T0377096

Modeling and Simulation of Chemical Process Systems

Modeling and Simulation of Chemical Process Systems

Nayef Ghasem

CRC Press is an imprint of the
Taylor & Francis Group, an **informa** business

CRC Press
Taylor & Francis Group
6000 Broken Sound Parkway NW, Suite 300
Boca Raton, FL 33487-2742

Printed on acid-free paper

International Standard Book Number-13: 978-1-1385-6851-8 (Hardback)
International Standard Book Number-13: 978-0-2037-0508-7 (eBook)

Library of Congress Cataloging-in-Publication Data

Names: Ghasem, Nayef, author.
Title: Modeling and simulation of chemical process systems / Nayef Mohamed Ghasem.
Description: Boca Raton, FL : CRC Press/Taylor & Francis Group, 2018. |
Includes bibliographical references and index.
Identifiers: LCCN 2018025236 | ISBN 9781138568518 (hardback : acid-free paper)
| ISBN 9780203705087 (ebook)
Subjects: LCSH: Chemical processes--Computer simulation. | Chemical
engineering--Data processing.
Classification: LCC TP184 .G483 2018 | DDC 660/.28--dc23
LC record available at https://lccn.loc.gov/2018025236

Visit the Taylor & Francis Web site at
http://www.taylorandfrancis.com

and the CRC Press Web site at
http://www.crcpress.com

Contents

Preface

Modeling and simulation of chemical process systems refer to translating the actual process behavior into mathematical expressions (process modeling) and solving that model numerically with the help of a computer (simulation). Modeling and simulation supports analysis, experimentation, and training and can facilitate understanding the behavior of the system. Modeling and simulation are valuable tools; it is safer and cheaper to perform tests on the model using computer simulations rather than carrying out repetitive experimentations and observations on the real system.

Modeling and Simulation of Chemical Process Systems covers modeling and simulation of both lumped parameter systems and distributed parameter systems. Lumped parameter systems include processes where the parameters of the system, such as temperature and concentration, are uniform throughout the process unit. The process model equations of the lumped parameters system originate from the transient material and energy balance equations. Lumped parameter systems, in general, use a single ordinary differential equation or a set of ordinary differential equations; in certain cases these equations can be simplified and solved manually. The students are also encouraged to solve the system model equations by using the MATLAB®/Simulink® software package. Students will be able to compare analytical and simulated results, which will give them self-confidence in their work.

This text also covers modeling and simulation of distributed parameter systems. The state variables of these systems, such as temperature, concentration, and momentum, generally have spatial variation. For these systems, students will learn not to start from shell momentum balance but from equations of change of heat, mass, and momentum. The equations to be simplified are based on the question information, and in general the resultant equations are partial differential equations. The generated simplified partial differential equations are solved by the COMSOL Multiphysics 5.3a software package, an effective tool for solving partial differential equations using the fine element method.

This textbook uses the transport phenomena approach to develop mathematical models of chemical process systems. Mathematical models contain equations that include known and unknown variables to be determined. Known variables are usually called parameters, and unknown variables are called decision variables. In this approach, chemical engineers and scientists start from the general three-dimensional conservation equations of mass, energy, and momentum. With the physical description of the system and with the assumptions based on the objectives of modeling, one sets several terms in general balance equations equal to zero to obtain the model equations of the system under consideration. Appropriate initial and boundary conditions are also to be stated. Once the process model is developed, it is essential to obtain the solution of the model equations to study the effect of system parameters and operating conditions on the performance of chemical process systems.

Students will practice using COMSOL and MATLAB software to obtain the solution of model equations. This approach and practice are also essential because it is frequently impossible to solve model equations analytically. Students should learn to use both numerical methods and available software tools. Numerical computational methods are also explained in the last two chapters of the book.

Nayef Ghasem

Acknowledgments

The author thanks Allison Shatkin, publisher of this book, and Camilla Michael, editorial assistant, for their help and cooperation. The author would also like to thank the engineers from COMSOL Multiphysics and MATLAB® for their help and cooperation. The comments and suggestions of the reviewers were highly appreciated.

Author

Nayef Ghasem is a professor of chemical engineering at the United Arab Emirates University, where he teaches undergraduate courses in process modeling and simulation, natural gas processing, and reactor design in chemical engineering, as well as graduate and undergraduate courses in chemical engineering. He has published primarily in the areas of modeling and simulation, bifurcation theory, polymer reaction engineering, advanced control, and CO_2 absorption in gas-liquid membrane contactor. He is the author of *Principles of Chemical Engineering Processes* (CRC Press, 2012) and *Computer Methods in Chemical Engineering* (CRC Press, 2009). He is a senior member of the American Institute of Chemical Engineers (AIChE).

1

Introduction

In this chapter, an introduction to process modeling and simulation of chemical process systems, tools, and methods of simulation are presented. Process simulation is based on mathematical models which describe a system using mathematical concepts and language. The mathematical model should reflect the degree of accuracy required by the application. Having a good knowledge of the modeling background is required for getting consistent conclusions and using software efficiently. The process modeling and simulation is an important subject in chemical engineering; basically, it deals with three aspects: modeling of chemical engineering processes, parameter estimations, and application of numerical methods for solution of models. This chapter includes description of the types of process modeling; lumped and distributed parameter systems; review of the general material and energy balance; and the basic approaches to measure bubble, dew, and flash points.

LEARNING OBJECTIVES

- Understand what process modeling and simulation are.
- Describe the types of process modeling.
- Distinguish between lumped and distributed parameter systems.
- Apply the general equation of material and energy balances.
- Measure bubble, dew, and flash points.

1.1 Background

Modeling and simulation is important in research. They are strictly related computer applications, which is a key factor in science and engineering. They help to reduce the cost and time needed for research. Modeling and simulation are useful tools for engineers to understand processes easily. The topic helps to understand the behavior of a dynamic system and how the various components of that system interact. Modeling and simulation are employed to test the reactions of the system in a cost-effective manner instead of carrying out experiments and observations on the real system. Modeling is the process of building mathematical models using different equations that can be analyzed and solved using computer software; simulation is the evaluation of the systems performance using a computer model. Modeling and simulation are vital in research; they represent the real systems either via physical reproductions on a smaller scale or via mathematical models that allow representing the dynamics of the system. Simulation permits exploring system behavior in a way that, in the real world, is frequently either not possible or too risky. A simulation of a system is the operation of a model, which is a representation of that system.

1.2 Mathematical Models

The process of system modeling is represented in Figure 1.1. The process of producing a model is referred to as modeling. A model can be described as a representation of the working and construction of some system of interest. Although the model often is very similar to the system it describes, the model is often a simpler presentation of the system. A model can help an

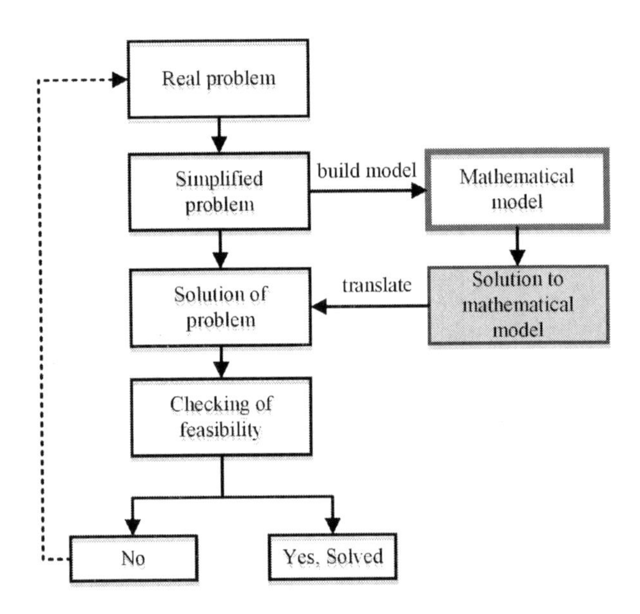

FIGURE 1.1
The modeling process.

analyst to predict the effect of changes to the system. Simulation is the reproduction of the real-world system or a process operation over time. Before simulating a real-world system or a process, the system or the process must first be modeled, where the model represents the behaviors/functions or the key characteristics of the selected system or process. A mathematical model can help in understanding the causes and effects among the modeled system and the complex physical interactions that occur within the system. Process modeling and simulation are the fastest and most cost-effective methods for attaining process optimization.

The process of building a dynamic model can be summarized as follows:

1. Specify the objectives of the mathematical models.
2. Develop the mathematical model of the process system.
3. Clearly state all the assumptions used in the developed mathematical model.
4. A partial differential equation will be required to describe the model if there are spatial variations of the process variables.
5. Write the appropriate mass, energy, and component conservation equations.
6. Include all required relations from transport phenomena, equipment geometry, reaction rate and thermodynamic relations.
7. Make sure that the model equations can be solved by performing a degree of freedom analysis.
8. Organize the equations to show clearly the independent and dependent variables.
9. Categorize inlet parameters as manipulated or disturbance variables.

1.3 Why Study Process Modeling and Simulation?

We study process modeling and simulation to enable us to model steady and dynamic behavior of chemical engineering systems, recognize the underlying mathematical problems, and develop some awareness of the available analytical and numerical solution techniques. In the discipline of chemical engineering, models can be of use in every part of the engineering process, starting from research and development to plant operation. Models are used to simulate the effects of several operating conditions and different parameters on the behavior of a process. In addition, models can also be utilized for the evaluation of equipment size, and for the startup and shutdown operation procedures of different plants, and to develop a better control structure and configuration.

1.4 Terminology of Process Modeling and Simulation

Models are developed to improve understanding and perceptions about some features of the real world. The basic concepts of modeling and simulation are presented in the following sections.

1.4.1 State Variables and State Equations

The state of a dynamic system can be defined as the smallest set of variables, which are referred to as sate variables. The knowledge of these variables at the initial state of the system ($t = t_0$) and the inputs to the system at $t \geq t_0$ can lead to finding out the system behavior at any point in time ($t \geq t_0$). The state variable of a dynamic system is a set of variables that, if known along with the equations describing the dynamics and the input functions of the system, will give the forthcoming state and the resulting output of the system. For a dynamic system, a set of state variables can describe the state of a system. The state variables describe the future response of a system, given the present state, the excitation inputs, and the equations describing the dynamics. The state variables of lumped parameter processes are described by a set of first-order differential equations written in terms of state variables. The dependent state variable (such as temperature or concentration) and independent state variables (such as time or spatial coordinate) should be connected through sets of equations that are extracted from mass, energy, and momentum balances that can describe the behavior of the system in connection to time and position. The basic structure of a computer simulation model of a formal system dynamic can be described mathematically by a system of couple, linear or nonlinear, first-order differential equations:

$$\frac{d}{dt}x(t) = f(x,p) \tag{1.1}$$

where:

- x is the state variables
- p is a set of parameters
- f is a nonlinear vector-valued function
- t is the independent variable

1.4.2 Steady State and Transient

When the processing system physical state does not change or vary with time, then the system is said to be at steady-state operation. Nevertheless, unsteady-state performance (transient) is considered when the physical state of processing the system varies with time. Transient and dynamic models are models that describe systems in unsteady-state situations. These models are a function of time and are often used to track system fluctuations and behavior changes with time. Such models are often described mathematically using differential equations. For example, if steam turbines in power plants operate at approximately constant temperature, constant flow, constant heat loss rates, and constant load hour after hour, then the operation of the steam turbines is considered steady-state operation. Heating or cooling operations are transient processes because the temperature of the system or device is constantly changing with time.

1.4.3 Lumped versus Distributed Parameters

Models where the dependent state variables have no spatial location obligation and are uniform over the entire system are called lumped parameter systems where time is the only independent variable. For lumped parameter systems, the process state variables are uniform over the entire system; that is, each state variable does not depend on the spatial variables (i.e., x, y, and z in Cartesian coordinates) but only on time t. In this case, the balance equation is written over the whole system using macroscopic modeling. By contrast, when the process dependent state variables vary with time and position, the process is measured as distributed parameter system. Modeling of distributed parameter systems must consider the variation of these variables throughout the entire system.

Lumped parameter system is a system where the relevant dependent variables are a function of time alone, and modeling of the system leads to first-order ordinary differential equations (ODEs). This means a system of a set of ordinary differential equations (ODEs) must be solved. The resulting process performance is presented in system outputs that are functions of time.

Distributed parameter system is a system where all the dependent variables are functions of one or more spatial variables and time. For the case of distributed systems, obtaining the solution for such systems requires solving partial differential equations (PDEs). Distributed system inputs can be only a function of time or functions of both time and one or more spatial coordinates. The resulting process performance is described in the system outputs as a function of both time and one or more spatial coordinates [1].

1.4.4 Model Verification

Model verification is different than model validation, where the verification of a model is ensuring that the developed model behaves in a similar way to what was intended from. In other words, it is the process of determining that the associated data with the model and the implementation of the model accurately represent the developer's conceptual description and conditions. The verification process is concerned mainly with correctly transforming the conceptual model into a program code or the simulation model to confirm that the program code correctly presents the performance that is implicitly presented in the conceptual model.

1.4.5 Model Validation

Compared to the model verification, which usually comes after building the model, the model validation is performed after the model has been developed and used. Model validation is the process where the results of the model are compared to the experimental or industrial data. Based on the difference between the model simulation results and the experimental data, the developed model may or may not present the real system behavior within an acceptable range of accuracy. The validation process is used to determine to which degree the developed model and its associated data can accurately represent the real-world behavior of the system intended to be modeled.

1.5 The Steps for Building a Mathematical Model

Mathematical model construction starts with identifying the system configuration, followed by identifying its nearby location and the modes of interface between them, then identifying the related state variables that designate the system and which processes take place inside the boundaries of the system. Building a mathematical model needs the know-how of the physical and chemical relations that take place within the borders of the structure. The knowledge of the essential simplifying assumption is essential throughout building the model. The model equations are derived from mass, energy momentum balances and energy suitable to the nature of the system. Validation of the developed mathematical model with the experimental results ensures the model reliability; the validation is also used to reevaluate the used assumptions, which may favor adding further simplifying assumptions or relax others.

1.6 Fundamental Balance Equations

The mathematical model of a system is formed from state equations, which define how the dependent variables (state variables) are related to the independent variables (i.e., time only for lumped systems, and time and spatial for distributed systems). The essential equations of a mathematical model are created from the total mass balance, component mass or mole balance, total energy balance, and momentum balance equations.

1.6.1 Material Balance

Based on the law of conservation of mass, matter is conserved, which means that mass cannot be created nor can it be destroyed. The law of conservation of mass is used as an overall mass balance equation, which accounts for the total mass involved in a process, including chemical reactions. The total mass balance can be written in the following form:

$$\begin{Bmatrix}\text{Rate of}\\\text{accumulation}\end{Bmatrix}=\begin{Bmatrix}\text{Rate of mass}\\\text{in}\end{Bmatrix}-\begin{Bmatrix}\text{Rate of mass}\\\text{out}\end{Bmatrix}$$

For a steady-state process, the mass entering the system is equal to the mass leaving it, which means that the accumulation of mass within the system is equal to zero. Thus

$$\text{Input} = \text{Output}$$

The steady-state mass balance equation is more suitable to present in the following form, especially, when applying the mass balance to a process with various inputs and outputs.

$$\sum \text{Masses entering via feed streams} = \sum \text{Masses exiting via product streams}$$

It is important to include all chemical species masses in every stream crossing the boundary of the system. The total mass balance equation applied to the batch and continuous processes can be written as follows

$$\sum \text{Mass in} = \sum \text{Mass out}$$

The consumption of feed and the formation of products must be considered if the process involves chemical reactions. One can present the mass balance of each chemical, and the formation and consumption of each chemical are determined as follows:

$$\begin{Bmatrix}\text{Mass}\\\text{in}\end{Bmatrix}+\begin{Bmatrix}\text{Mass formed}\\\text{by reaction}\end{Bmatrix}=\begin{Bmatrix}\text{Mass}\\\text{out}\end{Bmatrix}+\begin{Bmatrix}\text{Mass used}\\\text{by reaction}\end{Bmatrix}$$

This equation can be written more simply as follows:

$$\begin{Bmatrix}\text{Mass}\\\text{in}\end{Bmatrix}+\begin{Bmatrix}\text{Mass}\\\text{formed}\end{Bmatrix}=\begin{Bmatrix}\text{Mass}\\\text{out}\end{Bmatrix}+\begin{Bmatrix}\text{Mass}\\\text{consumed}\end{Bmatrix}$$

1.6.2 Total and Component Balances

Total mass balance is the result of using the total mass for each process stream, whereas a component balance is a mass balance written for each chemical species involved. For example, if a process unit involves three chemical components, one can write four balance equations: A total balance equation and three component balance equations. The sum of all component balances is the total mass balance of the system, which means that not all mass balances are independent. Accordingly, the total independent material balance equations are total mass balance plus total number of components minus one component balance.

Example 1.1: Steady State Mixing in Waste Storage Tank

A stream from a wastewater treatment plant at a volumetric flow rate of 1.0 m^3/s and comprising 5.0 mg/L of phosphorus combines with a second stream at flow rate of 25 m^3/s and comprising 0.010 mg/L of phosphorus compounds and both flow into a large waste storage tank (see Figure 1.2). Determine the resulting concentration of phosphorus (in units of mg/L) in the waste storage tank located downstream of the sewage outflow.

Solution

The phosphorus concentration in the resulting waste storage tank located downstream to the sewage outflow is calculated as follows. The first step is to perform a total mass balance calculation on the total mass of the river water. In this case, the concentration of the water stream is in the units of mass/volume:

$$\frac{dm}{dt} = \dot{m}_{\text{in}} - \dot{m}_{\text{out}}$$

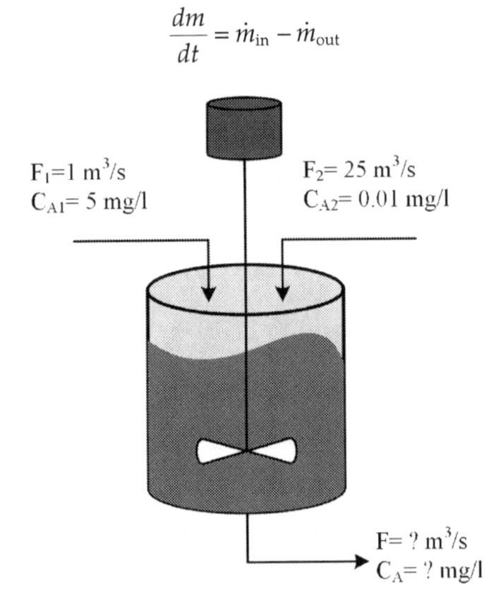

FIGURE 1.2
Schematic of wastewater storage tank.

Substitute $\dot{m}_{\text{in}} = C_{\text{in}}F_{\text{in}}$, $\dot{m}_{\text{out}} = C_{\text{out}}F_{\text{out}}$ in the general mass balance:

$$\frac{dm}{dt} = C_{\text{in}}F_{\text{in}} - C_{\text{out}}F_{\text{out}}$$

Here, the reaction term is set to zero because all chemical reactions are ignored: we are considering the steady-state operation of the mixing operation in the waste storage tank; in other words, the change of the mass of the system with time equals zero ($dm/dt = 0$).

Therefore, if the density (ρ) was considered as constant,

$$F_{\text{in}} = F_1 + F_2 = F_{\text{out}} = 1 + 25 = 26 \text{ m}^3/\text{s}$$

The exit concentration of phosphorus component at steady state is calculated as follows:

$$0 = C_{\text{in}}F_{\text{in}} - C_{\text{out}}F_{\text{out}}$$

Rearranging to find the exit concentration of the phosphorus compound, we have:

$$C_{\text{out}} = \frac{(C_{\text{in}}F_{\text{in}})}{F_{\text{out}}} = \frac{5 \text{ g/m}^3 \times 5 \text{ m}^3/\text{s} + 0.01 \text{ g/m}^3 \times 25 \text{ m}^3/\text{s}}{26 \text{ m}^3/\text{s}}$$

$$= 0.20 \text{ g/m}^3 = 0.20 \text{ mg/L}$$

1.6.3 Material Balance on Individual Components

The component material balance is designated in terms of molar concentration. For example, a rate mole balance of component A can be written as follows:

$$\left\{\begin{matrix}\text{Rate of}\\ \text{accumulation}\\ \text{of moles of A}\end{matrix}\right\} = \left\{\begin{matrix}\text{Rate of}\\ \text{moles of}\\ A \text{ in}\end{matrix}\right\} - \left\{\begin{matrix}\text{Rate of}\\ \text{moles of}\\ A \text{ out}\end{matrix}\right\} + \left\{\begin{matrix}\text{Rate of}\\ \text{generation of}\\ \text{moles of } A\end{matrix}\right\} - \left\{\begin{matrix}\text{Rate of}\\ \text{consumptions}\\ \text{of moles of } A\end{matrix}\right\}$$

Note that, unlike the total mass balance, the number of moles for A is not conserved, and the component mole balance equation will have rates of generation and consumption.

Example 1.2: Continuous Stirred Tank Reactor

Figure 1.3 shows a steady-state industrial waste continuous stirred tank reactor (CSTR) with a first-order reaction taking place. The industrial waste reactor uses a reaction that destroys the waste based on the first-order kinetics, which can be written as $r = kC$, where $k = 0.216 \text{ day}^{-1}$.

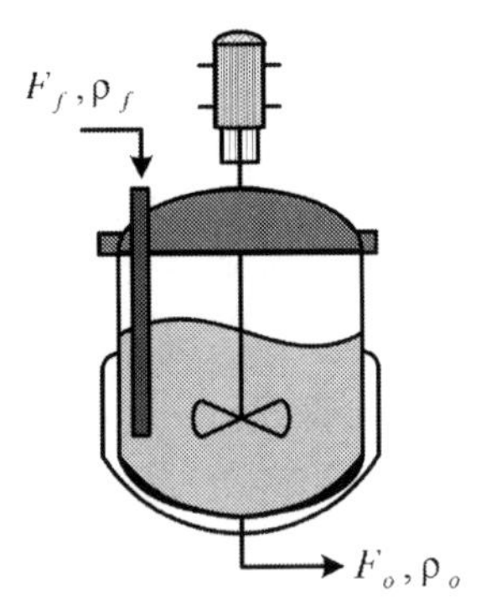

FIGURE 1.3
CSTR.

The volume of the reactor is 500 m^3, a volumetric flow rate of 50 m^3/day enters and leaves the rector, and the waste stream entering the reactor has a concentration of 100 g/m^3. Determine the concentration of the stream exiting the reactor at steady state.

Solution

The total mass balance equation is:

$$\frac{dm}{dt} = \dot{m}_{\text{in}} - \dot{m}_{\text{out}}$$

where $\dot{m}$ is the mass flow rate.

The component balance in terms of molar concentration is:

$$\frac{dn}{dt} = C_{\text{in}}F - CF - kCV$$

Rearranging to obtain the exit stream molar concentration of the waste material, we have:

$$C = \frac{C_{\text{in}}F}{1 + kV}$$

Substituting necessary values. The numerical solution is:

$$C = \frac{C_{\text{in}}F}{1 + kV} = \frac{100\ \text{g/m}^3 * 50\ \text{m}^3/\text{day}}{1 + (0.216\ \text{day}^{-1})\ (500\ \text{m}^3)} = 32\ \text{g/m}^3 = 32\ \text{mg/L}$$

1.6.4 Energy Balance

The energy accumulated within the system equals the rate of energy in minus the rate of energy out, plus the rate of energy generation minus the rate of energy consumption. The energy balance of the system can be written as follows:

$$\left\{\begin{matrix}\text{Rate of}\\ \text{Accumulation}\\ \text{of energy}\end{matrix}\right\} = \left\{\begin{matrix}\text{Rate of}\\ \text{energy}\\ \text{input}\end{matrix}\right\} - \left\{\begin{matrix}\text{Rate of}\\ \text{energy}\\ \text{output}\end{matrix}\right\} + \left\{\begin{matrix}\text{Rate of}\\ \text{energy}\\ \text{generated}\end{matrix}\right\} - \left\{\begin{matrix}\text{Rate of}\\ \text{energy}\\ \text{consumed}\end{matrix}\right\} \pm \left\{\begin{matrix}\text{Rate of energy}\\ \text{exchanged}\\ \text{with the}\\ \text{surrounding}\end{matrix}\right\}$$

Note that the "rate of energy generated" term does not violate the energy conservation principle; rather, it is just a presentation of the released heat of reaction with the condition that the reaction occurred inside the system, and the rate of energy consumed refers to the consumed heat of reaction by the component in the system as long as the reaction occurred within the boundary of the system.

1.6.5 Momentum Balance

The general momentum balance equation is:

$$\left\{\begin{matrix}\text{Rate of}\\ \text{accumulation}\\ \text{of momentum}\end{matrix}\right\} = \left\{\begin{matrix}\text{Rate of}\\ \text{momentum}\\ \text{in}\end{matrix}\right\} - \left\{\begin{matrix}\text{Rate of}\\ \text{momentum}\\ \text{out}\end{matrix}\right\} + \left\{\begin{matrix}\text{Rate of}\\ \text{momentum}\\ \text{generated}\end{matrix}\right\} - \left\{\begin{matrix}\text{Rate of}\\ \text{momentum}\\ \text{consumed}\end{matrix}\right\}$$

Newton's second law was used to write the momentum balance equation. The unit of momentum balance is $kg \cdot m/s^2$. Note that momentum is a vector quantity because the momentum is a function of the velocity, which is a vector quantity as well.

Example 1.3: Heating Water Tank

An electric water heater with a capacity of 50 L is used to heat tap water. The heater uses 5 kW of electricity. For a hot shower, someone must wait until the heater heats up the water in the tank. Determine the time necessary to wait for the temperature of the water in the tank to reach 55°C if the starting temperature of the water is 20°C (note that no water was removed from the heater tank once it is turned on).

Solution

By examining the water heater tank, we find out that there is a single energy input to the system, which is the electrical energy delivered to the heater. Note that the electrical heater is assumed to convert the electrical energy into thermal energy completely. Then from the energy balance equation, the rate change of the energy of the water in the tank can be presented as follows, in an equation for transient energy balance:

$$Q = \frac{d\left(\rho V C_p (T - T_{\text{ref}})\right)}{dt}$$

Simplifying, and assuming constant density and heat capacity, we have:

$$\frac{dT}{dt} = \frac{Q}{\rho V C_p}$$

Integrating the first-order differential equation, we have:

$$T = \frac{Q}{\rho V C_p} t + C$$

The arbitrary integration constant C is obtained from the initial conditions: at $t = 0$, $T = 20°\text{C}$, accordingly, $C = 20°\text{C}$

Substituting the value of the arbitrary integration constant, $C = 20°\text{C}$, we have:

$$T = \frac{Q}{\rho V C_p} t + 20$$

Substitute values of the known parameters and solving for time t, we have:

$$55 = \frac{5000\ \text{J/S} \times 60\ \text{s/min}}{1\ \text{kg/L}\ (50\ \text{L})\ (4184\ \text{J/kg} \cdot °\text{C})} t + 20$$

$$t = 24.4\ \text{min}$$

The time required for the electrical heater to heat the water from 20°C to 55°C is 24.4 min.

1.7 Process Classification

Based on how the state variables of the process change with time, the processes are differentiated between transient (unsteady-state) and steady-state processes. For a system undergoing a steady-state process, all variables of the system remain constant through time (they are independent in relation to time). However, for transient processes, the variables of the processes are functions of time. Based on how the processes operate, they are classified into continuous and batch [1,2].

1.7.1 Continuous Process

A continuous process can be defined as the process that has inlet streams and outlet streams transporting species continuously into and out of a process all the time. Nonstop production is known as continuous drift process because the materials that are being processed are constantly in motion, undergoing chemical reactions, or being subjected to mechanical or heat treatment.

Continuously operated systems work without interruption for real as well as commercial and financial purposes. In many continuous processes, shutting down and starting up results in off-quality product that must be reprocessed. Continuous processes utilize process control to automate and control operational variables such as flow rates, tank levels, pressures, temperatures, and machine speeds. Examples of continuous processes are oil refinery, pulp and paper, natural gas processing, wastewater treatment, and power generation.

1.7.2 Batch Process

A chemical that is needed in a small amount or only as desired is frequently made by a batch process, in which production does not go on all the time. Pharmaceutical medicines are made by batch processes. A batch procedure is a system in which the feed streams are fed to the process to get it started. When the system is charged with the feed, then the system carries on the required processes on the feed until all of the required processes are done and the products then are removed from the reactor. The batch process can be viewed as a closed system during the operation because no mass is allowed to leave or enter the system during operation. The batch process accepts mass in at the charging state and mass out after the entire process is completed. Examples of batch processes are cooking, brewing, and specialty chemicals. Table 1.1 summarizes some of the advantages and disadvantages of continuous and batch processes. You should be able to evaluate the advantages and disadvantages of each type of process, given relevant information.

TABLE 1.1

Comparing of Continuous and Batch Processes

Factor	Continuous Process	Batch Process
Equipment cost	High	Low
Production rate	High	Low
Shutdown	Occasional	Repeatedly
Labor	Few people needed	Several people needed
Ease of control	Relatively easy	Comparatively difficult

1.7.3 Semibatch Process

A semibatch process is not batch and not continuous; it is a hybrid of the continuous and the batch process. It acts as a continuous process for the streams entering the process and as a batch process when all the products are removed at once, and vice versa [3]. Like batch reactors or batch processes, semibatch reactors or processes uses similar reactor equipment, and they take a place in a single stirred tank. However, unlike the batch process, they can allow either the addition of reactants or removal of products with time similar to a continuous process.

The flexibility of adding more reactants over time through semibatch operation has several advantages over a batch reactor. It improves selectivity of a reaction and better control of exothermic reaction [4].

1.8 Types of Balances

1. Differential balance is applied to steady-state processes, where it is taken at a specific instant of time. In the differential balance, each term presents a stream, the chemicals forming it, and its mass flow rate.
2. Integral balance is applied to batch processes, and it is taken between two instants of time.

1.9 Procedure of Mass Balance

Chemical engineers use their knowledge of material and energy balance principles in building macroscopic and microscopic mathematical models. Each method is very beneficial for practical purposes. The macroscopic balances of momentum in chemical engineering structures are found out via energy balances or as the consequences of integration of the microscopic momentum balances. The following is a recommended systematic procedure to be followed in order solve material balance problems:

1. Construct and label the process flow diagram (PFD).
2. Inscribe the values of established streams, and allocate symbols to unknown stream variables.
3. Define a basis of calculation from the stream with given amounts or flow rates; if these are not available, then assumption can be made regarding an amount (basis) of a stream with known compositions.

4. Perform degree of freedom analysis.
5. Write material balance equations.
6. Solve the resulting equations for the unknown quantities to be determined.

1.9.1 Microscopic Balance

Considering the microscopic case of the balance equations, the equations are written for a differential element within the system, which in turn results in considering the state variable variations from different spots in the system, besides its variation with time. The differential element in the microscopic balance changes shape based on the coordinate system selected. For the case when Cartesian coordinates are used, a cube is used as a differential element, as shown in Figure 1.4. All system variables of the system are expected to vary on the three coordinates of x, y, and z and time, where it can be written for a state variable as $S_{var} = S_{var}(x, y, z, t)$. For the case when the cylindrical coordinates are used, a cylinder is used as a differential element, as shown in Figure 1.5; for the spherical coordinates, the sphere is the shape of the differential element that is selected. Care must be taken while selecting the appropriate coordinate system because the selection of the coordinate system depends on the system's

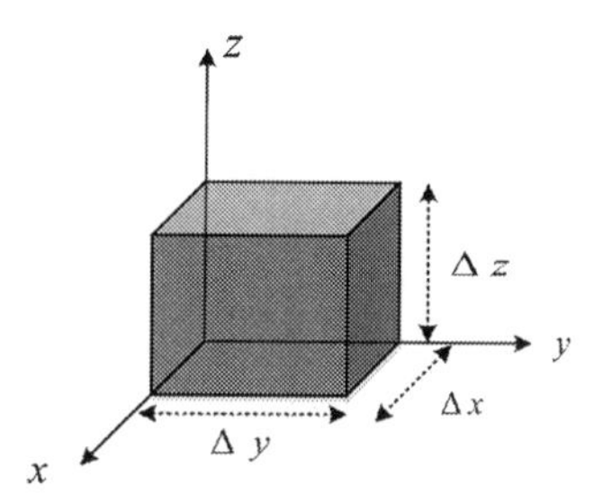

FIGURE 1.4
Cartesian coordinate system.

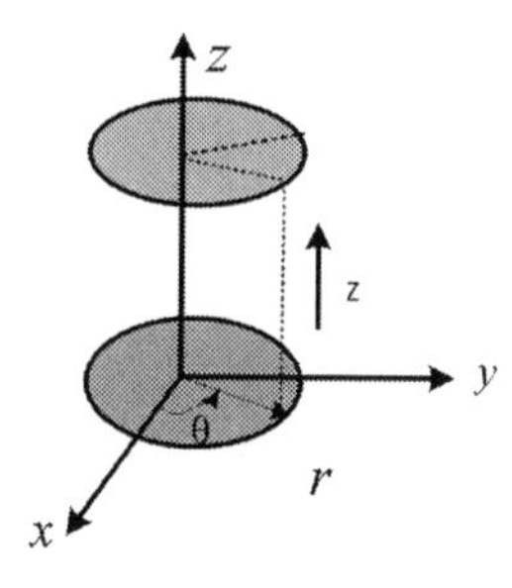

FIGURE 1.5
Cylindrical coordinate system.

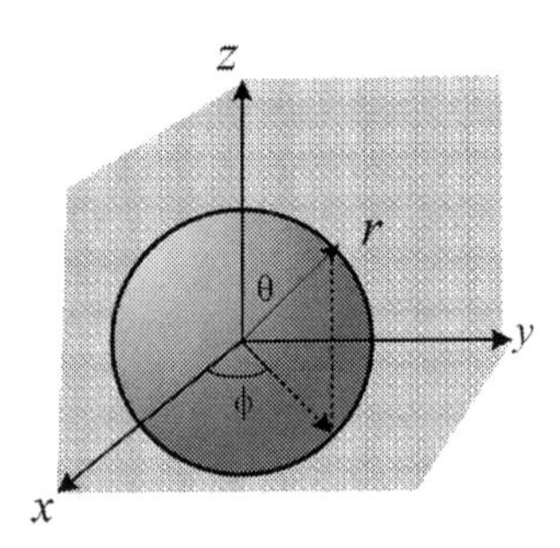

FIGURE 1.6
Spherical coordinate system.

geometry under consideration. Converting between coordinate systems is also possible and will be presented in more detail in Chapter 3 (Figure 1.6).

1.9.2 Macroscopic Balance

For situations where the system state variables do not vary with location through the entire system, these variables are independent on the special variables; however, it depends only on time. For such systems, the balance equations are written for the system through the principle of macroscopic modeling. If the process considered is modeled on a microscopic scale, the generated model is often in the form of PDEs where one or more spatial locations and time are the independent variables of the system. For a microscopic scale model of a steady-state system, the partial differential equation becomes independent of time (t) and the spatial spots become the only independent variables of the model. However, for a macroscopic scale model, the resulting model is in the form of ODEs. The microscopic and macroscopic modeling and balances of various chemical and physical processes are presented in detail in subsequent chapters.

The general balance equations, including the balance of mass, momentum, and energy, are often accompanied with many relations, which relate to thermodynamic relations. Some of these relationships are presented in the following sections.

1.10 Transport Rates

The fundamental quantities, including mass, energy, and momentum, take place through mainly two mechanisms, which are:

1. Transport due to bulk flow or convection
2. Transport due to potential difference or molecular diffusion

For many transport cases, two mechanisms can occur together, which leads to the conclusion that a flux due to the transference of any essential measure is the summation result of the flux billed to convection and diffusion. Values of the diffusion coefficient (D) are approximately $10^{-6} m^2/s$ for gases, and are much lower for liquids, around $10^{-9} m^2/s$. Because the average velocity of the random molecular motions depends on the kinetic energy of the molecules, the diffusion coefficient varies with temperature and the molecular weight of the diffusing molecule. As heat is introduced to a material and temperature increases, the thermal power is transformed to random kinetic strength of the molecules, and the molecules move quicker. This effects the diffusion coefficient with rising temperature. Gas molecules are free to move much larger distances before becoming motionless by bouncing into another molecule. By contrast, molecules of different molecular weight illustrate that, at a given temperature, a transfer of bigger molecules is slower; accordingly, with growing molecular weight, the diffusion coefficient decreases.

1.10.1 Mass Transport

Mass transport processes deal with shifting compounds through the air, floor water, or subsurface surroundings or through engineered systems. Those techniques are of interest to environmental engineers and scientists. The tactics circulate pollutants from the place at which they are generated to far from the pollutants' source. On the other hand, a few pollutants, which include sewage sludge, can be degraded in the environment if they are sufficiently diluted. For these pollutants, gradual shipping can result in excessively high pollutant concentrations, with ensuing extended detrimental effects.

Diffusion consequences from random motions are of two sorts:

1. The accidental motion of molecules in a fluid
2. The arbitrary vortexes that arise in turbulent current

Diffusion from the arbitrary molecular motion is known as molecular diffusion; diffusion from turbulent eddies is referred to as turbulent diffusion or eddy diffusion. The diffusion coefficient is the coefficient in Fick's first law, where J is the diffusion flux (amount of substance per unit area per unit time). Mass flux is the measurement of the amount of mass passing in or out of the control volume. The governing equation for calculating mass flux is the continuity equation. The mass flux is defined simply as mass flow per area.

The total flux n_{Au} of species A (kg/m^2s), where the density of species A $\rho_A(kg/m^3)$ moving with velocity v_u (m/s) in the u-direction, is:

$$\text{Total flux} = \text{bulk flux} + \text{diffusive flux}$$

$$n_{Au} = \rho_A v_u + j_{Au} \tag{1.2}$$

Fick's law can be used to write the diffusive flux $j_{Au}(\text{kg/m}^2\text{s})$ for a two-component mixture $A-B$ as follows:

$$j_{Au} = -\rho D_{AB}\frac{dw_A}{du} \tag{1.3}$$

The mass fraction (w_A) of species A:

$$w_A = \frac{\rho_A}{\rho} \tag{1.4}$$

where ρ is the density of the mixture in the units of kg/m^3 and the diffusivity coefficient of A in B, $D_{AB}(\text{m}^2/\text{s})$. The molar flux J_{A_u} $(\text{mol } A/\text{m}^2\text{s})$ is given by:

$$J_{Au}\left(\frac{\text{mol } A}{\text{m}^2\text{s}}\right) = -CD_{AB}\frac{dx_A}{du} \tag{1.5}$$

where the total molar concentration (C) of A and B $\left(\text{mol } (A+B)/\text{m}^3\right)$ and mole fraction of A in the mixture x_A:

$$x_A = \frac{C_A}{C} \tag{1.6}$$

For constant density, the flux becomes:

$$j_{Au}\left(\frac{\text{kg}\,A}{\text{m}^2\text{s}}\right) = -D_{AB}\frac{d\rho_A}{du} \tag{1.7}$$

In terms of concentration of specific component, we have:

$$J_{Au}\left(\frac{\text{mol}\,A}{\text{m}^2\text{s}}\right) = -D_{AB}\frac{dC_A}{du} \tag{1.8}$$

1.10.2 Momentum Transport

Convection and diffusion can also transport momentum. Compared to the scaler transportation of mass, however, linear momentum is a vector quantity, which means the transportation of momentum must be considered in all three directions of the three-dimensional space of a given process. Considering the transportation of the x-component of the momentum and also the y and z momentum components, the flux resulting from the convection of the x-component through the y-direction is as follows:

$$(\rho v_x)v_y\,(\text{kg}\cdot\text{m/s}^2) \tag{1.9}$$

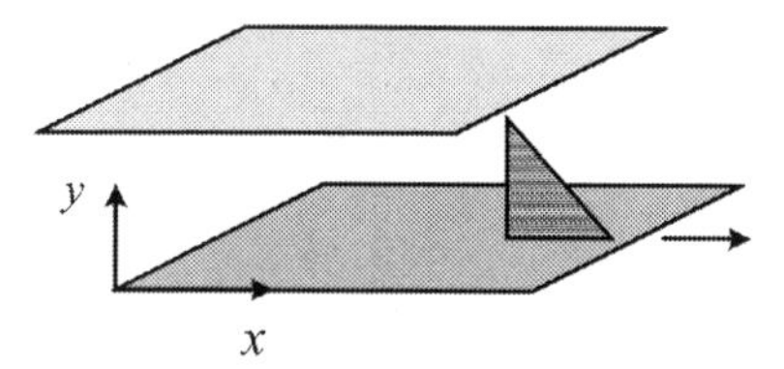

FIGURE 1.7
Fluid flow between two horizontal and parallel flat sheets.

The diffusion flux is denoted by Greek symbol τ with two subscripts of spatial directions; the first indicates the component of the momentum and the second specifies the direction. For example, τ_{xy} is the diffusion flux of the momentum x-component in the y-direction. Consider the case presented in Figure 1.7, where a fluid is flowing between two infinite parallel plates. At a moment in time, a constant force F_x is applied to the lower plate, which causes its movement, whereas the upper plate maintains its location (not moving). For the configuration and the position of the force F_x, this force is referred to as the shear force and it acts at the surface area A_y.

The shear stress is defined as the shear force per unit area:

$$\tau_{xy} = \frac{F_x}{A_y} \quad (\text{kg} \cdot \text{ms}^{-2}) \tag{1.10}$$

Due to the application of force F_x, the layer adjacent to the plate (at $y = 0$) moves with a fixed value of velocity ($V = v_x$). Due to the molecular transport phenomenon, the layer slightly higher than the adjacent layer to the plate moves with a slightly lower velocity, as shown by the velocity profile in Figure 1.7. As a result, there is a transport by diffusion of the momentum of the x-component in the y-direction. The shear stress τ_{xy} is in fact the flux of this diffusive transport. Consequently, the entire flux π_{yx} of the x-element in the y-path is the summation of the diffusive and convective terms:

$$\text{Momentum flux} = \text{diffusive flux} + \text{bulk flux}$$

$$\pi_{yx} = \tau_{yx} + (\rho v_x) v_y \tag{1.11}$$

For a fluid that follows the behavior of Newtonian fluids, the shear stress τ_{yx} is relative to the velocity gradient as follows:

$$\tau_{yx} = -\mu \frac{\partial v_x}{\partial y} \tag{1.12}$$

Here, μ is the viscosity of the fluid, which can have the units of $(\text{kg}/\text{m} \cdot \text{s})$.

1.10.3 Energy Transport

Energy flux is the rate of transfer of energy through a surface.

$$E_u (\mathrm{J/s \cdot m^2})$$

The total energy flux of a fluid at a static pressure moving with a velocity v_u that is in the u-direction is stated as follows:

$$\text{Energy flux} = \text{diffusive flux} + \text{bulk flux}$$

$$E_u = q_u + (\rho CpT)v_u \tag{1.13}$$

The heat flux by molecular diffusion q_u (the heat conduction in the u-direction) can be described by Fourier's law as follows:

$$q_u = -k\frac{\partial T}{\partial u} \tag{1.14}$$

Here, k is the thermal conductivity, which might have the units of $(\mathrm{J/s \cdot m \cdot K})$. From the three relations of mass, momentum, and energy transport, an analogy exists, which is that the diffusive flux in each case can be given as follows:

$$\text{Flux} = -\text{transport property} \times \text{potential difference (gradient)}$$

The term on the left side of the above equation presents the rate of transfer per unit area. The potential difference or the gradient indicates the transport property and the driving force is the proportionality constant. The transport laws for molecular diffusion are summarized in Table 1.2.

Note that for a process modeled on the macroscopic level, the flux can be expressed by a relation like the general relation of the diffusive flux definition. However, in this situation the gradient represents the variance between the bulk properties such as the temperature or concentration in two media that are in contact, and the transport property presents the overall transfer coefficient. For example, the molar flux of the mass transfer processes is stated in Equation 1.15.

TABLE 1.2

Transportation Laws for Molecular Dispersion in Single Dimension

Transport Type	Law	Flux	Transport Property	Gradient
Mass	Fick's	J_{Au}	D	$\frac{dC_A}{du}$
Heat	Fourier	q_u	k	$\frac{dT}{du}$
Momentum	Newton	τ_{ux}	μ	$\frac{dv_x}{du}$

$$J_A = K\Delta C_A \tag{1.15}$$

Here, K presents the overall mass transfer coefficient. The heat flux, on the other hand, is presented as follows:

$$q = U\Delta T \tag{1.16}$$

Here, q is the heat flux, U presents the overall heat transfer coefficient. For the case of the momentum balance, the pressure drop presents or is the gradient in this case, and the friction coefficient is the flux or equivalent to it. The following relation at a moment in time is generally used to represent the momentum laminar transport in a pipe:

$$f = \left(\frac{D}{2v^2 L\rho}\right)\Delta P \tag{1.17}$$

Here, f is the fanning friction factor, D is the pipe diameter, L is the pipe length, ΔP is the drop of pressure due to friction, and v is the fluid velocity.

Example 1.4: Thermal Pollution

Cooling water from a neighboring river is used to remove the excess heat from a power plant. The capacity of the power plant is 1000 MW of electricity. The overall efficiency of the power plant is 33%. The flow rate of the cooling water used for cooling is 100 m^3/s. What is the resulting increase in cooling water exit stream temperature?

Solution

We first need to determine the amount of energy removed by cooling water. Because of power plant cooling, a definite amount of heat is transferred to the cooling water and raises its temperature. The resulting temperature rise must be determined. The power plant produces 1000 MW of electricity, but it has only about one-third efficiency, meaning that it uses 3000 MW of fuel energy (1000/0.33=3000 MW).

The amount of heat transfer to the river is the amount of heat that is not converted to electricity, Q:

$$Q = 3000 - 1000 = 2000\,\text{MW}$$

Assign T_{in} to signify the upstream water temperature and T_{out} to the downstream temperature after heating:

$$\frac{dH}{dt} = m_{in}h_{in} - m_{out}h_{out} + Q$$

In terms of temperature at steady state, we have:

$$m_{in}C_p\left(T_{in} - T_{ref}\right) = m_{out}C_p\left(T - T_{ref}\right) + Q$$

Neglecting water loss due to evaporation, and assuming constant density and specific heat, we have:

$$mC_p(T_{\text{in}} - T) = Q$$

Rearranging, we have the following:

$$(T_{\text{in}} - T) = \frac{Q}{m_{\text{in}}C_p} = \frac{2\times 10^9\,\text{J/s}}{(100\,\text{m}^3/\text{s})(1000\,\text{kg/m}^3)(4184\,\text{J/kg}\cdot{}^\circ\text{C})} = 4.8^\circ\text{C}$$

1.11 Thermodynamic Relations

The relationship between thermodynamic properties for an ideal gas as well as for incompressible substances is provided by the equation of state. Nevertheless, simple equation of state is not sufficient to define the thermodynamic properties for systems with real and complex materials. Accordingly, more mathematical equations are required to represent realistic processes. Among the various thermodynamic properties, p, v, T are directly measurable whereas u, h, s are not directly measurable. This means that developing thermodynamic relations apart from the equation of states is necessary. An equation of state is that equations that links the volume of a fluid (V), the temperature (T), and the pressure (P). For ideal gasses, the ideal gas law is one of the simplest and most direct formulas, and it can be written as follows:

$$PV = nRT \tag{1.18}$$

The ideal gas low can be rearranged to find the density of an ideal gas as follows:

$$\rho = \frac{M_w P}{RT} \tag{1.19}$$

Here, M_w is the molecular weight. The variation of the liquid's density can be neglected except when huge changes in composition or temperature happen. The enthalpies of materials in both liquid and vapor phases can be calculated using the specific heat capacity at a constant pressure or when the pressure effect can be neglected, as follows:

$$h = C_p(T - T_{\text{ref}}) \tag{1.20}$$

$$H = C_p(T - T_{\text{ref}}) + \lambda \tag{1.21}$$

Here, h (lowercase h) is the specific enthalpy for the liquid phase, H is the specific enthalpy of vapor, C_p the average specific heat capacity, and λ presents the amount of energy required to completely evaporate one kilogram of the pure substance per unit mass and temperature and can be referred to latent heat of vaporization. It is important to note that there should be a reference temperature that the enthalpies are calculated relative to, and this temperature is presented in the enthalpy equations as T_{ref}. For cases where the mixing heat can be neglected, the enthalpy of the mixture is the summation of the specific enthalpies of the pure components multiplied by the mole fraction of each pure component. However, for cases where the temperature variation of the system can change the value of the specific heat capacity by a large margin, then a correlation in terms of the temperature can be used to find the value of the specific heat capacity, as follows:

$$C_p = a + bT + cT^2 + dT^3 \tag{1.22}$$

Then the specific enthalpy for liquid and vapor phases can be calculated using the following equations:

$$h = \int_{T_{ref}}^{T} C_p dT \tag{1.23}$$

$$H = \int_{T_{ref}}^{T} C_p dT + \lambda \tag{1.24}$$

The internal energy is one of the main fundamental thermophilic properties of materials, and even enthalpy is derived from the internal energy. However, when the variation of the pressure for liquids and solids is negligible, the internal energy has approximately the same values as the enthalpy, and this is the same for gases undergoing processes with small pressure changes.

1.12 Phase Equilibrium

In packed or tray towers, the different phases are brought into direct contact with each other. Several chemical processes include more than one stage. The interface between fluid phases is generally at an equilibrium when the transfer of either mass or energy occurs from one fluid phase to another. Looking at the case of heat transfer between phases, the equilibrium relationship states that, at the interface, the temperature is basically the same. By contrast, the situation is different for mass transfer. The mass transfer

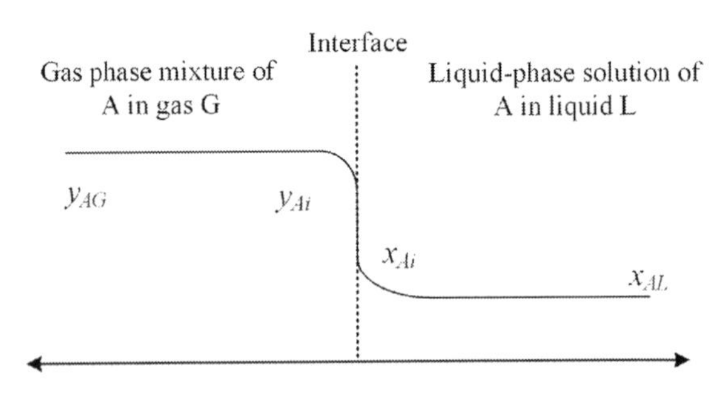

FIGURE 1.8
Mass transport between gas and liquid phases.

of species A from a liquid to a gas phase is shown in Figure 1.8. The bulk gas phase y_{AG} concentration decreases at the interface; on the other hand, from x_{AL} to x_{Ai}, the liquid concentration increases. Equilibrium appears at the interface, and y_{Ai} and x_{Ai} are related by a relation of the following form:

$$y_{Ai} = mx_{Ai} \tag{1.25}$$

Generally, the equilibrium relations are nonlinear. Nevertheless, results with an acceptable percentage of error can be obtained by using simpler relations, which are derived from the assumptions of ideal behavior.

For the case where the vapor phases are at low concentration, Henry's law can provide the following equilibrium relation:

$$p_A = Hx_A \tag{1.26}$$

Here, p_A is the partial pressure of species A in the vapor mixture, which is in the units of atm; x_A represents the mole fraction of A in the liquid; and H represents Henry's constant, which has the units of atm/mol. Another form of Henry's law can be obtaining through dividing Henry's law in terms of the partial pressure by the total pressure, and the resulting form of Henry's law is as follows:

$$y_A = Hx_A \tag{1.27}$$

Here, the constant (H) is pressure and temperature. For ideal vapor liquid mixtures, a suitable equilibrium law for such mixtures, called Raoult's law, is as follows:

$$y_A P = x_A P_A^s \tag{1.28}$$

Here, x_A presents the mole fraction of the liquid phase, y_A presents the mole fraction of the vapor phase, P_A^s presents the pure vapor pressure of A at the temperature of the system, and P is the total pressure of the gas phase side.

The relationship between the vapor pressure of pure A with the temperature can be presented as follows, in Antoine's equation:

$$\ln(P_A^s) = A - \frac{B}{C+T} \tag{1.29}$$

The parameters characterizing the fluid are A, B and C. For nonideal liquid and vapor behavior, Raoult's law can be adjusted to account for nonideal cases using the activity coefficients γ_i and $\varnothing_i$ of component i for liquid and vapor phases, respectively:

$$y_i \varnothing_i P = x_i \gamma_i P_i^s \tag{1.30}$$

for an ideal mixture ($\varnothing_i = \gamma_i = 1$), Equation 1.30 is reduced to Raoult's law.

In cases of liquid-liquid extraction, the transfer occurs in liquid-liquid phases. In liquid-solid phases such as ion exchange, the equilibrium relation is represented by Henry's law:

$$y_A = Kx_A \tag{1.31}$$

The equilibrium distribution coefficient (k) depends on temperature, concentration, and pressure. Phase equilibrium relations are used in dew point, bubble point, and flash calculations.

The calculations of bubble point consist of finding the molar vapor composition (y_i) and either P or T, for a given molar liquid compositions (x_i), when either T or P are specified. The calculations of the dew point depend on finding the molar liquid compositions (x_i) and either P or T, for a given molar vapor composition (y_i) and when either T or P is specified.

1.12.1 Flash Calculations

The flash calculations involve determining the liquid (x_i) and vapor compositions (y_i) via simultaneous solution of energy balance equations and component balance equations for known mixture compositions (z_i) and known (T) and (P).

$$Fz_i = y_i V + x_i L \tag{1.32}$$

In addition, the liquid and the vapor are assumed to be in equilibrium:

$$y_i = K_i x_i \tag{1.33}$$

The K-values, $K_i = K_i(T, P, x_i, y_i)$, are computed from the vapor liquid equilibria (VLE) model. In addition, we have the two relationships $\sum x_i = 1$ and $\sum y_i = 1$.

The simplest flash is usually to specify P and T (PT-flash), because K_i depends mainly on P and T. From the total mass balance:

$$F = L + V \tag{1.34}$$

The mass fraction is calculated from:

$$x_i = \frac{z_i}{1 + \frac{V}{L}(K_i - 1)} \tag{1.35}$$

Because the vapor split V/F is not known, we cannot directly calculate x_i. To find V/F, we may use the relationship $\sum x_i = 1$ or, alternatively, $\sum y_i = \sum K_i x_i = 1$.

1.13 Chemical Kinetics

Chemical kinetics is the study of chemical reactions with respect to reaction rates, re-arrangement of atoms, formation of intermediates, and effect of various variables. At the macroscopic level, the interest is in the amounts reacted, formed, and the rates of their formation. The following items are essential in discussing the mechanism of chemical reaction at the molecular or microscopic level:

1. During the chemical reactions, molecules or atoms of reactants must collide with each other.
2. To initiate the reaction, the molecules must have adequate energy.
3. The alignment of the molecules throughout the crash in certain cases must also be considered.

The rates of change in concentrations or amounts of either reactants or products are chemical reaction rates. The rate r in (mol/m^{3}s) of a chemical reaction is defined by:

$$r = \frac{1}{v_i V} \frac{dn_i}{dt} \tag{1.36}$$

where:

n_i is the moles of component i
v_i is the stochiometric coefficient of component i
V is the reactor holdup volume

The reaction rate (r) is a function of the reactant and product concentrations or partial pressures in addition to pressure and temperature. The necessity of the temperature comes from the reaction rate constant k given by Arrhenius law:

$$k = k_o e^{-\frac{E}{RT}} \tag{1.37}$$

where:
k_o is the pre-exponential factor
E is the activation energy
T is the absolute temperature
R is the ideal gas constant

The simple form of reaction rate consists of a function of temperature and concentration of reactants such as:

$$r = kC_1^{n1}C_2^{n2} \ldots C_r^n \tag{1.38}$$

The order of the reaction n_i is experimentally obtained, and the values are not essentially integers.

1.14 Process Control

The process involved in manipulation of an entity to preserve a parameter within a satisfactory deviation from an ideally required condition is known as process control. Feedback and feedforward control are the two basic process control schemes. Measuring the controlled variable and comparing it with a setpoint is considered feedback control. The discrepancy between the setpoint and the controlled variable is the error signal. The error signal is then utilized to diminish the deviation of the controlled variable from the setpoint. In any processing system, the presence of a control loop leads to an extra equation or relation for the process. By terminating the control loop, the control objective can be achieved. Examples of industrial control systems are.

1. ON–OFF control
2. Open loop control
3. Feedforward control
4. Closed loop control
5. Programmable logic controllers (PLCs)
6. Distributed control systems (DCSs)
7. Supervisory control and data acquisition (SCADA)

1.15 Number of Degrees of Freedom

A crucial stage in the model development and solution calculation is inspecting for the presence of an exact solution. After writing the equations and before attempting to solve them, the number of degrees of freedom of the model is to be checked. The number of degrees of freedom can be defined as the minimum number of independent coordinates that can specify the position of the system completely. For a process consisting of a set of N_e independent equations and N_v variables, the number of degrees of freedom F is:

$$F = N_v - N_e$$

According to the value of F three cases can be noteworthy:

1. If the model is fully defined ($F = 0$), number of independent equations equals the number of variables, and a single solution exists.
2. If the model is overdefined ($F < 0$), the number of equation is more than the number of variables.
3. If the model is underdefined ($F > 0$), the number of variables is more than the number of independent equations, and more relations or equations are needed.

While inspection the model consistency to avoid the conditions of over- or underspecified, perform the following:

1. Fix known quantities such as constant physical properties and equipment dimensions.
2. Define variables that can quantified by the external relations.

Example 1.5: Production of Orange Juice from Raw Orange

Oranges (0.4 kg/s, water content of an orange is 80%, density of the solids is 1.54 g/cm^3) are crushed and pressed to make orange juice, in the process shown in Figure 1.9. The crushed oranges enter a strainer. The filter can capture 90% of the solids; the residue exits with the orange juice as pulp. The mass flow rate of the orange juice product stream is 3925 kg/h. Do degree of freedom analysis.

System, crusher:

- Number of unknowns: 3 ($\dot{m}_1$, $\dot{m}_2$, and x_{S2})
- Number of independent mass balances: 2
- Number of degree of freedom: $3 - 2 = 1$

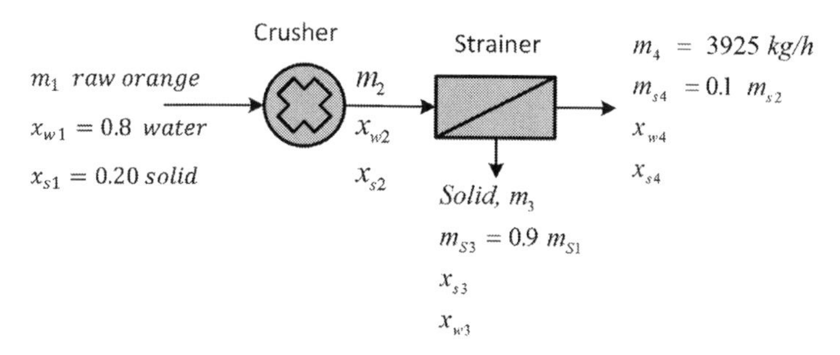

FIGURE 1.9
Schematic of orange juice maker.

System, strainer:

- Number of unknowns: 6 ($\dot{m}_2, X_{S2}, \dot{m}_3, X_{S3}, \dot{m}_4$, and X_{S4})
- Number of independent mass balances: 2 (one for each component)
- Number of relations: 2 (the conversion and the mass flow rate in the product stream)
- Number of degrees of freedom of the strainer are: $6 - 2 - 2 = 2$

Accordingly, the problem is underspecified.

Example 1.6: Hydrogenation of Ethylene into Ethane

Hydrogenation of ethylene into ethane takes place in a reactor, giving the following reaction:

$$2H_2 + C_2H_4 \rightarrow C_2H_6$$

The reaction goes to completion, and heat is released heat due to the exothermic nature of the reaction. The feed stream consists of 200 kg/h hydrogen and 584 kg/h ethylene gas. The outlet stream from the reactor contains 15% hydrogen by mass. The product stream from the reactor enters a membrane separator; the membrane exit steam consists of 100 kg/h, 5% hydrogen, and 93% ethane. The splitter recycled 70% and rejected 30% of the stream. Draw the process flow sheet and perform degrees of freedom analysis.

Solution

The labeled process flow sheet is shown in Figure 1.10.

Degrees of Freedom Analysis

System, reactor:

Number of unknowns: 6 ($\dot{m}_5, X_{A5}, X_{B5}, \dot{m}_3, X_{B3}, \xi$ (extent of reaction))
Number of equations: 3
Number degree of freedom: $6 - 3 = 3$

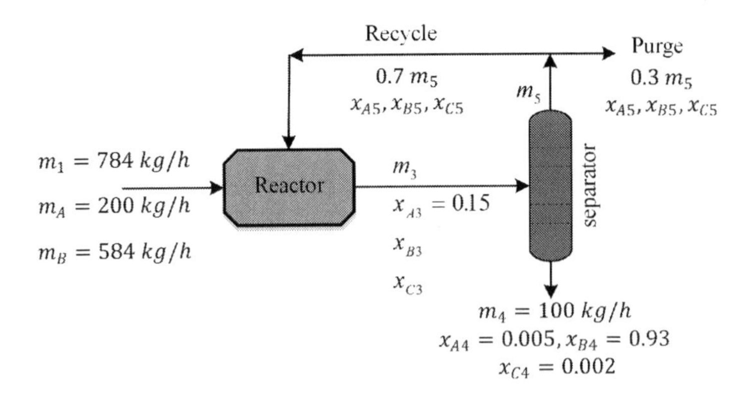

FIGURE 1.10
Process flow diagram for ethylene hydrogenation to ethane (A: H_2, B: C_2H_4, C: C_2H_6).

System, separator:

Number of unknowns: 5 ($\dot{m}_3, X_{B3}, \dot{m}_5, X_{A5}, X_{B5}$)
Number of independent equations: 3
Number degree of freedom: 5 – 3 = 2

System, splitter:

Number of unknowns: 3
Number of independent equations: 0 (all of them are used in labeling the chart)
Number degree of freedom: 3 – 0 = 3

System, overall:

Number of unknowns: 4 ($\dot{m}_5$,X_{A5},X_{B5},ξ(extent of reaction))
Number of independent equations: 3
Number degree of freedom: 3 – 3 = 1

Logically the following step after checking the solvability of the developed model is to solve the model.

1.16 Model Solution

The solution of the model often describes how the state variables of the system vary with time and spatial position or in general with the independent parameters. Another advantage of obtaining the solution of the model is that the it allows the parametric investigation of the system (modeled), which studies the effect of changing the value of a single or multiple parameter on the performance of the behavior of the modeled system, which can be very helpful in the process of optimizing the system. The ideal solution of a developed model is obtaining the analytical solution of that model. However, obtaining the ideal solution, which is in the form of an analytical solution, rarely occurs in chemical processes mainly because most of the chemical processes are nonlinear. Chemical processes are sets of nonlinear ODEs or nonlinear algebraic equations for the case when the processes are modeled through lumped parameter models, and they are sets of nonlinear PDEs for the case when the model is a distributed parameter model. In fact, the only type of differential equations where a complete analytical solution can be obtained is when the differential equations are linear ODE. A mathematical tool referred to as linearization can be used to linearize nonlinear differential equations to make it easier to obtain a linear solution for such models. However, the linearization tool is not always recommended (except for using it for control purposes). Most developed models of chemical processes are solved numerically by mathematical-based computer programs.

Example 1.7: Develop Mathematical Equations

Derive the mathematical equation that represents the mass balance of a pollutant of time-varying concentration C when there is a single inlet stream, flowrate Q at concentration C_{in}, and one outlet stream of equal flowrate Q, and there is no internal source and no internal decay. The volume of the system is V.

Solution

The component balance equation:

$$\left\{\begin{matrix}\text{Rate of}\\\text{accumulation}\\\text{of moles of } A\end{matrix}\right\}$$

$$=\left\{\begin{matrix}\text{Rate of}\\\text{moles of}\\A\text{ in}\end{matrix}\right\}-\left\{\begin{matrix}\text{Rate of}\\\text{moles of}\\A\text{ out}\end{matrix}\right\}+\left\{\begin{matrix}\text{Rate of}\\\text{generation of}\\\text{moles of } A\end{matrix}\right\}-\left\{\begin{matrix}\text{Rate of}\\\text{consumptions}\\\text{of moles of } A\end{matrix}\right\}$$

No reaction is taking place if there is no generation and no consumption; thus, the last two terms are dropped:

$$\frac{d(CV)}{dt}=C_{\text{in}}Q-QC+0-0$$

Since the inlet and exit flow rate are constant, the volume is constant:

$$V\frac{dC}{dt}=Q(C_{\text{in}}-C)$$

1.17 Model Evaluation

A crucial part of the modeling process is the evaluation of whether a given mathematical model describes a system accurately. The easiest part of model evaluation is checking whether a model fits experimental measurements or other empirical data. To implement a developed model, the model has to be validated with experimental data or compared against plant operating date, or at least published correlations. The importance of the model validation is enormous because using an invalidated model may lead to misleading results. If the developed model was far off the experimental data or another validation source, then the model parameters have to be adjusted in order to minimize the percentage difference with the experimental results or the validation data. However, it is important to keep in mind that most of the developed models are an approximation of the real world, which means difference and mismatch with the real data will always occur and it can be overlooked when the difference is within the acceptable range.

Example 1.8: Mathematical Models for Chemical Processes

For the liquid mixing tank shown in Figure 1.11, develop a mathematical model to investigate the variations of the tank holdup (i.e., volume of liquid in the tank) as a function of tank height and perform the degree of freedom analysis. Assume the tank is perfectly mixed (i.e., density of the effluent is the same as that of tank content). The tank is operating at constant temperature (isothermal).

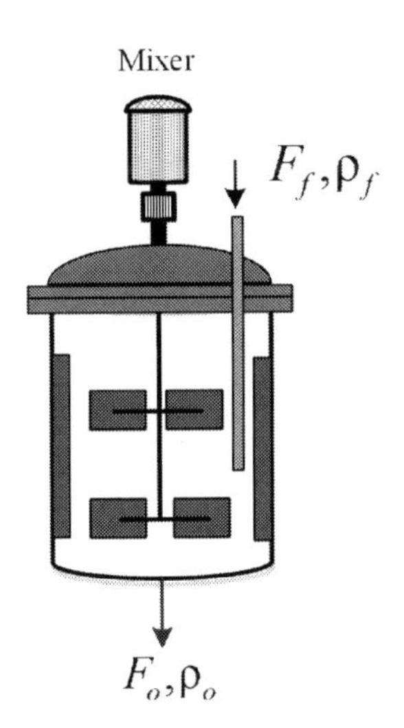

FIGURE 1.11
Liquid mixing tank.

Solution

Under the incompressible isothermal conditions, we assume that the density of the liquid is constant:

$$\frac{dV}{dt} = F_f - F_o$$

The tank is of constant cross-sectional area and hence $V = AL$:

$$A\frac{dL}{dt} = F_f - F_o$$

Degree of Freedom Analysis

Parameter of constant values: A
Variables for which values can be externally fixed (forced variable): F_f
Remaining variables: L and F_o
Number of equations: 1
Number of remaining variables, number of equations = 2 − 1 = 1

The extra relation is obtained by the relation between the effluent flow F_o and the level in open loop:

$$F_o = \alpha\sqrt{L}$$

Example 1.9: Isothermal CSTR

A liquid reaction is taking place in a CSTR at constant temperature (isothermal). Develop the system model equations for the variation of the volume of the reactor and the concentration of species A and B (Figure 1.12). The reaction taking place is:

$$A \rightarrow B$$

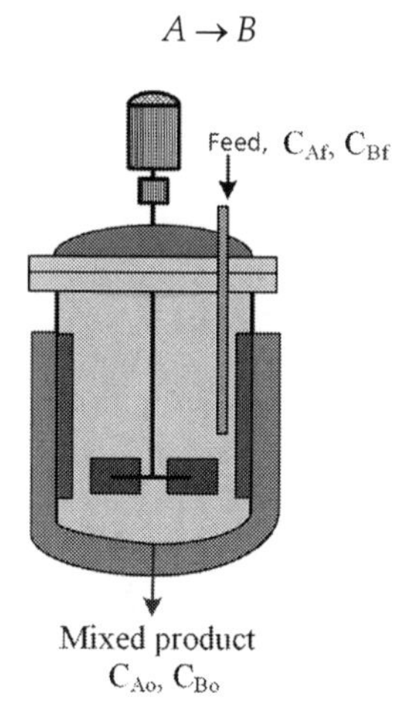

FIGURE 1.12
Isothermal CSTR with cooling jacket.

Solution

Component balance:
Flow of moles of A in: $F_f C_{Af}$
Flow of moles of A out: $F_o C_{Ao}$

Rate of accumulation:

$$\frac{dn}{dt} = \frac{d(CV)}{dt}$$

Rate of generation: rV

$$r = kC_A, \ r_A = -kC_A$$

where r(moles/m^{3}s) is the rate of reaction. The component A mole balance is:

$$\frac{d(VC_A)}{dt} = F_f C_{Af} - F_o C_A + r_A V$$

Because both V and C_A are state variables, the equation becomes:

$$\frac{d(VC_A)}{dt} = V\frac{d(C_A)}{dt} + C_A\frac{d(V)}{dt} = F_f C_{Af} - F_o C_A + r_A V$$

From the total mass balance:

$$\frac{dV}{dt} = F_f - F_o$$

Substitute the total mass balance equation:

$$\frac{d(VC_A)}{dt} = V\frac{d(C_A)}{dt} + C_A\left\{F_f - F_o\right\} = F_f C_{Af} - F_o C_A + r_A V$$

Rearranging, we have:

$$\frac{VdC_A}{dt} = F_f\left(C_{Af} - C_A\right) - kC_A V$$

Degree of Freedom Analysis

Parameter of constant values: A
Forced variable: F_f and C_{Af}
Remaining variables: V, F_o, and C_A
Number of equations: 2
The degree of freedom is: $f = 3 - 2 = 1$

The additional relation is obtained by the connection between the effluent flow F_o and the level in open loop such that:

$$F_o = \alpha\sqrt{L}$$

where α is the discharge coefficient.

Example 1.10: Deterioration of Chlorine

The degeneration rate of chlorine in a distribution system follows a first-order decay with a rate constant of 0.360 day^{-1} if the initial concentration of chlorine in a perfectly mixed liquid storage tank is 1.00 mg/L. Assuming no water flows outside of the storage reservoir, determine the concentration of chorine after one day.

Solution

The process can be considered as a batch process:

$$\frac{d(C_{cl}V)}{dt} = -kC_{cl}$$

Assuming constant volume and integration, we have:

$$\ln\frac{C_{cl}}{C_{cl0}} = -kt$$

Substitute values as follows:

$$\ln\frac{C_{cl}}{1 \text{ mg/L}} = -0.360\left(\frac{1}{\text{day}}\right)(1 \text{ day})$$

The concentration of chlorine after one day is 0.698 mg/L.

PROBLEMS

1.1 Sewage Wastewater Treatment

The municipality sewage with an impurity concentration, $C_{in} = 25\,\text{mg/L}$ enters a pool for treatment. The inlet and outflow flow rates for the pool is $Q = 4000\ \text{m}^3/\text{day}$. The volume of the pool is $20{,}000\,\text{m}^3$. The pool is used to treat sewage wastewater before it is discarded into a channel and goes to a waste storage lake. The purpose of the pond is to allow time for the deterioration of waste material to occur before being released into the environment. The degeneration rate in the pond is first-order with a rate constant equal to 0.25/day. Assuming a steady-state operation, what is the impurities concentration at the outflow of the pond, in units of mg/L?

Answer: 11 mg/L

1.2 Mixing of Two Gas Streams

For a mixture of two gas streams, the first gas stream has a pollutant concentration $C_1 = 5.00$ ppm_v (ppm = parts per million), the gas stream flow rate Q_1 is 0.010 L/min, the second gas stream is fresh air, and the total exit gas stream flow rate Q_{tot} is 1.000 L/min. What is the concentration of calibration gas after mixing (C_d)?

Answer: 50.0 ppb_v (ppb = parts per billion)

1.3 Removal of Waste Heat from Power Plant

A power plant (100 MW) is located next to a river and uses cooling water from the river to remove its waste heat. What is the resulting increase in river temperature? The power plant has an overall efficiency of 30%. All the waste heat from the power plant is removed with cooling water and added to the adjacent river. The river volumetric flow rate is 100 m^3/s.

Answer = 0.167 K

1.4 Leakage of Toxic Gas

A chemical engineer sitting in his office (volume = 1000 ft^3) in a chemical plant realized that a toxic gas has just started entering the room through a ventilation duct. The engineer knows the type of poison from its scent. She also knows that if the gas reaches a concentration of 100 mg/m^3 it will become fatal; it is safe if the concentration stays below 100 mg/m^3. If the ventilation air flow rate in the room is 100 ft^3/min and the incoming gas concentration is 200 mg/m^3, how long does the engineer have to escape?

Answer: 6.9 minutes

1.5 Sewage Waste Treatment

Sewage waste is added to a stream through a discharge pipe. The river's flow rate upstream of the discharge point is $F_u = 8.7\,m^3/s$. The discharge occurs at a flow of $F_d = 0.9\,m^3/s$ and has a dissolved oxygen concentration of 50.0 g/m^3. The upstream dissolved oxygen concentration is negligible. Dissolved oxygen is removed with a first-order decay rate constant $(dC_A/dt = -kC)$ equal to 0.0083 hr^{-1}.

1. What is the dissolved oxygen concentration just downstream of the discharge point?
2. If the stream has a cross-sectional area of 10 m^2, what would the dissolved oxygen concentration be 50 km downstream?

Answers: (1) 4.7 mg/L and (2) 4.2 mg/L

1.6 Mixing of Benzene and Toluene

A mixture containing 45% by mass benzene (B) and 55% by mass toluene (T) is fed to a distillation column. An overhead stream of 95 wt% B is produced, and 8% of the benzene fed to the column leaves in the bottom stream. The feed rate is 2000 kg/h. Determine the overhead flow rate and the mass flow rates of benzene and toluene in the bottom stream.

Answers: 72, 1060 kg/h

1.7 Power Required for Heating Water

How many watts of power would it take to heat 1 liter of water (weighing 1.0 kg, $C_p = 4.18\,\text{kJ/kg°C}$) by 10°C in 1.0 hour? Assume there is no loss of heat, which means that all the energy expended goes into heating the water.

Answer: 12 W

1.8 CSTRs

CSTRs are used to treat an industrial waste, using a reaction that destroys the waste per the first-order kinetics: $r = kC_A$, The reactor volume is 500 m^3; the volumetric flow rate of the inlet and exit streams is 2.1 m^3/h, $k = 0.009$ hr^{-1}; and the inlet waste concentration is 100 mg/L. What is the outlet concentration, assuming steady-state conditions?

Answer = 32 mg/L

1.9 Water-Gas Shift Reaction

The water-gas shift reaction proceeds to equilibrium at a temperature of 830°C:

$$CO + H_2O \Leftrightarrow CO_2 + H_2$$

The equilibrium relation:

$$K(830) = 1 = \frac{y_{CO_2}\, y_{H_2}}{y_{CO}\, y_{H_2O}}$$

where $K(T)$ is the reaction equilibrium constant. At $T = 830$°C, $K = 1.00$. The feed stream to the reactor contains 1.00 mol CO, 2.00 mol of H_2O, and no CO_2 or H_2, and the reaction mixture comes to equilibrium at $T = 830$°C. Calculate the fractional conversion of the limiting reactant.

Answer: $X_e = 0.667$

1.10 Evaporation Loss from a Lake

A lake contains $V = 2\times10^5$ m^3 of water and is fed by a stream discharging $Q_{in} = 9\times10^4$ m^3/year. Evaporation across the surface of the lake takes away $Q_{evap} = 1\times10^4$ m^3/year. The downstream flow rate $Q_{out} = 8\times10^4$ m^3/year exits the lake in the downstream stretch of the river. The upstream river is polluted, with a pollutant concentration of C = 6.0 mg/L. Inside the lake, this pollutant decays with rate $K = 0.12$/year. Calculate the concentration of pollutant in the downstream river.

Answer: 5.9 mg/L

References

1. Close, C. M., D. K. Frederick, 1995. *Modeling and Analysis of Dynamic Systems*, 2nd ed., New York: John Wiley.
2. Felder, R. M., R. W. Rousseau, 2005. *Elementary Principles of Chemical Processes*, 3rd ed., New York: John Wiley.
3. Ghasem, N., R. Henda, 2008. *Principles of Chemical Engineering Processes*, 2nd ed., New York: CRC Press.
4. Grau, M., J. Nougues, L. Puigjaner, 2000. Batch and semibatch reactor performance for an exothermic reaction. *Chemical Engineering and Processing: Process Intensification* 39(2): 141–148.

2

Lumped Parameter Systems

The states in lumped parameter systems are concentrated in single point and are not spatially distributed. Continuous models are described by differential equations, in which dependent variables (y) are defined over continuous range of independent variables (t). This chapter presents the process model construction methods for lumped parameter systems in which the dependent variables of interest are a function of time alone. The parameters in the model are constants and reflect the system's properties or composition. In general, this approach will lead to solving a single or a set of ordinary differential equations (ODEs). The simulation part (i.e., solution of ODEs) is performed via the MATLAB®/Simulink® software package.

LEARNING OBJECTIVES

- Develop mathematical formulation to derive an analytical solution.
- Obtain numerical solution of steady-state and unsteady-state problems related to chemical engineering systems.
- Solve the model equations of the lumped parameter systems using MATLAB/Simulink.

2.1 Introduction

Systems in which the dependent variables of interest are a function of time alone are called lumped systems. In general, this will lead to solving a set of ordinary differential equations (ODEs). For example, a tank full of one or more fluids of different concentrations and compositions is perfectly mixed, and there is no spatial variation in concentrations or temperature inside the tank; this considerably simplifies the model development. Models developed by ignoring spatial variation of a physical quantity of consideration are called lumped parameter models. These models are utilized for describing the steady-state or dynamic behavior of systems encountered in process industry. In the following subsections, we will study typical systems encountered in the process industry, which are modeled as lumped parameter systems.

2.2 Model Encountered in Material Balances Only

The following sections cover non-reactive isothermal systems such as storage tanks (Section 2.2.1), and isothermal reactive systems such as continuous stirred tank reactors (Section 2.2.2). Gas phase reactions in pressurized reactors are covered in Section 2.2.3. Heterogeneous systems, such as catalytic reactions with mass transfer, are studied in Section 2.2.4.

2.2.1 Material Balance Without Reactions

Consider the perfectly mixed storage tank shown in Figure 2.1. A liquid stream with volumetric rate $F_f(\text{m}^3/\text{s})$ and density ρ_f flow into the tank. The outlet stream has volumetric rate F_o and density ρ_o. Our objective is to develop a mathematical model for the variations of the tank holdup, that is, the volume of the fluid in the tank. The system is therefore the liquid in the tank. We will assume that the contents of the tank are perfectly mixed and that the density of the effluent is the same as that of tank content. We will also assume that the tank is isothermal; that is, there are no variations in the temperature. To model the tank, we need only to write a mass balance equation.

Since the system is perfectly mixed, the system properties do not vary with position inside the tank. The only variations are with time. The mass balance equation can be written on the whole system and not only on a differential element of it. This leads to a macroscopic model. The generation term is zero since the mass is conserved. The balance equation yields:

$$\frac{d(\rho V)}{dt} = \rho_f F_f - \rho_o F_o \tag{2.1}$$

Under isothermal conditions we can further assume that the density of the liquid is constant that is, $\rho_f = \rho_o = \rho$.

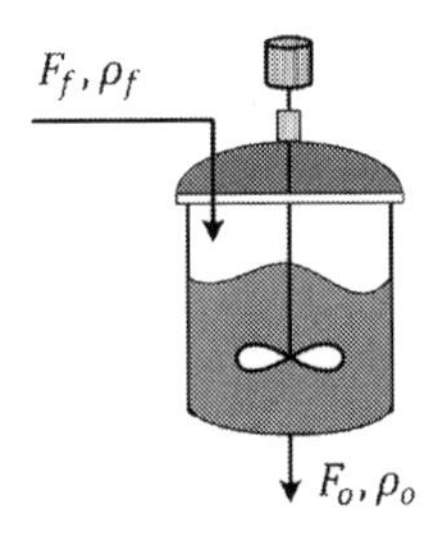

FIGURE 2.1
Liquid mixing tank.

In this case Equation 2.1 is reduced to:

$$\frac{dV}{dt} = F_f - F_o \tag{2.2}$$

The volume V is related to the height of the tank L and to the cross-sectional area A by:

$$V = AL \tag{2.3}$$

Since (A) is constant, we obtained the equation in terms of the state variable L:

$$A\frac{dL}{dt} = F_f - F_o \tag{2.4}$$

2.2.1.1 Degree of Freedom Analysis

The following information represents the system described by Equations 2.1 through 2.4:

- Constant values parameter: A
- Variables that have values that can be outwardly fixed: F_f
- Remaining variables: L and F_0
- Number of equations: 1

Therefore, the number degree of freedom can be written as:

Number of remaining variables − Number of independent Equations = 2 − 1 = 1

One additional equation is needed for the system to be exactly specified. This extra relation is obtained from practical engineering considerations. If the system is operated at an open loop (without control), then the outlet flow rate F_o is a function of the liquid level (L). Generally, a relation of the form can be used:

$$F_o = \alpha\sqrt{L} \tag{2.5}$$

The discharge coefficient is α.

Example 2.1: Dilution of Concentrated Solutions

A tank contains 3 m^3 of a solution consisting of 300 kg of sugar dissolved in water. Pure water is pumped into the tank at the rate of 15 L/s, and the mixture (kept uniform by stirring) is pumped out at the exact same rate. How long will it take for only 30 kg of sugar to remain in the tank? (See Figure 2.2.)

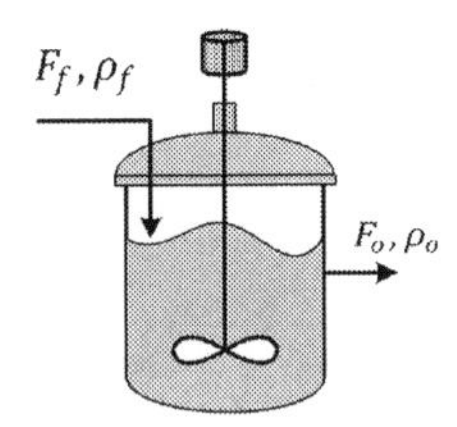

FIGURE 2.2
Sugar concentration tank.

Solution

Assume an isothermal process and a perfect mixing tank ($\rho_f = \rho_o = \rho$). After dividing both sides of the equation by ρ, the material balance equation is reduced to:

$$\frac{dV}{dt} = F_f - F_o$$

The inlet and exit volumetric flow rates are equal; hence:

$$F_f = F_o = 15\,\mathrm{L/s}$$

accordingly,

$$\frac{dV}{dt} = 0$$

The concentration of sugar in the tank is determined by component mass balance. The initial concentration of sugar in the tank is $c(0)$:

$$C(0) = \frac{300\,\mathrm{kg}}{3000\,\mathrm{L}} = 0.1\frac{\mathrm{kg}}{\mathrm{L}}$$

The final concentration is c_f:

$$C(f) = \frac{30\,\mathrm{kg}}{3000\,\mathrm{L}} = 0.01\frac{\mathrm{kg}}{\mathrm{L}}$$

Component Balance

The component balance of sugar (A) equation:

$$\frac{d(VC_A)}{dt} = F_f C_f - F_o C_A$$

Since the feed stream contains fresh water, the feed concentration of sugar (A) is zero; $c_f = 0$:

$$\frac{d(VC_A)}{dt} = -F_o C_A$$

Differentiation of the left side of the equation can be done via product rule:

$$V\frac{d(C_A)}{dt} + C_A\frac{d(V)}{dt} = -F_o C_A$$

Since volume of the tank from the total mass balance is constant, the equation is reduced to:

$$V\frac{dC_A}{dt} = -F_o C_A$$

Divide both sides of the equation by V:

$$\frac{dC_A}{dt} = -\frac{F_o C_A}{V}$$

Rearranging and integrating both sides of the equation, we have:

$$\int_{C_{Ao}}^{C_A} \frac{dC_A}{C_A} = \frac{-F_o}{V}\int_0^t dt$$

which leads to:

$$ln\frac{C_A}{C_{A0}} = \frac{-F_o}{V}t$$

Substitute the values of known parameters in the integrated equation:

$$\ln\frac{0.01}{0.1} = \frac{-15}{3000}t$$

Solve for time (the time required to achieve the final concentration):

$$t = 460.517 \text{ s}$$

The MATLAB/Simulink solution of the ODE is shown in Figure 2.3.

$$\frac{dC_A}{dt} = -\frac{F_o C_A}{V} = -\frac{15\left(\frac{L}{s}\right)C_A}{3000\,\text{L}}$$

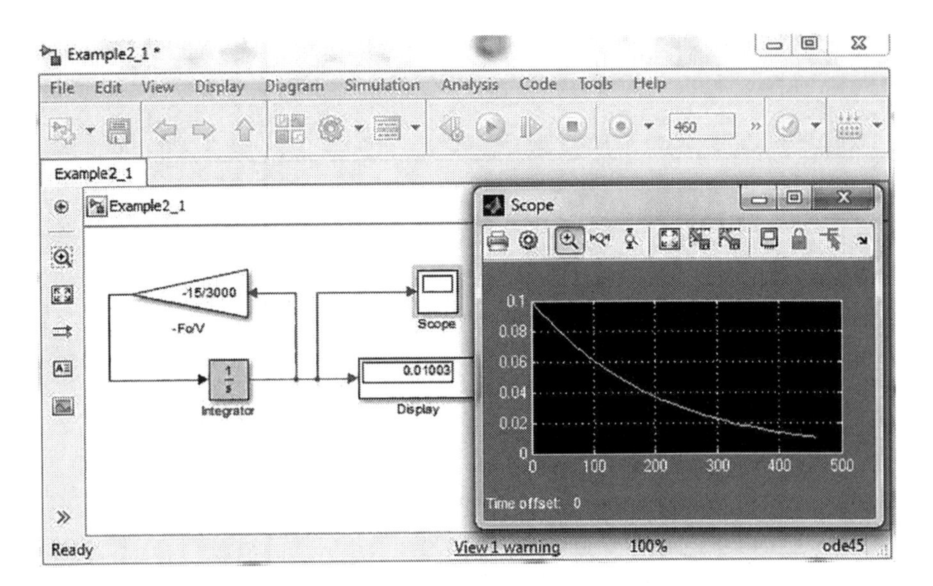

FIGURE 2.3
Simulink program, concentration versus time.

Initial conditions at

$$t = 0,\ C_A(0) = \frac{300\ \text{kg}}{3000\ \text{L}} = 0.3\ \text{kg/L}$$

The simulation predictions are the same as the manual calculations.

Example 2.2: Dilution of Brine Water

Pure water turns into a well-mixed tank filled with 100 L of brine. Water flows at a constant volumetric feed rate of 10 L/min. Initially, the brine has 7.0 kg of salt dissolved in the 100 L of water. The salt solution flows out of the tank at the same inlet volumetric flow rate of water. After 15 min of operation, calculate the amount of salt remaining in the tank. Plot the concentration of the salt versus time using Simulink (Figure 2.4).

Solution

The total mass balance of the tank contents is:

$$\frac{dm}{dt} = \dot{m}_{\text{in}} - \dot{m}_{\text{out}}$$

Because the content of the tank is assumed to be well mixed and the tank is under isothermal conditions (ρ is constant), then:

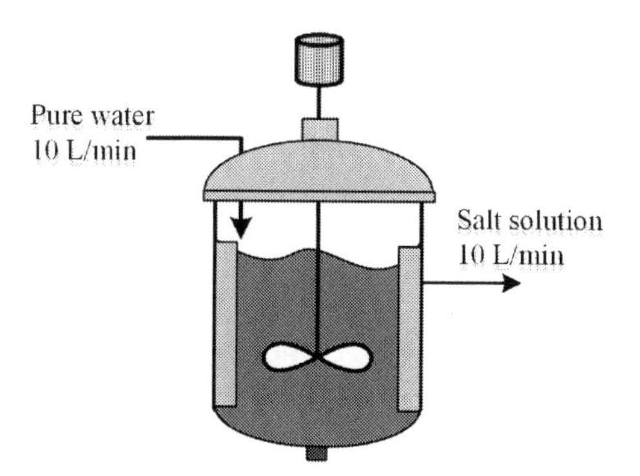

FIGURE 2.4
Mixed tank with brine and water.

$$\frac{dV}{dt} = F_{\text{in}} - F_{\text{out}}$$

where:

$F_{\text{in}}, F_{\text{out}}$ represents the inlet and exit volumetric flow rate, respectively
V is the tank holdup

Because the inlet volumetric flow rate to the tank is the same as the liquid volumetric flow rate leaving the tank (10 L/min), the volume of the tank is left constant, $V = 100$ L. Thus:

$$\frac{dV}{dt} = 0$$

The salt (A) component mass balance:

$$\frac{d\left(VC_{A,\text{tank}}\right)}{dt} = Fc_{A,\text{in}} - Fc_{A,\text{out}}$$

where V is the volume of the fluid in the tank, which is constant. The inlet stream consists of pure water, so the inlet salt concentration $C_{A,\text{in}} = 0$

Due to the assumption of a well-mixed tank, the exit salt concentration equals the concentration of the fluid in the tank, $C_{A,\text{out}} = C_{\text{tank}} = c_A$. Hence, the component mass balance equation is reduced to:

$$\frac{d\left(Vc_A\right)}{dt} = -FC_A$$

Since the volume is constant:

$$\int_{C_{Ao}}^{C_A} \frac{dC_A}{c_A} = -\frac{F}{V}\int_0^t dt$$

divide both sides by V, integrate, and rearrange:

$$\ln\frac{C_A}{C_{A0}} = \frac{-F}{V}t$$

Rearranging for the concentration, c_A, we have:

$$C_A = C_{A0}\exp\left(-\frac{F}{V}t\right)$$

The initial concentration in the tank, that is, the concentration at time zero is:

$$C_{A0} = \frac{7\ \text{kg}}{100\ \text{L}} = 0.07\ \text{kg/L}$$

Substitution of known parameters, the final concentration in the tank, C_A, is:

$$C_A = 0.07\ \text{kg/L} \times \exp\left(-\frac{10\,\text{L/min}}{100\,\text{L}} \times 15\ \text{min}\right) = 0.01562\ \text{kg/L}$$

The amount of the salt in the tank is:

$$m_{\text{salt}} = C_A \times V = 0.01562\ \text{kg/L} \times 100\ \text{L} = 1.562\ \text{kg}$$

It is important to note that the concentration of salt in the tank decreases with the time. After 15 min, the salt concentration is 0.015619 kg/L and the amount of it is 1.5619 kg. The Simulink solution of the ODE with initial conditions is shown in Figure 2.5.

$$\frac{dC_A}{dt} = -\frac{F}{V}C_A = -\frac{10\ \text{L/min}}{100\ \text{L}}C_A$$

Initial conditions:

$$C_A(0) = 0.07\ \text{kg/L}$$

The concentration of the mixture in the tank at the end of 15 min is 0.01562 kg/L, which is almost the same as the analytical solution.

2.2.2 Material Balance for Chemical Reactors

A perfectly mixed tank shown in Figure 2.6. It is used for processing the liquid phase chemical reactions: $A \to B$, $r = kc_A$.

The reaction is assumed to be irreversible and of first order. The feed that enters the reactor has the following parameters: volumetric feed rate,

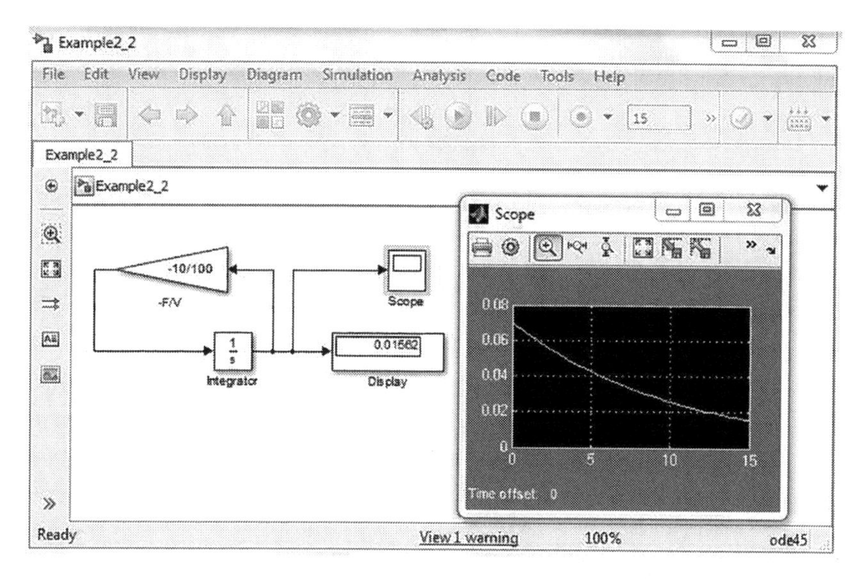

FIGURE 2.5
Simulation prediction with Simulink.

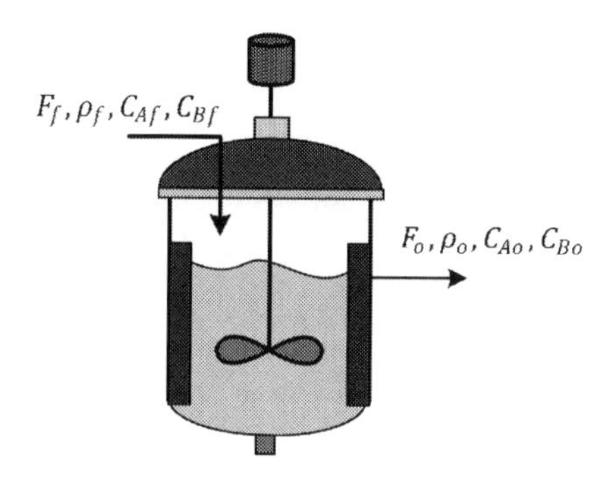

FIGURE 2.6
Isothermal CSTR.

F_f (m^3/s), density of feed stream, ρ_f (kg/m^3), and feed concentration C_{Af} (mol/m^3). The output that exits the reactor has the following parameters: volumetric outlet rate, F_o; density of outlet stream, ρ_o; and concentration of exit stream, C_{Ao} (mol/m^3) and c_{Bo} (mol/m^3). Assume the reactor is operating at an isothermal condition.

The goal is to develop a model for the variation of the volume of the reactor and the concentration of species A and B. The application of the component balance equation to the total number of moles ($n_A = c_AV$) is used to obtain the component balance on species A. Since the system is well mixed, the process concentration C_A and C_B is equal to the effluent concentration C_{Ao} and C_{Bo}. The rate of consumption of A is:

$$r_A = -rV$$

where r(mol/(m^{3}s)) is the rate of reaction. The general component mole balance equation is:

$$\frac{d(VC_A)}{dt} = F_f C_{Af} - F_o C_A + r_A V \tag{2.6}$$

We can check that all terms in the equation have the units (mol/s). Although a similar component balance on species B can be written, it is not needed because it will not represent an independent equation. A system of n species is accurately specified by a number of n independent equations. We can either write the total mass balance equations along with $(n - 1)$ component balance equations, or we can write n component balance equations. Applying the differential form of the product rule, Equation 2.6 can be written as follows:

$$\frac{d(VC_A)}{dt} = V\frac{d(C_A)}{dt} + C_A\frac{d(V)}{dt} = F_f C_{Af} - F_o C_A + r_A V \tag{2.7}$$

The total mass balance equation at constant fluid density is:

$$\frac{dV}{dt} = F_f - F_o \tag{2.8}$$

Using algebraic manipulations with Equations 2.8 into 2.7, we obtain:

$$V\frac{dC_A}{dt} = F_f(C_{Af} - C_A) + r_A V \tag{2.9}$$

The reaction rate is a first-order irreversible reaction:

$$r = kc_A \text{ and } r_A = -kc_A \tag{2.10}$$

Equations 2.8 and 2.9, which define the dynamic behavior of the reactor, can be solved if the system is exactly specified and if the initial conditions are given:

$$V(0) = V_i \text{ and } C_A(0) = C_{A_i}$$

The degrees of freedom analysis:

- The parameter of constant values: A
- The forced variables: F_f and C_{Af}
- The remaining variables: V, F_o, and C_A
- The number of independent equations: 2

The degree of freedom is then: 3 − 2 = 1.

The extra required relation is obtained by relating the effluent flow rate (F_o) with the level of fluid in the tank (h) in an open loop operation ($F_o = \alpha\sqrt{h}$). The steady-state behavior and the equal inlet and exit flow rates, $F_o = F_f$, can be obtained easily by setting the accumulation terms to zero. Accordingly, Equation 2.9 becomes:

$$F_f\left(C_{Af} - C_A\right) = r_A V \tag{2.11}$$

The same method can be used to model more complex situations. Consider the catalytic hydrogenation of ethylene:

$$A + B \rightarrow C$$

where A, B, C, represents ethylene, hydrogen, and ethane, respectively. The reaction happens in a continuous stirred tank reactor (CSTR), which can be observed in Figure 2.7. Two streams are feeding the reactor. A concentrated feed stream has a volumetric feed rate $F_1(\text{m}^3/\text{s})$ and concentration $C_{B1}(\text{mol/m}^3)$; the second stream is a dilute stream with flow rate $F_2(\text{m}^3/\text{s})$ and concentration $C_{B2}(\text{mol/m}^3)$. The effluent has flow rate $F_o(\text{m}^3/\text{s})$ and concentration $C_B(\text{mol/m}^3)$. The reactant A is assumed to be in excess; thus, it is not involved in the reaction rate.

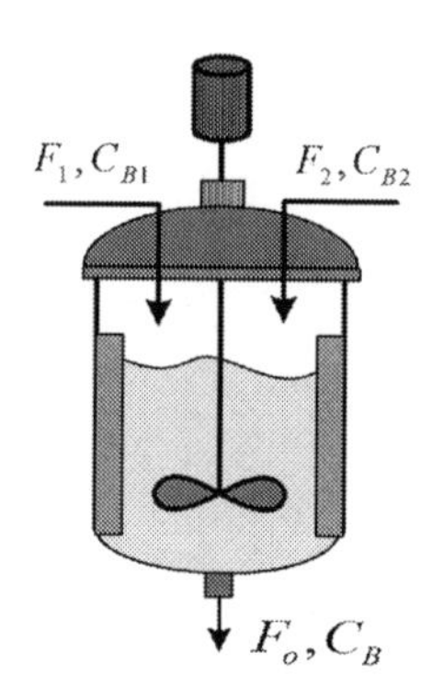

FIGURE 2.7
Reaction taking place in a stirred tank (CSTR).

The reaction rate is:

$$r\left(\frac{\text{mol}}{\text{m}^3\ \text{s}}\right)=\frac{k_1C_B}{\left(1+k_2C_B\right)^2} \tag{2.12}$$

where k_1 and k_2 are the reaction rate constant and the adsorption equilibrium constant, respectively. If we assume that the operation is isothermal and the density is constant, the following model can be obtained using the same procedure of the previous example:

Total mass balance at constant density of mixed fluids:

$$A\frac{dL}{dt}=F_1+F_2-F_o \tag{2.13}$$

The component (B) mole balance after assuming a constant volume holdup ($dV/dt=0$). Substituting $F_o=F_1+F_2$ in the component mole balance equation, the resultant equation is:

$$V\frac{dC_B}{dt}=F_1\left(c_{B1}-c_B\right)+F_2\left(c_{B2}-c_B\right)-rV \tag{2.14}$$

The degrees of freedom analysis:

- The parameters of constant values: A, k_1 and k_2
- The forced variables: F_1, F_2, c_{B1} and c_{B2}
- The remaining variables: V, F_o, and c_B
- The number of equations: 2
- The degree of freedom is therefore 3 − 2 = 1

The extra relation is between the effluent flow F_o and the liquid level L, as it was in the previous example.

Example 2.3: Liquid Phase Reaction

The elementary liquid-phase irreversible reaction is to be carried out in two continuous stirred tank reactors in series, both reactors having a volume of 15 L (Figure 2.8). The liquid phase reaction: $A+B\rightarrow C$.

We can describe the reaction inside the reactor by the following equation:

$$-r_A=kC_AC_B$$

The reaction rate constant, k, is:

$$k=0.3\frac{\text{L}}{\text{mol. min}}$$

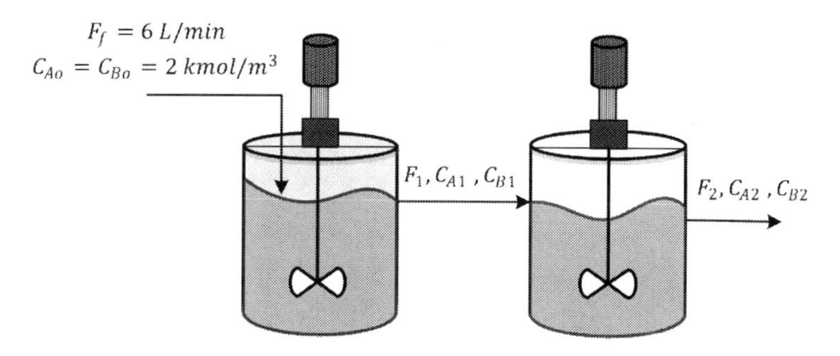

FIGURE 2.8
Two constant volume stirred stank reactors in series.

The feed stream is equimolar in A and B with concentration of 2 kmol/m^3 or 2 mol/L, and the volumetric flow rate is 6 L/min. Find the outlet concentrations for components A and B in each tank.

Assumption

The following assumption are considered:

- Isothermal operation.
- The contents of both tanks are perfectly mixed.
- Fluid is at constant density.

Total Mass Balance

Assuming constant density, the total mass balance around tank 1 is:

$$\frac{dV_1}{dt} = F_f - F_1$$

Since $F_f = F_1 = F_2$, so $\frac{dV_1}{dt} = \frac{dV_2}{dt} = 0$

Tank 1

Component (A) mole balance on tank 1 is:

$$\frac{d(V_1 C_{A1})}{dt} = F_f C_{A0} - F_1 C_{A1} - (kC_{A1} C_{B1}) V_1$$

Divide both sides of the equation by V_1 and $F_f = F_1$, then rearrange:

$$\frac{dC_{A1}}{dt} = \frac{F_f}{V_1}(C_{A0} - C_{A1}) - kC_{A1} C_{B1}$$

Component (B) mole balance:

$$\frac{d(V_1C_{B1})}{dt} = F_fC_{B0} - F_1C_{B1} - (kC_{A1}C_{B1})V_1$$

Divide both sides of the equation by V_1:

$$\frac{d(C_{B1})}{dt} = \frac{F_f}{V_1}(C_{B0} - C_{B1}) - (kC_{A1}C_{B1})$$

Tank 2

Component (A) mole balance around tank 2 is:

$$\frac{d(V_2C_{A2})}{dt} = F_1C_{A1} - F_2C_{A2} - (kC_{A2}C_{B2})V_2$$

Divide by V_2 and $F_f = F_2$, then rearrange:

$$\frac{d(C_{A2})}{dt} = \frac{F_f}{V_2}(C_{A1} - C_{A2}) - (kC_{A2}C_{B2})$$

Component (B) mole balance around tank 2 is:

$$\frac{d(VC_{B2})}{dt} = F_1C_{B1} - F_2C_{B2} - (kC_{A2}C_{B2})V_2$$

Divide both sides of the equation by V_2 and rearrange:

$$\frac{d(C_{B2})}{dt} = \frac{F_f}{V_2}(C_{B1} - C_{B2}) - kC_{A2}C_{B2}$$

Using MATLAB/Simulink software program, the following four ordinary differential equations are to be solved simultaneously.

Tank 1

$$\frac{dC_{A1}}{dt} = \frac{F_f}{V_1}(C_{A0} - C_{A1}) - kC_{A1}C_{B1}$$

$$\frac{dC_{B1}}{dt} = \frac{F_f}{V_1}(C_{B0} - C_{B1}) - kC_{A1}C_{B1}$$

Tank 2

$$\frac{dC_{A2}}{dt} = \frac{F_f}{V_2}(C_{A1} - C_{A2}) - kC_{A2}C_{B2}$$

$$\frac{dC_{B2}}{dt} = \frac{F_f}{V_2}(C_{B1} - C_{B2}) - kC_{A2}C_{B2}$$

Data provided:

$$F_f = 6\ \text{L/min}$$

$$C_{Ao} = C_{B0} = 2\ \text{mol/L}$$

$$k = 0.3\frac{\text{L}}{\text{mol. min}}$$

$$V_1 = V_2 = 15\ \text{L}$$

$$\frac{F_f}{V} = \frac{6}{15} = 0.4$$

The steady-state outlet concentrations are extracted from the Simulink solution in Figure 2.9:

$$C_{A1} = 1.097\ \text{mol/L}, C_{B1} = 1.097\ \text{mol/L}, C_{A2} = 0.7144\ \text{mol/L}, C_{B2} = 0.7144\ \text{mol/L}$$

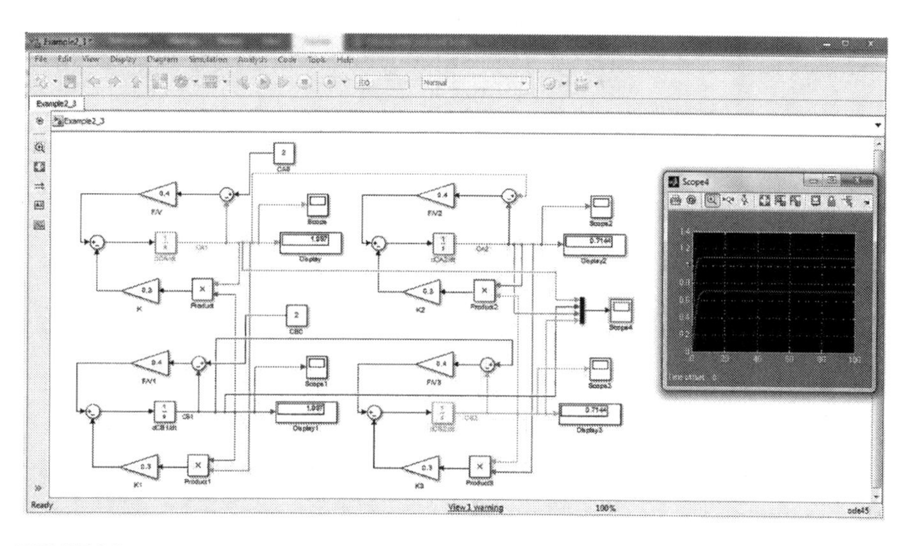

FIGURE 2.9
Simulink program for the two CSTR in series (Example 2.3).

Example 2.4: CSTR with Multiple Reactions

An isothermal CSTR is used for producing the desired component X. Unfortunately, side reactions take place in which components Y and Z are also produced forming a complex reaction. Changes inside the reactor can be described by following reactions:

$$A + B \xrightarrow{k_1} X$$

$$B + X \xrightarrow{k_2} Y$$

$$B + Y \xrightarrow{k_3} Z$$

The parameters and the related constant are shown in Table 2.1.

Solution

The schematic of the reactions process is shown in Figure 2.10.

If the reactant inside the tank is perfectly mixed and the volume of the reactant inside the reactor during the reaction is constant, the material balances inside the reactor can be used to derive the mathematical model of the system. All three reactions are assumed to follow second order kinetics. The component mole balance on A:

$$\frac{d(VC_A)}{dt} = (F_0 C_{Ao} - FC_A) - k_1 C_A C_B V$$

TABLE 2.1

Reactor Parameters and Constants

Parameters	Value	Units
k_1	5×10^{-4}	$m^3/kmol \cdot s$
k_2	5×10^{-2}	$m^3/kmol \cdot s$
k_3	2×10^{-2}	$m^3/kmol \cdot s$
c_{A0}	0.4	$kmol/m^3$
c_{B0}	0.6	$kmol/m^3$
c_{X0}, c_{Y0}, c_{Z0}	0	$kmol/m^3$
V	1	m^3
F	0.01	m^3/s

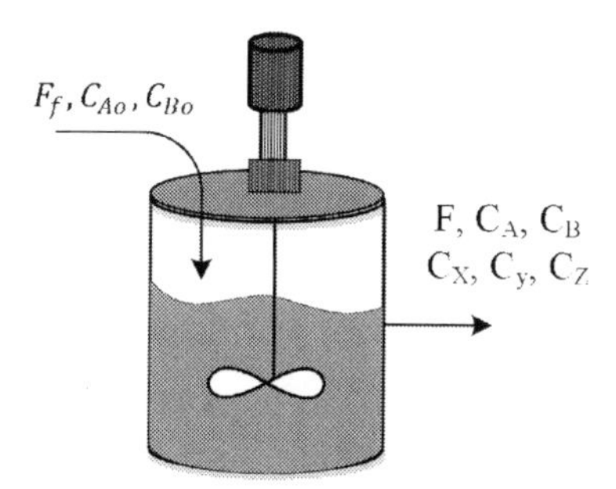

FIGURE 2.10
Continuous stirred tank reactor with multiple reactions.

Dividing by V, the tank holdup is constant:

$$\frac{dC_A}{dt} = \frac{(F_0 C_{Ao} - FC_A)}{V} - k_1 C_A C_B$$

Component mole balance on B

$$\frac{d(VC_B)}{dt} = (F_0 c_{Bo} - FC_B) - k_1 C_A C_B V - k_2 C_X C_B V - k_3 C_Y C_B V$$

Dividing all terms of the equation by V:

$$\frac{dC_B}{dt} = \frac{(F_0 C_{Bo} - FC_B)}{V} - k_1 C_A C_B - k_2 C_X C_B - k_3 C_Y C_B$$

Mole balance on component Z:

$$\frac{dC_Z}{dt} = \frac{(F_0 C_{Zo} - FC_Z)}{V} + k_3 C_Y C_B$$

Mole balance on component X:

$$\frac{dC_X}{dt} = \frac{(F_0 C_{Xo} - FC_X)}{V} + k_1 C_A C_B - k_2 C_X C_B$$

Mole balance on component balance on Y:

$$\frac{dC_Y}{dt} = \frac{(F_0 C_{Yo} - FC_Y)}{V} + k_2 C_X C_B - k_3 C_Y C_B$$

Mole balance on component Z:

$$\frac{dC_Z}{dt} = \frac{(F_0 C_{Zo} - FC_Z)}{V} + k_3 C_Y C_B$$

The previous sets of nonlinear ordinary differential equations illustrate the behavior of the state variables that are, in this case, concentrations of components $A, B, X, Y,$ and Z are $C_A, C_B, C_X, C_Y,$ and C_Z, in time t, respectively. We can say that this CSTR belongs to the class of nonlinear lumped-parameters systems. In the above equations, F represents volumetric flow rate, V represents the volume of the tank, k represents the reaction rate constants, and C represents the concentrations. Assume that the inlet flow rate is equal to the outlet flow rate. The industrial constants and parameters can be observed in Table 2.1. The steady-state analytical solution can be obtained. Steady-state analysis means computing the state variables in time ($t \to \infty$), where the value of this quantity is expected to be stable. From the computation point of view, it means that all derivatives with respect to time are equal to zero.

$$\frac{dC_A}{dt} = 0 = \frac{(F_0 C_{Ao} - FC_A)}{V} - k_1 C_A C_B$$

Rearranging, we have:

$$C_A = \frac{FC_{Ao}}{F + VK_1 C_B}$$

Similarly, at the steady state, we get the concentration of, C_A, C_B, C_X, C_Y, and C_Z as follows:

$$C_B = \frac{FC_{Bo}}{F + VK_1 C_A + VK_2 C_X + VK_3 C_Y}$$

$$C_X = \frac{FC_{Xo} + VK_1 C_A C_B}{F + VK_2 C_B}$$

$$C_Y = \frac{FC_{Yo} + VK_2 C_X C_B}{F + VK_3 C_B}$$

$$C_Z = \frac{FC_{Zo} + K_3 C_Y C_B}{F}$$

The polymath program and its solution are shown in Figures 2.11 and 2.12, respectively. The Simulink predictions are shown in Figure 2.13. Both results are in good agreement.

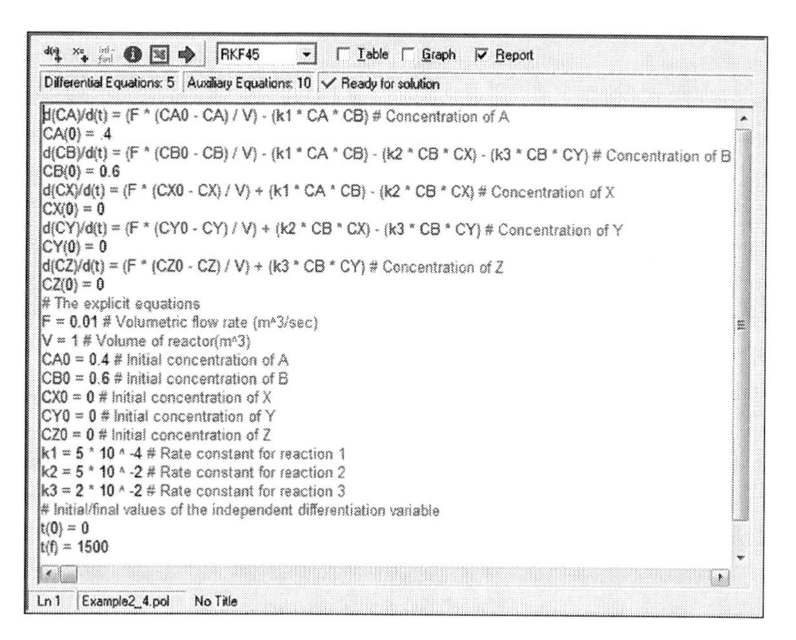

FIGURE 2.11
Polymath program for the solution of the set of ODEs.

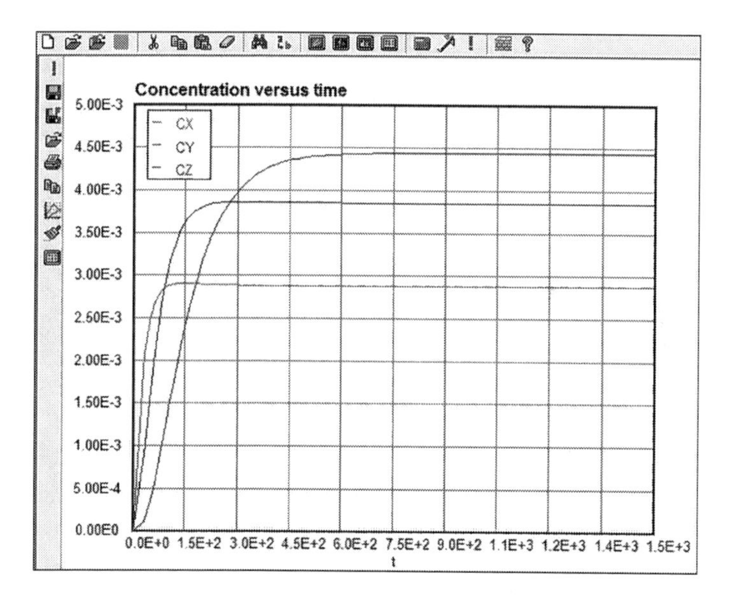

FIGURE 2.12
Concentration of X, Y, and Z versus time generated vial polymath.

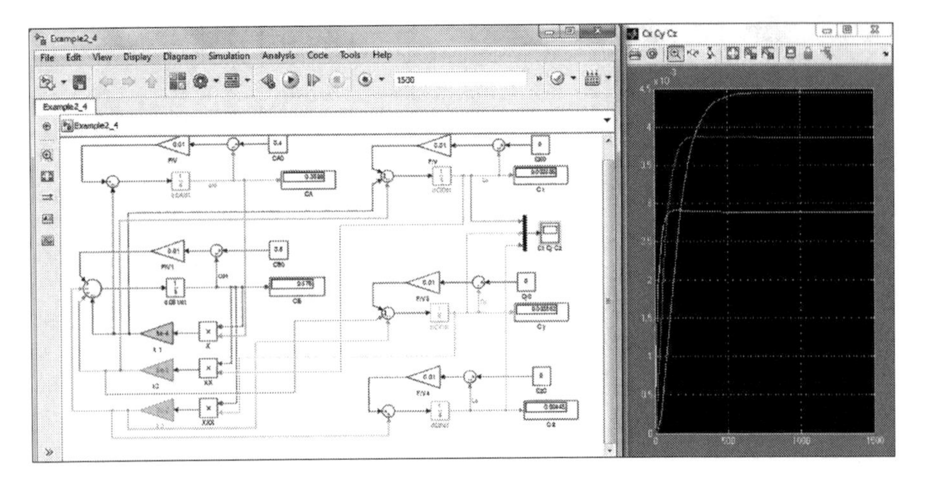

FIGURE 2.13
Concentration of X, Y, and Z versus time generated via Simulink.

2.2.3 Gas Phase Reaction in a Pressurized Reactor

So far, only liquid-phase reactions, where density can be taken as constant, have been considered. The following elementary reversible reaction is used to illustrate the effect of gas-phase chemical reaction on mass balance equation:

$$A \rightleftarrows 2B$$

The reaction is taking place in a perfectly mixed vessel, which is sketched in Figure 2.14. The influent to the vessel has volumetric rate F_f(m^3/s), density ρ_f(kg/m^3), and mole fraction y_f. Product comes out of the reactor with volumetric rate F_o, density ρ_o, and mole fraction y_o. The temperature and volume inside the vessel are constant. The reactor effluent passes through a control valve that regulates the gas pressure at constant pressure P_g.

Writing the macroscopic total mass balance around the vessel gives:

$$\frac{d(\rho V)}{dt} = \rho_f F_f - \rho_o F_o \tag{2.15}$$

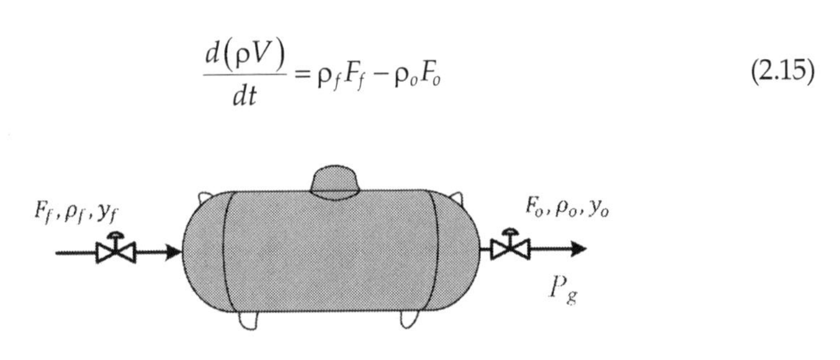

FIGURE 2.14
Gas pressurized reactor.

Since V is constant, we have:

$$V\frac{d(\rho)}{dt} = \rho_f F_f - \rho_o F_o \tag{2.16}$$

Writing the component mole balance, for fixed V, results in:

$$V\frac{dC_A}{dt} = F_f C_{Af} - F_o C_{Ao} + rV \tag{2.17}$$

The reaction rate for the reversible reaction is:

$$r_A = -k_1 C_A + k_2 C_B^2$$

The variations of density and molar concentration are defined in Equations 2.16 and 2.17. The equations can also be rewritten to define the behavior of the pressure (P) and mole fraction (y). The concentration can be expressed in terms of the density via the ideal gas law:

$$C_A = y_A P/RT \tag{2.18}$$

$$C_B = (1 - y_A)P/RT \tag{2.19}$$

Similarly, the density can be related to the pressure using the ideal gas law:

$$\rho = \frac{M_w P}{RT} = \frac{\left[M_{w,A} + M_{w,B}(1-y)\right]P}{RT} \tag{2.20}$$

where $M_{w,A}$ and $M_{w,B}$ are the molecular weights of A and B, respectively. Therefore, one can substitute Equations 2.18 through 2.20 into Equations 2.15 and 2.17 to write the latter two equations in terms of y and P.

Degrees of freedom analysis:

- The parameters: $V, k_1, k_2, R, T, M_{w,A}$ and $M_{w,B}$
- The forcing functions: F_f, c_{Af}, y_f
- The variables: c_A, c_B, y, P, ρ, F
- The Number of equations: 5

The degree of freedom is therefore 6 − 5 = 1. The extra relation relates the outlet flow to the pressure as follows:

$$F_o = C_v\sqrt{\frac{P - P_g}{\rho}} \tag{2.21}$$

where:

C_v is the valve-sizing coefficient

P_g is also assumed to be constant

2.2.4 Reaction with Mass Transfer

Chemical processes that incorporate diffusion usually involve chemical reactions. Often diffusion and reaction occur in the same region, and the two rate phenomena are coupled so closely that they must be treated simultaneously. An industrial example is the absorption of carbon dioxide via ethanolamine (EA) and diethanolamine (DEA), where diffusion and reaction are taking place simultaneously. Figure 2.15 shows a chemical reaction that takes place in a gas-liquid environment. The reactant A enters the reactor as a gas and the reactant B enters as a liquid. The gas dissolves in the liquid, where it chemically reacts to produce a liquid C. The effluent liquid flow rate, F_L, is used to drown off the product from the reactor. The unreacted gas vents from the top of the vessel. The reaction mechanism is given as follows:

$$A \rightarrow B \rightarrow C$$

2.2.4.1 Assumptions

1. Perfectly mixed reactor.
2. Isothermal operation.
3. Constant pressure, density, and holdup.
4. Negligible vapor holdup at the top of the column.

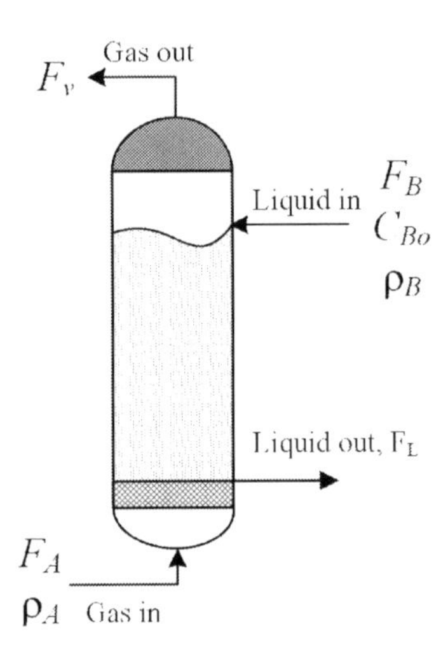

FIGURE 2.15
Reaction with mass transfer.

In cases where the two chemical phenomena (mass transfer and chemical reaction) occur together, the reaction process may become mass transfer dominant or reaction-rate dominant. If the mass transfer has a slower reaction rate, then mass transfer prevails, and vice versa. Due to the assumption of perfect mixing, macroscopic mass transfer of component A from the bulk gas to the bulk liquid is approximated by the following molar flux:

$$N_A = K_L\left(C_A^* - C_A\right)$$

where:
- K_L is the mass transfer coefficient
- C_A^* is the gas concentration at gas-liquid interface
- C_A is the gas concentration in bulk liquid

The macroscopic balance of the liquid phase where the chemical reaction takes place is derived to describe the process fully. This results in:

2.2.4.2 Liquid Phase

Total mass balance:

$$\frac{d(\rho V)}{dt} = \rho_B F_B + M_{w,A} A_m N_A - \rho F_L \tag{2.22}$$

Component balance on A:

$$V\frac{dC_A}{dt} = A_m N_A - F_L C_A + r_A V \tag{2.23}$$

Component balance on B:

$$V\frac{dC_B}{dt} = F_B C_{Bo} - F_L C_B + r_B V \tag{2.24}$$

2.2.4.3 Vapor Phase

Here, since vapor holdup is negligible, the steady-state total continuity equation can be written as follows:

$$\frac{d(\rho_A V_v)}{dt} = 0 = F_A \rho_A + M_{w,A} A_m N_A - F_y \rho_A \tag{2.25}$$

where:
- A_m is the total mass transfer area of the gas bubbles
- $M_{w,A}$ is the molecular weight of component A
- ρ_A is the vapor density
- V_v is the vapor volume

2.2.4.4 Degrees of Freedom Analysis

- The forcing variables: F_A, F_B, C_{B0}
- The parameters of constant values: $K_L, M_{w,A}, A_m, \rho, \rho_A, \rho_B$
- The remaining variables: C_B, N_A, C_A, F_v, V
- The number of equations: 5

It's important to note that the liquid flow rate, F_L, can be determined from the overall mass balance and that the reaction rate r should be defined.

2.3 Energy Balance

We reconsider the CSTR but for nonisothermal conditions. The reaction $A \rightarrow B$ is exothermic, and the heat generated in the reactor is removed via a cooling system, as shown in Figure 2.16. The effluent temperature is different from the inlet temperature due to the heat generation by the exothermic reaction.

If we assume that the density is constant, the macroscopic total mass balance and the component mole balance remain the same as before. However, one more ordinary differential equation will be produced from applying the conservation law for total energy balance. The dependence of the rate constant on the temperature (Arrhenius equation) should be emphasized:

$$k = k_o e^{-\frac{E}{RT}}$$

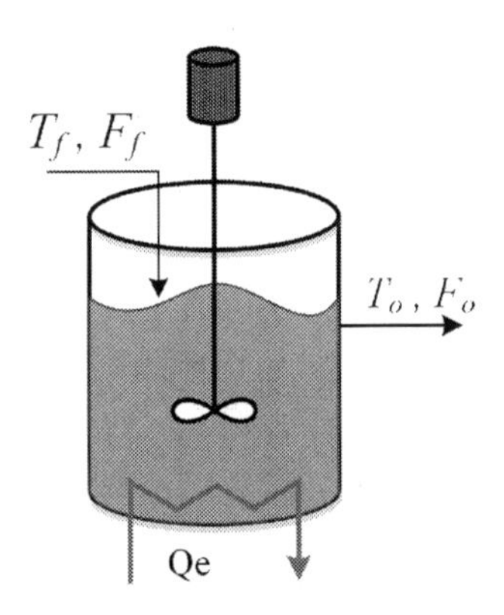

FIGURE 2.16
Non-isothermal CSTR.

The general energy balance equation (Equation 2.26) for macroscopic systems applied to the CSTR yields (assuming constant density and average heat capacity):

$$\rho C_p \frac{d(V(T - T_{\text{ref}}))}{dt} = \rho F_f C_p (T_f - T_{\text{ref}}) - \rho F_o C_p (T - T_{\text{ref}}) + Q_r - Q_c \qquad (2.26)$$

where:
Q_r(J/s) is the heat generated by the reaction
Q_c(J/s) the rate of heat removed by the cooling system

For simplicity and the use of the differentiation principles, we let $T_{\text{ref}} = 0$. Then Equation 2.26 can be rewritten as follows:

$$\rho C_p V \frac{dT}{dt} + \rho C_p T \frac{dV}{dt} = \rho F_f C_p (T_f - T_{\text{ref}}) - \rho F_o C_p (T - T_{\text{ref}}) + Q_r - Q_c \qquad (2.27)$$

Substituting ($dV/dt = F_f - F_o$) into Equation 2.27 and rearranging yields:

$$\rho C_p V \frac{dT}{dt} = \rho F_f C_p (T_f - T) + Q_r - Q_c \qquad (2.28)$$

The rate of heat exchanged Q_r due to the reaction is given by:

$$Q_r = -(\Delta H_r) V_r \qquad (2.29)$$

where ΔH_r(J/mol) is the heat of reaction, which has a negative value for an exothermic reaction and a positive value for an endothermic reaction. The nonisothermal CSTR is therefore modeled by three ODEs (total mass balance, component mole balance, and energy balance):

$$\frac{dV}{dt} = F_f - F_o \qquad (2.30)$$

$$V \frac{d(C_A)}{dt} = F_f (C_{Af} - C_A) - rV \qquad (2.31)$$

$$\rho C_p V \frac{dT}{dt} = \rho F_f C_p (T_f - T) + (-\Delta H_r) V_r - Q_c \qquad (2.32)$$

where the reaction rate (r) is given by:

$$r = k_o e^{-\frac{E}{RT}} c_A$$

If the system is exactly specified and if the initial conditions are given, then the system can be solved as follows: $V(0) = V_i$, $T(0) = T_i$ and $C_A(0) = C_{Ai}$.

Degrees of freedom analysis:

- The parameter of constant values: ρ, E, R, C_p, ΔH_r and k_o
- The forced variables: F_f, C_{Af} and T_f
- The remaining variables: V, F_o, T, c_A and Q_c
- The number of equations: 3

The degree of freedom is 5 – 3 = 2. Following the example of the previous analysis, it can be observed that the two extra relations are between the effluent stream (F_o) and the volume (V) on one hand and between the rate of heat exchanged (Q_c) and temperature (T) on the other hand, in either open loop or closed loop operations.

A more elaborate model of the CSTR would include the dynamic of the cooling jacket (Figure 2.17), with the assumption that the jacket is to be perfectly mixed with the constant volume V_j, density ρ_j, and constant average thermal capacity C_{pj}. We would then simply apply the macroscopic energy balance on the whole jacket, and the dynamic of the cooling jacket temperature can be modeled as follows:

$$\rho_j C_{p_j} V_j \frac{dT_j}{dt} = \rho_j F_c C_{p_j} \left(T_{jf} - T_j\right) + Q_c \quad (2.33)$$

Because V_j, ρ_j, C_{pf} and T_{jf} are constant or known, the addition of this equation introduces only one variable (T_j). The system is still exactly specified.

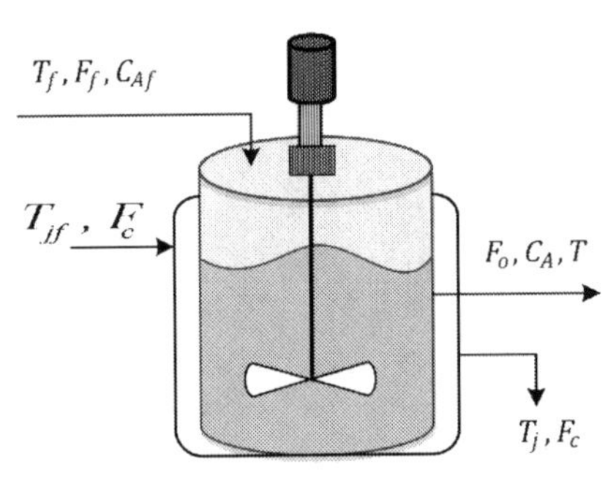

FIGURE 2.17
Jacketed non-isothermal CSTR.

Example 2.5: Jacketed Stirred Tank Reactor

Steam is employed to heat the contents of a jacketed stirred tank initially at 30°C. The steam temperature is fixed at 180°C. The jacket-tank contact surface area is 20 m^2. The volume of the tank is 100 L. The density of fluid in the tank is constant (1 kg/L), and the heat capacity is also constant $(4.187\,\mathrm{kJ}/(\mathrm{kg}\cdot{}^\circ\mathrm{C}))$. The inlet fluid temperature is 30°C, and the inlet stream flow rate is 250 L/min. At steady state, the fluid exit temperature is 60°C, and the exit steam flow rate is the same as the inlet stream flow rate (Figure 2.18). Derive the model equations that describe the temperature profile in the tank. If the inlet flow rate is raised to 300 L/min, what is the new tank steady-state temperature?

Solution

The flowing assumptions are considered:

- Lumped system (ρ is constant)
- Constant specific heat
- Nonisothermal process

The model equations that describe the temperature profile in the tank starting with the total mass balance:

$$\frac{dm}{dt} = \dot{m}_{\mathrm{in}} - \dot{m}_{\mathrm{out}}$$

Assuming constant density $(\rho = \rho_f = \rho_o)$, we have:

$$\rho\frac{dV}{dt} = \rho F_i - \rho F$$

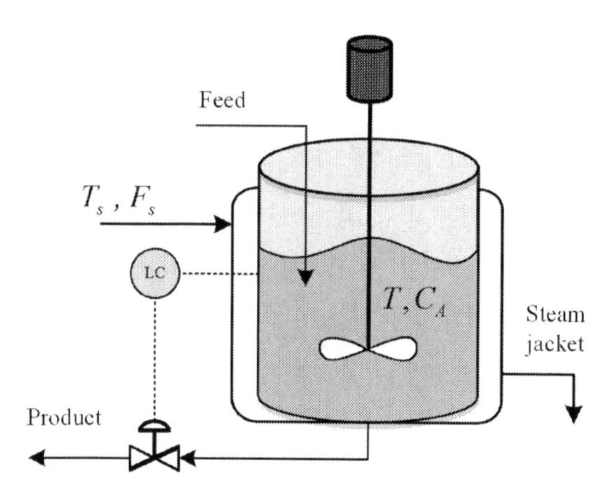

FIGURE 2.18
Stirred tank heated by steam jacket.

Since

$$F_i = F$$

The general energy balance equation is:

$$\frac{d(u)}{dt} = \sum \dot{m}_{\text{in}} h_{\text{in}} - \sum \dot{m}_{\text{out}} h_{\text{out}} + \dot{Q} - \dot{W}$$

For liquids: C_p and C_v are approximately equal. The heat added to the oil from the steam is: $Q = UA_e(T_s - T)$.

Volume of the tank, V, is:

$$V = 100\,\text{L} \times \frac{1\,\text{m}^3}{1000\,\text{L}} = 0.1\,\text{m}^3$$

Inlet and exit mass flow rate are equal, $\dot{m}$:

$$\dot{m} = 250\frac{\text{L}}{\text{min}} \times \frac{1\,\text{m}^3}{1000\,\text{L}} \times \frac{1000\,\text{kg}}{\text{m}^3} \times \frac{1\,\text{min}}{60\,\text{s}} = 4.17\frac{\text{kg}}{\text{s}}$$

After simplification and neglecting shaft work, the general energy balance equation can be written as follows:

$$\frac{d\left(mC_p(T - T_{\text{ref}})\right)}{dt} = \sum \dot{m}_{\text{in}} h_{\text{in}} - \sum \dot{m}_{\text{out}} h_{\text{out}} + \dot{Q}$$

To collect and separate variables, the equation can be rearranged as follows:

$$\frac{d\left(mC_p(T - T_{\text{ref}})\right)}{dt} = \dot{m}C_p(T_f - T_{\text{ref}}) - \dot{m}C_p(T - T_{\text{ref}}) + UA_e(T_s - T)$$

Divide both sides of the equation by mC_p:

$$\frac{dT}{dt} = \left\{\dot{m}C_p(T_f - T) + UA_e(T_s - T)\right\} / mC_p$$

Assuming steady-state conditions, we have:

$$\frac{dT}{dt} = 0$$

Substitute values:

$$0=\left(4.17\frac{\text{kg}}{\text{s}}\right)\left(4.187\frac{\text{kJ}}{\text{kg}\cdot{}^\circ\text{C}}\right)(30-60)^\circ\text{C}+(U)(20\ \text{m}^2)(180-60)^\circ\text{C}$$

$$0=-523.794\frac{\text{kJ}}{\text{s}}+2400^\circ\text{C}\cdot\text{m}^2(U)$$

$$523.794\frac{\text{kJ}}{\text{s}}=2400\,\text{m}^2\,^\circ\text{C}\,(U)$$

$$U=\frac{523.794\ \text{kJ/s}}{2400\,\text{m}^2\,^\circ\text{C}}=0.218248\frac{\text{kW}}{\text{m}^2\,^\circ\text{C}}=218.248\frac{\text{W}}{\text{m}^2\,^\circ\text{C}}$$

The new tank steady-state temperature with inlet flow rate equal to 300 L/min is obtained from the general energy balance equation at steady state:

$$0=\dot{m}C_p(T_f-T)+UA_e(T_s-T)$$

The new mass flow rate:

$$\dot{m}=300\frac{\text{L}}{\text{min}}\times\frac{1\ \text{m}^3}{1000\ \text{L}}\times\frac{1000\ \text{kg}}{\text{m}^3}\times\frac{1\ \text{min}}{60\ \text{s}}=5.0\frac{\text{kg}}{\text{s}}$$

Substitute variables:

$$0=\left(5\frac{\text{kg}}{\text{s}}\right)\left(4.187\frac{\text{kJ}}{\text{kg}\cdot{}^\circ\text{C}}\right)(30-T)^\circ\text{C}+\left(0.218\frac{\text{kW}}{\text{m}^2\,^\circ\text{C}}\right)(20\ \text{m}^2)(180-T)^\circ\text{C}$$

$$0=\left(-20.94\frac{\text{kJ}}{\text{s}^\circ\text{C}}\right)T+628.05\frac{\text{kJ}}{\text{s}}+784.8\ \text{kW}-4.36\frac{\text{kW}}{^\circ\text{C}}T$$

$$25.3\frac{\text{kJ}}{\text{s.}^\circ\text{C}}\times T=628.05\frac{\text{kJ}}{\text{s}}+784.8\frac{\text{kJ}}{\text{s}}$$

$$T=55.84^\circ\text{C}$$

The steady-state value can also be obtained using the dynamic model equation of the tank:

$$\frac{d(T)}{dt}=\frac{\{0-\dot{m}C_p(T-T_f)+UA_e(T_{\text{steam}}-T)\}}{mC_p}$$

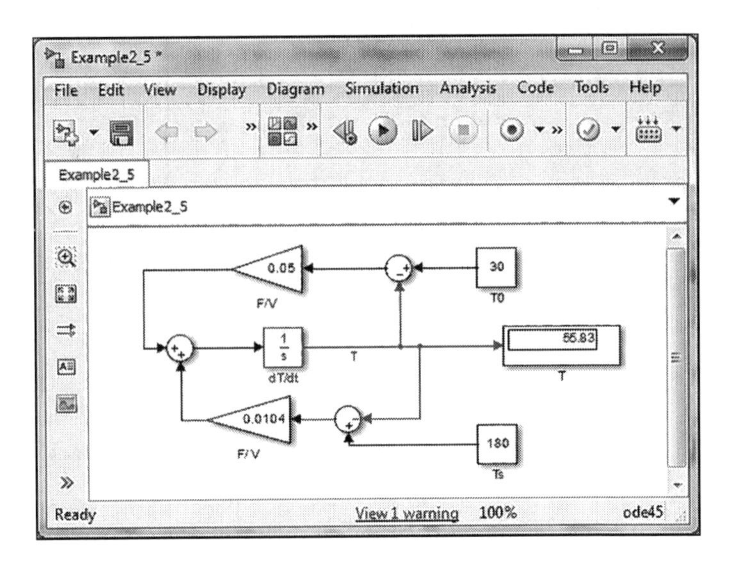

FIGURE 2.19
Solution using Simulink.

Substitute values and simplify:

$$\frac{d(T)}{dt} = \frac{\left\{\left(20.9\frac{\text{kJ}}{\text{s}\cdot{}^{\circ}\text{C}}\right)(30-T)+\left(4.36\frac{\text{kJ}}{\text{s}\cdot{}^{\circ}\text{C}}\right)(180-T)\right\}}{\left(418.7\frac{\text{kJ}}{{}^{\circ}\text{C}}\right)}$$

Simplifying further, we have:

$$\frac{d(T)}{dt} = (0.05)(30-T)+(0.01)(180-T)$$

The solution of the simplified dynamic equation and the calculation of T can be done in Simulink, as shown in Figure 2.19.

Example 2.6: Continuous Stirred Tank Reactors in Series

Consider the two CSTRs Figure 2.20. The following reaction is taking place in both reactors:

$$A \rightarrow B$$

The rate of reaction is first order:

$$-r_A = kC_A, k = 0.64 \text{ min}$$

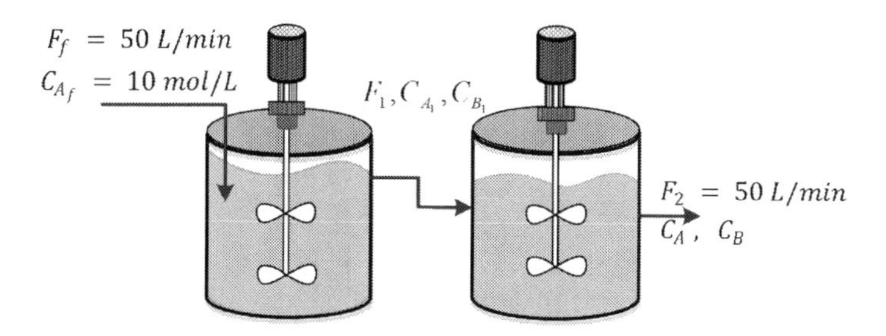

FIGURE 2.20
Continuous stirred tank reactor in series.

The inlet and exit volumetric flow rates are fixed at 50 L/min, volume of each reactor is 100 L, and the inlet concentration of component A to the first tank is 10 mol/L. Develop the system ordinary differential equation that described the concentration profile in the two tanks.

Solution

Assume that the contents of the tanks are well mixed (lumped system) and isothermal.

Mass balance on material in tank 1:

$$\frac{d(\rho V_1)}{dt} = F_f\rho - F_1\rho$$

Since the process is at constant density and constant flow rate, $F_f = F_1 = F_2$; consequently:

$$\frac{dV_1}{dt} = 0$$

Component mole balance (A):

$$\frac{d(V_1 c_{A1})}{dt} = C_{Af}F_f - C_{A1}F_1 - KC_{A1}V_1$$

Differentiating the accumulation term:

$$C_{A1}\frac{dV_1}{dt} + V_1\frac{dC_{A1}}{dt} = C_{Af}F_f - C_{A1}F_1 - KC_{A1}V_1$$

Substitute $dV/dt = 0$:

$$V_1\frac{dC_{A1}}{dt} = F(C_{Af} - C_{A1}) - kC_{A1}V_1$$

Mole balance on component B is:

$$\frac{d(V_1C_{B1})}{dt} = c_{Bf}F_f - C_{B1}F_1 + kc_{A1}V_1$$

Differentiating the accumulation term:

$$C_{B1}\frac{dV_1}{dt} + V_1\frac{dC_{B1}}{dt} = C_{Bf}F_f - C_{B1}F_1 + kC_{A1}V_1$$

Substitute: $dV_1/dt = 0$:

$$V_1\frac{dC_{B1}}{dt} = F(C_{Bf} - C_{B1}) + kC_{A1}V_1$$

Total mass balance on material in tank 2:

$$\frac{d(\rho V_2)}{dt} = F_1\rho - F_2\rho$$

If we assume constant density and $F_1 = F_2$, we have:

$$\frac{dV_2}{dt} = 0$$

Component mole balance on component A:

$$\frac{d(V_2C_{A2})}{dt} = C_{A1}F_1 - C_{A2}F_2 - kC_{A2}V_2$$

Use the differential product rule:

$$C_{A2}\frac{d(V_2)}{dt} + V_2\frac{d(C_{A2})}{dt} = C_{A1}F_1 - C_{A2}F_2 - kC_{A2}V_2$$

Since the volume is constant, $dV_2/dt = 0$:

$$V_2\frac{d(C_{A2})}{dt} = F(C_{A1} - C_{A2}) - kC_{A2}V_2$$

Component mole balance on component *B*:

$$\frac{d(V_2 C_{B2})}{dt} = C_{B1}F_1 - C_{B2}F + kC_{A_2}V_2$$

Use the differential product rule:

$$C_{B2}\frac{d(V_2)}{dt} + V_2\frac{d(C_{B2})}{dt} = C_{B1}F_1 - C_{B2}F_2 + kC_{A2}V_2$$

Substitute $dV_2/dt = 0$:

$$V_2\frac{d(C_{B2})}{dt} = F(C_{B1} - C_{B2}) + kC_{A2}V_2$$

Here is the resulting set of ODEs that will be solved using the Simulink software package:

$$\frac{d(C_{A1})}{dt} = \frac{F}{V_1}(C_{Af} - C_{A1}) - kC_{A1}$$

$$\frac{d(C_{B1})}{dt} = \frac{F}{V_1}(C_{Bf} - C_{B1}) + kC_{A1}$$

$$\frac{d(C_{A2})}{dt} = \frac{F}{V_2}(C_{A1} - C_{A2}) - kC_{A2}$$

$$\frac{d(C_{B2})}{dt} = \frac{F}{V_2}(C_{B1} - C_{B2}) + kC_{A2}$$

The parameters needed for solving the set of the four ODEs are:

$$C_{Af} = 10\ \text{kmol/L}$$

$$F = 50\ \text{L/min}$$

$$V_1 = V_2 = 100\ \text{L}$$

$$k = 0.64\ \text{min}$$

The four ODEs are solved using Simulink, and the results are depicted in Figure 2.21.

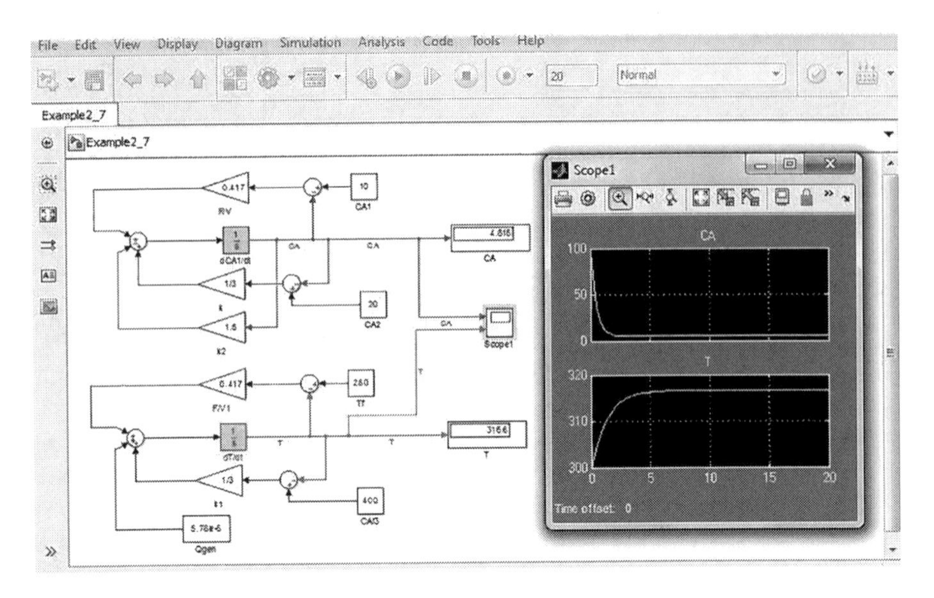

FIGURE 2.21
Concentration of component A and B in two CSTRs in series.

Example 2.7: Adiabatic CSTR

The contents of the adiabatic CSTR shown in Figure 2.22 is assumed to be perfectly mixed. Initially the tank contents are at a temperature of 300 K. Reactant A is fed to the reactor from two different inlets with constant flow rates, and it exits at a rate equal to the sum of both rates; thus, the volume of the tank holdup is kept constant. The initial concentration of reactant A in the tank is $100\,\text{kmol}/\text{m}^3$. The reaction rate is irreversible first order:

$$A \rightarrow B, \text{ where, } r_A = -1.5c_A$$

Due to the significant variations in temperature during the reactor operation, it is desirable to ensure that the process variables are kept

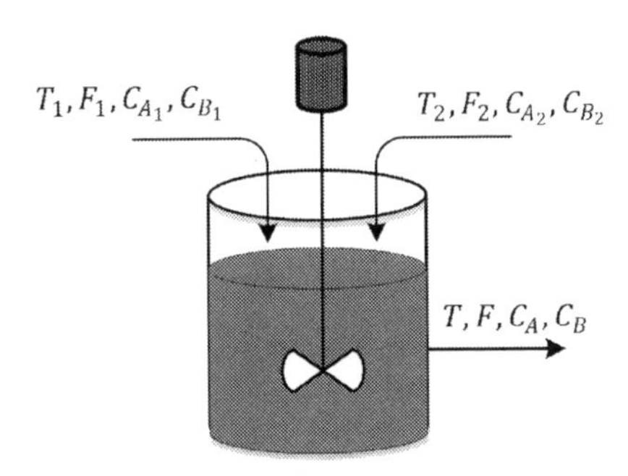

FIGURE 2.22
Continuous reactor with two feed streams.

TABLE 2.2

Parameters Used in the Example

Variable	Value
F_1 (m^3/hr)	250
F_2 (m^3/hr)	200
c_{A1} (kmol/m^3)	10
c_{B1} (kmol/m^3)	2
c_{A2} (kmol/m^3)	20
c_{B2} (kmol/m^3)	1
ρ (kg/m^3)	440
V (m^3)	600
T_1 (K)	250
T_2 (K)	400
C_p (kJ/kg)	2
Q_{gen} (kJ/h)	30

within reasonable limits. The constant parameters that will be used in the model are shown Table 2.2.

Solution

The following assumption are considered:

- Tank holdup is kept constant.
- The contents of the tank are well mixed, which indicates that rate of reaction (r_A) is the same throughout the reactor.

The total exit volumetric flow rate:

$$F_3 = F_1 + F_2$$

$$F_3 = 250 + 200 = 450 \text{ m}^3/\text{hr}$$

Mole balance on component A:

$$\frac{dC_A}{dt} = \frac{F_1}{V}(C_{A1} - C_A) + \frac{F_2}{V}(C_{A2} - C_A) - 1.5C_A$$

The overall energy balance equation is:

$$\frac{d\left(\rho VCp(T - T_{ref})\right)}{dt} = F_1\rho C_p(T_1 - T_{ref}) + F_2\rho C_p(T_2 - T_{ref}) - m_2C_p(T - T_{ref}) + Q_{gen}$$

Apply the assumptions of constant density and constant volume and divide by ρVC_p:

$$\frac{dT}{\partial t} = \frac{F_1}{V}(T_1 - T) + \frac{F_2}{V}(T_2 - T) + \frac{Q_{gen}}{\rho VC_p}$$

The resulting set of ODEs is simplified and the constants are substituted. Then the model equations are solved simultaneously via Simulink, as shown in Figure 2.23:

$$\frac{dC_A}{dt} = \frac{F_1}{V}(C_{A1} - C_A) + \frac{F_2}{V}(C_{A2} - C_A) - 1.5C_A$$

$$\frac{dT}{dt} = \frac{F_1}{V}(T_1 - T) + \frac{F_2}{V}(T_2 - T) + \frac{Q_{gen}}{\rho VCp}$$

Substitute parameter's value, we have:

$$\frac{dC_A}{dt} = \frac{250\ \text{m}^3/\text{h}}{600\ \text{m}^3}(10 - C_A) - \frac{200\ \text{m}^3/\text{h}}{600\ \text{m}^3}(20 - C_A) - 1.5C_A$$

Simplifying, we have:

$$\frac{dC_A}{dt} = 0.417\frac{1}{\text{h}}(10 - C_A) + 0.33\frac{1}{\text{h}}(20 - C_A) - 1.5C_A$$

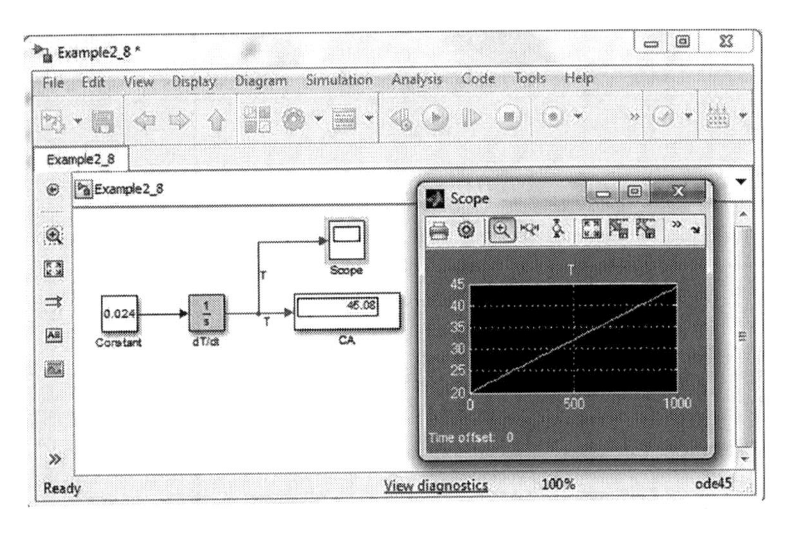

FIGURE 2.23
Concentration temperature profile.

$$C_A(0) = 100$$

The energy balance equation is:

$$\frac{dT}{\partial t} = \frac{250\,\text{m}^3/\text{h}}{600\,\text{m}^3}(250 - T) + \frac{200\,\text{m}^3/\text{h}}{600\,\text{m}^3}(400 - T) + \frac{30\,\text{kJ/h}}{440\,\text{kg/m}^3\,(600\,\text{m}^3)(2\,\text{kJ/kg}\cdot{}^\circ\text{C})}$$

Simplifying, we have:

$$\frac{dT}{\partial t} = 0.417\frac{1}{\text{h}}(250 - T) + \frac{1}{3}\frac{1}{\text{h}}(400 - T) + 5.78 \times 10^{-5}\,\frac{\text{C}}{\text{h}}$$

Initial conditions are:

$$T(0) = 300\ \text{K}$$

Example 2.8: Shower Heating Tank

A shower heater holds 15 kg of water (Figure 2.24). The water in thank is initially at 20°C. The heater is 1.5 kW power. The heater is switched on to start heating the water in the tank. Estimate the time in minutes until the temperature of the water gets to 45°C.

Solution

The following assumptions are considered:

- The contents of the tank are perfectly mixed.
- No reaction is taking place.
- Incompressible (constant density).
- Batch system.

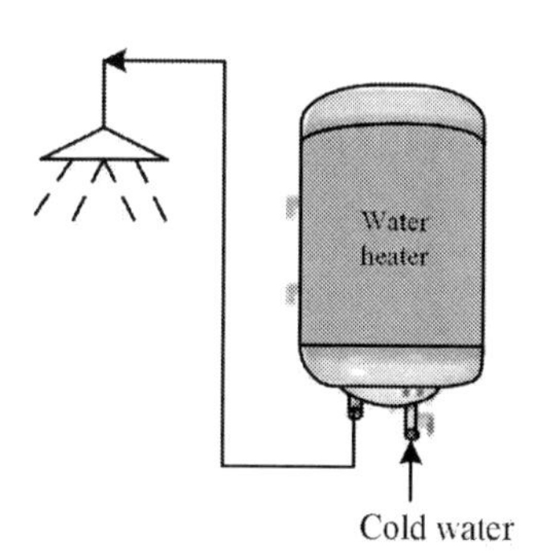

FIGURE 2.24
Schematic of shower heater.

Data

$$T_{\text{ref}} = T_o = 20°\text{C}$$

$$C_p = 4.18\ \text{J/g°C}$$

$$Q(\text{heat}) = 1500\ \text{J/s}$$

$$m\ (\text{mass of the water in the tank}) = 15\ \text{kg} = 15{,}000\ \text{g}$$

Solution

The energy balance of tank contents is:

$$\frac{d\left(mC_p(T - T_{\text{ref}})\right)}{dt} = \dot{m}_{\text{in}}C_p(T_f - T_{\text{ref}}) - \dot{m}_{\text{out}}C_p(T_o - T_{\text{ref}}) + Q_e - W_s$$

Take mC_p out of the integration (constant m, C_p). The tank is a batch system (i.e., no inlet or exit flow rate), and there is no shaft work. Hence, the energy balance equation is reduced to:

$$mC_p\frac{dT}{dt} = Q_e$$

Rearrange the equation by dividing both sides by mC_p:

$$\frac{dT}{dt} = \frac{Q_e}{mC_p}$$

Integrating leads to the following temperature profile equation with the new arbitrary integration constant, C_1:

$$T = \frac{Q_e}{mC_p}t + C_1$$

Using the initial conditions to find the value of C_1:

$$\text{at } t = 0,\ T = 20°\text{C}$$

From the initial conditions, $C_1 = 20$. Substituting C_1 in the temperature profile equation, we have:

$$T = \frac{Q_e}{mC_p}t + 20$$

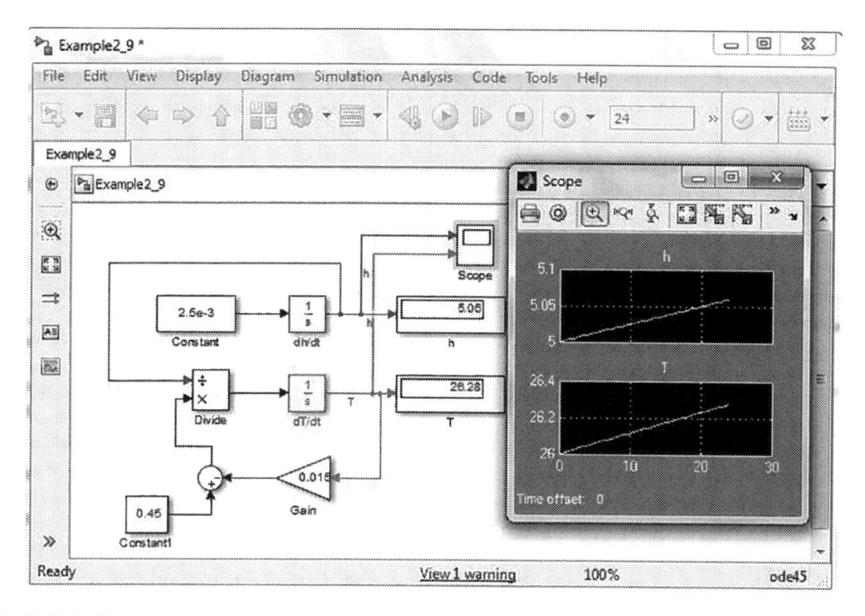

FIGURE 2.25
The time in seconds until the water gets to 45°C to have the shower.

Solve the equation by substituting the values of the known parameter:

$$45°\text{C} = \frac{1500\ \text{J/s}}{15{,}000\ \text{g} * 4.18\ \text{J/g°C}} t + 20°\text{C}$$

$$t = 1045\ \text{s} = 17.42\ \text{min}$$

The dynamic model equation can easily be solved using Simulink. After substituting the constant parameters, the ODE is reduced to:

$$\frac{dT}{dt} = \frac{Q_e}{mC_p} = \frac{1500\,\text{J/s}}{15{,}000\ \text{g} * 4.18\ \text{J/g°C}} = 0.024$$

The water is initially at temperature of 20°C ($T(0) = 20°\text{C}$). The model equation is solved by MATLAB/Simulink, as shown in Figure 2.25.

Example 2.9: Combined Mass and Energy Balance

Water from a string flows at a volumetric flow rate of 3 m^3/hr into a large storage concrete tank (Figure 2.26). The tank is used for different purposes in the nearby chemical plant. Initially the depth of the water in the tank is 5 m. The exit stream from the storage tank was constructed to

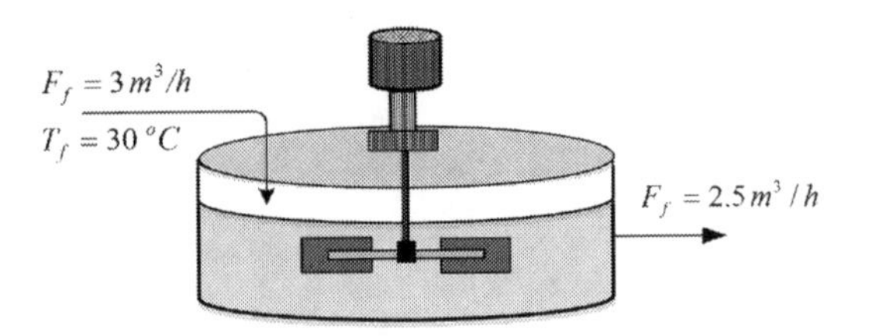

FIGURE 2.26
Schematic of a water storage tank.

reach various units in the plant nearby; the outlet stream carries around $2.5\ \text{m}^3/\text{hr}$ of water. The average temperature of the inlet water stream is around 30°C. Initially the storage tank temperature is 26°C. The storage tank covers a $200\ \text{m}^2$ piece of land and has a depth of 10 m. The density of water is $\rho = 1000\ \text{kg/m}^3$, and the heat capacity is $C_p = 4.187\ \text{kJ/(kg}\cdot°\text{C)}$. Develop a mathematical model to describe the change in depth and temperature of the water in the storage tank within one day.

Solution

The following assumptions are considered in developing the mathematical model of the tank's holdup:

- The contents of the tank are perfectly mixed.
- Transient system.
- Constant density.
- Tank with constant cross-sectional area.
- Neglect the evaporation loss of water.

The total mass balance:

$$\frac{dm}{dt} = \dot{m}_{\text{in}} - \dot{m}_{\text{out}}$$

Replace the mass flow rate ($\dot{m}$) with density (ρ) multiplied by volumetric flow rate (F):

$$\dot{m} = \rho F$$

Substitute $\dot{m} = \rho F$ in the total mass balance equation:

$$\rho \frac{dV}{dt} = \rho F_f - \rho F_o$$

Since the density is constant, divide both sides of the equation by ρ:

$$A\frac{dh}{dt} = F_f - F_o$$

Rearranging, we have:

$$\frac{dh}{dt} = \frac{(F_f - F_o)}{A} = \frac{(3-2.5)\ \text{m}^3/\text{h}}{200\ \text{m}^2} = 2.5\times 10^{-3}\ \frac{\text{m}}{\text{h}}$$

The integration generates the arbitrary integration constant, C_1:

$$h = 2.5\times 10^{-3}\ \frac{\text{m}}{\text{h}}\ t + C_1$$

To find the value of the integration constant, C_1, substitute the initial conditions. The initial condition: at $t = 0, h = 5$ m ($h(0) = 5$).

$$5 = 2.5\times 10^{-3}\ \frac{\text{m}}{\text{h}} \times 0 + C_1$$

The resultant value of C_1 is 5. Substitute the value of C_1. The change in the height of water in the lake with time is:

$$h = 2.5\times 10^{-3}\ \frac{\text{m}}{\text{h}}\ t + 5$$

Note that the depth of the water in the lake is increasing with time since the water downstream is less than the upstream water flow rate. The general total energy balance equation is:

$$\frac{d(H)}{dt} = \sum \dot{m}_{\text{in}} h_{\text{in}} - \sum \dot{m}_{\text{out}} h_{\text{out}} + \dot{Q} - \dot{W}_s$$

There is no shaft work. If we neglect heat loss from the lake, the energy balance equation is:

$$\frac{d\left(\rho A h C_p (T - T_{\text{ref}})\right)}{dt} = \rho F_f C_p (T_f - T_{\text{ref}}) - \rho F_o C_p (T - T_{\text{ref}})$$

Simplify the equation:

$$\rho A C_p \left[\frac{dh}{dt}(T - T_{\text{ref}}) + \frac{dT}{dt} h\right] = \rho F_f C_p (T_f - T_{\text{ref}}) - \rho F_o C_p (T - T_{\text{ref}})$$

Divide both sides of the equation by ρC_p:

$$A\left[\frac{dh}{dt}(T - T_{\text{ref}}) + \frac{dT}{dt}h\right] = F_f(T_f - T_{\text{ref}}) - F_o(T - T_{\text{ref}})$$

Substitute:

$$\frac{dh}{dt} = \frac{F_f - F_o}{A}$$

After substitution of dh/dt, we have obtained the following equation:

$$A\left[\frac{(F_f - F_o)}{A}(T - T_{\text{ref}}) + \frac{dT}{dt}h\right] = F_f(T_f - T_{\text{ref}}) - F_o(T - T_{\text{ref}})$$

Simplifying further, we have:

$$(F_f - F_o)(T - T_{\text{ref}}) + Ah\frac{dT}{dt} = F_f(T_f - T_{\text{ref}}) - F_o(T - T_{\text{ref}})$$

Rearranging, we have:

$$\frac{dT}{dt} = \frac{F_f(T_f - T)}{Ah}$$

From mass balance calculations, the change in height of water in the tank with time is:

$$\frac{dh}{dt} = 2.5 \times 10^{-3}\,\text{m/h}$$

The energy balance equation after the substitution of constants is:

$$\frac{dT}{dt} = \frac{3(30 - T)}{200\ \text{h}}$$

The simplified model describes the temperature profile in the tank:

$$\frac{dT}{dt} = \frac{1}{\text{h}}\left[0.45 - 0.015T\right]$$

The change in the height of water in the storage tank with time is:

$$\frac{dh}{dt} = 2.5 \times 10^{-3}\,\text{m/h}$$

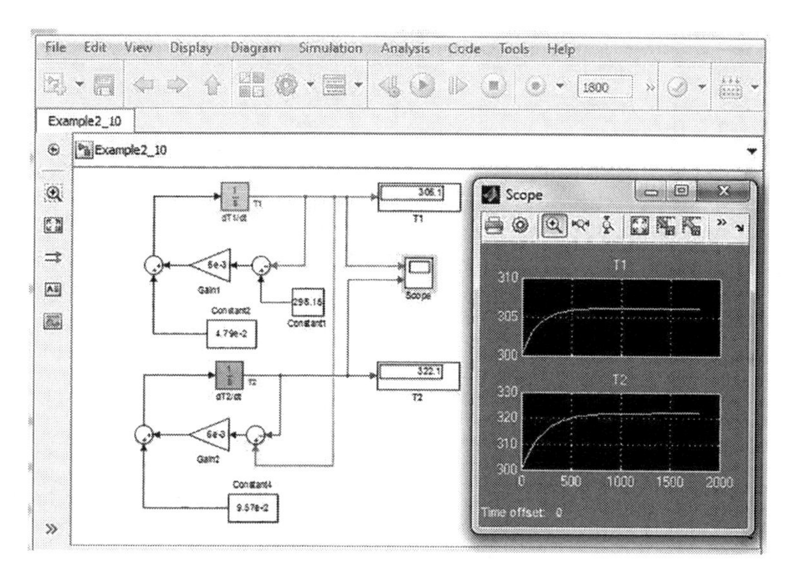

FIGURE 2.27
Tank height and temperature profile for a period of 24 hours.

The resulting two ordinary differential equations can be solved simultaneously using Simulink. Both h and T are variables and are difficult to solve manually. Note that the unit of the independent variable (t, time) in the following ODEs is in hours.

$$\frac{dh}{dt} = 2.5 \times 10^{-3} \frac{\text{m}}{\text{h}}, \quad h(0) = 5\text{ m}$$

$$\frac{dT}{dt} = \frac{1}{\text{h}}\left[0.45 - 0.015T\right], \quad T(0) = 26°\text{C}$$

The two ordinary differential equations are solved simultaneously using the MATLAB/Simulink software package, as shown in Figure 2.27.

Example 2.10: Heating Tanks in Series

Water in two perfectly mixed tanks in series is being heated by electrical heaters; the volume of both tanks is 5 m^3 each. Initially both tanks are full of water at a temperature of 26°C. Water is flowing into the first tank at a temperature of 25°C and at a flow rate of 0.03 m^3/s. Inlet and exit flow rates are kept constant through the heating process. The heat added to the first and second tank is 1000 kJ/s and 2000 kJ/s, respectively. Calculate the temperatures of the two tanks after one half hour of operation. The schematic diagram of the process is shown in Figure 2.28.

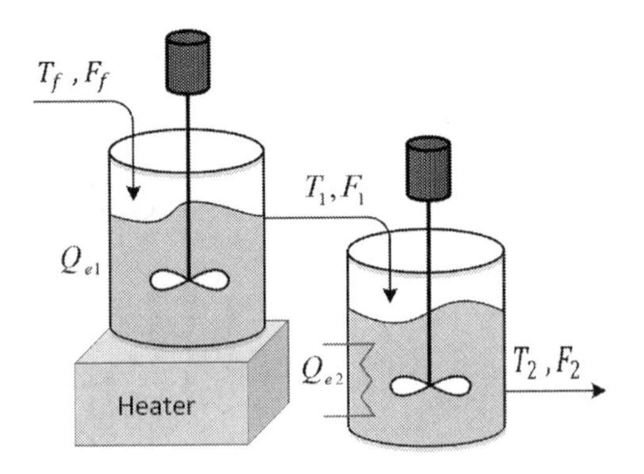

FIGURE 2.28
Schematic of two electrical heater in series.

Assumptions

The following assumptions are considered in building the mathematical model that represents the mass and energy balance in both tanks.

- No reactions are taking place in the tanks.
- The contents of the tanks are perfectly mixed (no spatial variation of temperature).
- Constant holdup (V_1, V_2 are constants).
- Incompressible fluid (constant density, $\rho_f = \rho_1 = \rho_2 = \rho$).

Mass Balance

The total mass balance around tank 1 is:

$$\frac{dV_1}{dt} = F_f - F_1 = 0$$

Since the tanks holdup are constants, we have:

$$F_f = F_1 = F_2 = F$$

Thus:

$$\frac{dV_1}{dt} = \frac{dV_2}{dt} = 0$$

The overall mass balance on tank 2 is:

$$\frac{dV_2}{dt} = F_1 - F_2 = 0$$

$$F_1 = F_2 = F$$

Hence:

$$\frac{dV_2}{dt} = 0$$

Energy Balance

The overall energy balance on tank 1 is:

$$\frac{d\left[\rho V_1 C_p (T_1 - T_{\text{ref}})\right]}{dt} = \rho F_f C_p (T_f - T_{\text{ref}}) - \rho F_1 C_p (T_1 - T_{\text{ref}}) + Q_{e1}$$

Simplify the energy balance equation:

$$\rho V_1 C_p \frac{d(T_1 - T_{\text{ref}})}{dt} = \rho F C_p (T_f - T_1) + Q_{e1}$$

Divide both sides of the equation by $\rho V_1 C_p$:

$$\frac{dT_1}{dt} = \frac{F}{V_1}(T_f - T_1) + \frac{Q_{e1}}{\rho V_1 C_p}$$

The overall energy balance on tank 2 is:

$$\frac{d\rho V_2 C_p (T_2 - T_{\text{ref}})}{dt} = \rho F_1 C_p (T_1 - T_{\text{ref}}) - \rho F_2 C_p (T_2 - T_{\text{ref}}) + Q_{e2}$$

Rearrange after replacing F_1 and F_2 by F:

$$\rho V_2 C_p \frac{d(T_2 - T_{\text{ref}})}{dt} = \rho F C_p (T_1 - T_2) + Q_{e2}$$

Divide both sides of the equation by $\rho V_2 C_p$:

$$\frac{dT_2}{dt} = \frac{F}{V_2}(T_1 - T_2) + \frac{Q_{e2}}{\rho V_2 C_p}$$

The Degree of Freedom Analysis

Known constants: $Q_{e1}, Q_{e2}, F, V_1, V_2, \rho, C_p, T_f$
Number of variables: $2(T_1, T_2)$
Number of independent equations: $2((dT_1/dt), (dT_2/dt))$
Number degree of freedom: $F = N_v - N_e = 2 - 2 = 0$

Thus, the system is fully specified. The following data are provided in the question:

$$T_f = 25 + 273.15 = 298.15\ \text{K}$$

$$V_1 = V_2 = 5\ \text{m}^3\ \text{(volume of tanks)}$$

$$C_p = 4.18\ \text{kJ/kg K (water specific heat)}$$

$$F_f = 0.03\ \text{m}^3/\text{s (tank feed rate)}$$

$$\rho = 1000\ \text{kg/m}^3\ \text{(density of water)}$$

$$Q_{e1} = 1000\ \text{kJ/s (heat supplied to first tank)}$$

$$Q_{e2} = 2000\ \text{kJ/s (heat provided to the second tank)}$$

The tanks are initially at 27°C:

$$T_1(0) = T_2(0) = 27 + 273.15 = 300.15\ \text{K}$$

The following two ODEs are simplified after adding the known parameters:

$$\frac{dT_1}{dt} = \frac{0.03\text{m}^3/\text{s}}{5\text{m}^3}(298.15 - T_1) + \frac{1000\ \text{kJ/s}}{1000\dfrac{\text{kg}}{\text{m}^3} \times 5\ \text{m}^3 \times 4.18\dfrac{\text{kJ}}{\text{kg}\cdot\text{K}}}$$

The energy balance equation on tank 1 is:

$$\frac{dT_1}{dt} = 6\times10^{-3}\frac{1}{\text{s}}(298.15 - T_1) + 4.79\times10^{-2}\frac{\text{K}}{\text{s}}$$

The energy balance equation on tank 2 is:

$$\frac{dT_2}{dt} = \frac{0.03\ \text{m}^3/\text{s}}{5\ \text{m}^3}(T_1 - T_2) + \frac{2000\ \text{kJ/s}}{1000\dfrac{\text{kg}}{\text{m}^3} \times 5\ \text{m}^3 \times 4.18\dfrac{\text{kJ}}{\text{kg}\cdot\text{K}}}$$

The equation is further simplified to:

$$\frac{dT_1}{dt} = 6\times10^{-3}\frac{1}{\text{s}}(T_1 - T_2) + 9.57\times10^{-2}\frac{\text{K}}{\text{s}}$$

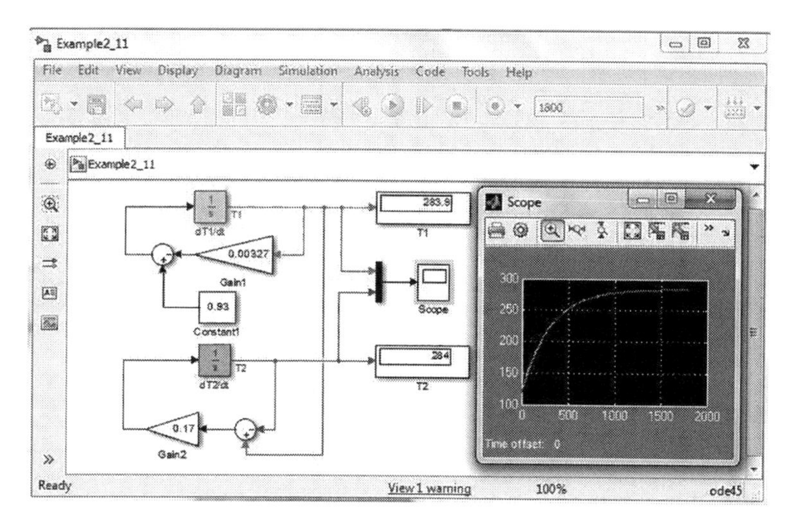

FIGURE 2.29
Simulink solution for the change of temperature versus time.

The results of the MATLAB/Simulink simulation of the resultant ODEs are depicted in Figure 2.29. Results reveal that, after one half hour of operation, the temperatures in both tanks are approximately $T_1 = 306$ K, $T_2 = 322$ K, respectively.

Example 2.11: Two Mixed Tanks in Series with One Tank Heated

The flow rate of oil in two perfectly mixed tanks in series is constant at 2.55 m^3/min. The density of the oil is kept at a constant flow rate of 640 kg/m^3, and its heat capacity is 2.5 kJ/kgK. The volume of the first tank is constant at 13 m^3, and the volume of the second tank holdup is kept constant at 2.50 m^3. The temperature of the oil entering the first tank is 65°C. The initial temperature in the two tanks is 120°C. The oil is heated in the first tank using steam. The rate of heat transferred to first tanks equal to 15,000 kJ/min.

Develop a transient mathematical model to describe the behavior of the temperature profile in both tanks as a function of time, and find the temperature of both tanks after 30 min [1]. The schematic diagram of the heating process is presented in Figure 2.30.

Solution

The following assumptions are considered in developing the transient model equations:

- Tanks' holdups are constants.
- The contents of the tanks are perfectly mixed.
- No reactions are taking place in the tanks.

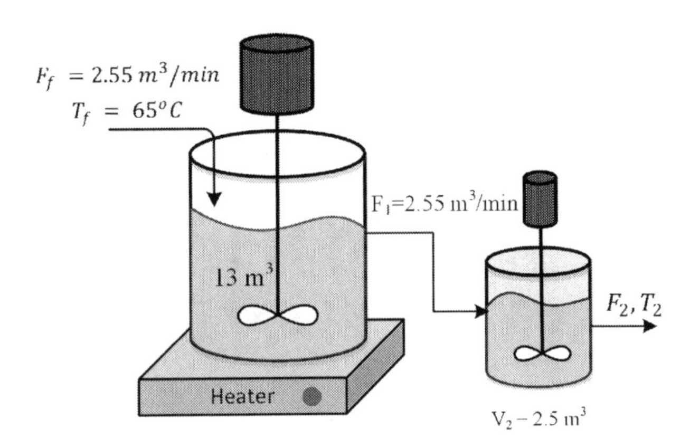

FIGURE 2.30
Two stirred oil tanks in series.

The general overall material balance equation for the incompressible fluid ($\rho_{in} = \rho_{out} = \rho$) is:

$$\begin{Bmatrix} \text{Rate of accumulation} \\ \text{of mass in the tank} \end{Bmatrix} = \begin{Bmatrix} \text{Rate of} \\ \text{mass in} \end{Bmatrix} - \begin{Bmatrix} \text{Rate of} \\ \text{mass out} \end{Bmatrix}$$

The total mass balance equation is:

$$\frac{dm}{dt} = \dot{m}_{in} - \dot{m}_{out}$$

Substitute $m = \rho V$ and $\dot{m} = \rho F$:

$$\frac{d(\rho V)}{dt} = \rho F_{in} - \rho F_{out}$$

$$\frac{dV}{dt} = 0 \text{ (constant flow rate } F_{in} = F_{out}\text{)}$$

The general overall energy balance equation:

$$\begin{Bmatrix} \text{Rate of energy} \\ \text{accumulation} \end{Bmatrix} = \begin{Bmatrix} \text{Rate of energy in} \\ \text{by convection} \end{Bmatrix} - \begin{Bmatrix} \text{Rate of energy out} \\ \text{by convection} \end{Bmatrix}$$

$$+ \begin{Bmatrix} \text{Net rate of heat addition} \\ \text{to the system from} \\ \text{the surroundings} \end{Bmatrix} + \begin{Bmatrix} \text{Net rate of work} \\ \text{performed on the system} \\ \text{by the surroundings} \end{Bmatrix}$$

Two energy balances are needed to model the system:

Tank 1

$$\frac{d\left(\rho C_p V_1(T_1 - T_{\text{ref}})\right)}{dt} = \rho C_p\left(F_0(T_0 - T_{\text{ref}}) - F_1(T_1 - T_{\text{ref}})\right) + Q_1$$

The volumetric flow rate is constant; that is, $F_{in} = F_{out} = F$, hence $dV/dt = 0$. The previous equation can be simplified and rearranged as follows:

$$\rho C_p V_1 \frac{dT_1}{dt} = \rho C_p F(T_0 - T_1) + Q_1$$

Tank 2

The energy balance around tank 2 is written as follows:

$$\frac{d\left(\rho C_p V_2(T_2 - T_{\text{ref}})\right)}{dt} = \rho C_p\left(F_1(T_1 - T_{\text{ref}}) - F_2(T_2 - T_{\text{ref}})\right)$$

Throughout the operation, the volumetric flow rates are constant, $F_0 = F_1 = F_2 = F$; thus, the densities, volumes, and heat capacities are constants. The previous equations can be simplified and rearranged to the following form:

$$\rho C_p V_2 \frac{dT_2}{dt} = \rho C_p F(T_1 - T_2)$$

The following set of ordinary differential equations are simplified and solved via Simulink:

$$\frac{dT_1}{dt} = \frac{F(T_0 - T_1)}{V_1} + \frac{Q_1}{\rho V_1 C_p}$$

Substitute values of known parameters:

$$\frac{dT_1}{dt} = \frac{2.55\dfrac{\text{m}^3}{\text{min}}\,1\dfrac{\text{min}}{60\text{ s}} \times (65°\text{C} - T_1)}{13\text{m}^3} + \frac{15{,}000\text{ kJ/s}}{640\dfrac{\text{kg}}{\text{m}^3} \times 13\text{ m}^3 \times 2.5\dfrac{\text{kJ}}{\text{kg}\cdot\text{K}}}$$

Rearranging, we have:

$$\frac{d(T_1)}{dt} = 0.00327(65 - T) + 0.72$$

Simplifying further, we have:

$$\frac{dT_1}{dt} = (285 - T_1)0.00327$$

Integrate using separable integration methods:

$$\frac{dT_1}{285 - T_1} = 0.00327dt$$

$$-\ln(285 - T_1) = 0.00327t + C_1$$

Use the initial conditions to find the arbitrary integration constant, C_1. At:

$$t(0),\ T_1 = 120°\text{C}$$

$$-\ln(285 - 0.00327 \times 120°\text{C}) = 0 + C_1$$

$$C_1 = -5.65$$

Accordingly, the temperature profile is:

$$-\ln(285 - T_1) = 0.00327t - 5.65$$

After 30 min (1800 s) the temperature in the first tank is:

$$-\ln(285 - T_1) = 0.00327(1800) - 5.65$$

$$T_1 = 283.73$$

Energy balance on tank 2

$$\frac{d(\rho C_p V_2 T_2)}{dt} = \rho C_p \left[V_2 \frac{d(T_2)}{dt} + T_2 \frac{d(V_2)}{dt} \right] = \rho C_p F(T_1 - T_2)$$

The volume of liquid in both tanks are constants. Thus, the equation is reduced to:

$$\rho C_p \left[V_2 \frac{d(T_2)}{dt} \right] = \rho C_p F(T_1 - T_2)$$

Divide both sides of the equation by $\rho C_p V_2$:

$$\frac{d(T_2)}{dt} = \frac{F}{V_2}(T_1 - T_2)$$

Substitute parameter values:

$$\frac{d(T_2)}{dt} = \frac{2.55\dfrac{m^3}{min} \times \dfrac{1\,min}{60\,s}(T_1 - T_2)}{2.50\ m^3}$$

Simplify:

$$\frac{dT_2}{dt} = 0.017(T_1 - T_2)$$

With the following initial condition

$$T_2(0) = 120°C$$

The degrees of freedom analysis of the system are calculated as follows:

The known parameter values are: ρ, C_p, V_1, V_2 and F.

Since there are two dependent variables (T_1 and T_2) as well as two equations, then the system is correctly specified.

The following two ordinary differential equations are solved simultaneously using MATLAB/Simulink software package, as shown in Figure 2.31.

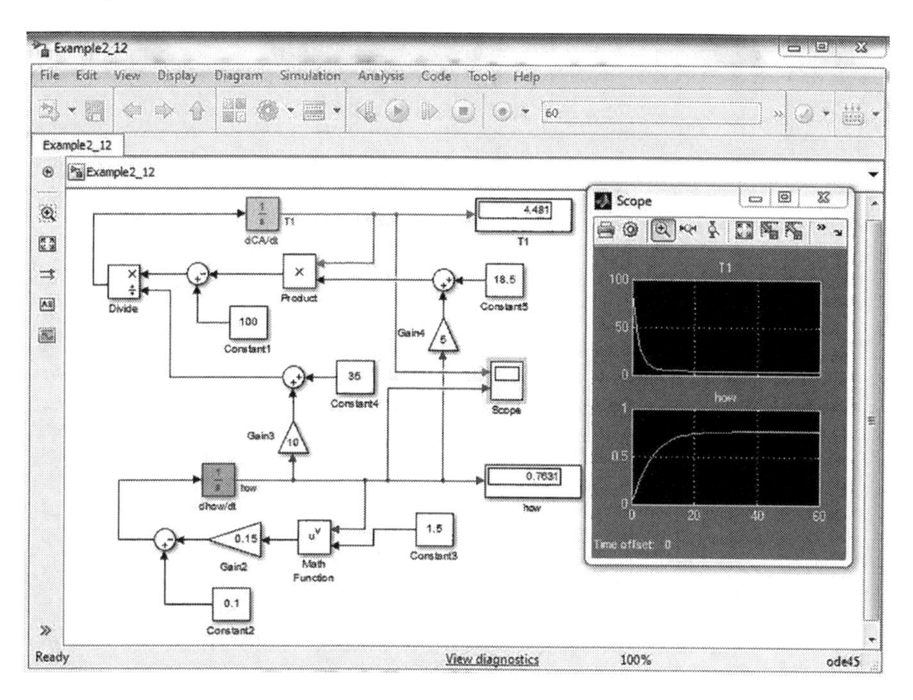

FIGURE 2.31
Plot of temperature versus time.

$$\frac{dT_1}{dt} = 0.93 - 0.00327T, \quad T_1(0) = 120°C$$

$$\frac{dT_2}{dt} = 0.017(T_1 - T_2), \quad T_2(0) = 120°C$$

Example 2.12: Isothermal Reactor

Figure 2.32 shows a perfectly mixed, isothermal stirred tank reactor where the following reaction is taking place:

$$A \rightarrow B$$

The isothermal CSTR has an outlet weir of height h_w. The initial concentration of A is 100 mol/m^3. The flow rate over the weir (F) is known to be proportional to the height of liquid over the weir, h_{ow} (m), to the 1.5 power ($F = h_{ow}^{1.5}$). The cross-sectional area of the cylindrical tank, A (m^2), is constant. Assume constant density, and derive a mathematical model for the given system if a first-order reaction is taking place in the tank.

Data

$$h_w = 3.5 \text{ m (liquid height)}$$

$$C_{Ao} = 100 \text{ mol/m}^3 \text{ (feed concentration)}$$

$$F_o = 1 \text{ m}^3\text{/s (inlet volumetric flow rate)}$$

$$k = 0.5 \text{ s}^{-1} \text{ (reaction rate constant)}$$

$$k_f = 1.5 \text{ (gain)}$$

$$A = 10 \text{ m}^2 \text{ (tank cross sectional area)}$$

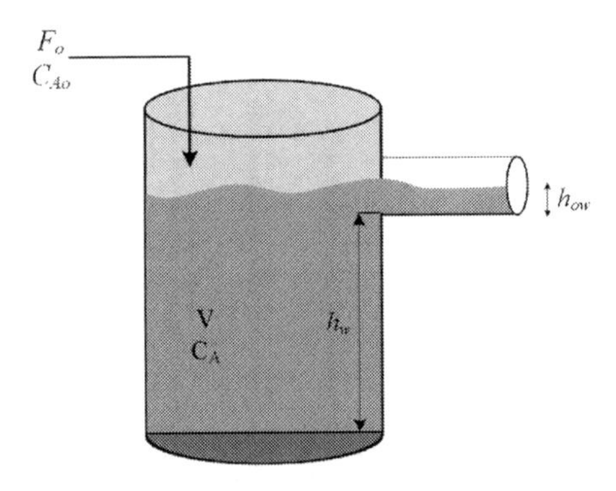

FIGURE 2.32
Schematic of CSTR.

Solution

The following assumptions are considered in developing the model equations:

- Perfectly mixed CSTR (i.e., intensive variables at any point in the tank are the same in the output stream).
- Isothermal reaction (i.e., T_{in} is equal to T_{out}).
- Constant density.
- First-order reaction.
- Assume constant volumetric flow rate $F_f = F$.
- Reaction rate of component A: $R_A = -kC_A$.

Symbols Description

$\dot{m}_f$: inlet mass flow rate (kg/s)
$\dot{m}_o$: outlet mass flow rate (kg/s)
ρ: density of the liquid (kg/m^3)
h: height of the tank (m) ($h_w + h_{ow}$)
A: cross-sectional area of the tank (m^2)
V: volume of holdup liquid in the tank (m^3)
k: constant rate of reaction (min^{-1})
C_A: concentration of component A (kg/m^3)

Total Component Mass Balance

[Rate of accumulation in system] = [rate of flow rate into system] – [rate of flow from system] + [rate of generation within system] – [rate of consumption within system], where the rate of generation within the system equals the rate of consumption within system. The total mass balance equation is:

$$\frac{d(m)}{dt} = \dot{m}_f - \dot{m}_o$$

In terms of density and volumetric flow rates, we have:

$$\frac{d(\rho V)}{dt} = F_f \rho_f - F\rho$$

If we assume constant density, then we have:

$$\rho \frac{d(V)}{dt} = \rho(F_f - F)$$

The volume of the tank holdup is equal to $V = hA$:

$$\frac{d(hA)}{dt} = F_f - F$$

For the exit flow rate, substitute $F = K_f\,(h_{ow})^{1.5}$:

$$A\frac{d(h_w + h_{ow})}{dt} = F_f - k_f(h_{ow})^{1.5}$$

The height of the liquid is h_w, which is constant and is deleted.

$$A\frac{d(h_w)}{dt} + A\frac{d(h_{ow})}{dt} = F_f - k_f\,(h_{ow})^{1.5}$$

The simplified model equation is now simplified to:

$$A\frac{d(h_{ow})}{dt} = F_f - k_f\,(h_{ow})^{1.5}$$

Component Balance (Component A)

The rate of accumulation of A = the rate of flow rate of A into system – the rate of flow of A from the system + the rate of generation of A within system – the rate of consumption of A within system. The rate of generation of A within system = 0. The mole balance of component A is:

$$\frac{d(C_A V)}{dt} = F_f C_{Af} - FC_A - kC_A V$$

Substitute $F = K_f(h_{ow})^{1.5}$:

$$\frac{d\big(C_A A(h_w + h_{ow})\big)}{dt} = F_f C_{Af} - k_f(h_{ow})^{1.5} C_A - kC_A V$$

Derive to simplify the differential equation:

$$A\left[\frac{dC_A}{dt}(h_w + h_{ow}) + C_A\frac{dh_{ow}}{dt}\right] = F_f C_{Af} - k_f(h_{ow})^{1.5} C_A - kC_A V$$

Simplify further, set $(dh_w/dt = 0)$ t, and replace dh_{ow}/dt by the following equation:

$$\frac{dh_{ow}}{dt} = \left\{F_f - k_o\,(h_{ow})^{1.5}\right\}/A$$

Then

$$A\left[\frac{dC_A}{dt}(h_w + h_{ow}) + C_A\left\{F_f - k_f\,(h_{ow})^{1.5}\right\}/A\right] = F_f C_{Af} - k_f(h_{ow})^{1.5} C_A - kC_A V$$

Rearranging, we have:

$$\frac{dC_A}{dt}(h_w + h_{ow}) = \left[F_f C_{Af} - k_f (h_{ow})^{1.5} C_A - kC_A V - C_A \left\{F_f - k_f (h_{ow})^{1.5}\right\}\right]/A$$

Simplifying further, we have:

$$\frac{dC_A}{dt}(h_w + h_{ow}) = \left[F_f C_{Af} - k_f (h_{ow})^{1.5} C_A - kC_A V - C_A F_f + C_A k_f (h_{ow})^{1.5}\right]/A$$

Rearranging again, we have:

$$\frac{dC_A}{dt} = \frac{F_f C_{Ao} - C_A F_f - kC_A V}{A(h_w + h_{ow})}$$

The following model ODEs are to be solved simultaneously:

$$A\frac{d(h_{ow})}{dt} = F_f - k_f (h_{ow})^{1.5}$$

$$\frac{dC_A}{dt} = \frac{F_f C_{Af} - C_A F_f - kC_A V}{A(h_w + h_{ow})}$$

Substitute values of known parameters and replace the volume by: $V = A(h_w + h_{ow})$:

$$\frac{dC_A}{dt} = \frac{1\dfrac{\text{m}^3}{\text{s}} \times 100\dfrac{\text{mol}}{\text{m}^3} - C_A\left(1\dfrac{\text{m}^3}{\text{s}}\right) - 0.5\, C_A\left(10\,\text{m}^2\right)\left(3.5\,\text{m} + h_{ow}\right)}{10\,\text{m}^2\left(3.5\,\text{m} + h_{ow}\right)}$$

Simplifying further, we have:

$$\frac{dC_A}{dt} = \frac{100\dfrac{\text{mol}}{\text{s}} - 18.5\, C_A - 5\, C_A h_{ow}}{35\,\text{m} + 10\, h_{ow}}$$

Streamlining, we have:

$$\frac{dC_A}{dt} = \frac{100 - C_A\left(5\, h_{ow} + 18.5\right)}{35 + 10\, h_{ow}}$$

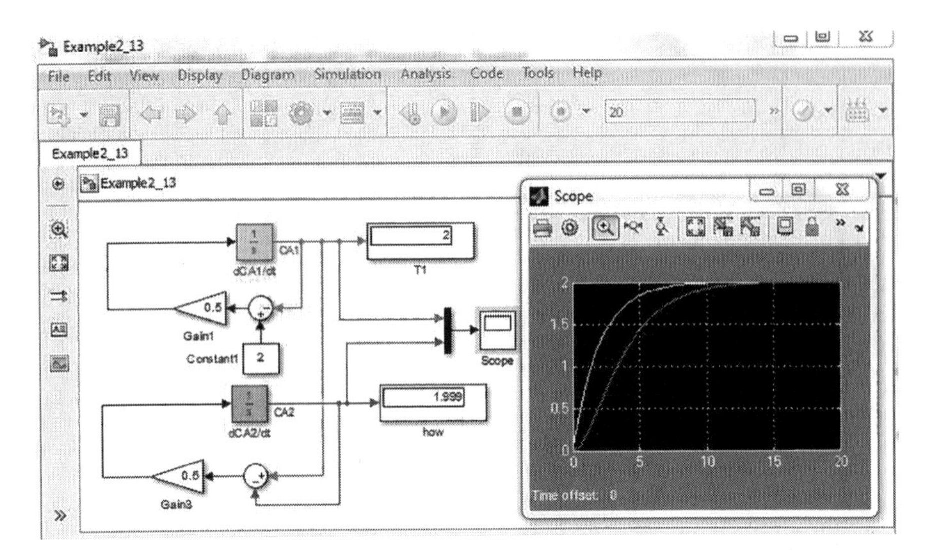

FIGURE 2.33
Simulink model for Example 2.12.

The ODE for the change in the h_{wo} with time is:

$$\frac{dh_{ow}}{dt} = \left(1 - 1.5\,(h_{ow})^{1.5}\right)/A$$

Finally, the following two equations are to be solved simultaneously using Simulink:

$$\frac{dC_A}{dt} = \frac{100 + C_A(5\,h_{ow} - 18.5)}{35 + 10\,h_{ow}},\quad C_A(0) = 100\ \text{mol/m}^3$$

$$\frac{dh_{ow}}{dt} = 0.1 - 0.15(h_{ow})^{1.5},\quad h_{ow}(0) = 0$$

The Simulink program and simulation results are shown in Figure 2.33.

Example 2.13: Multiple Stirred Tanks

Consider the two multiple stirred tanks connected in series as shown in Figure 2.34. The inlet concentration of component A to the first tank is 2 kmol/L, the volumetric feed rate is 10 L/min, and the volume of both tanks is 20 L. The initial concentration of component A in both tanks is zero. Derive the model equations that describe the concentration of component A in each tank and determine the time when the concentrations of A in both tanks are approximately identical.

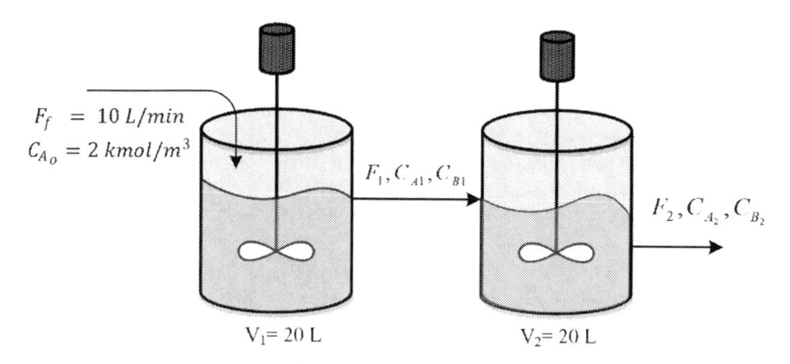

FIGURE 2.34
Two stirred tank reactors in series.

Solution

The following assumptions are considered in developing the concentration profile in both tanks:

- The system is perfectly mixed.
- No reactions are taking place.
- Isothermal operation.

The material balance on the holdup of tank 1 is:

$$\rho\frac{dV}{dt} = \rho F_f - \rho F_o = 0$$

Then, it will be:

$$\frac{dV}{dt} = 0 \text{ and } F_f = F_o$$

Component mole balance of A in tank 1:

$$\frac{d(C_A V)}{dt} = C_{A0}F_f - C_{A1}F$$

Using the differential form of the product rule, we have:

$$\frac{d(C_A V)}{dt} = V\frac{dC_A}{dt} + C_A\frac{dV}{dt} = C_{A0}F_f - C_{A1}F$$

Since the tank holdup is constant, we have:

$$\frac{dV}{dt} = 0$$

$$\frac{dC_{A1}}{dt} = \frac{F}{V}(C_{A0} - C_{A1})$$

Component balance on tank 2:

$$\frac{d(C_A V)}{dt} = C_{A1}F_f - C_{A2}F$$

$$V\frac{dC_A}{dt} + C_A\frac{dV}{dt} = C_{A1}F_f - C_{A2}F$$

The holdup of tank 2 is constant:

$$\frac{dV}{dt} = 0$$

Inlet and exit volumetric flow rate are constant:

$$F_f = F_o = F$$

Component mole balance of component A in tank 2:

$$\frac{dC_{A2}}{dt} = \frac{F}{V}(C_{A1} - C_{A2})$$

In general:

$$\frac{dC_{Ai}}{dt} = \frac{F}{Vi}(C_{Ai1} - C_{Ai})$$

Given information:

$$C_{Ao} = 2\ \text{kmol/L},\ F = 10\ \text{L/min},\ V_1 = V_2 = 20\ \text{L}$$

The following two ODEs are solved simultaneously using the MATLAB/Simulink software:

$$\frac{dC_{A1}}{dt} = \frac{F}{V}(C_{Ao} - C_{A1}) = 0.5(2 - C_{A1}),\ \ C_{A1}(0) = 2$$

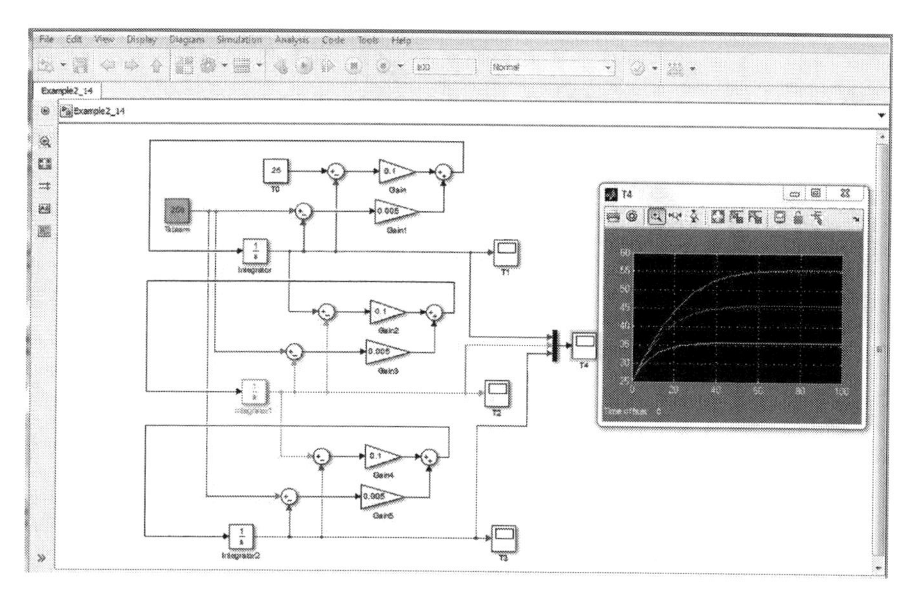

FIGURE 2.35
Simulink model for Example 2.13.

$$\frac{dC_{A2}}{dt} = \frac{F}{V}(C_{A1} - C_{A2}) = 0.5(C_{A1} - C_{A2}), \;\; C_{A2}(0) = 0$$

The simulation predictions reveal that the concentration of component A in both tanks is approximately equal within 20 s (Figure 2.35).

Example 2.14: Three Heated Tanks in Series

Chemical solutions are heated via three tanks in series as shown in Figure 2.36. Initially the tanks are filled with 15 kg of chemical solution at a temperature of 25°C. Saturated steam used to heat the oil at a temperature of 250°C condenses within the steam jacket surrounding the tanks. The oil is fed into the first tank at the rate of 1.5 kg/s and overflows into the second and the third tanks at the same flow rate. The temperature of the oil fed to the first tank is 25°C. The tanks are well mixed. The heat capacity, C_p, of the oil is 2.0 kJ/kg.

The overall heat transfer coefficient $U = 0.15$ kJ/s·°C and the area of the jacket for each tank is 1 m^2.

Develop the dynamic model for the concentration of the three tanks and solve the emerged set of ODEs using polymath software and the Simulink software package.

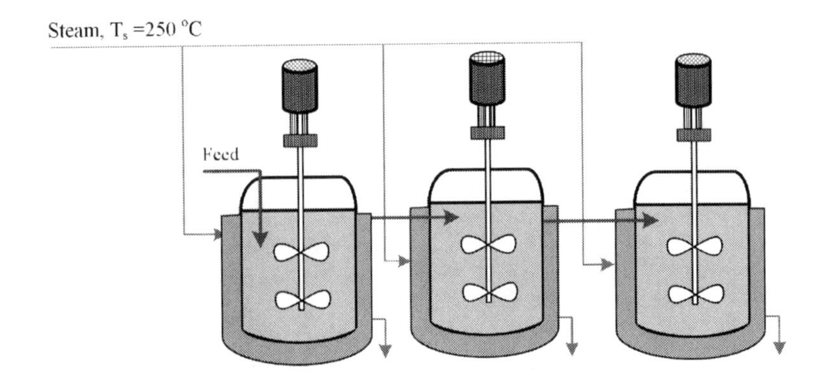

FIGURE 2.36
Schematic of three heated stirred tanks in series.

Solution

Total mass balance is:

$$\frac{dm}{dt} = \dot{m}_{\text{in}} - \dot{m}_{\text{out}}$$

where m is the initial mass of chemical solution in the tank and $\dot{m}$ is the inlet and outlet mass flow rate throughout the three thanks in series. The mass flow rates through the three tanks are constants (i.e., $dm/dt = 0$), and $\dot{m}_{\text{in}} = \dot{m}_{\text{out}} = \dot{m}$.

Energy Balance

The energy balance around tank 1:

$$\frac{d\left(mC_p(T_1 - T_{\text{ref}})\right)}{dt} = \dot{m}C_p(T_0 - T_1) - \dot{m}C_p(T_1 - T_{\text{ref}}) + UA(T_s - T_1)$$

Simplify, rearrange, and substitute all constants:

$$\frac{dT_1}{dt} = \frac{\dot{m}}{m}(T_o - T_1) + \frac{UA}{mC_p}(T_s - T_1) = 0.01\,(25 - T_1) + 0.005\,(250 - T_1)$$

Energy balance around tank 1:

$$\frac{dT_2}{dt} = \frac{\dot{m}}{m}(T_1 - T_2) + \frac{UA}{mC_p}(T_s - T_2) = 0.1\,(T_1 - T_2) + 0.005\,(250 - T_2)$$

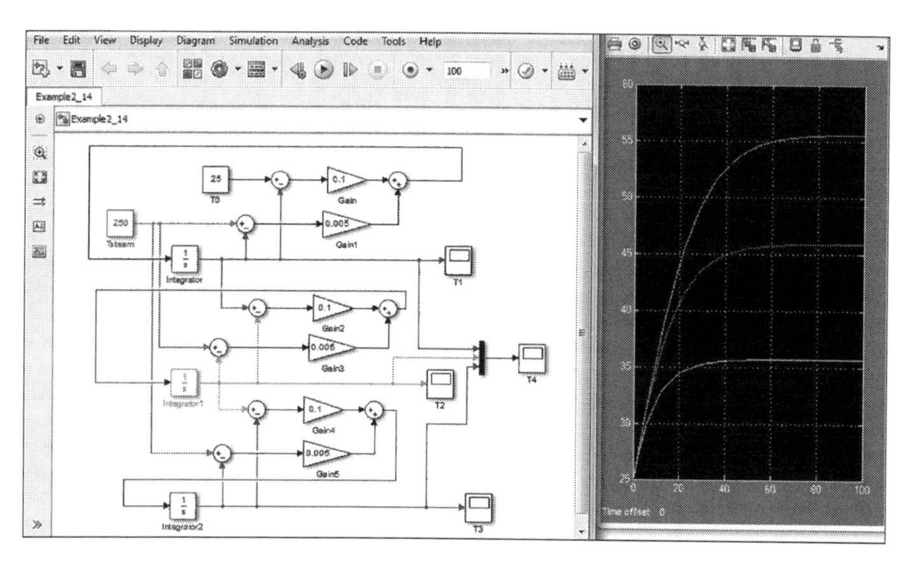

FIGURE 2.37
Simulink solution for the three heated tanks in series.

Energy balance around tank 3:

$$\frac{dT_3}{dt} = \frac{\dot{m}}{m}(T_2 - T_3) + \frac{UA}{mC_p}(T_s - T_3) = 0.1\,(T_2 - T_3) + 0.005\,(250 - T_3)$$

The three ODEs are solved simultaneously via MATLAB/Simulink, as shown in Figure 2.37.

Example 2.15: Heat Transfer from Sold Object

A hot solid block of mass m, average temperature T, and external surface area A is placed in and surrounded by fluid maintained at constant temperature, T_a. As the temperature of this solid object changes, its specific heat changes, but if the range of temperature variation is moderate, we can use an average value of the specific heat (C_p) and treat it as a constant for this analysis.

Solution

Now, assume that the mass is surrounded by a fluid at a temperature to which it loses heat, where U is the overall heat transfer coefficient between the mass solid and the surrounding air, and A is the surface area. The unsteady-state energy balance is:

$$\frac{d}{dt}\left(mC_p(T - T_{\text{ref}})\right) = -UA\,(T - T_a)$$

The mass, the specific heat, and the overall all heat transfer coefficient of the object are constants; thus:

$$mC_p\frac{dT}{dt} = -UA(T - T_a)$$

When T is greater than T_a, the mass is losing heat to the surroundings. Thus, the rate of addition is the negative of the rate of heat loss, which explains the minus sign on the right side. This equation is correct whether the mass is hotter than the surroundings or cooler than the surroundings, as you can verify for yourself. That is, we need to know the initial temperature of the mass:

$$\text{At } t = 0, T = T_0$$

Separate the differential equation and integrate both sides:

$$\frac{dT}{(T - T_a)} = -\frac{UA}{mC_p}dt$$

Integrate the unsteady-state energy balance:

$$\ln(T - T_a) = -\frac{UA}{mC_p}t + C_1$$

Apply the initial condition and find the value of C_1:

$$C_1 = \ln(T_0 - T_a)$$

Substituting C_1 and rearrange:

$$T = T_\infty + (T_0 - T_a)e^{-\frac{UA}{mC_p}t}$$

PROBLEMS

2.1 CSTR

CSTRs are common in the process industry. The CSTR is assumed to be perfectly mixed, with a single first-order exothermic and irreversible reaction $A \rightarrow B$, A the reactant converted to B, the product. The initial concentration of component A is C_{Ao}. Develop model equations that represent a jacketed non-adiabatic stirred tank reactor.

2.2 Jacketed Non-adiabatic CSTR

The reaction $A \rightarrow B$ takes place in a jacketed non-adiabatic CSTR reactor. The reaction is a single first-order exothermic and irreversible reaction. The reactant A is converted to B, the product. The initial concentration of component A is $C_{Ao} = 15\ \text{kmol/m}^3$. The inlet volumetric feed rate is $F_o = 200\ \text{m}^3/\text{h}$, the volume of the reactor is $V = 380\ \text{m}^3$, the inlet temperature is 250 K, the liquid density is $600\ \text{kg/m}^3$, and the fluid heat capacity is $C_p = 3\ \text{kJ/}(\text{kgn K})$. Develop model equations that represent a jacketed non-adiabatic stirred tank reactor and solve the resultant equations using Simulink.

2.3 Two Isothermal CSTR Reactors

An isothermal and two continuous flow stirred tank reactors connected in series are shown in Figure 2.38. The two systems are the liquid in each tank; $V_1 = V_2 = 1.05\ \text{m}^3$. Flow rates are constants and equal $0.085\ \text{m}^3/\text{min}$. Inlet feed concentration is $0.925\ \text{mol/m}^3$. The reaction rate is first order with a rate constant of $0.041/\text{min}$. Determine the steady-state concentrations in both reactors in mol/m^3.

Data

$$F = 0.085\ \text{m}^3/\text{min};\ C_{Ao} = 0.925\ \text{mol/m}^3.$$

The chemical reaction is first-order, $r_A = -kC_A$ with $k = 0.0401/\text{min}$.

2.4 Compound Dissolved in Water

Component A dissolves in water at a rate proportional to the product of the amount of an undissolved solid (Figure 2.39). The difference between the concentration in a saturated solution and the actual solution is $C_{sat} - C$. The dissolution rate is $0.257\ h^{-1}$. A saturated solution of this compound contains 0.4 g solid/g water. In a test run, start with 20 kg of undissolved compound

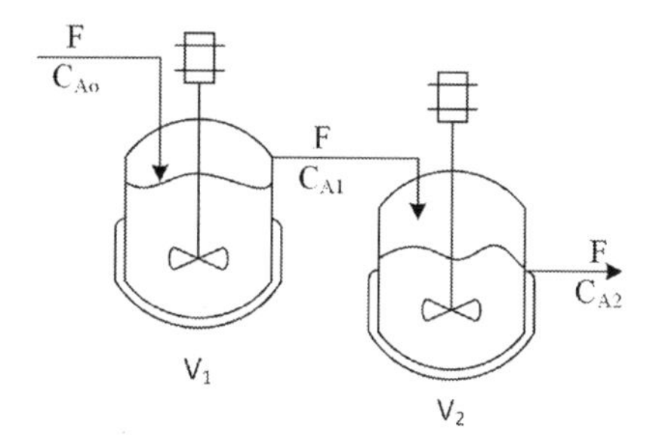

FIGURE 2.38
Two CSTR in series.

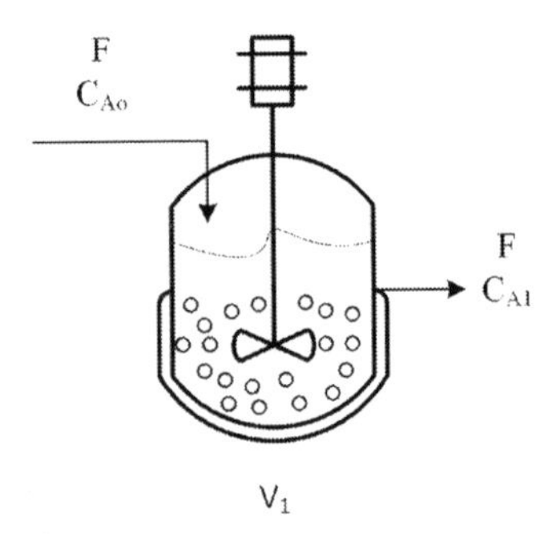

FIGURE 2.39
Schematic of the dissolved compound in a tank.

in 100 kg of water. How many kilograms of compound will remain undissolved after 10 h? Assume that the system is isothermal.

2.5 Leaking Stirred Tank

A stirred tank is used to heat 100 kg of a solvent (mass heat capacity $C_p = 2.5\ \text{J/g°C}$). An electrical coil delivers 2.0 kJ/s of power to the tank; the shaft work of the stirrer is 560 W. The solvent is initially at 25°C. The heat lost from the walls of the tank is 200 J/s. The tank is leaking at a rate of 0.2 kg/h. Calculate the temperature in the tank after 10 hours (Figure 2.40). Write the total material balance equations that describes the process.

2.6 Semibatch Tank

A stirred tank heater (Figure 2.41) initially contains 100 kg of a solvent; the initial temperature of the solvent in the tank is 25°C. The tank is heated with 2.0 kJ/s of power via a heating coil. The heat is transported to the tank using an electrical coil. The shaft work of the stirrer is 560 W. The heat lost from the walls of the tank is 200 J/s. The tank is continuously fed with a solvent at a

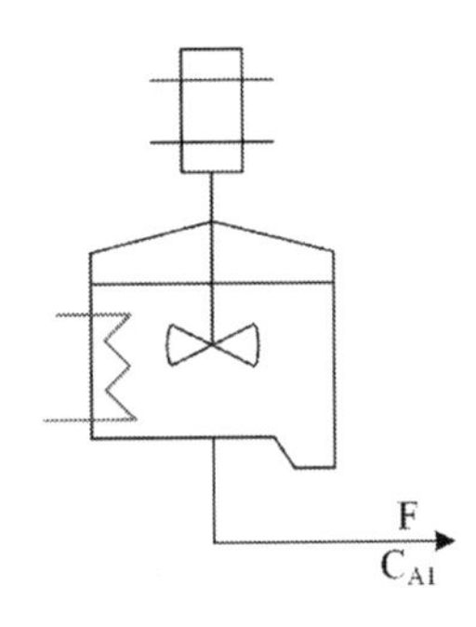

FIGURE 2.40
Schematic of the leaking tank.

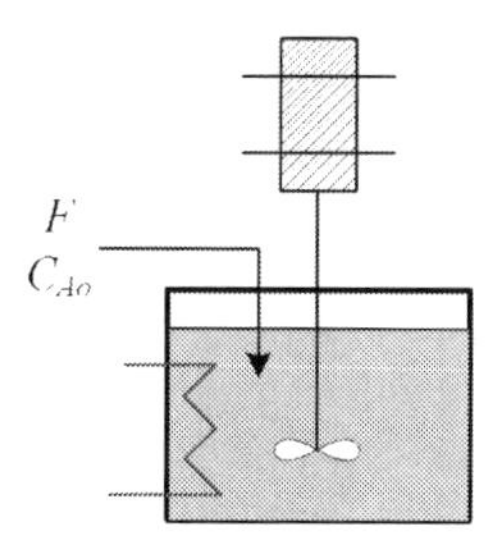

FIGURE 2.41
Semibatch heater tank.

rate of 0.2 kg/h and 25°C. The mass heat capacity of the solvent is 2.5 J/g°C. Calculate the temperature in the tank after 10 hours.

2.7 Continuous Heating Tank

Solvent is continuously fed to a well-mixed tank at a rate of 0.2 kg/h and 25°C (Figure 2.42). The tank initially contains 100 kg of a solvent (mass heat capacity 2.5 J/g°C). The solvent is initially at 25°C. An electrical coil delivers 1.80 kJ/s of power to the tank; the shaft work of the stirrer is 0.56 kW. There is no heat loss from the tank. The tank outlet flow rate is 0.2 kg/h. Calculate the temperature in the tank after 10 hours.

2.8 Continuous Stirred Tank Reactor

Consider CSTRs where the following first-order irreversible reaction takes place.

$$A \rightarrow B, -r_A = kC_A, k = 0.64 \text{ min}$$

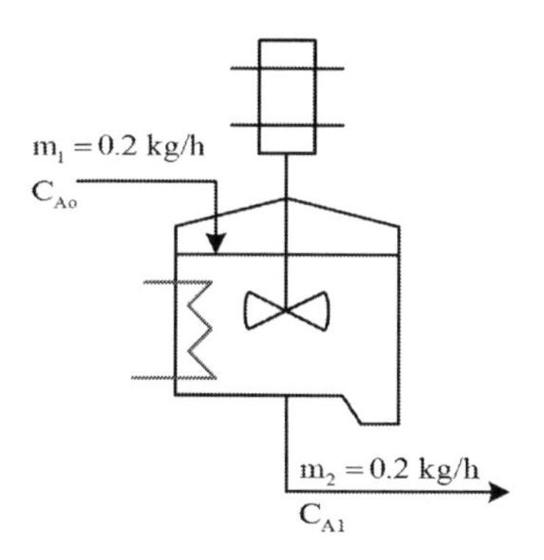

FIGURE 2.42
Schematic of the well-mixed tank.

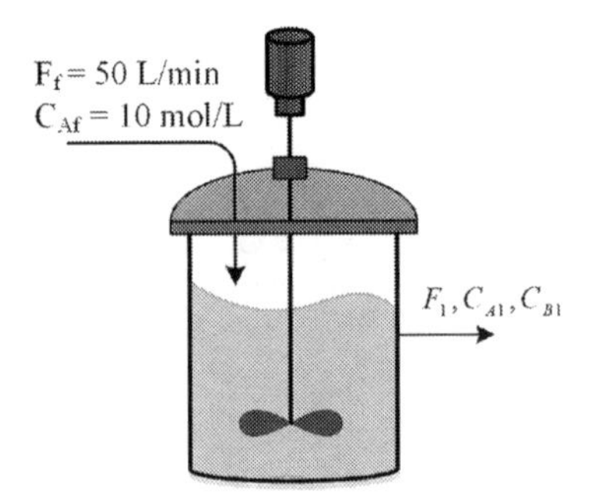

FIGURE 2.43
Continuous stirred tank reactor in series.

The inlet and exit volumetric flow rates are fixed at 50 L/min, the volume of the reactor is 100 L, and the inlet concentration of component A to the tank is 10 mol/L. The initial concentration of A in the tank is 5 mol/L. Develop the system ODEs that describe the concentration profile in the tank (Figure 2.43). Calculate the concentration of A and B after 30 min.

2.9 Isothermal CSTR

An isothermal CSTR is assumed to be perfectly mixed; initially it's at temperature 300 K (Figure 2.44). Reactant A is fed to the reactor from two different inlets with constant flow rates so the volume in the reactor tank is thus kept constant; 100 kmol/m^3 of reactant A is initially in the tank. The reaction is a single first-order exothermic and is irreversible:

$$A \rightarrow B, r_A = -1.5C_A$$

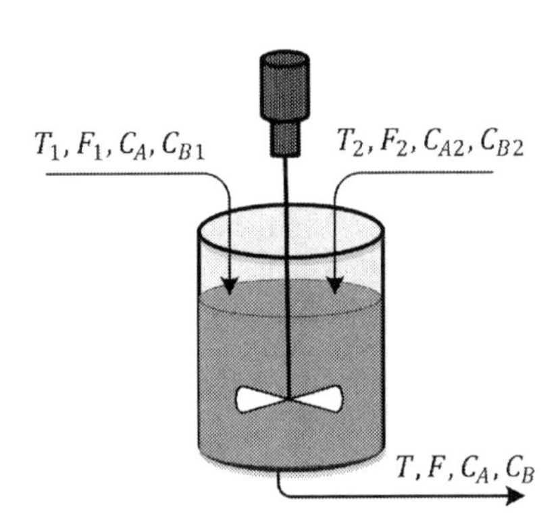

FIGURE 2.44
Continuous reactor with two feed streams.

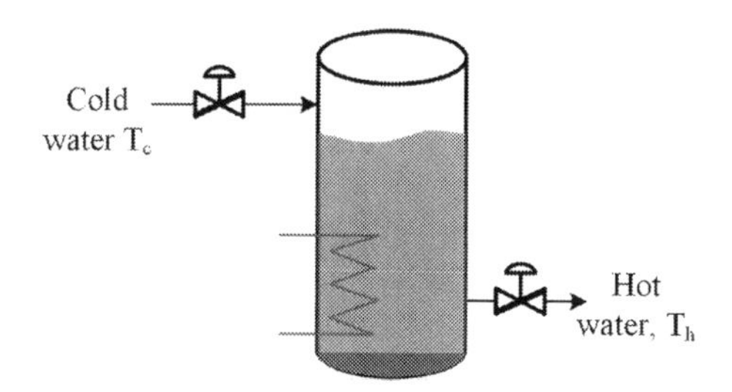

FIGURE 2.45
Schematic of shower heater.

Taking the significant variation of the tank temperature during operation of the reactor into account, it is desirable to ensure that this process variable is kept within reasonable limits.

2.10 Heating Water Tank

A heater holds 20 kg of water (Figure 2.45). The water initially is at 25°C. The heater is 3.0 kW power. The heater is switched on to start heating the water in it. Estimate the temperature of the water after 30 min. During heating, inlet or exit streams control valves are closed.

References

1. Kanse, N. G., P. B. Dhanke, T. Abhijit, 2012. Modeling and simulation study of the CSTR for complex reaction research. *Journal of Chemical Sciences* 2(4), 79–85.
2. William, L. L., 1973. *Process Modeling, Simulation and Control for Chemical Engineers*, New York: McGraw-Hill.
3. Michael, C., M. Schacham, 2008. *Problem Solving in Chemical and Biochemical Engineering with Polymath, Excel, and MATLAB®*, 2nd ed., New York: Prentice Hall.

3

Theory and Applications of Distributed Systems

Systems are called lumped, and mathematically described by ordinary differential equations, with time as the independent variable, when there are no important spatial variation in the system parameters. By contrast, distributed parameter systems vary continuously from one point to another, and the system state variables vary in space and in time. The mathematical description of a distributed system includes partial differential equations. The most important difference between the lumped and distributed systems is that, in the distributed system, the differential equation is in most cases not solvable in analytic form. Modeling of distributed parameter system is difficult but essential to simulation, control, and optimization. An appropriate mathematical model of the process is crucial to many applications such as system control design and optimization, analysis, and numerical simulation.

LEARNING OBJECTIVES

- Present systems of microscopic balances in one or two dimensions.
- Write the general balance equations in multidimensional systems.
- Extract the balance equations in Cartesian coordinates.
- Present the corresponding balance equations in cylindrical and spherical coordinates.

3.1 Introduction

Microscopic balances in distributed parameter systems will be studied in this chapter. The general balance equations in multidimensional conditions are considered. Once the equations are presented, we show through various examples how they can be used in a systematic way to model distributed systems. For that purpose, a background on transport phenomena is required to understand and interpret the details of materials processing and the synthesis and effects of processing parameters on the process. From this understanding, we can modify the process to produce materials in a shorter period at a reasonable cost with high quality. Knowledge of the basics of transport phenomena enables one to model the processes and do experiments in a computer environment. Transport phenomena is an engineering science with three main components [1–3]:

- Mass transport
- Fluid dynamics (momentum transport)
- Energy transport (heat transfer)

The three components are the key factors controlling the materials processing and synthesis and will be illustrated in the coming sections. Regarding boundary conditions, the shear stress is set to zero at a free liquid surface when flow is not driven by the adjoining gas dragging the liquid. In general, at a fluid-fluid interface, the velocity and the shear stress are continuous across the interface. The "no slip" boundary condition is the most common condition used, and it states that the fluid adjacent to a solid surface assumes the velocity of the solid.

3.2 Mass Transport

The definition of mass transfer is the net movement of mass from one location to another. The movement can be from component, stream, phase, or fraction. Mass transfer appears in multiple processes such as precipitation, absorption, evaporation, drying, membrane filtration, and distillation. Mass transfer is utilized by varying disciplines for different processes and mechanisms. The expression is usually used in engineering for physical processes that include diffusive and convective transport of chemical species within physical systems. Evaporation of water from a pond to the atmosphere is one of the most common examples of mass transfer processes; others are membrane separation such as the purification of blood in the kidneys and liver, the distillation of alcohol, and the absorption of carbon dioxide via gas liquid membrane contactor. Mass transfer operations in industrial processes include separation of chemical components in distillation columns, absorbers such as scrubbers or stripping, liquid-liquid extraction, and adsorbers such as activated carbon beds. Mass transfer is often tied to other transport processes; an example can be found in industrial cooling towers. These towers couple heat transfer to mass transfer by allowing hot water to flow in contact with hotter air and water evaporate as it absorbs the heat from the air cooling towers. Mass transfer also deals with multicomponent systems; an example is a binary two-component solution consisting of a solute in excess of chemically different solvent. Mass transfer also deals with situations involving chemical reactions, dissolution, or mixing phenomena such chemical absorption of carbon dioxide using aqueous diethanolamine (DEA) in conventional absorption towers.

In chemical engineering problems, mass transfer finds wide application. Mass transfer is used in reaction engineering, separations, heat transfer, and countless other subdisciplines of chemical engineering. The driving force for mass transfer is naturally the difference in chemical potential, although other thermodynamic gradients may couple to the flow of mass and drive it as well. A chemical species moves from areas of high chemical potential to areas of low chemical potential. Thus, the maximum theoretical extent of a given mass transfer is generally settled by the point at which the chemical potential is uniform [4–6].

3.2.1 Mass Transfer in Cartesian Coordinate

A Cartesian coordinate classification is a coordinate system that uses a pair of numerical coordinates to specify each point uniquely in a plane. The resultant equation in Cartesian coordinates for mass transfer is obtained by applying the law of conservation of mass to a differential control volume representing the system. The resulting equation is called the continuity equation and takes two forms:

1. Total continuity equation applies the law of conservation of mass to the total mass of the system.
2. Component continuity equation applies the law of conservation of mass to an individual component.

To derive the total continuity equation, consider the control volume, shown in Figure 3.1.

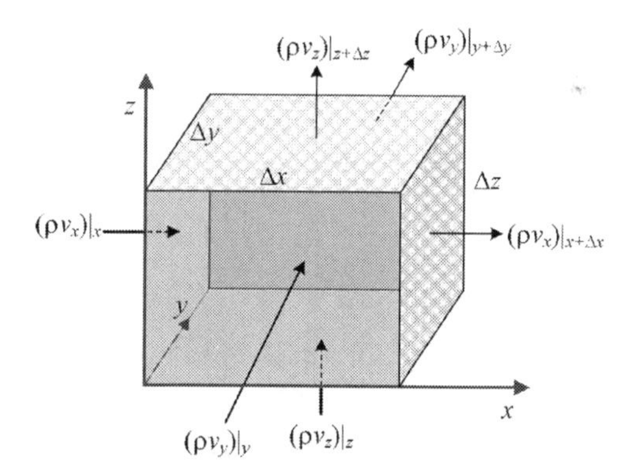

FIGURE 3.1
Total mass balance in Cartesian coordinates.

TABLE 3.1

Conservation of Total Mass to Control Volume in Cartesian Coordinate

Direction	In	Out	(In – Out) × Area
x	$\rho v_x \Delta y \Delta z\big\vert_x$	$\rho v_x \Delta y \Delta z\big\vert_{x+\Delta x}$	$\left(\rho v_x\big\vert_x - \rho v_x\big\vert_{x+\Delta x}\right)\Delta y \Delta z$
y	$\rho v_y \Delta x \Delta z\big\vert_y$	$\rho v_y \Delta x \Delta z\big\vert_{y+\Delta y}$	$\left(\rho v_y\big\vert_y - \rho v_y\big\vert_{y+\Delta y}\right)\Delta x \Delta z$
z	$\rho v_z \Delta x \Delta y\big\vert_z$	$\rho v_z \Delta x \Delta y\big\vert_{z+\Delta z}$	$\left(\rho v_z\big\vert_z - \rho v_z\big\vert_{z+\Delta z}\right)\Delta x \Delta y$

To write the mass balance around the volume, we need to consider the total inlet mass flow rate, the total outlet mass flow rate, and the accumulated mass flow rate. The total inlet mass flow rate is the mass entering in the three directions x, y, and z: the mass entering in the x-direction at the cross-sectional area ($\Delta y \Delta z$), the mass entering in the y-direction at the cross-sectional area ($\Delta x \Delta z$), and the mass entering in the z-direction at the cross-sectional area ($\Delta x \Delta y$). The total mass flow rate leaving is the mass exiting in the x-direction at the cross-sectional area ($\Delta y \Delta z$), the mass exiting the y-direction at the cross-sectional area ($\Delta x \Delta z$), and the mass exiting the z-direction at the cross-sectional area ($\Delta y \Delta z$), plus the accumulated mass with the control volume $dM/dt = \rho(\Delta x \Delta y \Delta z)$. Apply the law of conservation of mass on this control volume as shown in Table 3.1.

$$\text{in} - \text{out} = \text{accumulation}$$

Write the above terms in the overall equation [in – out = accumulation]:

$$\left[\rho v_x\big\vert_x \Delta y \Delta z + \rho v_y\big\vert_y \Delta x \Delta z + \rho v_z\big\vert_z \Delta x \Delta y\right] - \left[\rho v_x\big\vert_{x+\Delta x} \Delta y \Delta z + \rho v_y\big\vert_{y+\Delta y} \Delta x \Delta z + \rho v_z\big\vert_{z+\Delta z} \Delta x \Delta y\right] = \frac{dm}{dt} \tag{3.1}$$

$$\text{Accumulation} = \frac{\partial m}{\partial t} = \frac{\partial \rho \Delta x \Delta y \Delta z}{\partial t} = \Delta x \Delta y \Delta z \frac{\partial \rho}{\partial t}$$

Divide each term in the equation by $\Delta x \Delta y \Delta z$:

$$\frac{\rho v_x\big\vert_x - \rho v_x\big\vert_{x+\Delta x}}{\Delta x} + \frac{\rho v_y\big\vert_y - \rho v_y\big\vert_{y+\Delta y}}{\Delta y} + \frac{\rho v_z\big\vert_z - \rho v_z\big\vert_{z+\Delta z}}{\Delta z} = \frac{d\rho}{dt} \tag{3.2}$$

Take the limit as Δx, Δy, and Δz approach zero:

$$\frac{\partial \rho}{\partial t} = -\frac{\partial \rho v_x}{\partial x} - \frac{\partial \rho v_y}{\partial y} - \frac{\partial \rho v_z}{\partial z} \tag{3.3}$$

Equation 3.3 is the general total continuity equation. This is the general form of the mass balance in Cartesian coordinates. If the fluid is incompressible, then the density is assumed constant, both in time and position. That means the partial derivatives of the density (ρ) are all zero. The total continuity Equation 3.3 is equivalent to:

$$0 = \rho\left[\frac{\partial v_x}{\partial x} + \frac{\partial v_y}{\partial y} + \frac{\partial v_z}{\partial z}\right] \tag{3.4}$$

or simply:

$$0 = \frac{\partial v_x}{\partial x} + \frac{\partial v_y}{\partial y} + \frac{\partial v_z}{\partial z} \tag{3.5}$$

3.2.2 Component Continuity Equation

Whenever composition changes are to be investigated, a component balance equation must be written. Most separator or reactor problems involve a component balance. Compositions are usually conveyed in terms of mole fractions. The component continuity equation takes two forms depending on the units of concentration:

1. Mass continuity equation (mass units)
2. Molar continuity equation (molar units)

The importance of the component differential equation of mass transfer is to describe the concentration profiles and the flux or other parameters of engineering interest within a diffusing system. Do not forget that, in the total continuity equation, there is no generation or consumption terms.

3.2.2.1 Component Mass Continuity Equation

To determine the component mass continuity equation, consider the control volume, $\Delta x \Delta y \Delta z$ (Figure 3.2).

We consider a fluid consisting of species *A*, *B*, …, where a chemical reaction is generating the species *A* at a rate r_A (kg/m^{3}s). The fluid is in motion with mass-average velocity $v = n_t/\rho$ (m/s), where $n_t = n_A + n_B + \ldots$ (kg/m^{2}s) is the total mass flux and ρ (kg/m^{3}s) is the density of the mixture. Our objective is to establish the component mass balance equation of *A* as it diffuses in all

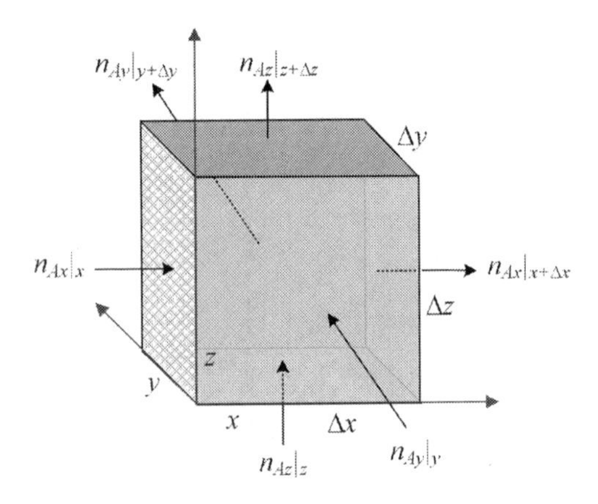

FIGURE 3.2
Component mass balance of component A.

directions x, y, z. Apply the law of conservation of mass for component balance to a control volume (Table 3.2):

$$\text{in} - \text{out} + \text{generation} - \text{consumption} = \text{accumulation}$$

$$\text{Accumulation of component } A = \frac{\partial m_A}{\partial t} = \frac{\partial \rho_A \Delta x \Delta y \Delta z}{\partial t} = \Delta x \Delta y \Delta z \frac{\partial \rho_A}{\partial t}$$

A is produced within the control volume by a chemical reaction at a rate of:

$$r_A \frac{\text{mass}}{(\text{volume})(\text{time})}$$

$$\text{Rate of production of } A \text{ (generation)} = r_A \Delta x \Delta y \Delta z$$

Place all terms in the equation in terms of:

$$\text{in} - \text{out} + \text{generation} - \text{consumption} = \text{accumulation}$$

TABLE 3.2
Conservation of Component Mass Balance to Control Volume

Direction	In	Out	(In – Out) × Area
x	$n_{Ax}\Delta y \Delta z\big|_x$	$n_{Ax}\Delta y \Delta z\big|_{x+\Delta x}$	$\left(n_{Ax}\big|_x - n_{Ax}\big|_{x+\Delta x}\right)\Delta y \Delta z$
y	$n_{Ay}\Delta x \Delta z\big|_y$	$n_{Ay}\Delta x \Delta z\big|_{y+\Delta y}$	$\left(n_{Ay}\big|_y - n_{Ay}\big|_{y+\Delta y}\right)\Delta x \Delta z$
z	$n_{Az}\Delta x \Delta y\big|_z$	$n_{Az}\Delta x \Delta y\big|_{z+\Delta z}$	$\left(n_{Az}\big|_z - n_{Az}\big|_{z+\Delta z}\right)\Delta x \Delta y$

$$\left\{n_{Ax}\Big|_x(\Delta y\Delta z)+n_{Ay}\Big|_y(\Delta x\Delta z)+n_{Az}\Big|_z(\Delta x\Delta y)\right\}$$
$$-\left\{n_{Ax}\Big|_{x+\Delta x}(\Delta y\Delta z)+n_{Ay}\Big|_{y+\Delta y}(\Delta x\Delta z)+n_{Az}\Big|_{z+\Delta z}(\Delta x\Delta y)\right\} \tag{3.6}$$
$$+\left\{r_A(\Delta x\Delta y\Delta z)\right\}=(\Delta x\Delta y\Delta z)\frac{d\rho_A}{dt}$$

Divide each term in the above equation by $\Delta x\Delta y\Delta z$:

$$\frac{n_{Ax}\Big|_x-n_{Ax}\Big|_{x+\Delta x}}{\Delta x}+\frac{n_{Ay}\Big|_y-n_{Ay}\Big|_{y+\Delta y}}{\Delta y}+\frac{n_{Az}\Big|_z-n_{Az}\Big|_{z+\Delta z}}{\Delta z}+r_A=\frac{d\rho_A}{dt} \tag{3.7}$$

Take the limit as Δx, Δy, and Δz approach zero:

$$-\frac{\partial n_{Ax}}{\partial x}-\frac{\partial n_{Ay}}{\partial y}-\frac{\partial n_{Az}}{\partial y}+r_A=\frac{\partial \rho}{\partial t} \tag{3.8}$$

Equation 3.8 is the component mass continuity equation, and it can be written in the following form:

$$\frac{\partial \rho}{\partial t}+\frac{\partial n_{Ax}}{\partial x}+\frac{\partial n_{Ay}}{\partial y}+\frac{\partial n_{Az}}{\partial y}=r_A \tag{3.9}$$

3.2.2.2 Component Molar Continuity Equation

The mass flux n_{Ax} is the sum of a term due to convection ($\rho_A v_{Ax}$) and a term due to diffusion $j_{Ax}(\text{kg/m}^2\text{s})$:

$$n_{Ax}=\rho_A v_x+j_{Ax} \tag{3.10}$$

Substituting the flux in Equation 3.9 gives:

$$\frac{\partial \rho}{\partial t}=-\frac{\partial(\rho_A v_x)}{\partial x}-\frac{\partial(\rho_A v_y)}{\partial y}-\frac{\partial(\rho_A v_z)}{\partial z}-\frac{\partial j_{Ax}}{\partial x}+\frac{\partial j_{Ay}}{\partial y}+\frac{\partial j_{Az}}{\partial z}+r_A \tag{3.11}$$

For a binary mixture (A, B), Fick's law gives the flux in the u-direction as:

$$j_{Au}=-\rho D_{AB}\frac{\partial w_A}{\partial u} \tag{3.12}$$

where the mass fraction $w_A=\rho_A/\rho$.

Expanding Equation 3.11 and substituting for the fluxes yield:

$$\frac{\partial \rho_A}{\partial t} = -\rho_A\left(\frac{\partial v_x}{\partial x}+\frac{\partial v_y}{\partial y}+\frac{\partial v_z}{\partial z}\right)-\left(v_x\frac{\partial \rho_A}{\partial x}+v_y\frac{\partial \rho_A}{\partial y}+v_z\frac{\partial \rho_A}{\partial z}\right)$$
$$+\left(\frac{\partial}{\partial x}\left(\frac{\partial \rho D_{AB} w_A}{\partial x}\right)+\frac{\partial}{\partial y}\left(\frac{\partial \rho D_{AB} w_A}{\partial y}\right)+\frac{\partial}{\partial z}\left(\frac{\partial \rho D_{AB} w_A}{\partial z}\right)\right)+r_A \tag{3.13}$$

Species A in Equation 3.13 represents the general component balance or equation of continuity. This equation can be further reduced according to the nature of properties of the fluid involved. If the binary mixture is a dilute liquid and can be considered incompressible, then density ρ and diffusivity D_{AB} are constant. Substituting the continuity Equation 3.5 into Equation 3.13 leads the component mass to continuity Equation 3.14:

$$\frac{\partial \rho_A}{\partial t} = -\left(v_x\frac{\partial \rho_A}{\partial x}+v_y\frac{\partial \rho_A}{\partial y}+v_z\frac{\partial \rho_A}{\partial z}\right)+D_{AB}\left(\frac{\partial^2 \rho_A}{\partial x^2}+\frac{\partial^2 \rho_A}{\partial y^2}+\frac{\partial^2 \rho_A}{\partial z^2}\right)+r_A \tag{3.14}$$

This equation can also be written in molar units by dividing each term by the molecular weight M_A to yield:

$$\overbrace{\frac{\partial C_A}{\partial t}}^{\text{accum.}} = -\left(\overbrace{v_x\frac{\partial C_A}{\partial x}+v_y\frac{\partial C_A}{\partial y}+v_z\frac{\partial C_A}{\partial z}}^{\text{Convection}}\right)+\overbrace{D_{AB}\left(\frac{\partial^2 C_A}{\partial x^2}+\frac{\partial^2 C_A}{\partial y^2}+\frac{\partial^2 C_A}{\partial z^2}\right)}^{\text{Diffusion}}+\overbrace{R_A}^{\text{reaction}} \tag{3.15}$$

The general differential equation consists of the transient term, a convective term, a diffusive term, and a reaction term, which are the main terms of the component balance equation, where C_A is the molar concentration in mol/m^3, and rate of reaction R_A in $\text{mol/(m}^3\text{s)}$.

3.3 Fluid Dynamics

Fluid dynamics is defined as the study of the behavior of fluids when they are in motion. Fluids can flow either steadily or turbulently. The fluid maintains a steady velocity while passing a given point in a steady flow process. For turbulent flow, the speed and or the direction of the flow varies. The motion in steady flow can be represented with streamlines showing the direction in which the fluid flows in different areas. The density of the streamlines increases as the velocity increases. Fluids can be compressible or incompressible. Liquids are generally incompressible, which means that their volume does not drastically change in response to a pressure change. Gases, on the other hand, are compressible, their volume response to a change in pressure. The application of the law of conservation of mass to a small volume element within a flowing fluid results in developing the equation of continuity. In addition, the equation of continuity is frequently used in conjunction with the momentum equation. The general momentum equation is also called the equation of motion, or the Navier-Stoke's equation. The following two equations can be used to describe most fluid flow problems mathematically:

- The equation of continuity
- The momentum equation

To develop the equation of motion, we start with Newton's law of conservation of energy:

$$
\begin{aligned}
\text{Rate of momentum accumulation} = & \text{ transport rate of momentum in} \\
& - \text{transport rate of momentum out} \\
& + \text{sum of forces acting on element}
\end{aligned}
$$

Consider a fluid flowing with a velocity $v(t, x, y, z)$ in the cube of Figure 3.3. The flow is assumed laminar. The momentum is transferred through convection (bulk flow) and by molecular transfer (velocity gradient).

The momentum is a vector that has three components: the equation for the conservation of the x-component, the y-component, and the z-component of the momentum, all of which are obtained in a similar way. To establish the momentum balance for the x-component, we need to consider its transfer

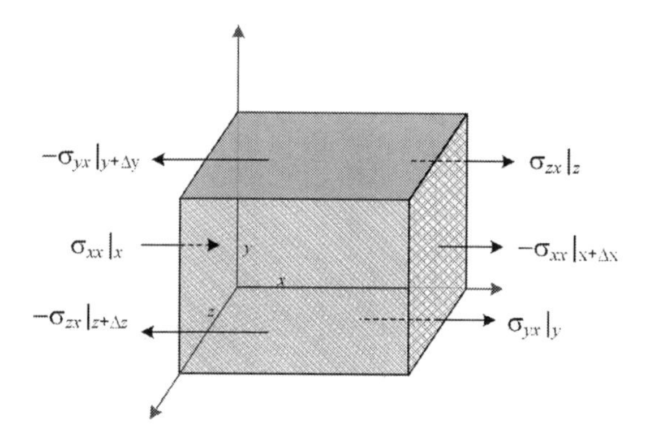

FIGURE 3.3
Momentum balance of the x-component.

in the x-direction, y-direction, and z-direction. Convective transport of momentum is by means of the kinetic energy of the fluid mass moving in and out of the six faces of our cubical element. Thus, the convective transport of momentum in the x-direction (x-momentum) consists of three terms:

$$(\rho v_x)v_x,\ (\rho v_y)v_x,\ (\rho v_z)v_x$$

The summary of the convective transport of momentum in the x-direction is shown in Table 3.3.

Diffusive transport of momentum also takes place at the six surfaces of the cubical element (Figure 3.3) by means of the viscous shear stresses (Table 3.4).

The fluid pressure force on the volume element in the x-direction is: $\left(P|_x - P|_{x+\Delta x}\right)\Delta x \Delta y \Delta z$.

The net gravitational force in the x-direction is: $\left(\rho g|_x\right)\Delta x \Delta y \Delta z$

Accumulation is: $\left(\rho v_x|_{t+dt} - \rho v_x|_t\right)\Delta x \Delta y \Delta z$

Substituting all these equations in the Newton law of conservation of energy, we have:

$$\begin{aligned}\text{Rate of momentum accumulation} &= \text{transport rate of momentum in}\\ &\quad - \text{transport rate of momentum out}\\ &\quad + \text{sum of forces acting on element}\end{aligned}$$

TABLE 3.3

Convective Transport of Momentum in the x-Direction

Direction	In	Out	(In – Out) × Area
x	$(\rho v_x)v_x\|_x$	$(\rho v_x)v_x\|_{x+\Delta x}$	$\left((\rho v_x)v_x\|_x - (\rho v_x)v_x\|_{x+\Delta x}\right)\Delta y \Delta z$
y	$(\rho v_y)v_x\|_y$	$(\rho v_y)v_x\|_{y+\Delta y}$	$\left((\rho v_y)v_x\|_y - (\rho v_y)v_x\|_{y+\Delta y}\right)\Delta x \Delta z$
z	$(\rho v_z)v_x\|_z$	$(\rho v_z)v_x\|_{z+\Delta z}$	$\left((\rho v_z)v_x\|_z - (\rho v_z)v_x\|_{z+\Delta z}\right)\Delta x \Delta y$

TABLE 3.4

Diffusive Transport of Momentum in Cartesian Coordinate

Direction	In	Out	(In – Out) × Area
x	$\tau_{xx}\|_x$	$\tau_{xx}\|_{x+\Delta x}$	$\left(\tau_{xx}\|_x - \tau_{xx}\|_{x+\Delta x}\right)\Delta y \Delta z$
y	$\tau_{yx}\|_y$	$\tau_{yx}\|_{y+\Delta y}$	$\left(\tau_{yx}\|_y - \tau_{yx}\|_{y+\Delta y}\right)\Delta x \Delta z$
z	$\tau_{zx}\|_z$	$\tau_{zx}\|_{z+\Delta z}$	$\left(\tau_{zx}\|_z - \tau_{zx}\|_{z+\Delta z}\right)\Delta x \Delta y$

Divide each term by $\Delta x \Delta y \Delta z$ and take the limit as $\Delta x, \Delta y, \Delta z$ goes to zero:

$$\frac{\partial(\rho v_x)}{\partial t} = -\left(\frac{\partial(\rho v_x v_x)}{\partial x} + \frac{\partial(\rho v_x v_y)}{\partial y} + \frac{\partial(\rho v_x v_z)}{\partial z}\right)$$
$$-\left(\frac{\partial \tau_{xx}}{\partial x} + \frac{\partial \tau_{yx}}{\partial y} + \frac{\partial \tau_{zx}}{\partial z}\right) - \frac{\partial P}{\partial x} + \rho g_x \quad (3.16)$$

Expand the partial derivative and rearrange:

$$v_x \frac{\partial \rho}{\partial t} + \rho \frac{\partial v_x}{\partial t} = -v_x \left(\frac{\partial(\rho v_x)}{\partial x} + \frac{\partial(\rho v_y)}{\partial y} + \frac{\partial(\rho v_z)}{\partial z}\right)$$
$$-\rho\left(v_x \frac{\partial v_x}{\partial x} + v_y \frac{\partial v_x}{\partial y} + v_z \frac{\partial v_x}{\partial z}\right) \quad (3.17)$$
$$-\left(\frac{\partial \tau_{xx}}{\partial x} + \frac{\partial \tau_{yx}}{\partial y} + \frac{\partial \tau_{zx}}{\partial z}\right) - \frac{\partial P}{\partial x} + \rho g_x$$

Use the equation of continuity for incompressible fluid (constant density) and substitute using the assumption of Newtonian fluid:

$$\tau_{xx} = -\mu \frac{\partial v_x}{\partial x}, \ \tau_{yx} = -\mu \frac{\partial v_x}{\partial y}, \ \tau_{zx} = -\mu \frac{\partial v_x}{\partial z}$$

Equation 3.17 yields the momentum balances in the x-direction:

$$\overbrace{\rho \frac{\partial v_x}{\partial t}}^{\text{accum.}} = \overbrace{-\rho\left(v_x \frac{\partial v_x}{\partial x} + v_y \frac{\partial v_x}{\partial y} + v_z \frac{\partial v_x}{\partial z}\right)}^{\text{transport by bulk flow}} + \overbrace{\mu\left(\frac{\partial^2 v_x}{\partial x^2} + \frac{\partial^2 v_x}{\partial y^2} + \frac{\partial^2 v_x}{\partial z^2}\right)}^{\text{transport by viscous forces}} \overbrace{- \frac{\partial P}{\partial x} + \rho g_x}^{\text{generation}} \quad (3.18)$$

The momentum balances in the y-direction:

$$\overbrace{\rho \frac{\partial v_y}{\partial t}}^{\text{accum.}} = \overbrace{-\rho\left(v_x \frac{\partial v_y}{\partial x} + V_y \frac{\partial v_y}{\partial y} + V_z \frac{\partial v_y}{\partial z}\right)}^{\text{transport by bulk flow}} + \overbrace{\mu\left(\frac{\partial^2 v_y}{\partial x^2} + \frac{\partial^2 v_y}{\partial y^2} + \frac{\partial^2 v_y}{\partial z^2}\right)}^{\text{transport by viscous forces}} \overbrace{- \frac{\partial P}{\partial x} + \rho g_y}^{\text{generation}} \quad (3.19)$$

and the momentum balances in the z-direction:

$$\overbrace{\rho \frac{\partial v_z}{\partial t}}^{\text{accum.}} = \overbrace{-\rho\left(v_x \frac{\partial v_z}{\partial x} + v_y \frac{\partial v_z}{\partial y} + v_z \frac{\partial v_z}{\partial z}\right)}^{\text{transport by bulk flow}} + \overbrace{\mu\left(\frac{\partial^2 v_z}{\partial x^2} + \frac{\partial^2 v_z}{\partial y^2} + \frac{\partial^2 v_z}{\partial z^2}\right)}^{\text{transport by viscous forces}} \overbrace{- \frac{\partial P}{\partial x} + \rho g_z}^{\text{generation}} \quad (3.20)$$

The following steps are useful for solving fluid-mechanics problems:

1. Read and sketch the system.
2. Choose system coordinates: Cartesian, cylindrical, or spherical.
3. Perform a total mass balance.
4. Perform and simplify the momentum balance equation.
5. Solve the differential equation by applying the suitable boundary conditions.

3.4 Energy Transport

The transfer of heat is normally from a high-temperature object to a lower-temperature object. Heat is transferred by three mechanisms—conduction, radiation, and convection—when there is physical contact between systems at different temperatures. Through molecular motion, conduction heat transfer occurs [5,6]. Heat transfer through solids can be achieved by conduction alone, whereas the heat may be transferred from a solid surface to a fluid partly by conduction and partly by convection. Heat transfer by radiation can occur between solid surfaces. The energy that radiates from hot surfaces as electromagnetic waves is called thermal radiation. It does not need a medium for its dissemination. Solids radiate over a wide range of wavelengths, while some gases can only release and absorb radiation on certain wavelengths. The energy flux emitted by an ideal radiator is proportional to the fourth power of its absolute temperature.

The flow of energy per unit of area per unit of time is the heat flux. In SI, its units are watts per square meter ($W \cdot m^{-2}$). The heat flux (q_y) in the y-direction is $q_y = -k\partial T/\partial y$. The thermal conductivity (k) of a substance is defined as the heat flow per unit area per unit time when the temperature decreases by one degree in unit distance. The thermal conductivity is a material property that reflects the relative ease or difficulty of the transfer of energy through the material depending on the bonding and structure of the material. Heat transfer within a fluid is called convection. Convection involves the gross motion of the fluid itself. Fluid motion can occur due to differences in density, as in free convection. Density differences are a direct result of temperature differences between the fluid and the solid wall surface. In forced convection, mechanical means are used to produce fluid motion. An example of mechanical means is a domestic fan-heater that blows the air across an electric element. When a moving fluid at one temperature encounters a solid at a different temperature, heat is exchanged between the solid and the fluid by conduction at a rate given by Fourier's law. Under such cases the heat transfer coefficient (h) can be used to determine the distribution of temperature within the fluid and the heat flux at the wall. Heat flux at the wall, $q_o = h(T_s - T_f)$, where T_s is the surface temperature, and T_f, is the bulk fluid temperature. Energy balance is essential whenever the temperature within a system changes; temperature is almost always inside the derivative.

3.4.1 Energy Transport in Cartesian Coordinates

The equation for energy balance is derived by the analogy that exists between mass and energy transport [1,2], assuming constant heat capacity, density, and thermal conductivity for the incompressible fluid. The fluid is assumed at constant pressure. Consider the energy balance in Cartesian coordinates shown in Figure 3.4.

The total energy flux is the sum of heat flux and bulk flux: $E = q + \rho v C_p T$. The total energy in and out of a system in Cartesian coordinates in x, y, and z directions is shown in Table 3.5.

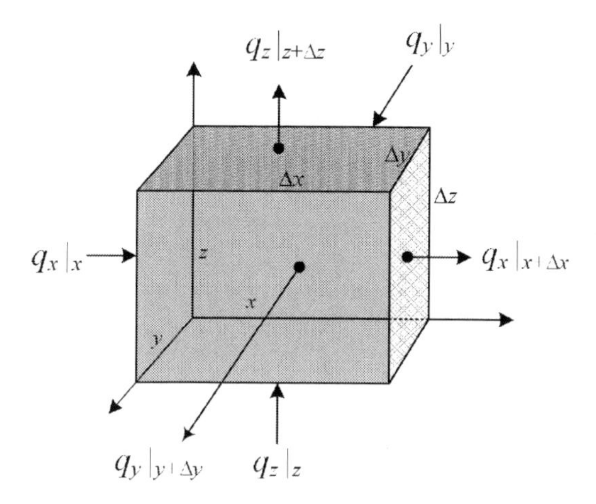

FIGURE 3.4
Energy balance in Cartesian coordinate.

TABLE 3.5
Total Energy Transport In and Out In a Cartesian Coordinate System

Direction	In	Out
x	$\left(q_x + \rho v_x C_p T\right)\Big\|_x$	$\left(q_x + \rho v_x C_p T\right)\Big\|_{x+\Delta x}$
y	$\left(q_y + \rho v_y C_p T\right)\Big\|_y$	$\left(q_y + \rho v_y C_p T\right)\Big\|_{y+\Delta y}$
z	$\left(q_z + \rho v_z C_p T\right)\Big\|_z$	$\left(q_z + \rho v_z C_p T\right)\Big\|_{z+\Delta z}$

The total energy entering in the x-direction y and z directions at boundaries x, y, and z is:

$$\left(q_x + \rho v_x C_p T\right)\Big|_x \Delta y \Delta z + \left(q_y + \rho v_y C_p T\right)\Big|_y \Delta x \Delta z + \left(q_z + \rho v_z C_p T\right)\Big|_z \Delta x \Delta y \quad (3.21)$$

The total energy leaving the x, y, and z directions is:

$$\left(q_x + \rho v_x C_p T\right)\Big|_{x+\Delta x} \Delta y \Delta z + \left(q_y + \rho v_y C_p T\right)\Big|_{y+\Delta y} \Delta x \Delta z + \left(q_z + \rho v_z C_p T\right)\Big|_{z+\Delta z} \Delta x \Delta y \quad (3.22)$$

The energy accumulated is approximated by $(\partial(\rho C_p T)/\partial t)\Delta x \Delta y \Delta z$.

The rate of generation Φ includes all the sources of heat generation, that is, reaction, pressure forces, fluid friction, and gravity forces. Substitute all these terms in the general energy equation, divide the equation by the term $\Delta x \Delta y \Delta z$, and let each of these terms approach zero [7,8]:

$$\frac{\partial(\rho C_p T)}{\partial t} = -\frac{\partial(\rho v_x C_p T)}{\partial x} - \frac{\partial(\rho v_y C_p T)}{\partial y} - \frac{\partial(\rho v_z C_p T)}{\partial z} - \frac{\partial q_x}{\partial x} - \frac{\partial q_y}{\partial y} - \frac{\partial q_z}{\partial z} + \Phi \quad (3.23)$$

Expanding the partial derivative yields:

$$\begin{aligned} \rho C_p \frac{\partial T}{\partial t} + C_p T \frac{\partial \rho}{\partial t} &= -\rho C_p \left(v_x \frac{\partial T}{\partial x} + v_y \frac{\partial T}{\partial y} + v_z \frac{\partial T}{\partial z} \right) \\ &\quad - \rho C_p T \left(\frac{\partial \rho v_x}{\partial x} + \frac{\partial \rho v_y}{\partial y} + \frac{\partial \rho v_z}{\partial z} \right) \\ &\quad - \frac{\partial q_x}{\partial x} - \frac{\partial q_y}{\partial y} - \frac{\partial q_z}{\partial z} + \Phi \end{aligned} \quad (3.24)$$

Using the equation of continuity for incompressible fluids reduces the equation to:

$$\rho C_p \frac{\partial T}{\partial t} = -\rho C_p \left(v_x \frac{\partial T}{\partial x} + v_y \frac{\partial T}{\partial y} + v_z \frac{\partial T}{\partial z} \right) - \frac{\partial q_x}{\partial x} - \frac{\partial q_y}{\partial y} - \frac{\partial q_z}{\partial z} + \Phi \quad (3.25)$$

Using Fourier's law, $q_u = -k(dT/du)$ into Equation 3.25 gives:

$$\overbrace{\rho C_p \frac{\partial T}{\partial t}}^{\text{accum.}} = \overbrace{-\rho C_p \left(v_x \frac{\partial T}{\partial x} + v_y \frac{\partial T}{\partial y} + v_z \frac{\partial T}{\partial z} \right)}^{\text{transport by bulk flow}} + \overbrace{k \left(\frac{\partial^2 T}{\partial x^2} + \frac{\partial^2 T}{\partial y^2} + \frac{\partial^2 T}{\partial z^2} \right)}^{\text{transport by thermal diffusion}} + \overbrace{\Phi}^{\text{gen.}} \quad (3.26)$$

The energy balance includes as before a transient term, convection term, diffusion term, and generation term. For solids, the density is constant and has no velocity; that is, $v = 0$. The equation is reduced to:

$$\underbrace{\rho C_p \frac{\partial T}{\partial t}}_{\text{accum.}} = \overbrace{k\left(\frac{\partial^2 T}{\partial x^2} + \frac{\partial^2 T}{\partial y^2} + \frac{\partial^2 T}{\partial z^2}\right)}^{\text{transport by thermal diffusion}} + \overbrace{\Phi}^{\text{gen.}} \tag{3.27}$$

3.4.2 Conversion Between the Coordinates

The equations of energy change in Cartesian coordinates have been derived in the previous sections. In the same way, the equations of change can be written for other coordinate systems such as the cylindrical or spherical coordinates. Alternatively, one can transform the equation of change written in Cartesian coordinates to the others through the following transformation expressions. The relations between Cartesian coordinates (x, y, z) and cylindrical coordinates (r, z, θ) (Figures 3.3 and 3.4) are the following relations:

$$x = r\cos(\theta),\ y = r\sin(\theta),\ z = z$$

Therefore, $r = \sqrt{x^2 + y^2}$, $\theta = \tan^{-1}(y/x)$

The relations between Cartesian coordinates (x, y, z) and spherical coordinates $(r, \theta, \varnothing)$ (Figures 3.3 and 3.5) are the following relations:

$$x = r\sin(\theta)\cos(\varnothing),\ y = r\sin(\theta)\sin(\varnothing),\ z = r\cos(\theta)$$

Therefore, $r = \sqrt{x^2 + y^2 + z^2}$, $\theta = \tan^{-1}\left(\sqrt{x^2 + y^2}/z\right)$, $\varnothing = \tan^{-1}(y/x)$

The general balance equations in the three coordinates are listed in Section 3.5. These equations are written under the assumptions mentioned in Sections 3.2 and 3.3. For the more general case, where density is considered variable, the reader can refer to the books listed in the references [1–5].

3.5 Introduction of Equations of Change

In the following sections the equations of change in Cartesian, cylindrical, and spherical coordinate are presented. Equations of continuity, energy and equations of motion are listed for each coordinate system.

3.5.1 Equations of Change in Cartesian Coordinates

The equation of continuity states that, for an incompressible fluid flowing in a tube of varying cross-section, the mass flow rate is the same everywhere in the tube. The rate at which mass flows past a given point is called the mass flow rate, which means that it can be computed by dividing the total mass flowing past a given point by the time interval. In this section the continuity equation, component balance equation, energy balance equation, and momentum equation are listed as follows [6,9]:

Continuity equation

$$0 = \frac{\partial v_x}{\partial x} + \frac{\partial v_y}{\partial y} + \frac{\partial v_z}{\partial z} \tag{3.28}$$

Component balance for component A in binary mixture with chemical reaction rate R_A:

$$\frac{\partial C_A}{\partial t} = -\left(v_x \frac{\partial C_A}{\partial x} + v_y \frac{\partial C_A}{\partial y} + v_z \frac{\partial C_A}{\partial z} \right) + D_{AB} \left(\frac{\partial^2 C_A}{\partial x^2} + \frac{\partial^2 C_A}{\partial y^2} + \frac{\partial^2 C_A}{\partial z^2} \right) + R_A \tag{3.29}$$

Energy balance

$$\rho C_p \frac{\partial T}{\partial t} = -\rho C_p \left(v_x \frac{\partial T}{\partial x} + v_y \frac{\partial T}{\partial y} + v_z \frac{\partial T}{\partial z} \right) + k \left(\frac{\partial^2 T}{\partial x^2} + \frac{\partial^2 T}{\partial y^2} + \frac{\partial^2 T}{\partial z^2} \right) + \Phi_H \tag{3.30}$$

Equation of motion in rectangular coordinates for Newtonian fluids for constant density and viscosity (the x, y, and z components are listed below, respectively):

x component:

$$\rho \frac{\partial v_x}{\partial t} = -\rho \left(v_x \frac{\partial v_x}{\partial x} + v_y \frac{\partial v_x}{\partial y} + v_z \frac{\partial v_x}{\partial z} \right) + \mu \left(\frac{\partial^2 v_x}{\partial x^2} + \frac{\partial^2 v_x}{\partial y^2} + \frac{\partial^2 v_x}{\partial z^2} \right) - \frac{\partial p}{\partial x} + \rho g_x \tag{3.31}$$

y component:

$$\rho \frac{\partial v_y}{\partial t} = -\rho \left(v_x \frac{\partial v_y}{\partial x} + v_y \frac{\partial v_y}{\partial y} + v_z \frac{\partial v_y}{\partial z} \right) + \mu \left(\frac{\partial^2 v_y}{\partial x^2} + \frac{\partial^2 v_y}{\partial y^2} + \frac{\partial^2 v_y}{\partial z^2} \right) - \frac{\partial p}{\partial y} + \rho g_y \tag{3.32}$$

z component:

$$\rho \frac{\partial v_z}{\partial t} = -\rho \left(v_x \frac{\partial v_z}{\partial x} + v_y \frac{\partial v_z}{\partial y} + v_z \frac{\partial v_z}{\partial z} \right) + \mu \left(\frac{\partial^2 v_z}{\partial x^2} + \frac{\partial^2 v_z}{\partial y^2} + \frac{\partial^2 v_z}{\partial z^2} \right) - \frac{\partial p}{\partial z} + \rho g_z \tag{3.33}$$

3.5.2 Equations of Change in Cylindrical Coordinates

The following form of the continuity or total mass-balance equation in cylindrical coordinates is expressed in terms of the mass density ρ, which can be nonconstant and mass-average velocity component.

Continuity equation:

$$\rho\left(\frac{1}{r}\frac{\partial r v_r}{\partial r}+\frac{1}{r}\frac{\partial v_\theta}{\partial \theta}+\frac{\partial v_z}{\partial z}\right)=0 \tag{3.34}$$

Component balance for component A in binary mixture (A–B) with reaction rate R_A:

$$\begin{aligned}\frac{\partial C_A}{\partial t}&=-\left(v_r\frac{\partial C_A}{\partial r}+v_\theta\frac{1}{r}\frac{\partial C_A}{\partial \theta}+v_z\frac{\partial C_A}{\partial z}\right)\\&\quad+D_A\left(\frac{1}{r}\frac{\partial}{\partial r}\left(r\frac{\partial C_A}{\partial r}\right)+\frac{1}{r^2}\frac{\partial^2 C_A}{\partial \theta^2}+\frac{\partial^2 C_A}{\partial z^2}\right)+R_A\end{aligned} \tag{3.35}$$

Energy balance:

$$\begin{aligned}\rho C_p\frac{\partial T}{\partial t}&=-\rho C_p\left(v_r\frac{\partial T}{\partial r}+\frac{v_\theta}{r}\frac{\partial T}{\partial \theta}+v_z\frac{\partial T}{\partial z}\right)+\\&\quad k\left(\frac{1}{r}\frac{\partial}{\partial r}\left(r\frac{\partial T}{\partial r}\right)+\frac{1}{r^2}\frac{\partial^2 T}{\partial \theta^2}+\frac{\partial^2 T}{\partial z^2}\right)+Q\end{aligned} \tag{3.36}$$

Equation of motion in cylindrical coordinates for Newtonian fluids for constant density and viscosity (the r, θ, and z components are listed below, respectively):

r component:

$$\begin{aligned}\rho\frac{\partial v_r}{\partial t}&=-\rho\left(v_r\frac{\partial v_r}{\partial r}+\frac{v_\theta}{r}\frac{\partial v_r}{\partial \theta}+\frac{v_\theta^2}{r}+v_z\frac{\partial v_r}{\partial z}\right)\\&\quad+\mu\left(\frac{\partial}{\partial r}\left(\frac{1}{r}\frac{\partial (rv_r)}{\partial r}\right)+\frac{1}{r^2}\frac{\partial^2 v_r}{\partial \theta^2}+\frac{2}{r^2}\frac{\partial v_\theta}{\partial \theta}+\frac{\partial^2 v_r}{\partial z^2}\right)-\frac{\partial P}{\partial r}+\rho g_r\end{aligned} \tag{3.37}$$

θ component:

$$\begin{aligned}\rho\frac{\partial v_\theta}{\partial t}&=-\rho\left(v_r\frac{\partial v_\theta}{\partial r}+\frac{v_\theta}{r}\frac{\partial v_\theta}{\partial \theta}+\frac{v_r v_\theta}{r}+v_z\frac{\partial v_\theta}{\partial z}\right)\\&\quad+\mu\left(\frac{\partial}{\partial r}\left(\frac{1}{r}\frac{\partial r v_\theta}{\partial r}\right)+\frac{1}{r^2}\frac{\partial^2 v_\theta}{\partial \theta^2}+\frac{2}{r^2}\frac{\partial v_r}{\partial \theta}+\frac{\partial^2 v_\theta}{\partial z^2}\right)-\frac{1}{r}\frac{\partial p}{\partial \theta}+\rho g_\theta\end{aligned} \tag{3.38}$$

z component

$$\rho\frac{\partial v_z}{\partial t}=-\rho\left(v_r\frac{\partial v_z}{\partial r}+\frac{v_\theta}{r}\frac{\partial v_z}{\partial \theta}+v_z\frac{\partial v_z}{\partial z}\right)$$
$$+\mu\left(\frac{1}{r}\frac{\partial}{\partial r}\left(r\frac{\partial v_z}{\partial r}\right)+\frac{1}{r^2}\frac{\partial^2 v_z^2}{\partial \theta^2}+\frac{\partial^2 v_z}{\partial z^2}\right)-\frac{\partial P}{\partial z}+\rho g_z \quad (3.39)$$

3.5.3 Equations of Change in Spherical Coordinates

Continuity equation in spherical coordinate:

$$\rho\left(\frac{1}{r^2}\frac{\partial(r^2 v_r)}{\partial r}+\frac{1}{r\sin(\theta)}\frac{\partial(v_\theta\sin(\theta))}{\partial\theta}+\frac{1}{r\sin(\theta)}\frac{\partial v_\varnothing}{\partial\varnothing}\right)=0 \quad (3.40)$$

Component balance for component A in binary mixture (A–B) with reaction rate R_A:

$$\frac{\partial C_A}{\partial t}=-\left(v_r\frac{\partial C_A}{\partial r}+\frac{v_\theta}{r}\frac{\partial C_A}{\partial\theta}+\frac{v_\varnothing}{r\sin(\theta)}\frac{\partial C_A}{\partial\varnothing}\right)+D_{AB}\left(\frac{1}{r^2}\frac{\partial}{\partial r}\left(r^2\frac{\partial C_A}{\partial r}\right)+\right.$$
$$\left.\left(\frac{1}{r^2\sin(\theta)}\frac{\partial}{\partial\theta}\left(\sin(\theta)\frac{\partial C_A}{\partial\theta}\right)+\frac{1}{r^2\sin^2(\theta)}\frac{\partial^2 C_A}{\partial\varnothing^2}\right)\right)+R_A=0 \quad (3.41)$$

Energy balance:

$$\rho C_p\frac{\partial T}{\partial t}=-\rho C_p\left(v_r\frac{\partial T}{\partial r}+\frac{v_\theta}{r}\frac{\partial T}{\partial\theta}+\frac{v_\varnothing}{r\sin(\theta)}\frac{\partial T}{\partial\varnothing}\right)$$
$$+k\left(\frac{1}{r^2}\frac{\partial}{\partial r}\left(r^2\frac{\partial T}{\partial r}\right)+\frac{1}{r^2\sin(\theta)}\frac{\partial}{\partial\theta}\left(\sin(\theta)\frac{\partial T}{\partial\theta}\right)+\frac{1}{r^2\sin^2(\theta)}\frac{\partial^2 T}{\partial\varnothing^2}\right)+\varnothing=0 \quad (3.42)$$

Equation of motion in spherical coordinates for Newtonian fluids for constant density and viscosity (the r, θ, and $\varnothing$ components are listed below, respectively):

r component:

$$\rho\frac{\partial v_r}{\partial t}=-\rho\left(v_r\frac{\partial v_r}{\partial r}+\frac{v_\theta}{r}\frac{\partial v_r}{\partial \theta}+\frac{v_\varnothing}{r\sin(\theta)}\frac{\partial v_r}{\partial \varnothing}-\frac{v_\theta^2+v_\varnothing^2}{r}\right)+\mu\left(\frac{1}{r^2}\frac{\partial^2}{\partial r^2}(r^2 v_r)+\left(\frac{1}{r^2\sin(\theta)}\frac{\partial}{\partial\theta}\left(\sin(\theta)\frac{\partial v_r}{\partial\theta}\right)+\frac{1}{r^2\sin^2(\theta)}\frac{\partial^2 v_r}{\partial\varnothing^2}\right)-\frac{\partial P}{\partial r}+\rho g_r \tag{3.43}$$

θ component:

$$\rho\frac{\partial v_\theta}{\partial t}=-\rho\left(v_r\frac{\partial v_\theta}{\partial r}+\frac{v_\theta}{r}\frac{\partial v_\theta}{\partial\theta}+\frac{v_\varnothing}{r\sin(\theta)}\frac{\partial v_\theta}{\partial\varnothing}+\frac{v_r v_\theta}{r}-\frac{v_\varnothing^2\cot(\theta)}{r}\right)$$
$$+\mu\left(\frac{1}{r^2}\frac{\partial}{\partial r}\left(r^2\frac{\partial v_\theta}{\partial r}\right)+\frac{1}{r^2}\frac{\partial}{\partial\theta}\left(\sin(\theta)\frac{\partial\sin(\theta)v_\theta}{\partial\theta}\right)+\frac{1}{r^2\sin^2(\theta)}\frac{\partial^2 v_\theta}{\partial\varnothing^2}\right) \tag{3.44}$$
$$+\frac{2\mu}{r^2}\left(\frac{\partial v_r}{\partial\theta}-\frac{\cos(\theta)}{\sin^2(\theta)}\frac{\partial v_\varnothing}{\partial\varnothing}\right)-\frac{1}{r}\frac{\partial P}{\partial\theta}+\rho g_\theta$$

$\varnothing$ component:

$$\rho\frac{\partial v_\varnothing}{\partial t}=-\rho\left(v_r\frac{\partial v_\varnothing}{\partial r}+\frac{v_\theta}{r}\frac{\partial v_\varnothing}{\partial\theta}+\frac{v_\varnothing}{r\sin(\theta)}\frac{\partial v_\varnothing}{\partial\varnothing}+\frac{v_r v_\varnothing}{r}-\frac{v_\theta v_\varnothing\cot(\theta)}{r}\right)$$
$$+\mu\left(\frac{1}{r^2}\frac{\partial}{\partial r}\left(r^2\frac{\partial v_\varnothing}{\partial r}\right)+\frac{1}{r^2}\frac{\partial}{\partial\theta}\left(\frac{1}{\sin(\theta)}\frac{\partial\sin(\theta)v_\varnothing}{\partial\theta}\right)+\frac{1}{r^2\sin^2(\theta)}\frac{\partial^2 v_\varnothing}{\partial\varnothing^2}\right) \tag{3.45}$$
$$+\mu\left(\frac{2}{r^2\sin(\theta)}\frac{\partial v_r}{\partial\varnothing}+\frac{2\cos(\theta)}{r^2\sin^2(\theta)}\frac{\partial v_\theta}{\partial\varnothing}\right)-\frac{1}{r\sin(\theta)}\frac{\partial P}{\partial\varnothing}+\rho g_\varnothing$$

3.6 Applications of the Equations of Change

This section presents a few examples and applications that clarify the fluid dynamics, mass transfer, and energy transfer in distributed parameter systems. The examples will illustrate the use of the equations of change to obtain mathematical models. The approach will show how the general partial differential equations can be simplified to solvable differential equations.

Example 3.1: Liquid Flow in a Pipe

Develop a velocity profile of the one-dimensional plug flow of fluid through a pipe. The fluid is an incompressible liquid. State all assumptions and boundary conditions. The schematic of the pipe is shown in Figure 3.5.

Solution

Consider the following assumptions in developing the velocity profile in a horizontal pipe:

1. Steady, incompressible flow (no changes with time, constant ρ and μ).
2. Fully developed flow.
3. Velocity is a function of r only $(v = v(r))$.
4. No radial or circumferential velocity components ($v_\theta = 0$).
5. Pressure changes linearly with z and is independent of r (constant pressure gradient).

The incompressible fluid with constant density and viscosity have the continuity equation in cylindrical coordinates:

$$\frac{1}{r}\frac{\partial r v_r}{\partial r} + \frac{\partial v_\theta}{r\partial \theta} + \frac{\partial v_z}{\partial z} = 0$$

The plug flow assumptions imply that $v_r = v_\theta = 0$, and the continuity equation is reduced to:

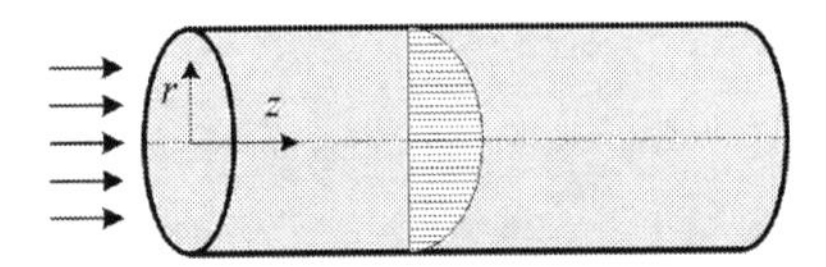

FIGURE 3.5
Liquid flow in a pipe.

$$\frac{\partial v_z}{\partial z} = 0$$

The z-component of equation of motion in cylindrical coordinate is:

$$\rho\frac{\partial v_z}{\partial t} + \rho\left(v_r\frac{\partial v_z}{\partial r} + \frac{v_\theta}{r}\frac{\partial v_z}{\partial \theta} + v_z\frac{\partial v_z}{\partial z}\right) = \mu\left(\frac{1}{r}\frac{\partial}{\partial r}\left(r\frac{\partial v_z}{\partial r}\right) + \frac{1}{r^2}\frac{\partial^2 v_z^2}{\partial \theta^2} + \frac{\partial^2 v_z}{\partial z^2}\right) - \frac{\partial P}{\partial z} + \rho g_z$$

If we assume steady state, incompressible fluid, plug flow, and isothermal flow, the equation is simplified to:

$$0 + \rho(0 + 0 + 0) = \mu\left(\frac{1}{r}\frac{\partial}{\partial r}\left(r\frac{\partial v_z}{\partial r}\right) + 0 + 0\right) - \frac{\partial P}{\partial z} + 0$$

Rearrange the equation to the following form:

$$\frac{\partial P}{\partial z} = \mu\left(\frac{1}{r}\frac{\partial}{\partial r}\left(r\frac{\partial v_z}{\partial r}\right)\right)$$

Since the pressure gradient is constant (assumption 4), we can integrate once:

$$r\frac{\partial v_z}{\partial r} = \frac{r^2}{2\mu}\frac{\partial p}{\partial z} + C_1$$

By symmetry, we have:

$$\text{at } r = 0, \quad \frac{\partial v_z(r)}{\partial r} = 0, \quad \text{so } C_1 = 0$$

with:

$$\frac{\partial v_z}{\partial r} = \frac{r}{2\mu}\frac{\partial p}{\partial z} + 0$$

The second integration generates the second arbitrary integration constant, C_2:

$$v_z = \frac{r^2}{4\mu}\frac{\partial p}{\partial z} + C_2$$

The no slip condition at $r = R$ is $v_z = 0$:

$$C_2 = -\frac{R^2}{4\mu}\frac{\partial p}{\partial z}$$

Substitute C_1 and C_2 to produce the following expression for velocity profile in horizontal pipe:

$$v_z(r) = -\frac{R^2}{4\mu}\frac{\partial p}{\partial z}\left(1 - \frac{r^2}{R^2}\right)$$

Example 3.2: Diffusion with Chemical Reaction in a Spherical Catalyst

Consider the diffusion with chemical reaction of species A that reacts at the surface of catalytic sphere (Figure 3.6). The fluid properties are assumed to be constant. Compound A reacts quickly to form compound B according to the reaction $A \rightarrow B$. The concentration of A at the outer edge of the gas film is C_{A0}. Develop the concentration profile inside the catalyst.

Solution

The following assumptions are considered in deriving the component balance equation:

- The fluid properties are constant.
- The reaction rate to produce the product B at the surface of the sphere is very fast compared to the diffusion; thus, the reaction rate is completely controlled by the diffusion process. This is a phenomenon that might be detected at high temperature where mass transfer is often the main factor that is controlling the reaction rate.
- Neglect concentration gradient in the θ, $\varnothing$ direction.
- One-dimensional, spherical coordinate.

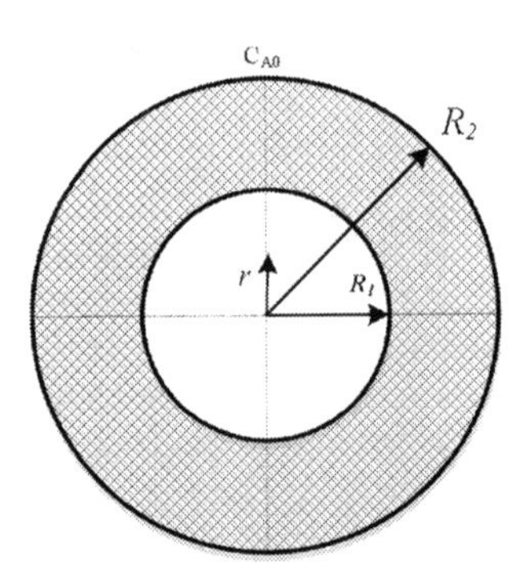

FIGURE 3.6
Diffusion through a catalytic sphere.

The general differential equation of mole balance in spherical coordinates is:

$$\frac{\partial C_A}{\partial t} = -\left(v_r \frac{\partial C_A}{\partial r} + \frac{v_\theta}{r}\frac{\partial C_A}{\partial \theta} + \frac{v_\varnothing}{\mathrm{rsin}(\theta)}\frac{\partial C_A}{\partial \varnothing}\right) + D_{AB}\left(\frac{1}{r^2}\frac{\partial}{\partial r}\left(r^2\frac{\partial C_A}{\partial r}\right)\right.$$

$$\left. + \frac{1}{r^2 \sin(\theta)}\frac{\partial}{\partial \theta}\left(\sin(\theta)\frac{\partial C_A}{\partial \theta}\right) + \frac{1}{r^2 \sin^2(\theta)}\frac{\partial^2 C_A}{\partial \varnothing^2}\right) + R_A = 0$$

We can infer the proceeding assumptions. Because the system is at steady state, we have, $\partial C_A/\partial t = 0$. If we assume that there is no bulk flow, then, $v_r = v_\theta = v_\varnothing = 0$. For diffusion in the r-direction only, the following holds:

$$D_{AB}\left(\frac{1}{r^2}\frac{\partial}{\partial r}\left(r^2\frac{\partial C_A}{\partial r}\right)\right) = 0$$

The equation is then reduced to:

$$\frac{\partial}{\partial r}\left(r^2\frac{\partial C_A}{\partial r}\right) = 0$$

Boundary conditions:

- Boundary condition 1, B.C.1: at $r = R_1$, $C_A = 0$, very fast reaction
- Boundary condition 2, B.C.2: at $r = R_2$, $C_A = C_{A0}$, concentration at surface of catalyst.

Integrating the equation yields:

$$\frac{\partial C_A}{\partial r} = \frac{C_1}{r^2}$$

Taking the second integral, the expression for the concentration of A:

$$C_A = -\frac{C_1}{r} + C_2$$

The expressions of the two arbitrary integration constants are obtained from the following two boundary conditions.

$$\text{B.C.1: } 0 = -\frac{C_1}{R_1} + C_2$$

$$\text{B.C.2: } C_{A0} = -\frac{C_1}{R_2} + C_2$$

We have two equations and two unknowns; thus, we can solve for C_1 and C_2 by subtracting one from the other:

$$C_1 = \frac{R_1 R_2}{R_2 - R_1} C_{A0}$$

$$C_2 = \frac{R_2}{R_2 - R_1} C_{A0}$$

Substitute C_1 and C_2 in the resulting double integrated equation and obtain the expressions for the concentration of A along the radius of the catalyst:

$$\frac{C_A}{C_{A0}} = \frac{R_2}{R_2 - R_1}\left(1 - \frac{R_1}{r}\right)$$

Note that $R_2 > r > R_1$.

Example 3.3: Nonisothermal Plug Flow Reactor

Consider the exothermic reaction ($A \rightarrow B$) taking place in the double pipe plug flow reactor shown in Figure 3.7. The fluid is incompressible. Cooling water is circulated in the shell side of the reactor to absorb the excess heat generated for the exothermic reaction. Develop the model equations that describes the concentration and temperature distributing in the plug flow reactor.

Solution

The component mole balance equation can be used to model the nonisothermal plug flow reactor. The plug flow conditions imply that $v_r = v_\theta = 0$ and $\partial C_A/\partial r = \partial C_A/\partial \theta = 0$.

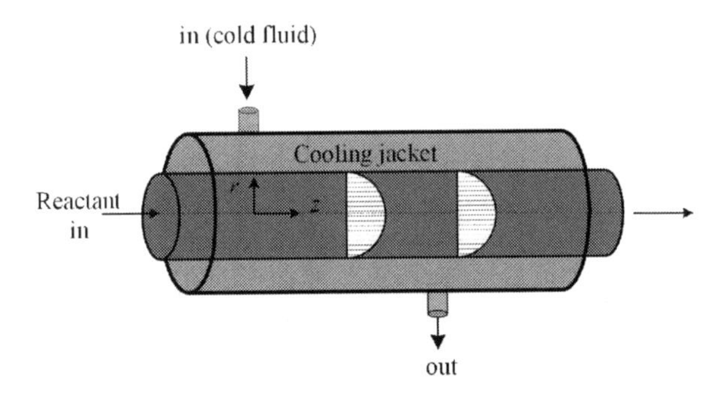

FIGURE 3.7
Non-isothermal plug flow reactor.

The component mole balance equation in cylindrical coordinate:

$$\frac{\partial C_A}{\partial t}+\left(v_r\frac{\partial C_A}{\partial r}+v_\theta\frac{1}{r}\frac{\partial C_A}{\partial \theta}+v_z\frac{\partial C_A}{\partial z}\right)=D_A\left(\frac{1}{r}\frac{\partial}{\partial r}\left(r\frac{\partial C_A}{\partial r}\right)+\frac{1}{r^2}\frac{\partial^2 C_A}{\partial \theta^2}+\frac{\partial^2 C_A}{\partial z^2}\right)+R_A$$

After applying the plug flow assumptions, the component mole balance equation is reduced to the following form:

$$\frac{\partial C_A}{\partial t}=-v_z\frac{\partial C_A}{\partial z}+D_{AB}\frac{\partial^2 C_A}{\partial z^2}+R_A$$

The equation can be simplified further by assuming that the convective term $\left(v_z\,\partial(C_A/\partial z)\right)$ is much greater than the diffusion term $(D_{AB}\,(\partial^2 C_A/\partial z^2))$:

$$v_z\frac{\partial C_A}{\partial z}\gg D_{AB}\frac{\partial^2 C_A}{\partial z^2}$$

The equation is reduced to:

$$\frac{\partial C_A}{\partial t}=-v_z\frac{\partial C_A}{\partial z}+R_A$$

For the nonisothermal plug flow reactor, the energy balance in cylindrical coordinate is obtained for fluid with constant properties and constant pressure:

$$\rho C_p\frac{\partial T}{\partial t}=-\rho C_p\left(v_r\frac{\partial T}{\partial r}+\frac{v_\theta}{r}\frac{\partial T}{\partial \theta}+v_z\frac{\partial T}{\partial z}\right)+k\left(\frac{\partial^2 T}{\partial r^2}+\frac{1}{r}\frac{\partial T}{\partial r}+\frac{1}{r^2}\frac{\partial^2 T}{\partial \theta^2}+\frac{\partial^2 T}{\partial z^2}\right)+\Phi_H$$

Using the plug-flow assumptions, $v_r=v_\theta=0$, $\partial T/\partial r$, $\partial T/\partial\theta$, and neglecting the viscous forces, consider the term Φ_H, which includes the heat generation by reaction rate R_A and heat exchanged with the cooling jacket, $hA(T-T_w)$. Thus, the equation is reduced to:

$$\rho C_p\frac{\partial T}{\partial t}=-\rho C_p v_z\frac{\partial T}{\partial z}-\Delta H_r k_o e^{-\frac{E}{RT}}C_A-h\frac{\pi DL}{\pi D^2 L/4}\left(T-T_w\right)$$

Each term in the equation is in the unit of energy per unit volume (W/m^3), which is why the last term is multiplied by heat transfer area and divided by the volume of the cylindrical plug flow reactor. When T is greater than T_w, the mass is losing heat to the surroundings. Thus, the rate of addition is the negative of the rate of heat loss, which explains the minus sign before the convective term. This equation is correct whether the mass is hotter or cooler than the surrounding environment.

Example 3.4: Steady-State Energy Transport with Heat Generation

Heat is being generated from an electrical heater at a uniform rate of $\Phi_H(\text{J/m}^2\text{.s})$ in a solid cylinder of radius R. The heat is removed from the system, which maintains its surface temperature at the constant value T_w (Figure 3.8) using a cooling system. Our objective is to derive the temperature variations in the cylinder. We assume that the solid is of constant density, heat capacity, and thermal conductivity.

Solution

This is a distributed parameter system since the temperature can vary with time and with all positions in the cylinder. We will use then the equation of energy change in cylindrical coordinates. Several assumptions can be made:

The system is at steady state i.e. $\partial T/\partial t = 0$

The variation of temperature is only allowed in radial directions. Therefore, the terms $\partial^2 T/\partial z^2$ and $\partial^2 T/\partial \theta^2$ are zero.

For a solid object, there is no convective heat; thus, the equation of energy change is reduced to the following form:

$$\frac{\partial T}{\partial t} = \frac{k}{\rho C_p}\left(\frac{\partial^2 T}{\partial r^2} + \frac{1}{r}\frac{\partial T}{\partial r} + \frac{1}{r^2}\frac{\partial^2 T}{\partial \theta^2} + \frac{\partial^2 T}{\partial z^2}\right) + \frac{\Phi_H}{\rho C_p}$$

Employing the assumptions, the general energy balance equation is reduced to:

$$0 = \frac{k}{\rho C_p}\left(\frac{d^2 T}{dr^2} + \frac{1}{r}\frac{dT}{dr}\right) + \frac{\Phi_H}{\rho C_p}$$

Boundary conditions:

B.C.1: The temperature at the wall is constant: at $r = R,\ T = T_w$.

B.C.2: The maximum temperature will be reached at the center $(r=0)$, therefore at:

$$r = 0, \frac{dT}{dr} = 0$$

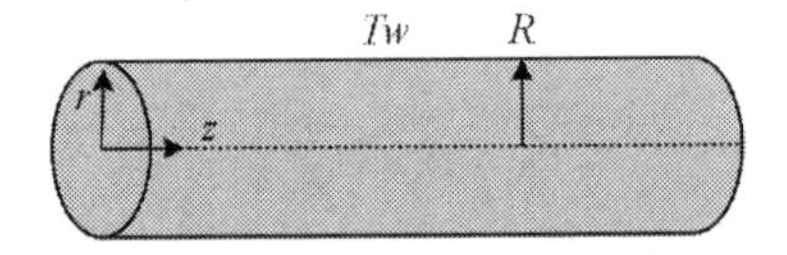

FIGURE 3.8
Heat transfer in solid rod.

The equation can be rewritten as follows:

$$-\Phi_H = k\frac{1}{r}\frac{dT}{dr}\left(r\frac{dT}{dr}\right)$$

The left-hand side represents the rate of heat production per unit volume, while the right-hand side presents the rate of diffusion of heat per unit volume.

As an exercise solve the simplified equation by using the boundary conditions and obtain the temperature profile as a function of r.

Example 3.5: Momentum Transport in a Circular Tube

Derive the steady-state velocity profile equation for the laminar flow inside a horizontal circular tube.

Solution

We will see how the model can be obtained using the momentum equation of change. Assume that the fluid is incompressible and Newtonian. The z component of equation of motion in the cylindrical coordinates is:

$$\rho\frac{\partial v_z}{\partial t} = -\rho\left(v_x\frac{\partial v_z}{\partial x} + v_y\frac{\partial v_z}{\partial y} + v_z\frac{\partial v_z}{\partial z}\right) + \mu\left(\frac{\partial^2 v_z}{\partial x^2} + \frac{\partial^2 v_z}{\partial y^2} + \frac{v_z}{\partial z^2}\right) - \frac{\partial p}{\partial z} + \rho g_z$$

Assumptions:

The flow is only in the z direction; that is, $v_r = v_\theta = 0$.

The flow is at steady state, $\partial v_z/\partial t = 0$.

The steady-state momentum equation in cylindrical coordinates and velocity in the z-direction is reduced to:

$$\rho v_z\frac{\partial v_z}{\partial z} = \mu\left(\frac{1}{r}\frac{\partial}{\partial r}\left(r\frac{\partial v_z}{\partial r}\right) + \frac{1}{r^2}\frac{\partial^2 v_z}{\partial \theta^2} + \frac{\partial^2 v_z}{\partial z^2}\right) - \frac{\partial p}{\partial z} + \rho g_z$$

Use the continuity equation; $v_r = v_\theta = 0$ gives $dv_z/dz = 0$. There is no variation of the velocity with θ because the flow is symmetrical around the z-axis; thus:

$$\frac{\partial^2 v_z}{\partial \theta^2} = 0$$

The cylinder is horizontal, $\rho g_z = 0$. Neglect gravitational effects. The momentum equation in the z direction is then reduced to:

$$\mu\left(\frac{1}{r}\frac{\partial}{\partial r}\left(r\frac{\partial v_z}{\partial r}\right)\right)=\frac{\partial p}{\partial z}$$

This equation advocates that dp/dz is constant because the left-hand side depends only on r. The right-hand side is the rate of momentum due to pressure drop. The left-side term is the rate of momentum diffusion per unit volume. Therefore:

$$\frac{\partial p}{\partial z}=\frac{\Delta p}{\Delta z}=\frac{\Delta p}{L}$$

where Δp is the pressure drop across the tube. The equation reduces to the following:

$$\frac{\partial}{\partial r}\left(r\frac{\partial v_z}{\partial r}\right)=\frac{\Delta p}{\mu L}r$$

The first integral generated the arbitrary integration constant C_1:

$$\left(r\frac{dv_z}{dr}\right)=\frac{\Delta p}{\mu L}\frac{r^2}{2}+C_1$$

Boundary condition 1: The axial symmetry at $r=0$, $dv_z/dr=0$, leads to $C_1=0$. Divide each term by r and integrate for the second time:

$$v_z=\frac{\Delta p}{\mu L}\frac{r^2}{4}+C_2$$

Boundary conditions 2: The assumption of no-slip conditions at $r=R, v_z=0$; thus:

$$C_2=-\frac{\Delta p}{\mu L}\frac{R^2}{4}$$

Substitute C_2 in the general velocity profile equation and rearrange to get the velocity distribution inside a horizontal cylindrical pipe:

$$v_z=-\frac{\Delta p}{4\mu L}\left(1-\left(\frac{r}{R}\right)^2\right)$$

Example 3.6: Unsteady-State Heat Generation

Consider the temperature variations of the reactor with time as well. This may be needed to compute the heat transferred during startup or shutdown operations. Keep the same assumptions as in Example 3.4 (except the steady-state assumption).

Solution

The unsteady-state energy balance in cylindrical coordinates yields:

$$\rho C_p \frac{\partial T}{\partial t} = k \frac{1}{r} \frac{d}{dr}\left(r \frac{dT}{dr} \right) + \frac{\Phi_H}{k}$$

with the following initial conditions:

$$t = 0,\ T(r,0) = T_0$$

The boundary conditions:

at $r = 0,\ dT/dr = 0$ (axial symmetry)
at $r = R,\ T = T_w$ (wall temperature)

Example 3.7: Heat Transfer with Constant Wall Temperature

Consider a fluid flowing at constant velocity v_z into an adiabatic horizontal cylindrical tube, as shown in Figure 3.9. The fluid enters with uniform temperature T_i. The wall is assumed to be at constant temperature T_w. We would like to model the variations of the fluid velocity and temperature distribution inside the tube.

Solution

Apply the general differential energy balance equation in cylindrical coordinates. Assume that the fluid is an incompressible, Newtonian

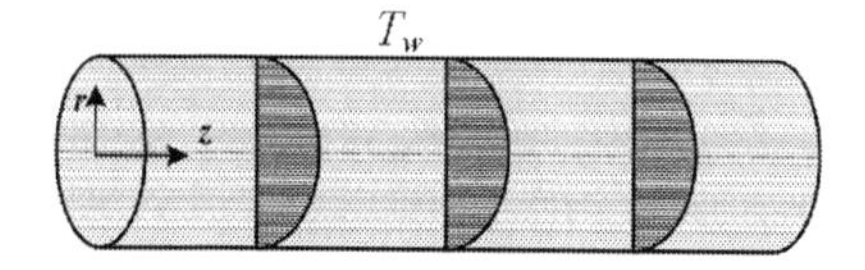

FIGURE 3.9
Heat transfer with constant wall temperature.

fluid and that it is of constant thermal conductivity. Start with the energy equation in cylindrical coordinates:

$$\rho C_p \frac{\partial T}{\partial t} = -\rho C_p \left(v_r \frac{\partial T}{\partial r} + v_\theta \frac{1}{r} \frac{\partial T}{\partial \theta} + v_z \frac{\partial T}{\partial z} \right) + k \left(\frac{1}{r} \frac{\partial}{\partial r} \left(r \frac{\partial T}{\partial r} \right) + \frac{1}{r^2} \frac{\partial^2 T}{\partial \theta^2} + \frac{\partial^2 T}{\partial z^2} \right) + Q$$

Since the system is at steady state, we have $\partial T/\partial t = 0, dv_z/d = 0$. The flow is one-dimensional (in the z-direction), $v_r = v_\theta = 0$. Since the temperature is symmetrical then: $\partial T/\partial \theta = \partial^2 T/\partial \theta^2$.

In some cases, we can neglect the conduction term $k(\partial^2 T/\partial z^2)$ compared to the convective term $v_z(\partial T/\partial z)$. The energy balance equation is reduced and described by the following equation:

$$0 = -\rho C_p \left(0 + 0 + v_z \frac{\partial T}{\partial z} \right) + k \left(\frac{1}{r} \frac{\partial}{\partial r} \left(r \frac{\partial T}{\partial r} \right) + 0 + 0 \right) + 0$$

Accordingly:

$$\rho C_p \left(v_z \frac{\partial T}{\partial z} \right) = k \left(\frac{1}{r} \frac{\partial}{\partial r} \left(r \frac{\partial T}{\partial r} \right) \right)$$

The velocity profile of the system is described by the following momentum equation as derived in the previous examples:

$$\mu \frac{d}{dr} \left(\frac{r dv_z}{dr} \right) = \frac{\Delta P}{L}$$

The energy balance and momentum equations are therefore coupled by v_z, and both equations need to be solved simultaneously. The boundary conditions for the momentum equation are:

B.C.1: at $r = R$, $v_z = 0$ (no-slip conditions)
B.C.2: at $r = 0$, $dv_z/dr = 0$ (axial symmetry)

The boundary conditions of the energy balance equation are:

B.C.3: at $z = 0, T = T_i$ (temperature of inlet stream)
B.C.4: at $r = 0, dT/dr = 0$ (axial symmetry)
B.C.5: at $r = R, T = T_w$ (wall temperature)

Example 3.8: Laminar Flow and Mass Transfer

Consider the case of a fluid flowing through a horizontal pipe with constant velocity v_z. The pipe wall is made of a solute of constant concentration C_{Aw} that is dissolved in the fluid (Figure 3.10). The concentration of the fluid at the entrance $z = 0$ is C_{Ao}. The regime is assumed laminar and at steady state. Assume that the fluid properties are constant. Model the variation of the concentration of A along the z axis of the pipe.

Solution

The component balance for A in cylindrical coordinates is described by the following equation:

$$v_r \frac{\partial C_A}{\partial r} + v_\theta \frac{1}{r}\frac{\partial C_A}{\partial \theta} + v_z \frac{\partial C_A}{\partial z} = D_A\left(\frac{1}{r}\frac{\partial}{\partial r}\left(r\frac{\partial C_A}{\partial r}\right) + \frac{1}{r^2}\frac{\partial^2 C_A}{\partial \theta^2} + \frac{\partial^2 C_A}{\partial z^2}\right)$$

The flow is in the z direction; hence, $v_r = v_\theta = 0$. The balance equation is reduced to:

$$v_z \frac{\partial C_A}{\partial z} = D_{AB}\left(\frac{1}{r}\frac{\partial}{\partial r}\left(r\frac{\partial C_A}{\partial r}\right) + \frac{\partial^2 C_A}{\partial z^2}\right)$$

The convection term ($v_z \partial C_A/\partial z$) is much higher than diffusion term ($D_{AB}\partial^2 C_A/\partial z^2$); thus, the diffusion term in the z-direction is negligible compared to convection term. The equation is simplified to the following form:

$$v_z \frac{\partial C_A}{\partial z} = D_{AB}\left(\frac{1}{r}\frac{\partial}{\partial r}\left(r\frac{\partial C_A}{\partial r}\right)\right)$$

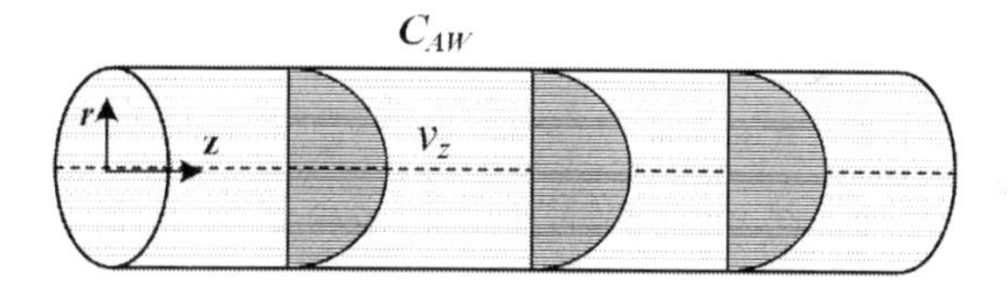

FIGURE 3.10
Laminar flow and mass transfer in horizontal cylinder.

where v_z is to be obtained from the momentum equation:

$$\mu\frac{d}{dr}\left(\frac{rdv_z}{dr}\right)=\frac{\Delta p}{L}$$

The two equations are coupled through v_z. The boundary conditions are similar to Example 3.7 for velocity profile in the pipe:

at $r = R$, $v_z = 0$ (no-slip conditions)
at $r = 0$, $dv_z/dr = 0$ (axial symmetry)

Boundary conditions for the component mole balance:

at $z = 0$, $C_A = C_{Ao}$ (feed concentration)
at $r = 0$, $dC_A/dr = 0$ (axial symmetry)
at $r = R$, $C_A = C_{Aw}$ (concentration of A at the walls)

Note the similarity between this example and the heat transfer case of the previous example.

Example 3.9: Axial Flow of Incompressible Fluid in a Vertical Cylinder

Newtonian liquid is flowing downward in a vertical circular pipe. Derive the steady-state velocity profile of the axial flow of an incompressible fluid in a circular tube of length L and radius R (Figure 3.11).

Solution

To solve a flow problem, first, select the continuity equation and the equation of motion in the appropriate coordinate system and for the appropriate symmetry—in this case, cylindrical coordinates. Discard all terms that have the value of zero. Integrate the resultant differential

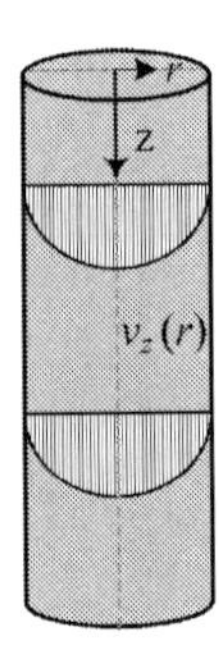

FIGURE 3.11
Fluid flow in vertical cylinder.

equation using appropriate boundary conditions and at the end check your assumptions.

Continuity equation:

$$\rho\left(\frac{1}{r}\frac{\partial r v_r}{\partial r}+\frac{1}{r}\frac{\partial v_\theta}{\partial \theta}+\frac{\partial v_z}{\partial z}\right)=0$$

Set $v_\theta = v_r = 0$ along the z-axis; v_z is not a function of θ because of cylindrical symmetry:

$$\frac{\partial v_z}{\partial z}=\frac{\partial^2 v_z}{\partial z^2}=0$$

Consider the z-component of the equation of motion:

$$\rho\frac{\partial v_z}{\partial t}+\rho\left(v_r\frac{\partial v_z}{\partial r}+\frac{v_\theta}{r}\frac{\partial v_z}{\partial \theta}+v_z\frac{\partial v_z}{\partial z}\right)=\mu\left(\frac{1}{r}\frac{\partial}{\partial r}\left(r\frac{\partial v_z}{\partial r}\right)+\frac{1}{r^2}\frac{\partial^2 v_z^2}{\partial \theta^2}+\frac{\partial^2 v_z}{\partial z^2}\right)-\frac{\partial p}{\partial z}+\rho g_z$$

Simplifying the z-component of the equation of motion based on the given assumptions:

$$0+\rho(0+0+0)=\mu\left(\frac{1}{r}\frac{\partial}{\partial r}\left(r\frac{\partial v_z}{\partial r}\right)+0+0\right)-\frac{\partial p}{\partial z}+\rho g_z$$

The equation is simplified and rearranged to:

$$\frac{1}{r}\frac{\partial}{\partial r}\left(r\frac{\partial v_z}{\partial r}\right)=\frac{\left(\frac{\partial p}{\partial z}-\rho g\right)}{\mu}=K$$

Integrate twice with respect to r using the boundary conditions:

$v_z = 0$ at $r = R$ and v_z finite $(dv_z/dr = 0)$ at $r = 0$

The resultant velocity profile in the vertical pipe is described by the following equation:

$$v_z=\left(\frac{KR^2}{4\mu L}\right)\left(1-\left(\frac{r}{R}\right)^2\right)$$

Example 3.10: Transient Heat Transfer in a Finite Slab

Heat conduction problems encountered in engineering applications involve time as an independent variable. Consider a solid slab material of length L (Figure 3.12). The material has an initial temperature of T_0 at

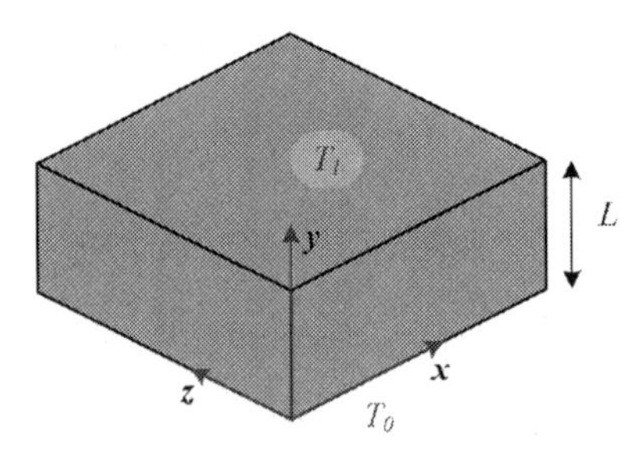

FIGURE 3.12
Heat transfer in finite slab.

time $t=0$; the surface at $y=L$ is maintained at temperature T_1 and maintained at that temperature for $t>0$. Simplify the energy balance equation and state the initial and boundary conditions. Determine the variation of the temperature as a function of time and position $T(y,t)$ within the heat-conducting object.

Solution

Consider the following assumption

- Heat is flowing in only one coordinate direction (y direction). Thus, there is no temperature gradient in the other two directions, and the two partials associated with these directions are equal to zero ($\partial T/\partial x=0$, $\partial T/\partial z=0$).
- If there are no heat sources within the system, then the term $\Phi_H=0$.

The microscopic energy balance equation in the y-direction in a Cartesian coordinate states:

$$\rho C_p \frac{\partial T}{\partial t} + \rho C_p \left(v_x \frac{\partial T}{\partial x} + v_y \frac{\partial T}{\partial y} + v_z \frac{\partial T}{\partial z} \right) = k \left(\frac{\partial^2 T}{\partial x^2} + \frac{\partial^2 T}{\partial y^2} + \frac{\partial^2 T}{\partial z^2} \right) + \Phi_H$$

Solutions to the above equation must be obtained and must satisfy the initial and boundary conditions. Since heat is transferred through the material by conduction, there is no heat transfer by convection and no heat generation source. There is no heat transfer in the x and z directions. These assumptions simplify the equation such that:

$$\rho C_p \frac{\partial T}{\partial t} + \rho C_p (0+0+0) = k \left(0 + \frac{\partial^2 T}{\partial y^2} + 0 \right) + 0$$

Rearranging the simplified differential equation leads to the following form:

$$\rho C_p \frac{\partial T}{\partial t} = k\left(\frac{\partial^2 T}{\partial y^2}\right)$$

The initial and boundary conditions state the following:

Initial conditions:

$$\text{at } t = 0,\ T(y,0) = T_o$$

Boundary conditions,

$$\text{at } y = 0,\ T(0,t) = T_0$$
$$\text{at } y = L,\ T(L,t) = T_1$$

The equation can be solved using COMSOL software packages.

Example 3.11: Flow Through a Double Pipe Heat-Exchanger

A double pipe heat exchanger consists of two pipes, one pipe inside another larger pipe (Figure 3.13). In this configuration of heat exchangers, one fluid flows through the inside pipe and the other flows through the annulus between the two pipes. The heat transfer surface is the wall of the inner pipe. To make the overall unit more compact, the pipes are usually doubled back multiple times. The advantage of a double pipe heat exchanger is that it can be operated in a counterflow pattern. This gives the highest overall heat transfer coefficient for the double pipe heat exchanger design. Consider the heat exchanger as shown in Figure 3.13, with an inner radius of R_1 and outer radius of R_2 with z pointing to the right. Fluid is moving to the right (positive z-direction) in the outer tube,

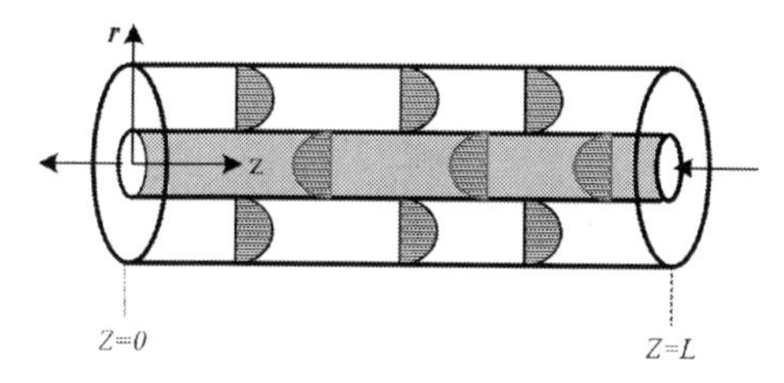

FIGURE 3.13
Double pipe heat exchanger.

and the fluid is moving to the left in the inner tube (countercurrent in the negative z direction). Assume that the pressure gradient is linear and that $R_1 < r < R_2$.

Solution

The system is represented by cylindrical coordinates, and velocity is only in the z direction. Assume constant physical properties (ρ and μ). Set up the Navier-Stokes equation in the z direction. The double pipe heat exchanger is horizontal:

$$\rho\frac{\partial v_z}{\partial t}+\rho\left(v_r\frac{\partial v_z}{\partial r}+\frac{v_\theta}{r}\frac{\partial v_z}{\partial \theta}+v_z\frac{\partial v_z}{\partial z}\right)=\mu\left(\frac{1}{r}\frac{\partial}{\partial r}\left(r\frac{\partial v_z}{\partial r}\right)+\frac{1}{r^2}\frac{\partial^2 v_z^2}{\partial \theta^2}+\frac{\partial^2 v_z}{\partial z^2}\right)-\frac{\partial p}{\partial z}+\rho g_z$$

The flow is at steady state, $\partial v_z/\partial t = 0$. We also note that because the flow is symmetrical around the z-axis, we have necessarily no variation of the velocity with θ; that is:

$$\frac{\partial^2 v_z}{\partial \theta}=0$$

Recall that ρg is not necessary since gravity is not playing a role in a horizontal pipe. Thus, the equation is simplified to the following:

$$0+\rho(0+0+0)=\mu\left(\frac{1}{r}\frac{\partial}{\partial r}\left(r\frac{\partial v_z}{\partial r}\right)+0+0\right)-\frac{\partial p}{\partial z}+0$$

Rearrange to obtain the following equation:

$$\frac{1}{r}\frac{\partial}{\partial r}\left(r\frac{\partial v_z}{\partial r}\right)=\frac{\partial p}{\partial z}$$

The left-hand side depends only on r. This equation suggests that dp/dz is constant; therefore:

$$\frac{\partial P}{\partial z}=\frac{\Delta p}{\Delta z}$$

where ΔP is the pressure drop across the tube. The equation is equivalent to:

$$\left(\frac{1}{r}\frac{\partial}{\partial r}\left(r\frac{\partial v_z}{\partial r}\right)\right)=\frac{\Delta p}{\mu \Delta z}=\frac{\Delta p}{\mu L}$$

The left term is the rate of momentum diffusion per unit volume, and the right-hand side is in fact the rate of production of momentum due to pressure. Double integration of the simplified differential equation generates two arbitrary integration constants (C_1 and C_2). The expressions of the integration constants are determined by two boundary conditions.

$$v_z = \frac{\Delta p}{4\mu L} r^2 + C_1 \ln r + C_2$$

B.C.1: at $r = R_1, v_z = 0$ (no slip) since it is a solid-liquid boundary, so:

$$0 = \frac{\Delta p}{4\mu L} R_1^2 + C_1 \ln R_1 + C_2$$

B.C.2: at $r = R_2, v_z = 0$ (no slip) since it is a solid-liquid boundary, so:

$$0 = \frac{\Delta p}{4\mu L} R_2^2 + C_1 \ln R_2 + C_2$$

Solve for C_1 and C_2, and rearrange to obtain the following equation for velocity profile between the two coaxial cylinders:

$$v_z = \frac{\Delta p}{4\mu L}\left[\frac{\ln \frac{r}{R_2}}{\ln \frac{R_2}{R_1}}(R_2^2 - R_1^2) + (R_2^2 - r^2)\right]$$

Example 3.12: Flow Between Concentric Cylinders

Consider the flow in an annular space between two fixed, concentric cylinders, as shown in Figure 3.14. The fluid has constant density (ρ) and viscosity (μ). The outer cylinder is fixed, and the inner cylinder rotates at an angular velocity (ω_i). There is no axial motion or end effects (i.e., $v_z = 0$) and no change in velocity in the direction of θ (i.e., $\partial v_\theta / \partial\theta = 0$). The inner and the outer cylinders have radii r_i and r_o, respectively, and

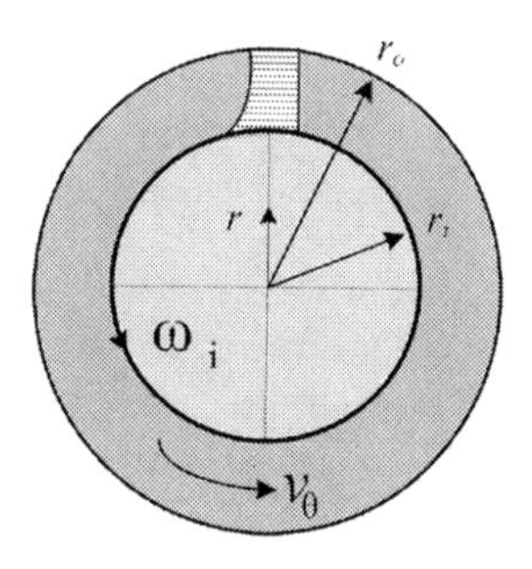

FIGURE 3.14
Flow through annulus.

the velocity varies between r_i and r_o. The velocity varies in the direction of r only. Determine the velocity profile in a fluid situated between two coaxial cylinders [1].

Solution

The continuity equation may be written in cylindrical coordinates, as follows:

$$\rho\left(\frac{1}{r}\frac{\partial r v_r}{\partial r}+\frac{1}{r}\frac{\partial v_\theta}{\partial \theta}+\frac{\partial v_z}{\partial z}\right)=0$$

Note that v_θ does not vary with θ, and at the inner and outer radii, there is no velocity. So the equation of motion can be treated as purely circumferential so that $v_r = 0$, and $v_\theta = v_\theta(r)$:

$$\rho\left(\frac{1}{r}\frac{\partial r v_r}{\partial r}+0+0\right)=0$$

From the continuity equation, we have:

$$\frac{1}{r}\frac{\partial r\, v_r}{\partial r}=0$$

The θ-momentum component equation may be written as follows:

$$\rho\frac{\partial v_\theta}{\partial t}+\rho\left(v_r\frac{\partial v_\theta}{\partial r}+\frac{v_\theta}{r}\frac{\partial v_\theta}{\partial \theta}+\frac{v_r v_\theta}{r}+v_z\frac{\partial v_\theta}{\partial z}\right)$$
$$=\mu\left(\frac{\partial}{\partial r}\left(\frac{1}{r}\frac{\partial r v_\theta}{\partial r}\right)+\frac{1}{r^2}\frac{\partial^2 v_\theta}{\partial \theta^2}+\frac{2}{r^2}\frac{\partial v_r}{\partial \theta}+\frac{\partial^2 v_\theta}{\partial z^2}\right)-\frac{\partial p}{\partial \theta}+\rho g_\theta$$

Considering the nature of the present problem, most of the terms in this equation will be cancelled when applying the assumptions. The basic equation for the flow between rotating cylinders becomes a linear second-order ordinary differential equation:

$$0=\mu\left(\frac{\partial}{\partial r}\left(\frac{1}{r}\frac{\partial r\, v_\theta}{\partial r}\right)\right)$$

Double integrating the simplified differential equation generates two arbitrary integration constants, C_1 and C_2:

$$v_\theta = C_1 r+\frac{C_2}{r}$$

The constants appearing in the solution of v_θ are found by no-slip conditions at the inner and outer cylinders;

B.C.1: at $r = r_o, v_\theta = 0, \rightarrow v_\theta = C_1 r_o + C_2/r_o$
B.C.2: at $r = r_i$, $v_\theta = \omega_i r_i, \rightarrow \omega_i r_i = C_1 r_i + C_2/r_i$

Solve for C_1 and C_2.

The second integration constant (C_1) is obtained using boundary conditions 1:

$$C_1 = \frac{\omega_i}{\left(1 - \frac{r_o^2}{r_i^2}\right)}$$

The second integration constant (C_2) is obtained using boundary conditions 2:

$$C_2 = \frac{\omega_i}{\left(\frac{1}{r_i^2} - \frac{1}{r_o^2}\right)}$$

Substitute the integration constant. The final solution for velocity distribution is given by:

$$v_\theta = \omega_i r_i \left[\frac{\left(\frac{r_o}{r}\right) - \left(\frac{r}{r_o}\right)}{\left(\frac{r_o}{r_i}\right) - \left(\frac{r_i}{r_o}\right)}\right]$$

Example 3.13: Flow of Fluid Down an Inclined Plane

Newtonian fluid is flowing down an inclined plane with angle β, as shown in Figure 3.15. The fluid is assumed to be incompressible. The film of water flows downward because of gravitational force. Develop the velocity profile for the fluid film flowing on the inclined plane.

Solution

For laminar flow of a Newtonian fluid down an incline, the following assumptions are valid:

- Fluid is Newtonian (constant viscosity), laminar (no velocity in the x- or y-directions).
- Shear stress at interface between the fluid and air is negligible.
- At the solid-liquid interface, no slip conditions are applied at wall.
- The flow is at steady state.

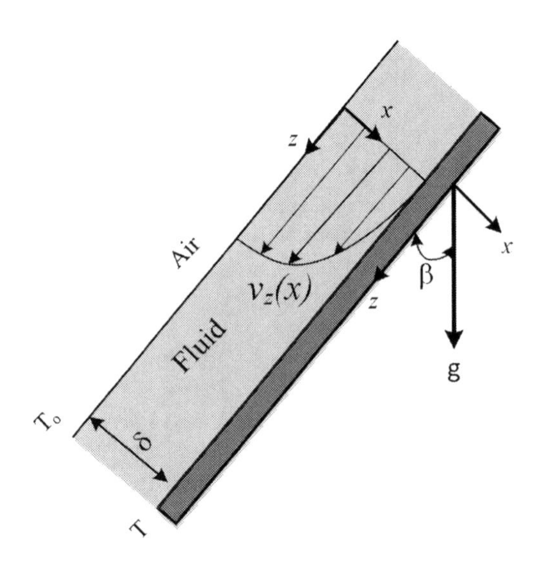

FIGURE 3.15
Flow of fluid down an inclined plane.

- In regions far from the entrance, fluid flow in circular pipe is well developed.
- No edge effects in the y-direction (width).
- Constant density.

The equation of continuity, constant density, and Cartesian coordinates:

$$0 = \frac{\partial v_x}{\partial x} + \frac{\partial v_y}{\partial y} + \frac{\partial v_z}{\partial z}$$

Thus, the velocity is in the z-direction, which means that $v_x = v_y = 0$, and $\partial v_z / \partial z = 0$.

The equation of motion in Cartesian coordinates:

$$\rho \frac{\partial v_z}{\partial t} + \rho \left(v_x \frac{\partial v_z}{\partial x} + v_y \frac{\partial v_z}{\partial y} + v_z \frac{\partial v_z}{\partial z} \right) = \mu \left(\frac{\partial^2 v_z}{\partial x^2} + \frac{\partial^2 v_z}{\partial y^2} + \frac{\partial v_z}{\partial z^2} \right) - \frac{\partial p}{\partial z} + \rho g_z$$

Applying the given assumptions:

$$0 + \rho(0 + 0 + 0) = \mu \left(\frac{\partial^2 v_z}{\partial x^2} + 0 + 0 \right) - 0 + \rho g \cos\beta$$

Note that $dp/dz = 0$ because there is no applied pressure gradient to drive the flow. Flow is driven by gravity alone. Rearranging, we have:

$$\mu\left(\frac{\partial^2 v_z}{\partial x^2}\right) = -\rho g\cos\beta$$

Double integration generates two arbitrary integration constants:

$$v_z = -\frac{\rho g\cos\beta}{2\mu}x^2 + C_1x + C_2$$

Boundary conditions

B.C.1: at $x = 0, dv_z/dx = 0$ (stress matches at the boundary):

$$\frac{dv_z}{dx} = -\frac{\rho g\cos\beta}{\mu}x + C_1$$

Use boundary condition 1:

$$0 = -\frac{\rho g\cos\beta}{\mu}(0) + C_1 \rightarrow C_1 = 0$$

B.C.2: at $x = H, v_z = 0$ (no slip conditions at the liquid-wall interface):
Applying boundary condition 2, we have:

$$0 = -\frac{\rho g\cos\beta}{2\mu}H^2 + 0 + C_2 \rightarrow C_2 = \frac{\rho g\cos\beta}{2\mu}H^2$$

Substitute C_1 and C_2:

$$v_z = -\frac{\rho g\cos\beta}{2\mu}x^2 + \frac{\rho g\cos\beta}{2\mu}H^2$$

Rearranging, we have:

$$v_z(x) = \frac{\rho g\cos\beta}{2\mu}(H^2 - x^2)$$

where:

v_z is the velocity in the z direction.
ρ is the density.
μ is the dynamic viscosity.
g is the acceleration due to gravity.
β is the angle of the plane to the vertical.
H is the thickness of the film.

Example 3.14: Temperature Profile in an Inclined Plane

A liquid is falling over an inclined plane (angle of β with the vertical). The temperature of the film at the fluid-solid interface ($x = 0$) is kept at temperature T_o, and the temperature at the liquid-air interface ($x = \delta$) is at temperature T (Figure 3.16). Derive the temperature profile for the liquid film. Although convective terms are important, it will be neglected in this example to simplify the problems. Assume that the temperature across the liquid film varies due to the conduction term in the y-direction.

Solution

The equation of energy can be taken for Cartesian coordinates and reduced to a solvable form by using the following assumptions:

1. Steady state.
2. The temperature, T, is only a function of y, convection is much greater in the x direction than conduction, accordingly, conduction term $(\partial^2 T)/(\partial x^2)$ can be dropped.
3. Conduction is in the y direction only.
4. Convection and viscous dissipation are negligible, to make the problem simpler.
5. No heat generation; thus, the generation term is neglected.

The general differential equation of energy transport in Cartesian coordinates is:

$$\rho C_p \frac{\partial T}{\partial t} + \rho C_p \left(v_x \frac{\partial T}{\partial x} + v_y \frac{\partial T}{\partial y} + v_z \frac{\partial T}{\partial z} \right) = k \left(\frac{\partial^2 T}{\partial x^2} + \frac{\partial^2 T}{\partial y^2} + \frac{\partial^2 T}{\partial z^2} \right) + \Phi_H$$

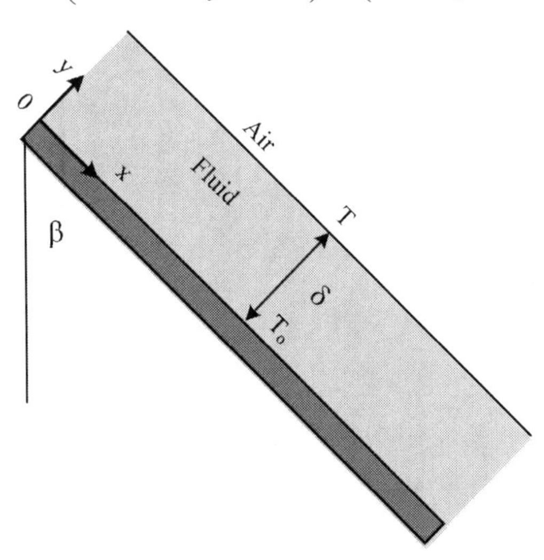

FIGURE 3.16
Temperature profile of water flowing in an inclined plane.

Applying the assumptions, we have:

$$0+\rho C_p(0+0+0)=k\left(0+\frac{\partial^2 T}{\partial y^2}+0\right)+0$$

The equation is reduced to:

$$\frac{\partial^2 T}{\partial y^2}=0$$

Integrating twice, we have:

$$T=C_1 y+C_2$$

Boundary conditions:

B.C.1: at $y=0, T=T_o, \xrightarrow{\text{yields}} C_2=T_o$

B.C.2: at $y=\delta, T=T, \xrightarrow{\text{yields}} C_1=\frac{T-T_o}{\delta}$

Substitute the integrating constants, C_1 and C_2:

$$T=\frac{T-T_o}{\delta}y+T_o$$

Example 3.15: A Strip Moves Upward at Constant Velocity

A strip close to a liquid film moves upward at velocity V, constructing a film of viscous liquid of thickness, h, as shown in Figure 3.17. Near the strip, the film moves upward. The other side of the liquid film moves downward due to gravity. Consider the case where the only nonzero velocity is v_z. With zero shear stress at the outer film edge, derive a formula for the velocity distribution $v(x)$.

Solution

The assumption of parallel flow, $v_x=v_y=0$ and $v_z=v(x)$, satisfies continuity and makes the x and y components of the momentum equations irrelevant. The rest is the z-momentum equation. The equation of continuity at constant density and Cartesian coordinates is:

$$0=\frac{\partial v_x}{\partial x}+\frac{\partial v_y}{\partial y}+\frac{\partial v_z}{\partial z}$$

The velocity is in the z-direction; thus, $v_x=v_y=0$, and $\partial v_z/\partial z=0$.

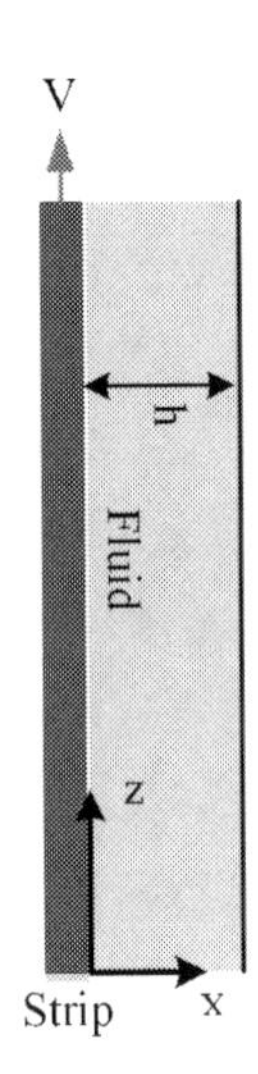

FIGURE 3.17
A strip moves upward at velocity *V*.

The equation of motion in Cartesian coordinates is:

$$\rho\frac{\partial v_z}{\partial t}+\rho\left(v_x\frac{\partial v_z}{\partial x}+v_y\frac{\partial v_z}{\partial y}+v_z\frac{\partial v_z}{\partial z}\right)=\mu\left(\frac{\partial^2 v_z}{\partial x^2}+\frac{\partial^2 v_z}{\partial y^2}+\frac{\partial v_z}{\partial z^2}\right)-\frac{\partial p}{\partial z}+\rho g_z$$

After applying the appropriate assumptions, the equation of motion is simplified to:

$$0+\rho(0+0+0)=\mu\left(\frac{\partial^2 v_z}{\partial x^2}+0+0\right)-0+\rho g_z$$

There is no convective acceleration. In addition, the pressure gradient is negligible because of the free surface. Finally, a second-order linear differential equation for $v_z(x)$ remains as follows:

$$\left(\frac{\partial^2 v_z}{\partial x^2}\right)=\frac{\rho g}{\mu}$$

Perform double integrating again:

$$v_z=\frac{\rho g}{2\mu}x^2+C_1x+C_2$$

B.C.1: at free surface, $x=h,\ dv_z/dx=0$, hence, $C_1=-\rho gh/\mu$
B.C.2: at the wall, $x=0,\ v_z=V$, hence, $C_2=V$

Substitution of C_1 and C_2 leads to the following solution:

$$v_z = V - \frac{\rho g h}{\mu} x + \frac{\rho g}{2\mu} x^2$$

PROBLEMS

3.1 Fluid Flowing in an Inclined Pipe

Develop the velocity profile for fluid flowing in an inclined pipe. The radius of the pipe is R, the length of the pipe is L, and the pressure drop across the pipe is ΔP. The pipe is inclined with at a degree of θ with the vertical coordinate. The quantity P is known as the hydrodynamic pressure or simply the dynamic pressure. Its gradient is zero in a stationary fluid, and therefore it is uniform when the fluid is not in motion.

Answer: $v_z(r) = \Delta P \frac{R^2}{4\pi L}\left[1 - \frac{r^2}{R^2}\right]$ were $\Delta P = P(0) - P(L) + \rho g \cos\theta$

3.2 Fluid Flow Between Wide Parallel Plates

Develop the velocity profile for fluid flow between wide parallel plates that are separated by a distance $2H$. Consider flow driven by a pressure difference between the inlet and exit of a channel formed two horizontal parallel plates. The plate's length is L and its width W. The maximum velocity occurs at the centerline.

Answer: $V_z(y) = (\Delta P.H^2)/4\pi L[1 - y^2/H^2]$

3.3 Heat Transfer in a Cylindrical Pipe Wall

Consider a straight circular pipe of inner radius r_1, outer radius r_2, and length L. The temperature is specified at both the inner and outer pipe wall surfaces. Let the steady temperature of the inner surface be T_1 and that of the outer surface be T_2. Assume that the temperature varies only in the radial direction. Develop the temperature profile between the inner and outer surface of pipe.

Answer: $(T_1 - T)/(T_1 - T_2) = \ln(r/r_1)/\ln(r_2/r_1)$

3.4 Heat Generation in a Cylinder

Heat is generated because of an electrical current passing through wire and the resistance to the flow of electricity in the wire. The heat generation can be characterized as a volumetric source of heat $Q(W/m^3)$. The wire is cylindrical, with diameter D and length L, and contains a uniform heat source Q. The wire is surrounded by air temperature T_0, and the wire steady-state surface temperature is T_s. The cylinder is assumed to be long with insulated ends, so that heat transfer through the cylinder and to the air occurs only in the radial direction. Let the heat transfer coefficient between the surface of the cylinder and the surrounding air be h_o. The thermal conductivity of the wire is K.

Answer: $T = T_s + \frac{Q}{4K}\left(R^2 - r^2\right)$

3.5 Separation of Helium from Natural Gas

Pyrex glass tubes can be used for separation of helium from natural gas because the diffusivity of helium in Pyrex glass is very high in comparison to the other gases. Suppose a natural gas mixture is flowing in the Pyrex tube whose inner radius is r_1 and whose outer radius is r_2. Obtain an expression for the rate at which helium will separate out in terms of diffusivity of helium through the wall of the Pyrex tube, the interfacial concentration of helium in the Pyrex tube wall, and the dimensions of the tube.

Answer: $N_{He} = D_{AB}/r\ln(r_2/r_1)(C_{He1} - C_{He2})$

3.6 Heat Transfer in a Slab

Consider a wall of infinite extent in the y and z directions and with temperature at each side being fixed. There are two boundary conditions for $T(x)$: $T(0) = T_0$ and $T(L) = T_L$. Assume a steady solution, with no flow and and no heat generation.

Answer: $T = T_0 + (T_L - T_0)x/L$

3.7 Temperature Profile in Rectangular Film

Consider a fin with length L in the x-direction and thickness b in the z-direction. The width of the fin extended in the y-direction is assumed to be infinite. The air surrounding the fin is at a temperature T_a. The left-hand base of the fin is maintained at a fixed temperature, T_o. Assume the end of the fin is insulated. Develop the temperature profile for heat transfer in the x-direction.

Answer: $(T - T_a)/(T_w - T_a) = \cosh(2h/bk)(1 - x/L)/\cosh(2h/bk)$

3.8 Heat Transfer in Cylindrical Rod Fin

A fin is a cylindrical rod shape where the left end of the rod is fixed at a temperature T_1. The right end of the rod is fixed at a temperature T_2. The rod is surrounded by ambient air at a temperature T_a. There is also a heat source given by Φ. Find the temperature distribution in the rod. Assume that the rod has a radius R and is oriented along the z axis in cylindrical coordinates. The length of the rod is given by L.

3.9 Diffusion with Heterogeneous Chemical Reaction

Consider a gas A diffusing into a liquid B. As it diffuses, the reaction $A + B \rightarrow P$ occurs. You can ignore the small amount of P that is present. The concentration of A at the surface of the liquid is C_{Ao}. The reaction rate is first order:

$$r = kC_A$$

Develop the concentration profile of A through the liquid phase.

Answer: $C_A/C_{A0} = (\cosh((KL^2/D_{AB})^{0.5}(1 - z/L))/\cosh(KL^2/D_{AB})^{0.5}$

3.10 Diffusion into a Falling Liquid Film

Water is flowing in laminar motion down a vertical wall is used of gas absorption. The thickness of the film in the x-direction is δ, and the water is falling along the wall in the z-direction. The width of the film is W in the y-direction. The concentration of liquid-gas interface is C_{Ao}.

Answer: $C_A/C_{A0} = 1 - \mathrm{erf}(x/(4D_{AB}\, z/v_{z,\max})^{0.5})$

References

1. Bird, R. B., W. E. Stewart, E. N. Lightfoot, 2002. *Transport Phenomena*, 2nd ed., New York: John Wiley & Sons.
2. Brenan, K. E., S. L. Campbell, L. R. Petzold, 1989. *Numerical Solution of Initial-Value Problems in Differential-Algebraic Equations*, New York: Elsevier Science Publishers.
3. Villadsen, J., M. L. Michelsen, 1978. *Solution of Differential Equation Models by Polynomial Approximation*, Upper Saddle River, NJ: Prentice-Hall.
4. Geankoplis, C. J., 2009. *Transport Processes and Separation Process Principles*, 4th ed., Upper Saddle River, NJ: Prentice-Hall.
5. McCabe, W. L., J. C. Smith, P. Harriott, 2001. *Unit Operations of Chemical Engineering*, 6th ed., New York: McGraw-Hill.
6. Seader, J. D., E. J. Henley, 2006. *Separation Process Principles*, 2nd ed., New York: John Wiley & Sons.
7. Mills, A. F., 1999. *Heat Transfer*, 2nd ed., Upper Saddle River, NJ: Prentice-Hall.
8. Holman, J. P., 2010. *Heat Transfer*, 10th ed., New York: McGraw-Hill.
9. Griskey, R. G., 2002. *Transport Phenomena and Unit Operations, a Combined Approach*, New York: John Wiley & Sons.

4

Computational Fluid Dynamics

The branch of fluid mechanics that uses numerical analysis to solve and analyze problems that include fluid flows is called computational fluid dynamics (CFD). The main task in fluid dynamics is to find the velocity field describing the flow in a specific domain. Computers are used to perform the calculations required to simulate the interaction of liquids and gases with surfaces defined by boundary conditions. With high-performance computing (HPC) via supercomputers, faster and better solutions can be achieved. In this chapter, mathematical models are developed for fluid flow in distributed parameter systems, that is, those systems designated by partial differential equations (PDEs). COMSOL Multiphysics is used in solving the generated model equations.

LEARNING OBJECTIVES

- Develop mathematical models for fluid flow in distributed parameter systems.
- Rephrase the system by partial differential equations (PDEs).
- Solve manually the developed PDEs when possible.
- Compare manual results with predictions from COMSOL software.

4.1 Introduction

Distributed parameter systems are modeled by sets of PDEs, boundary conditions, and initial conditions that describe the evolution of the state variables in several independent coordinates (space and time). Most distributed parameter models are derived from first principles (conservation of mass, energy, and momentum). The solution of PDEs is a difficult task. Equations can be simplified under suitable assumptions; the PDEs are reduced to a one-dimensional equation and steady-state conditions. The difference between lumped and distributed parameter models depends sometimes on the assumptions put forward by the model designer. The generated models are solved manually when possible and compared with the simulation results generated by COMSOL Multiphysics. Models are to be validated by experimental data to be used for further investigations outside the range of the experimental results.

4.2 Equations of Motion

The equation of conservation of momentum and the equation of motion are listed in the following sections without going through detailed derivation because this can be found in Chapter 3 or from other resources [1]. The equation of motion in rectangular coordinates (Cartesian coordinate), cylindrical, and spherical coordinates are presented in the following sections.

4.2.1 Cartesian Coordinate

The equations of motion in Cartesian coordinate and Newtonian fluids of constant density and viscosity for the x, y, and z component of momentum are obtained, respectively. The continuity equation for constant density is:

$$\rho\left(\frac{\partial v_x}{\partial x}+\frac{\partial v_y}{\partial y}+\frac{\partial v_z}{\partial z}\right)=0 \tag{4.1}$$

The Navier-Stokes equation for the x component is:

$$\rho\frac{\partial v_x}{\partial t}+\rho\left(v_x\frac{\partial v_x}{\partial x}+v_y\frac{\partial v_x}{\partial y}+v_z\frac{\partial v_x}{\partial z}\right)=\mu\left(\frac{\partial^2 v_x}{\partial x^2}+\frac{\partial^2 v_x}{\partial y^2}+\frac{\partial^2 v_x}{\partial z^2}\right)-\frac{\partial p}{\partial x}+\rho g_x \tag{4.2}$$

The Navier-Stokes equation for the y component is:

$$\rho\frac{\partial v_y}{\partial t}+\rho\left(v_x\frac{\partial v_y}{\partial x}+v_y\frac{\partial v_y}{\partial y}+v_z\frac{\partial v_y}{\partial z}\right)=\mu\left(\frac{\partial^2 v_y}{\partial x^2}+\frac{\partial^2 v_y}{\partial y^2}+\frac{\partial^2 v_y}{\partial z^2}\right)-\frac{\partial p}{\partial y}+\rho g_y \tag{4.3}$$

The Navier-Stokes equation for the z component is:

$$\rho\frac{\partial v_z}{\partial t}+\rho\left(v_x\frac{\partial v_z}{\partial x}+v_y\frac{\partial v_z}{\partial y}+v_z\frac{\partial v_z}{\partial z}\right)=\mu\left(\frac{\partial^2 v_z}{\partial x^2}+\frac{\partial^2 v_z}{\partial y^2}+\frac{v_z}{\partial z^2}\right)-\frac{\partial p}{\partial z}+\rho g_z \tag{4.4}$$

4.2.2 Cylindrical Coordinates

The equations of motion for Newtonian fluid in cylindrical coordinates at constant density ρ and viscosity μ for the r, θ, and z components are shown in equations 4.6, 4.7, and 4.8, respectively.

$$\rho\left(\frac{1}{r}\frac{\partial(rv_r)}{\partial r}+\frac{1}{r}\frac{\partial v_\theta}{\partial \theta}+\frac{\partial v_z}{\partial z}\right)=0 \tag{4.5}$$

The r-component of the Navier-Stokes equation is:

$$\rho\frac{\partial v_r}{\partial t}+\rho\left(v_r\frac{\partial v_r}{\partial r}+\frac{v_\theta}{r}\frac{\partial v_r}{\partial \theta}+\frac{v_\theta^2}{r}+v_z\frac{\partial v_r}{\partial z}\right)$$
$$=\mu\left(\frac{\partial}{\partial r}\left(\frac{1}{r}\frac{\partial(rv_r)}{\partial r}\right)+\frac{1}{r^2}\frac{\partial^2 v_r}{\partial \theta^2}+\frac{2}{r^2}\frac{\partial v_\theta}{\partial \theta}+\frac{\partial^2 v_r}{\partial z^2}\right)-\frac{\partial p}{\partial r}+\rho g_r \tag{4.6}$$

The θ-component of the Navier-Stokes equation is:

$$\rho\frac{\partial v_\theta}{\partial t}+\rho\left(v_r\frac{\partial v_\theta}{\partial r}+\frac{v_\theta}{r}\frac{\partial v_\theta}{\partial \theta}+\frac{v_r v_\theta}{r}+v_z\frac{\partial v_\theta}{\partial z}\right)$$
$$=\mu\left(\frac{\partial}{\partial r}\left(\frac{1}{r}\frac{\partial(rv_\theta)}{\partial r}\right)+\frac{1}{r^2}\frac{\partial^2 v_\theta}{\partial \theta^2}+\frac{2}{r^2}\frac{\partial v_r}{\partial \theta}+\frac{\partial^2 v_\theta}{\partial z^2}\right)-\frac{\partial p}{\partial \theta}+\rho g_\theta \tag{4.7}$$

The z-component of the Navier-Stokes equation is:

$$\rho\frac{\partial v_z}{\partial t}+\rho\left(v_r\frac{\partial v_z}{\partial r}+\frac{v_\theta}{r}\frac{\partial v_z}{\partial \theta}+v_z\frac{\partial v_z}{\partial z}\right)$$
$$=\mu\left(\frac{1}{r}\frac{\partial}{\partial r}\left(r\frac{\partial v_z}{\partial r}\right)+\frac{1}{r^2}\frac{\partial^2 v_z^2}{\partial \theta^2}+\frac{\partial^2 v_z}{\partial z^2}\right)-\frac{\partial p}{\partial z}+\rho g_z \tag{4.8}$$

4.2.3 Spherical Coordinates

Since spherical coordinates are rarely found in chemical engineering process units, it will not be discussed here. Those who are interested in spherical coordinate systems can refer to Chapter 3 or other references (see 1–5 in the References section).

4.2.4 Solving Procedure

Solving for velocity and stress fields when using the microscopic balance can be achieved using the following steps:

1. Sketch the system.
2. Choose suitable coordinates for the system.
3. Simplify the continuity equation (mass balance).
4. Simplify the three components of the equation of motion (momentum balance).

5. Solve the differential equations for velocity and pressure if applicable.
6. Apply boundary conditions.
7. Calculate any engineering values of interest (average velocity, flow rate).

4.3 Fluid Dynamic Systems

Fluid dynamics is a branch of fluid mechanics that describes the flow of fluids (liquids and gases). Fluid dynamics has a wide range of applications, including determining the mass flow rate of petroleum through pipelines and predicting weather patterns. In the flowing sections, fluid dynamics systems are considered [6].

4.3.1 Velocity Profile in a Triangular Duct

How to find a velocity profile of an incompressible fluid in a triangular duct (Figure 4.1).

For unidirectional laminar flow, the x-component of velocity v_x depends on coordinates y and z. The components of velocity in the other directions are identically zero.

$$\rho\frac{\partial v_x}{\partial t}+\rho\left(v_x\frac{\partial v_x}{\partial x}+v_y\frac{\partial v_x}{\partial y}+v_z\frac{\partial v_x}{\partial z}\right)=\mu\left(\frac{\partial^2 v_x}{\partial x^2}+\frac{\partial^2 v_x}{\partial y^2}+\frac{\partial^2 v_x}{\partial z^2}\right)-\frac{\partial p}{\partial x}+\rho g_x \quad (4.9)$$

The steady-state Navier-Stokes equation is reduced to Equation 4.10:

$$0=\mu\left(\frac{\partial^2 v_x}{\partial y^2}+\frac{\partial^2 v_x}{\partial z^2}\right)-\frac{\partial p}{\partial x} \quad (4.10)$$

Since v_x depends only on y and z and p does not, it follows that the partial derivative of the pressure with respect to x must be a constant: $(1/\mu)(\partial p/\partial x)=G$. Thus, the equation is reduced to the following form:

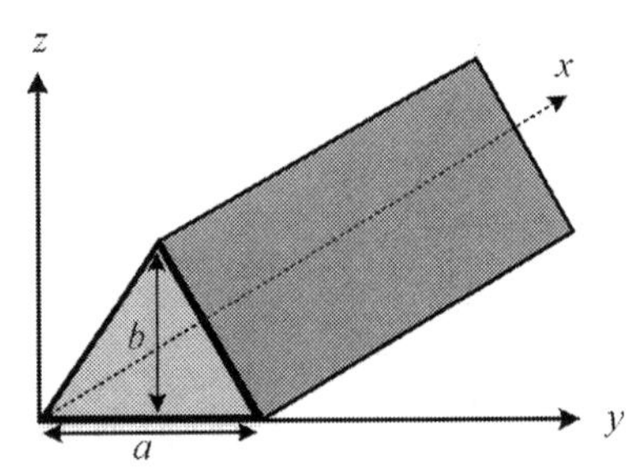

FIGURE 4.1
Fluid flow in a triangular duct.

$$G = \frac{\partial^2 v_x}{\partial y^2} + \frac{\partial^2 v_x}{\partial z^2} \tag{4.11}$$

There is no closed-form solution to this problem on the specified triangular region in terms of elementary functions. Approximate numerical solutions can be obtained. The most common approaches for problems of this type are finite element accessible via COMSOL Multiphysics.

4.3.2 Fluid Flow in a Nuzzle

Develop of velocity profile in a nozzle is shown in Figure 4.2. Gas enters at velocity 0.1 m/s.

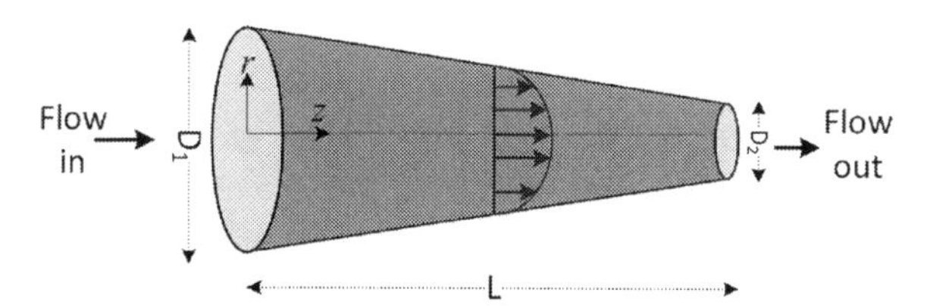

FIGURE 4.2
Velocity profile in a nuzzle.

The Navier-Stokes equation for the z component is:

$$\rho \frac{\partial v_z}{\partial t} + \rho \left(v_r \frac{\partial v_z}{\partial r} + \frac{v_\theta}{r} \frac{\partial v_z}{\partial \theta} + v_z \frac{\partial v_z}{\partial z} \right)$$

$$= \mu \left(\frac{1}{r} \frac{\partial}{\partial r} \left(r \frac{\partial v_z}{\partial r} \right) + \frac{1}{r^2} \frac{\partial^2 v_z^2}{\partial \theta^2} + \frac{\partial^2 v_z}{\partial z^2} \right) - \frac{\partial P}{\partial z} + \rho g_z \tag{4.12}$$

For unidirectional laminar flow, the z-component of velocity v_z depends on coordinates r and direction z. The components of velocity in the angular directions are zero. The steady state Navier-Stokes equation is reduced to:

$$\rho \left(v_z \frac{\partial v_z}{\partial z} \right) = \mu \left(\frac{1}{r} \frac{\partial}{\partial r} \left(r \frac{\partial v_z}{\partial r} \right) + \frac{\partial^2 v_z}{\partial z^2} \right) - \frac{\partial p}{\partial z} \tag{4.13}$$

There is no closed-form solution to this problem, but an approximate numerical solution can be obtained. The most common approaches for problems of this type are finite element working in COMSOL Multiphysics.

4.3.3 Fluid Flow Past a Stationary Sphere

A stainless-steel sphere is being tested in a water tunnel. In the tunnel, the water is forced past the sphere at a uniform fluid velocity, where r_o is the

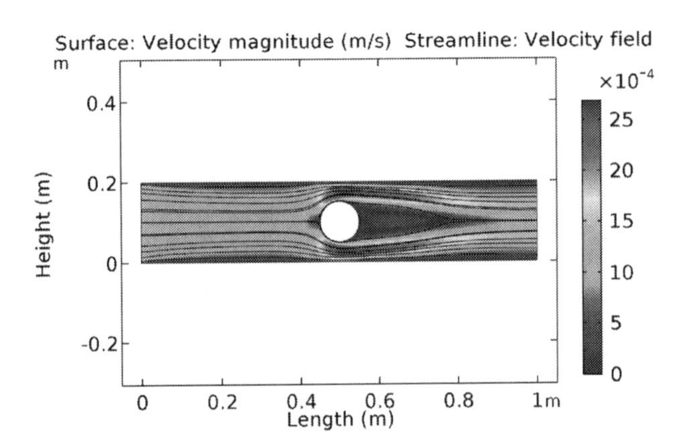

FIGURE 4.3
Fluid flow past a stationary solid sphere.

radius of the sphere and v_∞ is the free stream velocity of the fluid in the z-direction. The velocity profiles in the r-direction and θ-direction are [2]:

$$v_r = v_\infty \left[1 - \frac{3}{2}\left(\frac{r_o}{r}\right) + \frac{1}{2}\left(\frac{r_o}{r}\right)^3 \right] \cos\theta \tag{4.14}$$

$$v_\theta = v_\infty \left[1 - \frac{3}{4}\left(\frac{r_o}{r}\right) + \frac{1}{4}\left(\frac{r_o}{r}\right)^3 \right] \sin\theta \tag{4.15}$$

4.3.3.1 COMSOL Simulation

Numerical solution of this case is calculated using COMSOL Multiphysics (Figure 4.3). The rectangle is (0.2, 1), and the radius of the circle inside the pipe is 0.05 m. The inlet velocity is $v_\infty = 0.001\ \mathrm{m/s}$. The physical properties of water (density and viscosity) are built into COMSOL. The diagram shows the velocity streamlines around the sphere. The velocity is a maximum between the sphere and walls of the pipe due to the narrow area; it is almost zero right behind the sphere. The model builder is shown in Figure 4.4.

4.3.4 Incompressible Fluid Flows Past a Solid Flat Plate

An incompressible viscous fluid with density ρ flows past a solid flat plate that has width b (in the z-direction). The flow initially has a uniform velocity V before contacting the plate for $y \le \delta$ and $u = \mathrm{V}$ for $y \ge \delta$, where δ is the boundary layer thickness (Figure 4.5).

The continuity equation for steady, incompressible flow over a flat plate is:

$$\rho\left(\frac{\partial v_x}{\partial x} + \frac{\partial v_y}{\partial y} + \frac{\partial v_z}{\partial z}\right) = 0 \tag{4.16}$$

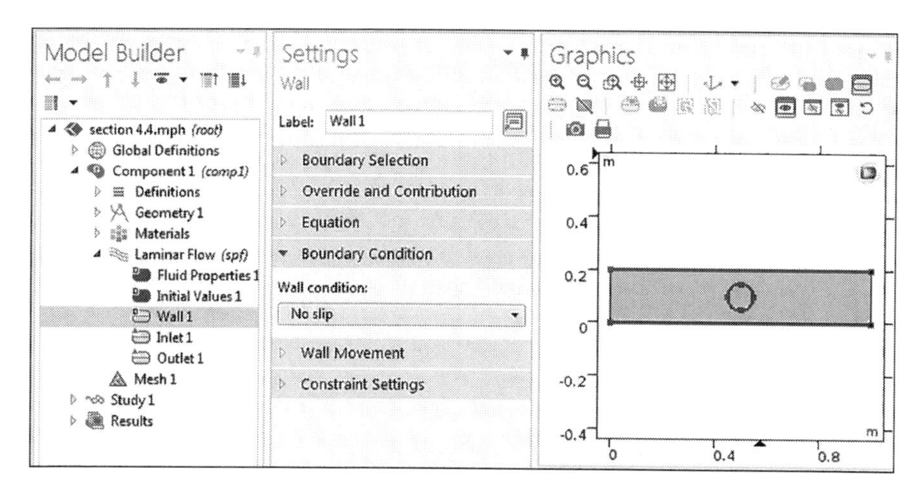

FIGURE 4.4
Model builder geometry, physics, and boundary conditions.

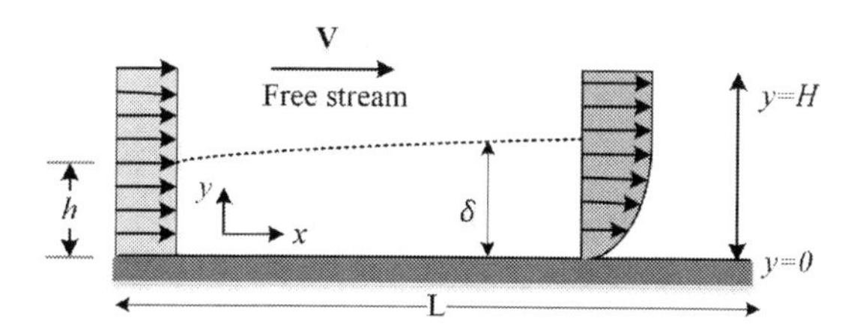

FIGURE 4.5
Schematic of flow past a flat plate, depth b into page.

There is no velocity change in the z-direction (i.e., the depth vertical to paper), $\partial v_z/\partial z = 0$; thus:

$$\frac{\partial v_x}{\partial x} + \frac{\partial v_y}{\partial y} = 0 \tag{4.17}$$

No velocity in z-direction

$$\rho\left(\frac{\partial v_x}{\partial x} + \frac{\partial v_y}{\partial y} + \frac{\partial v_z}{\partial z}\right) = 0$$

The momentum conservation equation in the x direction, also called the Navier-Stokes equation, for the change of velocity in x-direction is:

$$\rho\frac{\partial v_x}{\partial t} + \rho\left(v_x\frac{\partial v_x}{\partial x} + v_y\frac{\partial v_x}{\partial y} + v_z\frac{\partial v_x}{\partial z}\right) = \mu\left(\frac{\partial^2 v_x}{\partial x^2} + \frac{\partial^2 v_x}{\partial y^2} + \frac{\partial^2 v_x}{\partial z^2}\right) - \frac{\partial P}{\partial x} + \rho g_x \tag{4.18}$$

After applying the assumptions, the momentum equation is simplified to:

$$\rho\left(v_x\frac{\partial v_x}{\partial x}+v_y\frac{\partial v_x}{\partial y}\right)=\mu\left(\frac{\partial^2 v_x}{\partial x^2}+\frac{\partial^2 v_x}{\partial y^2}\right)-\frac{\partial p}{\partial x} \tag{4.19}$$

Steady state

No change in the velocity in the z-direction (2D flow)

No gravity in the x-direction

$$\rho\frac{\partial v_x}{\partial t}+\rho\left(v_x\frac{\partial v_x}{\partial x}+v_y\frac{\partial v_x}{\partial y}+v_z\frac{\partial v_x}{\partial z}\right)=\mu\left(\frac{\partial^2 v_x}{\partial x^2}+\frac{\partial^2 v_x}{\partial y^2}+\frac{\partial^2 v_x}{\partial z^2}\right)-\frac{\partial P}{\partial x}+\rho g_x$$

No change in the velocity in the z-direction (2D flow)

Boundary conditions

B.C.1: at $y=0,\ v_x=0$

B.C.2: at $y=0,\ v_y=0$

B.C.3: at $y=\infty,\ v_x=V$

B.C.4: at $x=0,\ v_x=V$

The analytical solution of the set of partial differential equations is difficult. The approximate solution of the velocity profile at location x is estimated to have a parabolic shape [2]. The flow initially has a uniform velocity V before contacting the plate. The velocity profile at location x is estimated to have a parabolic shape:

$$v_x=V\left[\left(\frac{2y}{\delta}\right)-\left(\frac{y}{\delta}\right)^2\right] \tag{4.20}$$

The parabolic shape is valid for $y\le\delta$ and $u=V$ for $y\ge\delta$, where δ is the boundary layer thickness. The upstream height from the plate h of a streamline that has height δ at the downstream location can be obtained from the continuity equation.

Mass flow rate in = mass flow rate out:

$$\int_0^h \rho Vbdy=\int_0^\delta \rho v_x bdy \tag{4.21}$$

Assume incompressible fluid (constant density):

$$Vbh=\int_0^\delta V\left[\left(\frac{2y}{\delta}\right)-\left(\frac{y}{\delta}\right)^2\right]dy=bV\left(\frac{y^2}{\delta}-\frac{y^3}{3\delta^2}\right)_0^\delta=bV\left(\frac{2\delta}{3}\right) \tag{4.22}$$

Simplify:

$$h=\frac{2}{3}\delta \tag{4.23}$$

4.3.4.1 COMSOL Solution

The numerical solution is much easier with COMSOL Multiphysics (Figure 4.6). The length of the arrows is proportional to the magnitude of the velocity.

The snapshot of COMSOL model builder is shown in Figure 4.7. The boundary conditions are as follows:

Top boundary: slip wall.

Bottom boundary: no slip conditions.

Inlet boundary: normal inflow velocity ($V = 0.001\ \text{m/s}$).

Outlet boundary is atmospheric pressure ($p_a = 0$), where p_a is gauge pressure.

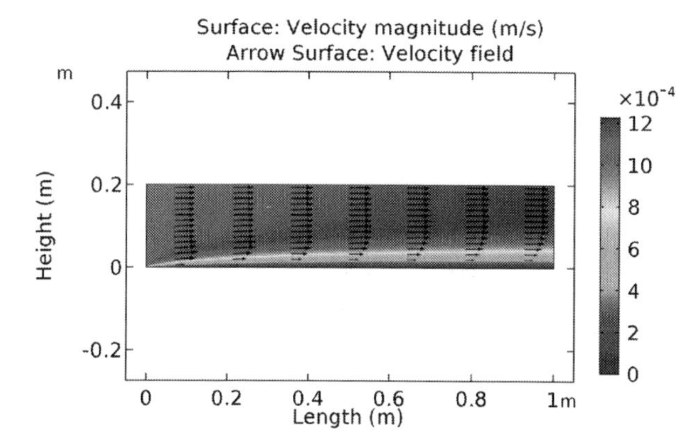

FIGURE 4.6
Velocity profile for fluid flow along a flat sheet.

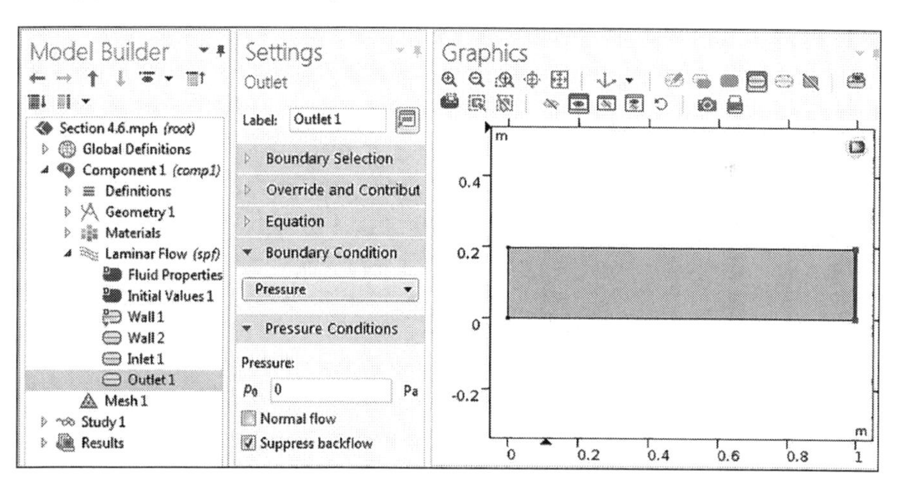

FIGURE 4.7
Boundary conditions.

4.4 Application to Fluid Dynamics

CFD has applications in many different processing industries, including air flow in clean rooms, ovens, and chillers; flow of foods in continuous-flow systems; and convection patterns during in-container thermal processing. Fluid dynamics is a subdiscipline of fluid mechanics that describes the flow of fluids, liquids, and gases. The following are worked examples to illustrate the application of fluid dynamics.

Example 4.1: Velocity Profile Between Two Vertical Parallel Plate

Develop a model for the velocity profile for a laminar flow of a liquid water flowing upward between two parallel vertical plates (Figure 4.8). The inlet velocity is 0.02 m/s. The distance between the two plates is 0.02 m, and the height of the plates is 0.05 m. The pressure drop is constant and approximately 490.8 Pa. The density of water is 1000 kg/m^3 and the viscosity is 0.01 Pa.s.

Solution

For flow between two vertical parallel plates, Cartesian coordinates are selected. Since the fluid flow is in the y-direction, $v_x = 0$, $v_z = 0$, and only v_y exists. Further, the pressure gradient is held constant. The continuity equation is:

$$\rho\left(\frac{\partial v_x}{\partial x}+\frac{\partial v_y}{\partial y}+\frac{\partial v_z}{\partial z}\right)=0$$

Since the density is assumed constant, it will be omitted and the velocity in v_x and v_z are set to zero; after scraping these two terms, the velocity gradient in the y-direction is zero.

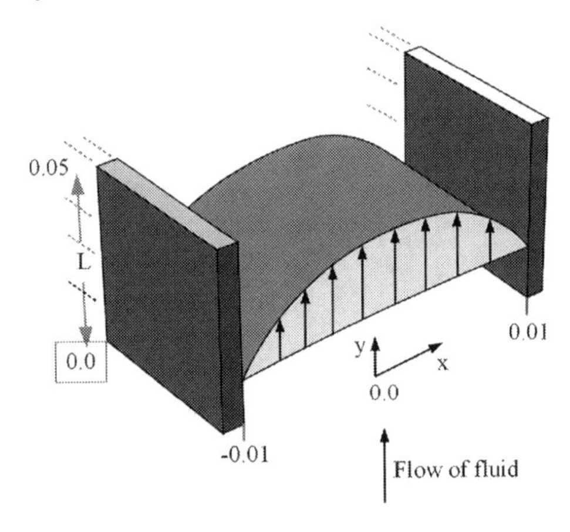

FIGURE 4.8
Schematic of fluid flow between two parallel plates.

$$\frac{\partial v_y}{\partial y} = 0$$

Fluid flows in the y-direction

$v_x = 0$

$$\rho\left(\frac{\partial v_x}{\partial x} + \frac{\partial v_y}{\partial y} + \frac{\partial v_z}{\partial z}\right) = 0$$

$v_z = 0$

The momentum conservation equation in the y direction (Navier-Stokes equation for the change of velocity in the y-direction) is:

$$\rho\frac{\partial v_y}{\partial t} + \rho\left(v_x\frac{\partial v_y}{\partial x} + v_y\frac{\partial v_y}{\partial y} + v_z\frac{\partial v_y}{\partial z}\right)$$

$$= \mu\left(\frac{\partial^2 v_y}{\partial x^2} + \frac{\partial^2 v_y}{\partial y^2} + \frac{\partial^2 v_y}{\partial z^2}\right) - \frac{\partial p}{\partial y} + \rho g_y$$

Assuming steady state, there is no change for the velocity in the y direction with time. The term $\rho(\partial v_y/\partial t)$ from the Navier-Stokes equation will be omitted. The force of gravity ρg_y is replaced by $-\rho g$; the minus sign is attributed to the opposite direction of force of gravity to direction of the fluid flow path. The Navier-Stokes equation is simplified as follows:

Steady state; Fluid flows in the y-direction; $v_x = 0$ and $v_z = 0$; No change in the velocity v_y in the z-direction; $-\rho g$

$$\rho\frac{\partial v_y}{\partial t} + \rho\left(v_x\frac{\partial v_y}{\partial x} + v_y\frac{\partial v_y}{\partial y} + v_z\frac{\partial v_y}{\partial z}\right) = \mu\left(\frac{\partial^2 v_y}{\partial x^2} + \frac{\partial^2 v_y}{\partial y^2} + \frac{\partial^2 v_y}{\partial z^2}\right) - \frac{\partial p}{\partial y} + \rho g_y$$

From continuity $\frac{\partial v_y}{\partial y} = 0$ From continuity $\frac{\partial v_y}{\partial y} = 0$

From the continuity equation, $\partial v_y/\partial y = \partial^2 v_y/\partial y^2 = 0$, so it will be discarded and there is no change in the velocity v_y with the z direction, $\partial^2 v_y/\partial z^2 = 0$. The pressure gradient is constant, consequently:

$$\mu\left(\frac{\partial^2 v_y}{\partial x^2}\right) = \frac{\partial p}{\partial y} + \rho g$$

Let:

$$\frac{\partial p}{\partial y} + \rho g = \varphi$$

Substitute φ in the velocity equation:

$$\frac{d^2 v_y}{dx^2} = \frac{\varphi}{\mu}$$

Doing the integration twice:

$$\frac{dy}{dx} = \frac{\varphi x}{\mu} + C_1$$

The second integration leads to the velocity distribution profile:

$$v_y = \frac{\varphi x^2}{2\mu} + C_1 x + C_2$$

B.C.1: At the mid-distance between two plates (axial symmetry):

$$\text{at} \qquad x = 0, \qquad \frac{dv_y}{dx} = 0$$

Substituting this boundary condition leads to: $C_1 = 0$

Using boundary condition 2 to find the value of C_2, we have:

$$\text{at} \qquad x = w, \qquad v_y = 0$$

Using the boundary condition 2, we have:

$$0 = \frac{\varphi w^2}{2\mu} + C_2$$

Accordingly:

$$C_2 = -\frac{\varphi w^2}{2\mu}$$

Substitute the expression of C_1 and C_2 in the generated velocity profile equation:

$$v_y = \frac{\varphi x^2}{2\mu} - \frac{\varphi w^2}{2\mu}$$

Arrange terms by taking $\varphi/2\mu$ as a common factor:

$$v_y = \frac{\varphi}{2\mu}(x^2 - w^2)$$

The velocity profile for fluid flow upward in the x-direction between two parallel plates with distance w between the plates is:

$$v_y = \frac{\varphi}{2\mu}(x^2 - w^2)$$

Pressure Drops

The pressure drop is approximately 490.8 Pa; thus, the value of k is:

$$\varphi = \frac{\Delta p}{\Delta y} + \rho g = \frac{-490.8\ \text{kg.m/s}^2\text{m}^2}{0.05\ \text{m}} + 1000\ \text{kg/m}^3 \times 9.81\ \text{m/s}^2 = -6\ \text{kg/m}^2\text{s}^2$$

Substitute values in the velocity profile equation to obtain the following equation as a function of x

$$v_z = \frac{\varphi}{2\mu}(x^2 - w^2) = \frac{-6\ \text{kg/m}^2\text{s}^2}{2 \times 0.01\ \text{kg / m.s}}\left(x^2 - (0.01)^2\right)$$

The analytical solution for the velocity versus distance is shown in Figure 4.9. The plot reveals that the velocity is maximum at the middle of two plates and zero at the surface of both plates.

Solution Using COMSOL

Start COMSOL and click on Model Wizard; select 2D under Select Space Dimension. Under Select Physics use:

Fluid Flow > Single phase flow> Laminar flow, then click "Add" (Figure 4.10).

Click "Study." Then click on "Stationary" under preset studies, then click "Done." Under model builder, right-click on Geometry and select "Rectangle" button. Left-click on "Rectangle" and type the values below in the respective edit fields for the rectangle dimensions. Under Base select corner and place the values of −0.01 in the cell near x and 0 for y. The snapshot of the model builder is shown in Figure 4.11.

The fluid density is 1000 kg/m^3 and viscosity is 0.01Pa.s. The volume force vector, $F = (F_x, F_y, F_z)$, describes a distributed force field such as gravity.

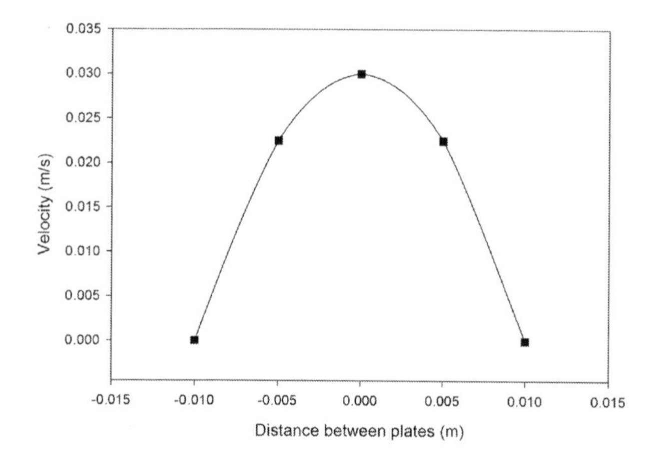

FIGURE 4.9
Analytical solution for velocity versus distance.

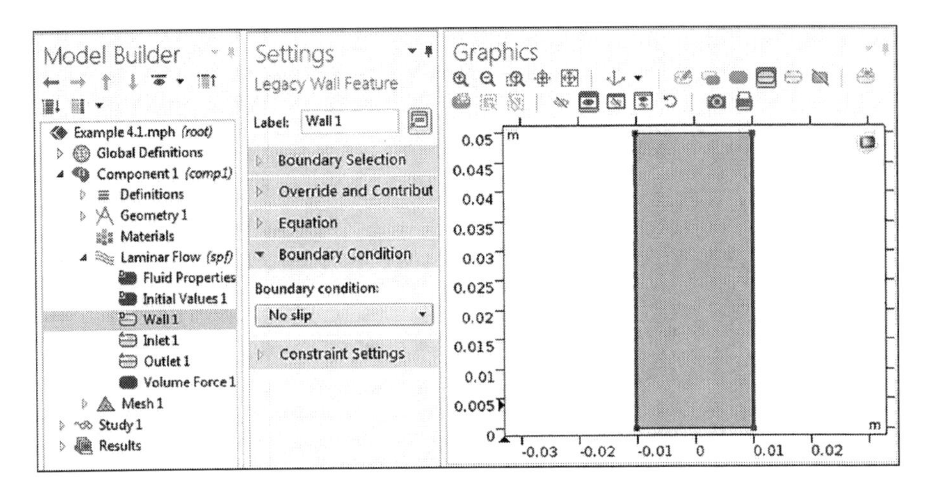

FIGURE 4.10
Physics selection, fluid flow, laminar flow.

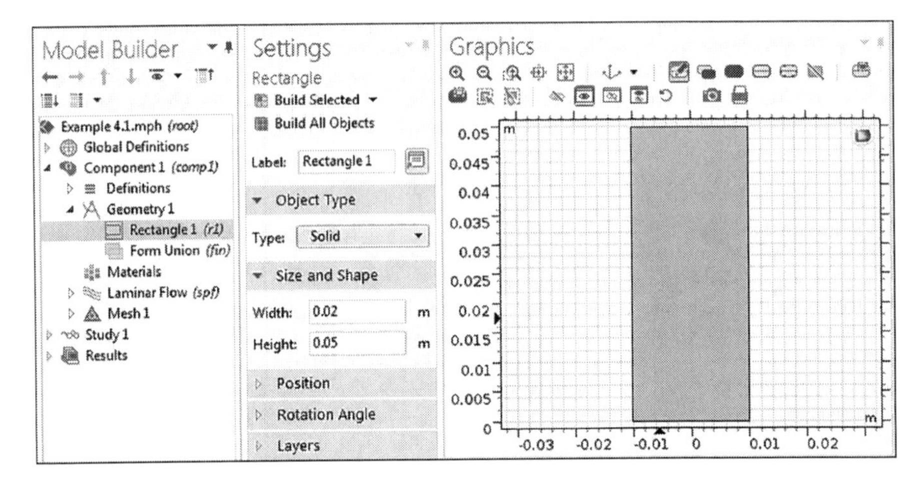

FIGURE 4.11
Dimension of the rectangular geometry.

The unit of the volume force is force/volume. In this example, the gravity is in the negative y-direction. Right-click on laminar flow and add "Volume force." Left-click on volume force, and in the force field enter the value of $-\rho g = -1000 \times 9.81\ \text{m/s}^2$ to your constants (Figure 4.12).

The surface velocity profile between the two plates, considering effect of gravity, is shown in Figure 4.13.

The surface plot of the pressure profile between two plates, considering the effect of gravity, is shown in Figure 4.14. The diagram shows

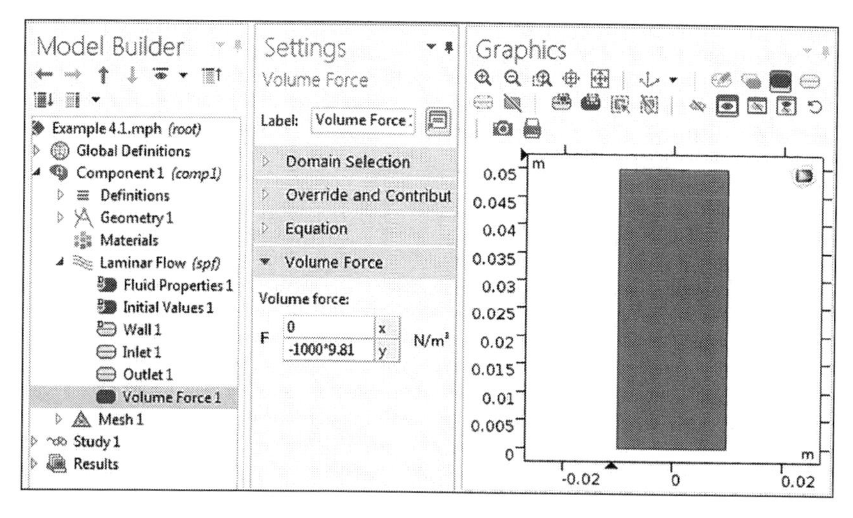

FIGURE 4.12
Selection of volume force representing gravitational force.

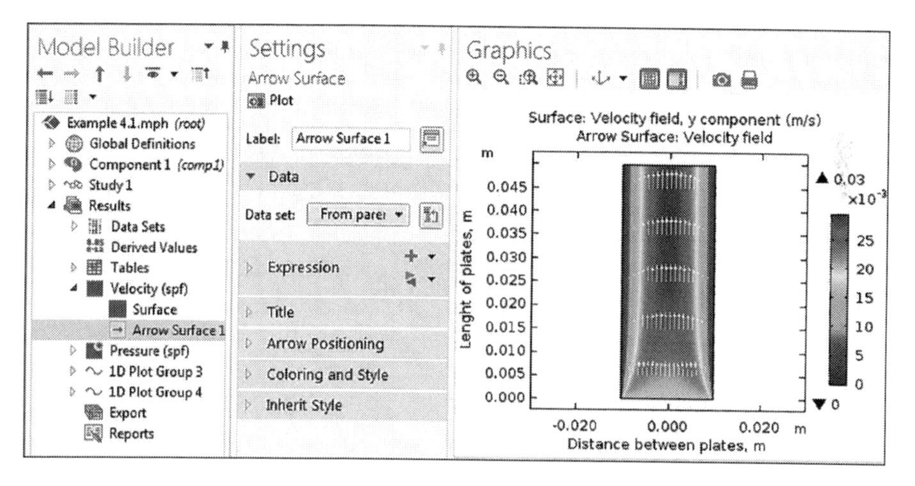

FIGURE 4.13
Velocity profile between the two parallel plates.

that the pressure is maximum at the bottom of the tube and decreases upward along the plates.

Comparison of the model simulation results and the analytical solution is shown in Figure 4.15. The figure depicts the predicted velocity profile at a distance 0.03 m of the length of the plates. A comparison between manual calculations and COMSOL predictions were in good agreement.

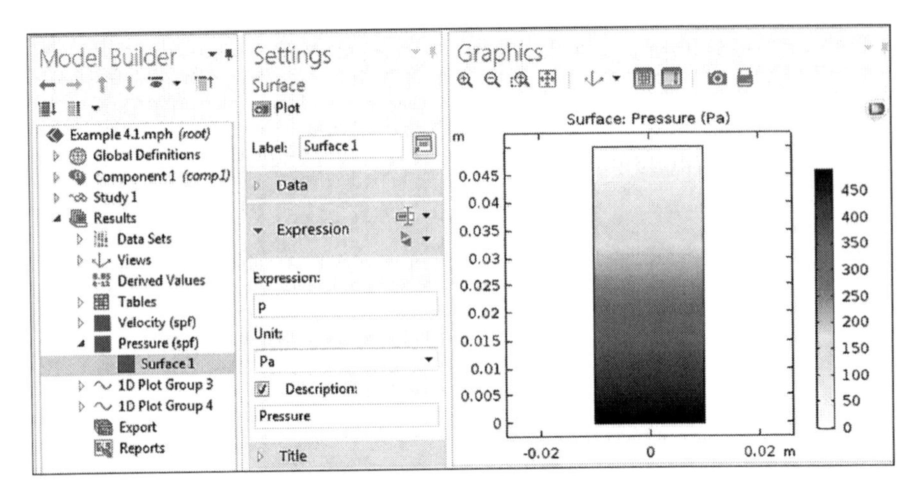

FIGURE 4.14
Pressure drop across the length between two plates.

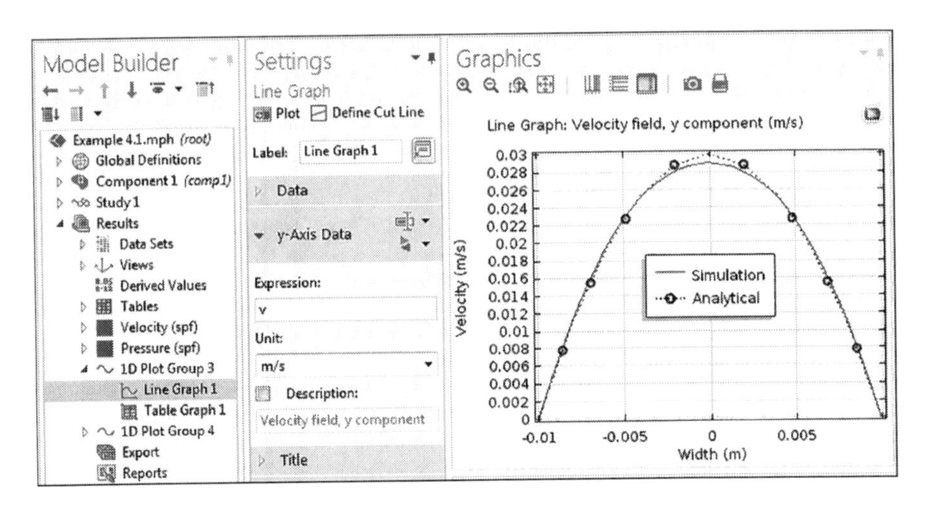

FIGURE 4.15
Velocity profile through the distance between the two plates.

Example 4.2: Fluid Flow Between Two Horizontal Parallel Plates

Consider the flow between two horizontal and parallel plates separated by a distance $2H$. The Newtonian fluid (liquid water) flows in steady laminar flow between the two parallel and flat surfaces of length $L = 0.5$ m; the distance between the two plates is 0.02 m (Figure 4.16). The velocity profile is fully developed between the two surfaces with constant velocity in the x direction (v_x). The fluid is flowing under the influence of constant pressure gradient. The density of water is

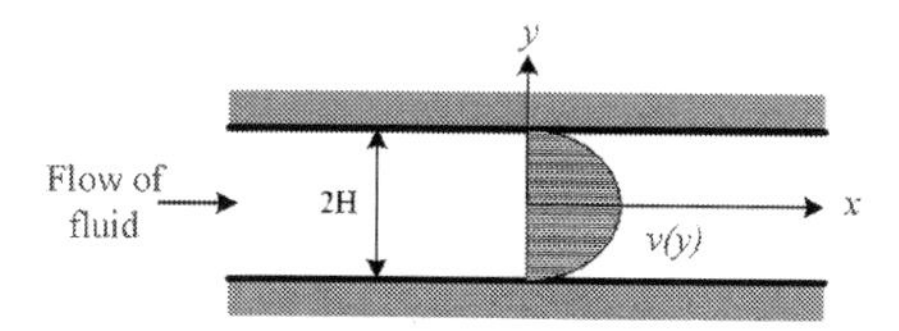

FIGURE 4.16
Fluid flow between two horizontal parallel plates.

1000 kg/m^3 and the viscosity is 0.01 Pa.s. The pressure gradient in the x-direction is 0.29 Pa/m. Determine the velocity profile and compare with COMSOL predictions.

Solution

Consider the following assumptions in the development of the velocity profile between two horizontal and parallel plates:

- Steady, laminar flow
- Incompressible, constant properties (means constant density)
- Fully developed (means no end effects)
- Newtonian fluid (means constant viscosity)
- Negligible gravity effects

The only nonzero velocity component is v_x, and it depends on the coordinate y. Therefore, $v_x = v_x(y)$. For flow between two parallel plates, Cartesian coordinates are chosen. Since the fluid flow is in the x-direction, $v_y = 0$, $v_z = 0$, and only v_x exists. Furthermore, the pressure gradient is held constant. The continuity equation is used and simplified as follows:

$$\rho\left(\frac{\partial v_x}{\partial x} + \frac{\partial v_y}{\partial y} + \frac{\partial v_z}{\partial z}\right) = 0 \qquad v_y = 0,\; v_z = 0$$

Since the density is constant, it will be eliminated, and the velocity in v_y and v_z direction is zero, so the final equation obtained after canceling these two terms of the velocity in the y and z directions is: $\partial v_x / \partial x = 0$.

The velocity profile is obtained from the Navier-Stokes equation for the change of velocity in x-direction:

$$\rho\frac{\partial v_x}{\partial t} + \rho\left(v_x\frac{\partial v_x}{\partial x} + v_y\frac{\partial v_x}{\partial y} + v_z\frac{\partial v_x}{\partial z}\right) = \mu\left(\frac{\partial^2 v_x}{\partial x^2} + \frac{\partial^2 v_x}{\partial y^2} + \frac{\partial^2 v_x}{\partial z^2}\right) - \frac{\partial p}{\partial x} + \rho g_x$$

Assuming steady state, there is no change for the velocity in the x direction with time. The term $\rho(\partial v_x/\partial t)$ from the Navier-Stokes equation will be eliminated. The Navier-Stokes equation is reduced to:

Steady state — No change in the velocity in the z-direction (2D flow) — No gravity in the x-direction

$$\rho\frac{\partial v_x}{\partial t}+\rho\left(v_x\frac{\partial v_x}{\partial x}+v_y\frac{\partial v_x}{\partial y}+v_z\frac{\partial v_x}{\partial z}\right)=\mu\left(\frac{\partial^2 v_x}{\partial x^2}+\frac{\partial^2 v_x}{\partial y^2}+\frac{\partial^2 v_x}{\partial z^2}\right)-\frac{\partial P}{\partial x}+\rho g_x$$

From continuity $\frac{\partial v_x}{\partial x}=0$ — $v_y = 0$ — From continuity $\frac{\partial v_x}{\partial x}=0$ — No change in the velocity in the z-direction (2D flow)

From the continuity equation, $\partial v_x/\partial x=0$; hence, $\partial^2 v_x/\partial x^2=0$. So these two terms will be detached and there is no change in the x-velocity with the z-direction. The pressure gradient is constant and the gravity g_x is zero because there is negligible gravitational acceleration effect in the x direction on the flow of fluid; consequently, the Navier-Stokes equation is reduced to:

$$\mu\left(\frac{\partial^2 v_x}{\partial y^2}\right)=\frac{\partial p}{\partial x}$$

Define the pressure gradient, $\partial p/\partial x=\Delta P/L=\left(p(L)-p(0)\right)/L$:

$$\frac{d^2 v_x}{dy^2}=\frac{\Delta P}{\mu L}$$

Performing double integration generates two arbitrary integration constants, C_1 and C_2:

$$v_x=\frac{\Delta P}{2\mu L}y^2+C_1 y+C_2$$

To evaluate the two integration arbitrary constants C_1 and C_2, we need two boundary conditions. We can use the no-slip conditions at each wall or the axial symmetry at the centerline. The first boundary condition (B.C.1): At the mid distance between two plates, the concept of fully developed velocity profile is used (axis symmetry):

B.C.1: at $y=0$, $\frac{dv_x}{\partial y}=0$

Applying boundary condition 1:

$$\frac{dv_x}{dy}=\frac{\Delta P}{\mu L}y+C_1\Rightarrow 0=\frac{\Delta P}{2\mu L}(0)+C_1\Rightarrow C_1=0$$

The second boundary condition (B.C.2) is a no-slip condition at the top or bottom flat plate:

B.C.2: at $y=H$, $v_x=0$ (no slip at the top wall)

Substitute the second boundary conditions:

$$v_x = \frac{\Delta P}{2\mu L} y^2 + 0 + C_2 \Rightarrow 0 = \frac{\Delta P}{2\mu L} H^2 + C_2 \Rightarrow C_2 = -\frac{\Delta P}{2\mu L} H^2$$

Substitute the expression of C_1 and C_2 in the general velocity profile equation:

$$v_x = \frac{\Delta P}{2\mu L} y^2 - \frac{\Delta P}{2\mu L} H^2$$

Take $\Delta PH^2/2\mu L$ as common factor so that we can write the velocity distribution as follows:

$$v_x = -\frac{\Delta PH^2}{2\mu L}\left(1 - \left(\frac{y}{H}\right)^2\right)$$

The velocity profile for fluid flow in the x-direction between two parallel and horizontal plates with distance $2H$ between the parallel plates is:

$$v_x = -\frac{\Delta PH^2}{2\mu L}\left(1 - \left(\frac{y}{H}\right)^2\right)$$

Just as in a circular tube, this is a parabolic velocity profile that is symmetric about the centerline. The maximum velocity (v_{max}) occurs at the centerline, $y = 0$.

$$v_{max} = -\frac{\Delta PH^2}{2\mu L}$$

The differential cross-sectional area for flow is wdy, and the differential volumetric flow rate would be $dQ = v_x(y)wdy$. By integrating this result across the cross-section, we obtain the volumetric flow rate Q as:

$$Q = -\frac{2\Delta PwH^3}{3\mu L}$$

The average velocity (v_{avg}) is obtained by dividing the volumetric flow rate, Q by the cross-sectional area $A_c = 2wH$:

$$v_{avg} = -\frac{\Delta PH^2}{3\mu L}$$

We can see that the ratio of the average to the maximum is two-thirds for steady laminar flow between wide parallel plates. Thus, the velocity

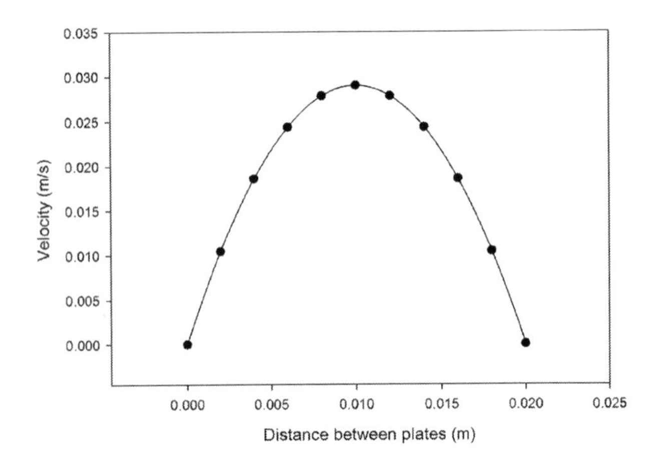

FIGURE 4.17
Analytical solution of velocity profile.

profile is parabolic. For $\Delta P/L$ that is -0.29 Pa/m, the analytical solution of velocity profile is shown in Figure 4.17.

COMSOL Simulation

In this example the gravitational force is zero; thus, the addition of volume force is not needed. The geometry of the fluid flow between two horizontal parallel plates is as shown in Figure 4.18.

The surface velocity profile is shown in Figure 4.19. The arrows in the surface diagram show that the velocity at the entrance of the pipe is not fully established, and the velocity profile is developed gradually along

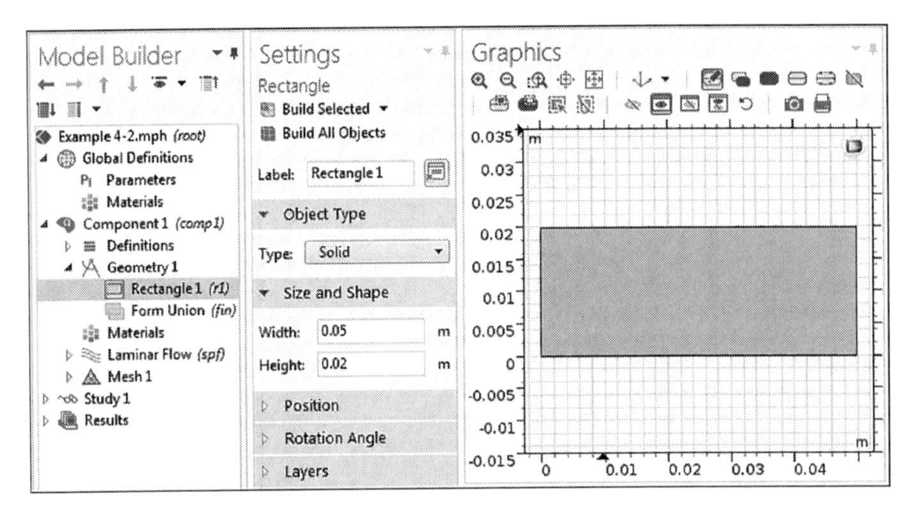

FIGURE 4.18
System geometry of fluid flow between two horizontal and parallel plates.

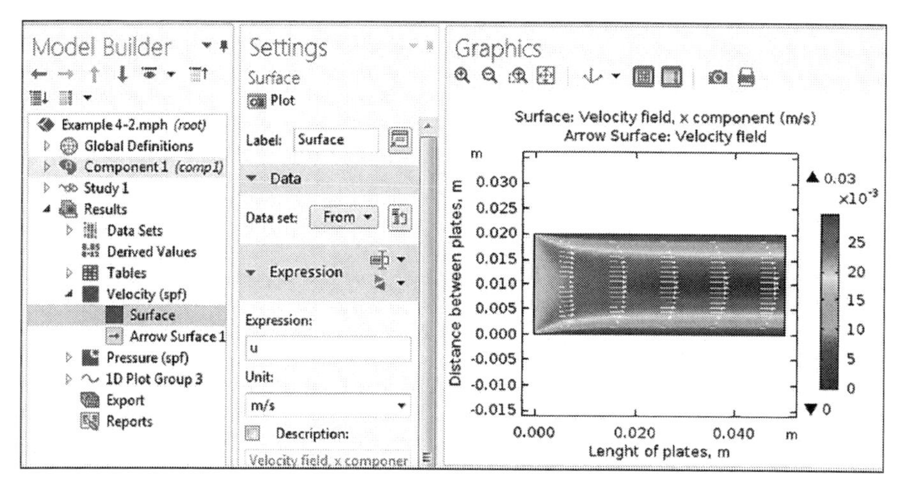

FIGURE 4.19
Surface plot velocity profile with arrows.

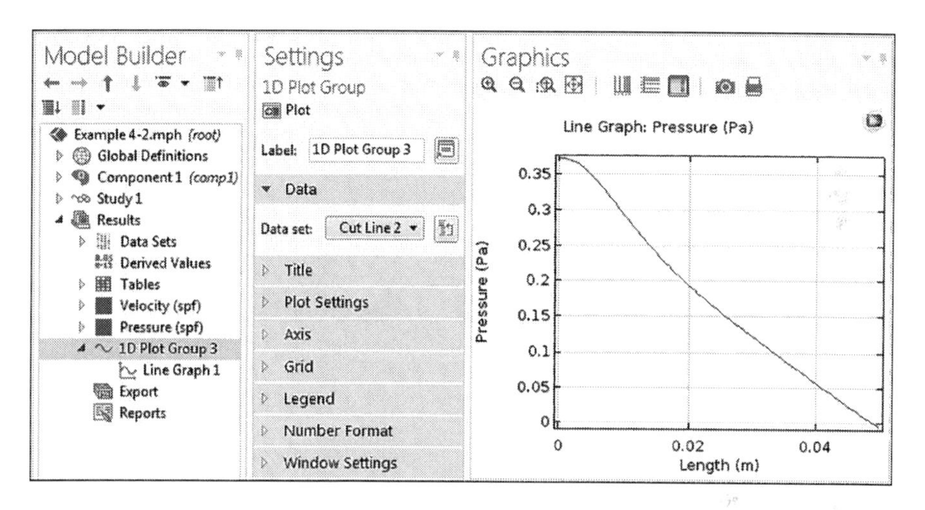

FIGURE 4.20
Gauge pressure along the length of the plates.

the length of the pipe until the profile is fully developed after approximately 0.03 m from the entrance of the pipe.

The pressure drop is shown in the Figure 4.20. The pressure gradually drops along the length of the pipe until it reaches atmospheric pressure. The pressure in the figure is the gauge pressure. This means the pressure drop is around 0.36 Pa.

The velocity profile between the two plates, where the line cut is taken at a half distance of the plates length, 0.025 m, is shown in Figure 4.21. The figure shows that simulation results and analytical solution are in good agreement.

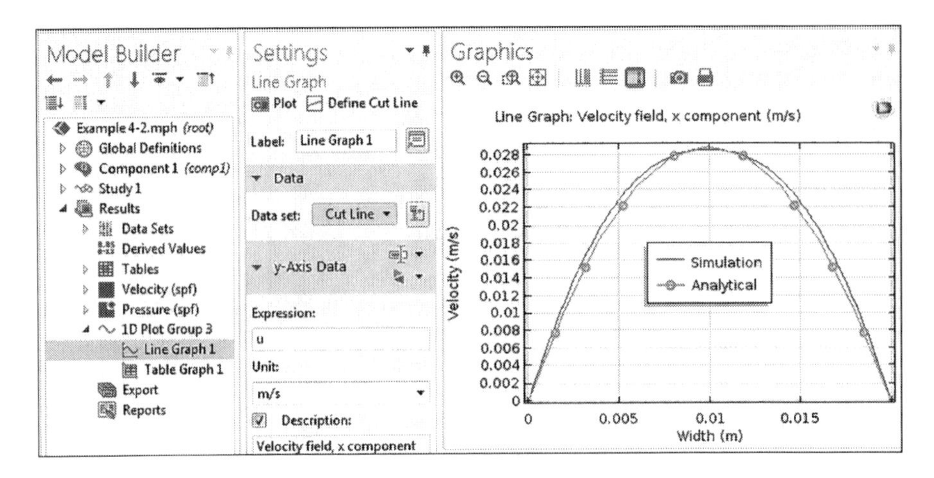

FIGURE 4.21
Velocity profile (v_x) at $x = 0.025\,\text{m}$.

Example 4.3: Velocity Profile Inside a Horizontal Pipe

Consider steady, incompressible, laminar flow of a Newtonian fluid in a horizontal, long, round pipe as shown in Figure 4.22. Water (density $1000\,\text{kg/m}^3$, viscosity 8.9×10^{-4} Pa.s). The fluid flows in a horizontal pipe of length 1 m and diameter 0.1 m. The pressure drop is 0.45 Pa. Find the velocity profile in the pipe at steady state. The fluid is flowing is in the z-direction and the fluid is moving under the influence of constant pressure gradient, where z_1 and z_2 are two arbitrary locations along the z-axis, and P_1 and P_2 are the pressures at those two locations. A constant negative pressure gradient $\partial P/\partial z$ is applied in the z-direction. The pressure gradient may be caused by pump and/or gravity. Derive an expression for the velocity field in the annular space in the pipe.

Solution

Consider the following assumptions in the development of the velocity profile of fluid flow in a horizontal pipe:

- The flow is steady, laminar, and parallel ($v_r = 0$).
- The fluid is incompressible and Newtonian.

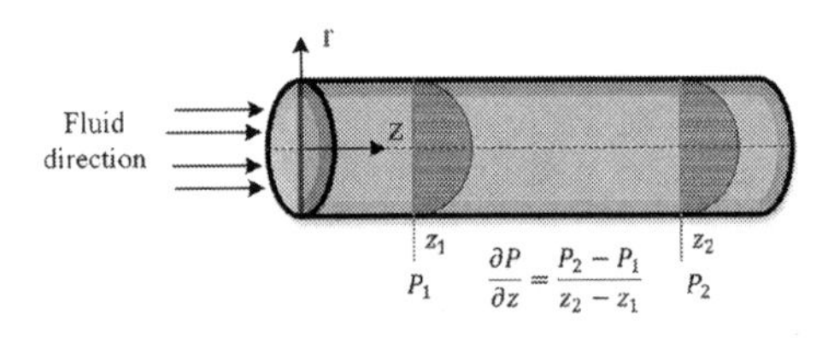

FIGURE 4.22
Fluid flowing inside a horizontal pipe.

- A constant negative pressure gradient is applied in the z-direction.
- The velocity field is axisymmetric with no swirl, so $v_\theta = 0$ and the partial derivative with respect to θ will be zero.
- The gravity effects are neglected.

For axial flow in cylindrical coordinates, the partial differential continuity equation is expressed and simplified by:

Velocity in the r-direction equals to zero $(v_r = 0)$

$$\rho\left(\frac{1}{r}\frac{\partial(rv_r)}{\partial r} + \frac{1}{r}\frac{\partial v_\theta}{\partial \theta} + \frac{\partial v_z}{\partial z}\right) = 0$$

Angular velocity can be neglected $(v_\theta = 0)$

where the angular velocity can be neglected ($v_\theta = 0$) and the velocity in the r-direction equals to zero ($v_r = 0$), so the continuity equation ends with $\partial v_z/\partial z = 0$. Consequently:

$$\frac{\partial^2 v_z}{\partial z^2} = 0$$

The general partial differential equation of motion for fluid moving in the z-component in cylindrical coordinate is:

$$\rho\frac{\partial v_z}{\partial t} + \rho\left(v_r\frac{\partial v_z}{\partial r} + \frac{v_\theta}{r}\frac{\partial v_z}{\partial \theta} + v_z\frac{\partial v_z}{\partial z}\right)$$
$$= \mu\left(\frac{1}{r}\frac{\partial}{\partial r}\left(r\frac{\partial v_z}{\partial r}\right) + \frac{1}{r^2}\frac{\partial^2 v_z^2}{\partial \theta^2} + \frac{\partial^2 v_z}{\partial z^2}\right) - \frac{\partial p}{\partial z} + \rho g_z$$

Here we will assume that the flow is fully developed; that is, it is not influenced by the entrance effects. In this case the term dp/dz is constant and hence we have:

$$\frac{\partial p}{\partial z} = \frac{\Delta p}{\Delta z} = \frac{p_2 - p_1}{z_2 - z_1}$$

Simplify the Navier stocks equation after eliminating the velocity in the radial and angular directions:

Velocity in the r-direction equals to zero $(v_r = 0)$ | From continuity $\frac{\partial v_z}{\partial z} = 0$ | From continuity $\frac{\partial v_z}{\partial z} = 0$

$$\rho\frac{\partial v_z}{\partial t} + \rho\left(v_r\frac{\partial v_z}{\partial r} + \frac{v_\theta}{r}\frac{\partial v_z}{\partial \theta} + v_z\frac{\partial v_z}{\partial z}\right) = \mu\left(\frac{1}{r}\frac{\partial}{\partial r}\left(r\frac{\partial v_z}{\partial r}\right) + \frac{1}{r^2}\frac{\partial^2 v_z^2}{\partial \theta^2} + \frac{\partial^2 v_z}{\partial z^2}\right) - \frac{\partial p}{\partial z} + \rho g_z$$

Steady state | Angular velocity can be neglected $(v_\theta = 0)$ | v_z does not change with θ | No gravity in the z-direction

Accordingly, the equation is reduced to:

$$0 = \mu\left(\frac{1}{r}\frac{\partial}{\partial r}\left(r\frac{\partial v_z}{\partial r}\right)\right) - \frac{\Delta p}{\Delta z}$$

Assuming fluid is Newtonian means that the shear stress is proportional to the velocity gradient and hence the viscosity is constant. The system is described by the second order ordinary differential equation. This equation can be integrated with the following boundary conditions; owing to symmetry, the velocity profile reaches a maximum at the center of the tube:

B.C.1: at $r = 0$, $dv_z/dr = 0$ (because of symmetry about the centerline)

The velocity is zero at the wall of the tube:

B.C.2: at $r = R$, $v_z = 0$ (the no-slip condition at the pipe surface)

First integration results in C_1:

$$r\frac{\partial v_z}{\partial r} = \frac{\Delta p}{\mu L} \times \frac{r^2}{2} + C_1$$

$L = \Delta z$; the tube length,

Divide both sides of the equation by r:

$$\frac{\partial v_z}{\partial r} = \frac{\Delta p}{\mu L}\frac{r^2}{2r} + \frac{C_1}{r}$$

Simplify:

$$\frac{\partial v_z}{\partial r} = \frac{\Delta p}{\mu L}\frac{r}{2} + \frac{C_1}{r}$$

Second integration results in the second arbitrary integration constant, C_2:

$$v_z = \frac{\Delta p}{\mu L}\frac{r^2}{4} + C_1 \ln r + C_2$$

From the first boundary condition B.C.1 $\rightarrow C_1 = 0$

Substitute the boundary conditions 2:

$$0 = \frac{\Delta p}{\mu L}\frac{R^2}{4} + (0)\ln r + C_2$$

Then the expression of the integration constant C_2 is:

$$C_2 = -\frac{\Delta p}{4\mu L} R^2$$

Finally, substitute C_1 and C_2, the velocity profile for the fluid can be expressed by:

$$v_z = \frac{\Delta p}{4\mu L} r^2 - \frac{\Delta p}{4\mu L} R^2$$

Rearrange by taking a common factor ($\Delta p R^2/4\mu L$):

$$v_z = -\frac{\Delta p R^2}{4\mu L}\left[1 - \left(\frac{r}{R}\right)^2\right]$$

The change in fluid velocity as a function of pipe radius is:

$$v_z = \frac{(p_o - p_L)R^2}{4\mu L}\left[1 - \left(\frac{r}{R}\right)^2\right]$$

Accordingly, in a horizontal pipe, the velocity profile in a fully developed laminar flow is parabolic with a minimum velocity (zero) at the pipe wall and maximum at the centerline. Also, the axial velocity v_z is positive for any r, and thus the axial pressure gradient dp/dx must be negative (i.e., pressure decreases in the flow direction because of viscous effects). The maximum velocity ($v_{z,\max}$) of the fluid can be obtained at $r = 0$.

$$v_{z,\max} = -\frac{\Delta p R^2}{4\mu L} = \frac{(p_o - p_L)R^2}{4\mu L}$$

The pressure drop is negative because the pressure is decreasing in the z-direction, so the overall value of velocity is a positive value.

Simulation Using COMSOL

The water is flowing inside a horizontal pipe of length 1 m and 0.1 m diameter. The pressure drop is 0.45 Pa. Neglect inertial term in COMSOL to make the fluid creeping form match the simplified analytical solution. The surface velocity profile is predicted in Figure 4.23, and the comparison with the manual calculations is shown in Figure 4.24. In simulating the fluid flow in a horizontal pipe, the force of gravity is not included (negligible), the inlet pressure from the top is set to 0.45 Pa and zero for exit pressure at the bottom. Figure 4.24 reveals that the simulation predictions and manual calculations are in good agreement.

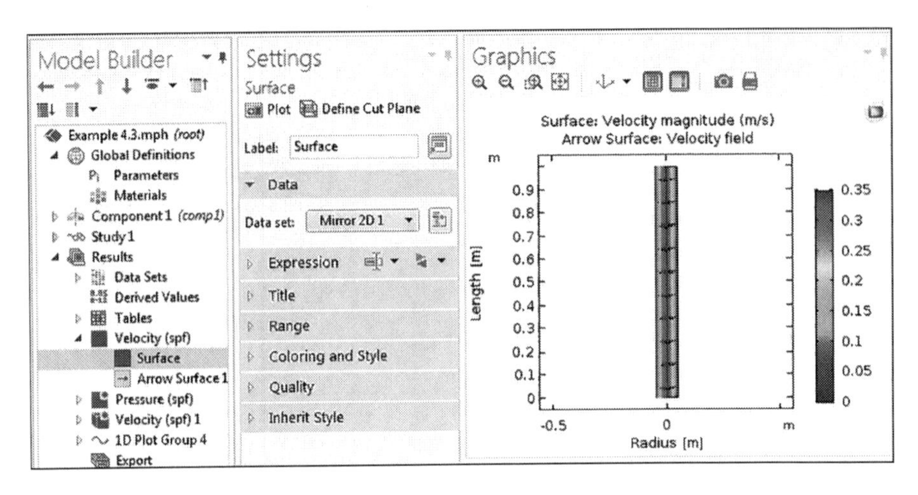

FIGURE 4.23
Fluid flowing inside a horizontal pipe.

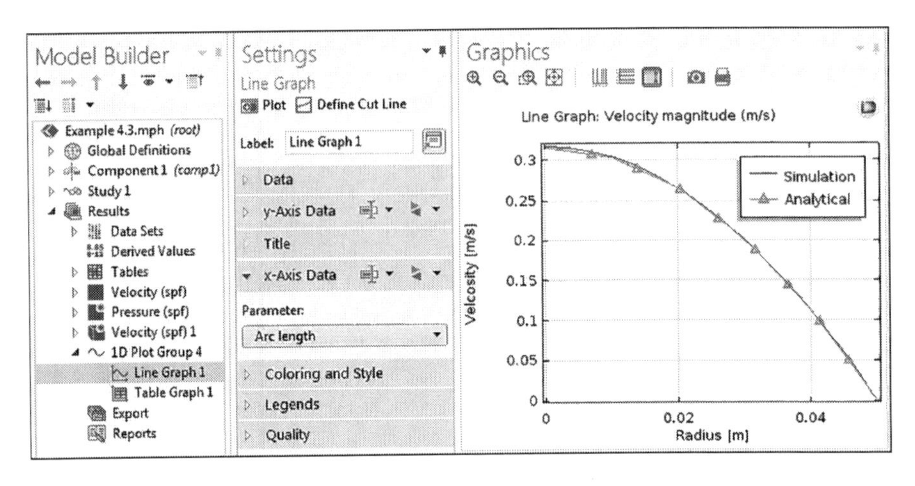

FIGURE 4.24
Fluid velocity versus radius of the horizontal pipe.

Example 4.4: Velocity Profile Inside an Inclined Pipe

Consider an incompressible laminar flow fluid (of density $\rho = 1000$ kg/m^3 and viscosity $\mu = 0.001$ Pa.s), and flow inside an inclined circular pipe ($\beta = 60°$) of radius $R = 0.1$ m and length $L = 1$ m. End effects may be neglected because the tube length L is relatively large compared to the tube radius R (Figure 4.25). The fluid flows under the influence of a pressure difference Δp, gravity, or both [1]. Determine the steady-state velocity distribution for a Newtonian fluid (of constant viscosity μ) and the maximum velocity.

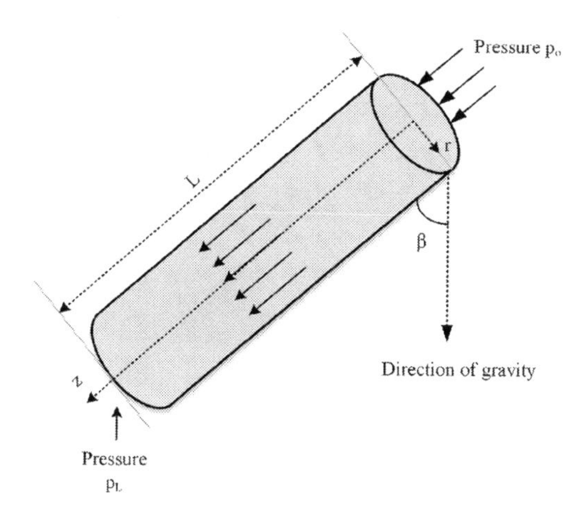

FIGURE 4.25
Fluid flowing in an inclined pipe.

Solution

The relationship between the shear stress (τ_{rz}) and the velocity gradient is presented in the Newtonian constitutive model:

$$\tau_{rz} = \mu \frac{dv_z}{dr}$$

where (μ) is the dynamic viscosity of the fluid. The shear stress is zero at the centerline and its magnitude is a maximum at the tube wall. It varies linearly across the cross-section. We can explain why it is negative. For axial flow in cylindrical coordinates, the differential equation for the continuity equation is expressed by:

$$\rho\left(\frac{1}{r}\frac{\partial(rv_r)}{\partial r} + \frac{1}{r}\frac{\partial v_\theta}{\partial \theta} + \frac{\partial v_z}{\partial z}\right) = 0$$

Velocity in the r-direction
equals to zero ($v_r = 0$)

$$\rho\left(\frac{1}{r}\frac{\partial(rv_r)}{\partial r} + \frac{1}{r}\frac{\partial v_\theta}{\partial \theta} + \frac{\partial v_z}{\partial z}\right) = 0$$

Angular velocity can be
neglected ($v_\theta = 0$)

where the angular velocity (v_θ) can be neglected and the velocity in the r-direction equals zero ($v_r = 0$), so the result of the continuity equation ends with $dv_z/dz = 0$; hence:

$$\frac{\partial^2 v_z}{\partial z^2} = 0$$

For axial flow in cylindrical coordinates, the differential equation for the momentum flux and if the fluid flows in the circular tube under the effect of the gravity and constant pressure gradient, for fluid flowing in the z-direction:

$$\rho\frac{\partial v_z}{\partial t}+\rho\left(v_r\frac{\partial v_z}{\partial r}+\frac{v_\theta}{r}\frac{\partial v_z}{\partial \theta}+v_z\frac{\partial v_z}{\partial z}\right)$$

$$=\mu\left(\frac{1}{r}\frac{\partial}{\partial r}\left(r\frac{\partial v_z}{\partial r}\right)+\frac{1}{r^2}\frac{\partial^2 v_z^2}{\partial \theta^2}+\frac{\partial^2 v_z}{\partial z^2}\right)-\frac{\partial p}{\partial z}+\rho g_z$$

The force of gravity on the fluid is given as the product of the mass of the fluid and the (vector) acceleration due to gravity g_z, which acts vertically downward. We need to use the component of this force in the z-direction, where g is the magnitude of the (vector) acceleration due to gravity g_z.

Simplify the equation after setting the velocities in the r and θ component to zero; the change of velocity in the z-direction is also zero based on the continuity equation (i.e., $dv_z/dz=0$ and $\partial^2 v_z/\partial z^2=0$). The assumptions are simplified as follows:

Velocity in the r-direction equals to zero ($v_r = 0$) — From continuity $\frac{\partial v_z}{\partial z}=0$ — From continuity $\frac{\partial v_z}{\partial z}=0$

$$\rho\frac{\partial v_z}{\partial t}+\rho\left(v_r\frac{\partial v_z}{\partial r}+\frac{v_\theta}{r}\frac{\partial v_z}{\partial \theta}+v_z\frac{\partial v_z}{\partial z}\right)=\mu\left(\frac{1}{r}\frac{\partial}{\partial r}\left(r\frac{\partial v_z}{\partial r}\right)+\frac{1}{r^2}\frac{\partial^2 v_z^2}{\partial \theta^2}+\frac{\partial^2 v_z}{\partial z^2}\right)-\frac{\partial p}{\partial z}+\rho g_z$$

Steady state — Angular velocity can be neglected ($v_\theta = 0$) — v_z does not change with θ — $+\rho g \cos\beta$

The steady-state velocity distribution for the fluid can be determined by:

$$0=\mu\left(\frac{1}{r}\frac{\partial}{\partial r}\left(r\frac{\partial v_z}{\partial r}\right)\right)-\frac{\partial p}{\partial z}+\rho g\cos\beta$$

Rearrange:

$$\mu\left(\frac{1}{r}\frac{\partial}{\partial r}\left(r\frac{\partial v_z}{\partial r}\right)\right)=\frac{\partial p}{\partial z}-\rho g\cos\beta$$

The integral of $\partial P/\partial z$:

$$\int_{p_0}^{p_L}\frac{\partial p}{\partial z}=\frac{p_L-p_0}{L}$$

Hence:

$$\mu\left(\frac{1}{r}\frac{\partial}{\partial r}\left(r\frac{\partial v_z}{\partial r}\right)\right)=\frac{(p_L-p_0-\rho g L\cos\beta)}{L}$$

The fluid flows under the influence of a pressure difference Δp and gravity; the right-hand side may be written in a compact and convenient way by introducing the modified pressure P, which is the sum of the pressure and gravitational terms. The general definition of the modified pressure is $P=p+\rho gh$, where h is the distance upward from a reference plane of choice in the direction opposite to gravity. The fluid may flow because of a pressure difference, gravity, or both. Here, h is negative since the z-axis points downward, giving $h=-L\cos\beta$ and therefore:

$$P_o = p_o \text{ at } z=0 \text{ and } P_L = p_L - \rho g L\cos\beta \text{ at } z = L.$$

The quantity P is known as the hydrodynamic pressure or simply the dynamic pressure. Its gradient is zero in a stationary fluid and therefore it is uniform when the fluid is not in motion:

$$\frac{p_L - p_o - \rho g L\cos\beta}{L} = \frac{(p_L - \rho g L\cos\beta) - p_o}{L} = \frac{P_L - P_o}{L} = \frac{\Delta P}{L}$$

where L is the length of the tube.

$$\frac{1}{r}\frac{\partial}{\partial r}\left(r\frac{\partial v_z}{\partial r}\right) = \frac{\Delta P}{\mu L}$$

We can integrate the differential equation for the velocity profile by noting that the right side is just a constant while the left side is the derivative of $r\tau_{rz}$. The result of this integration is:

$$r\frac{\partial v_z}{\partial r} = \frac{\Delta P}{\mu L}\frac{r^2}{2} + C_1$$

where C_1 is an arbitrary constant of integration that needs to be determined. We can divide each term in the previous equation by r, and we can rewrite this result as:

$$\frac{\partial v_z}{\partial r} = \frac{\Delta P}{\mu L}\frac{r^2}{2r} + \frac{C_1}{r}$$

According to this result, the velocity gradient approaches infinity as the radial coordinate $r \to 0$. This is not physically possible. Therefore, we must choose the arbitrary constant of integration $C_1 = 0$ and write the velocity distribution in the tube as follows:

$$\frac{\partial v_z}{\partial r} = \frac{\Delta P}{\mu L}\frac{r}{2}$$

This equation can be integrated to yield:

$$v_z = \frac{\Delta P}{\mu L}\frac{r^2}{4} + C_2$$

where a new arbitrary constant of integration C_2 has been introduced. To determine this constant, we must enforce a boundary condition on the velocity field. This is the no-slip boundary condition at the tube wall $r = R$.

$$0 = \frac{\Delta P}{\mu L}\frac{R^2}{4} + C_2$$

Use of this condition leads to the result such that the constant C_2 is:

$$C_2 = -\frac{\Delta P}{4\mu L}R^2$$

Finally, substitute the integration constant C_2. The velocity profile for the fluid in the inclined tube can be expressed by:

$$v_z = \frac{\Delta P}{4\mu L}r^2 - \frac{\Delta P}{4\mu L}R^2$$

Rearrange to get the velocity profile in the following form:

$$v_z = -\frac{\Delta P R^2}{4\mu L}\left[1 - \left(\frac{r}{R}\right)^2\right]$$

Since $\Delta P = P_L - P_o$, the velocity profile equation can be written as:

$$v_z = \frac{(P_o - P_L)R^2}{4\mu L}\left[1 - \left(\frac{r}{R}\right)^2\right]$$

The maximum velocity of the fluid can be attained at $r = 0$. Accordingly, the maximum velocity of fluid in the z direction is:

$$v_{z,\max} = -\frac{\Delta P R^2}{4\mu L}(1 - 0)$$

Note that ΔP is negative, and hence maximum velocity is positive:

$$v_{z,\max} = \frac{(P_o - P_L)R^2}{4\mu L}$$

We can calculate the volumetric flow rate through the tube by multiplying the velocity at a given r by the differential cross-sectional area $2\pi r dr$ to yield dQ (the differential volumetric flow rate) and then integrating across the cross-section:

$$Q = \int_0^Q dQ = \int_0^R 2\pi r\, v_z(r)dr = -\frac{\pi\Delta P R^2}{2\mu L}\int_0^R r\left(1 - \frac{r^2}{R^2}\right)dr = -\frac{\pi \Delta P R^4}{8\mu L}$$

The average velocity in the tube, defined as the volumetric flow rate per unit area of cross-section, is:

$$v_{\text{avg}} = \frac{Q}{\pi R^2} = -\frac{\pi \Delta P R^4}{8\mu L}\frac{1}{\pi R^2} = -\frac{\Delta P R^2}{8\mu L}$$

In a circular tube, the average velocity is one-half the maximum velocity. The ΔP is negative as the pressure is decreasing in the z-direction, the overall value of the velocity is positive. The simulation surface plot with

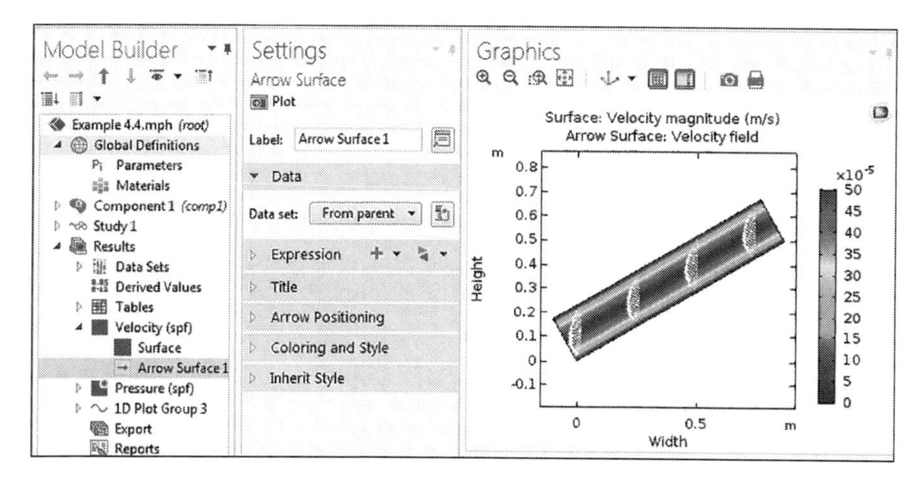

FIGURE 4.26
COMSOL fluid flowing in inclined pipe.

arrows proportional to velocity magnitude is calculated using the software program COMSOL, as shown in Figure 4.26. The analytical solution of the velocity profile is:

$$v_z = \frac{(P_o - P_L)R^2}{4\mu L}\left[1-\left(\frac{r}{R}\right)^2\right]$$

Substitution of known value: $v_z = \dfrac{(1.5\times10^{-4})(0.1)^2}{4\times(0.00089)\times0.1}\left[1-\left(\dfrac{r}{0.1}\right)^2\right]$

Comparison between the analytical solution and the simulation predictions are shown in Figure 4.27. The results are in good agreement;

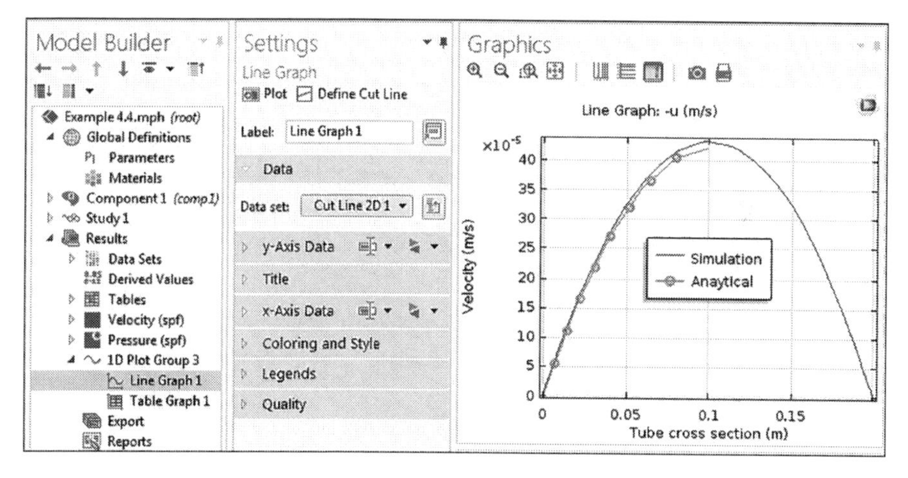

FIGURE 4.27
Velocity profile across the tube width.

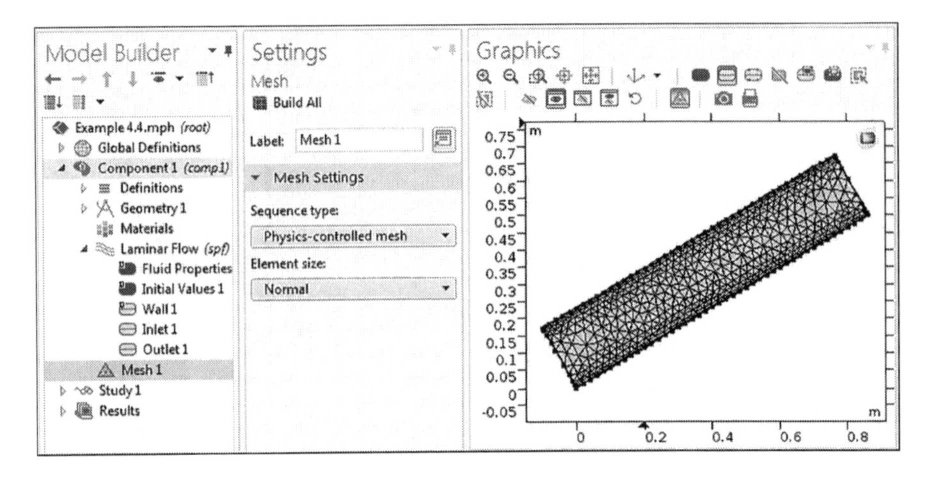

FIGURE 4.28
Model builder and normal mesh size.

the pressure drop in the analytical solution is taken from COMSOL (1.5×10^{-4} Pa).

The Model builder is shown in Figure 4.28.

Example 4.5: Fluid Flowing Between a Stationary and Moving Plates

Consider the steady-state Newtonian incompressible fluid flowing between two parallel and horizontal plates shown in Figure 4.29. The bottom surface is stationary, whereas the top surface is moving horizontally at a constant velocity of 0.01 m/s. Determine the velocity field in the channel (assume fully developed flow).

Solution

The following assumptions are considered in developing the simplified model equations

- Density is constant.
- Steady state.
- Two-dimensional flow of fluid.
- Incompressible fluid.

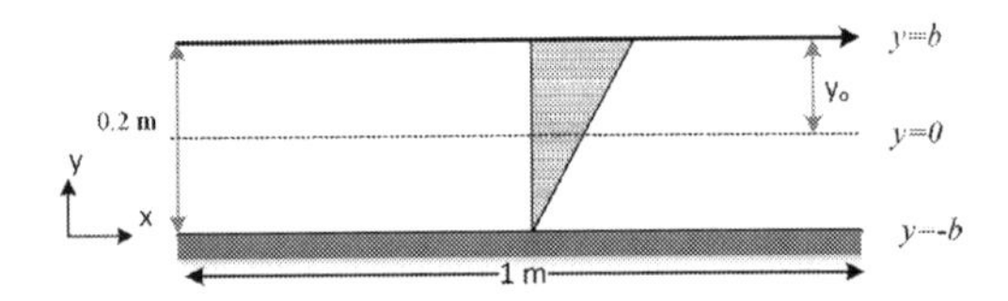

FIGURE 4.29
Schematic of two parallel plates; the top one is moving.

Start with continuity equation in Cartesian coordinates:

$$\frac{\partial v_x}{\partial x}+\frac{\partial v_y}{\partial y}+\frac{\partial v_z}{\partial z}=0$$

The fluid is moving in the x-direction; thus, the velocities in the y- and z-directions are neglected ($v_z = v_y = 0$) and change of velocities in the y- and z-directions equal zero:

The velocities in the y and z directions are neglected $(v_z = v_y = 0)$

$$\frac{\partial v_x}{\partial x}+\frac{\partial v_y}{\partial y}+\frac{\partial v_z}{\partial z}=0$$

The top surface is moving horizontally at constant velocity. Substitute values of $v_y = v_z = 0$ in the continuity equation, and simplify it to the following form: $\partial v_x/\partial x = 0$.

The Navier-Stokes equations are the fundamental PDEs that describe the flow of incompressible fluids. The x-component Navier-Stokes equation is:

$$\rho\frac{\partial v_x}{\partial t}+\rho\left(v_x\frac{\partial v_x}{\partial x}+v_y\frac{\partial v_x}{\partial y}+v_z\frac{\partial v_x}{\partial z}\right)$$

$$=\mu\left(\frac{\partial^2 v_x}{\partial x^2}+\frac{\partial^2 v_x}{\partial y^2}+\frac{\partial^2 v_x}{\partial z^2}\right)-\frac{\partial p}{\partial x}+\rho g_x$$

At steady state ($\partial v_x/\partial t = 0$), flow in the horizontal rectangular channel is $g_x = 0$, also no external pressure was applied, $\partial p/\partial x = 0$, substitute all in the x-component of the Navier Stocks equation. After applying the earlier assumptions, the equation is reduced as follows:

Steady state $(v_z = v_y = 0)$ No external pressure was applied

$$\rho\frac{\partial v_x}{\partial t}+\rho\left(v_x\frac{\partial v_x}{\partial x}+v_y\frac{\partial v_x}{\partial y}+v_z\frac{\partial v_x}{\partial z}\right)=\mu\left(\frac{\partial^2 v_x}{\partial x^2}+\frac{\partial^2 v_x}{\partial y^2}+\frac{\partial^2 v_x}{\partial z^2}\right)-\frac{\partial p}{\partial x}+\rho g_x$$

No gravity in the x-direction

From continuity $\frac{\partial v_x}{\partial x}=0$ From continuity $\frac{\partial v_x}{\partial x}=0$ No change in the velocity v_x in the z-direction (2D flow)

The equation is reduced to:

$$\mu\frac{d^2 v_x}{dy^2}=0$$

Perform double integration. The general velocity profile equation is:

$$v_x = C_1 y + C_2$$

where C_1 and C_2 are the arbitrary integration constants and need to be determined by enforcing the flowing boundary conditions:

B.C.1: at $y = -b, v_x = 0$
B.C. 2: at $y = b, v_x = V$

Substitute the first and second boundary conditions:

$$0 = -C_1 b + C_2$$

$$V = C_1 b + C_2$$

Subtract one from the other:

$$V = C_1(b + b)$$

$C_1 = V/2b$, and $C_2 = C_1 b = V/2$

Substitute C_1 and C_2 in the general velocity profile equation to obtain the following velocity profile for flow between moving and stationary parallel and horizontal flat plates:

$$v_x = \frac{V}{2b} y + \frac{V}{2} = \frac{V}{2}\left(\frac{y}{b} + 1\right)$$

COMSOL Simulation

Simply select new case in COMSOL, and select 2D coordinate and stationary case study. Select laminar flow physics and the suitable boundary conditions as shown in Figure 4.30. The surface velocity profile is shown in Figure 4.31. For the top wall, select sliding wall

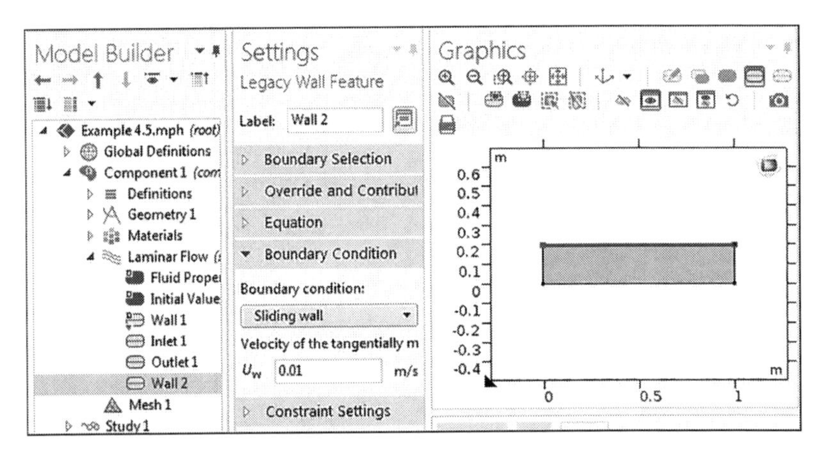

FIGURE 4.30
Model builder and sliding wall boundary.

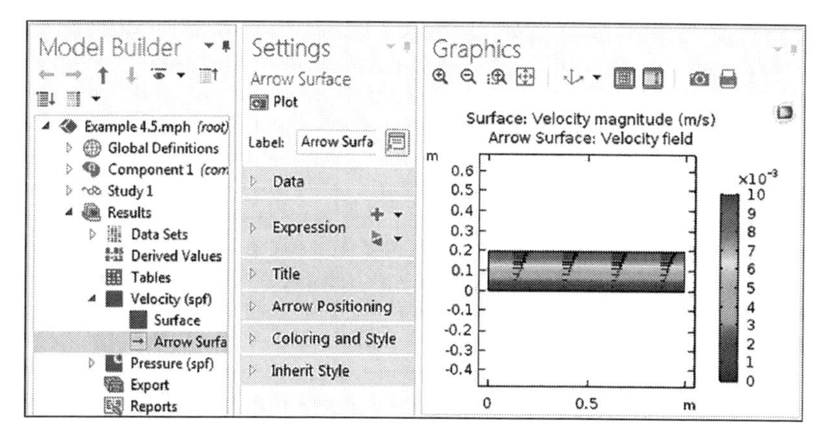

FIGURE 4.31
Surface plot for the velocity magnitude.

and set the velocity of 0.01 m/s. The velocity surface profile and velocity magnitude are shown in Figure 4.31.

Example 4.6: Coloring Process

A coloring process tank is used to paint a metal rod of radius kR inside a tank of radius R. The rod is moving with constant velocity V as shown in Figure 4.32. Find the velocity profile of the fluid in the dyeing tank for the case where the fluid is under the influence of a constant pressure gradient and for the case where there is no pressure gradient.

Solution

The assumptions that were implemented to derive the model equation for this example are:

- Steady state.
- Incompressible fluid.
- Laminar flow.
- Cylindrical coordinates will be used to model this system.

For axial flow in cylindrical coordinates, the differential equation for the continuity equation is expressed by:

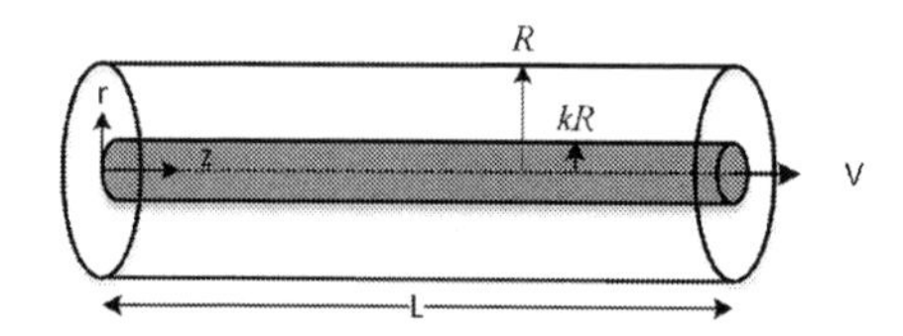

FIGURE 4.32
Schematic diagram of the rod dyeing process.

$$\rho\left(\frac{1}{r}\frac{\partial(rv_r)}{\partial r}+\frac{1}{r}\frac{\partial v_\theta}{\partial \theta}+\frac{\partial v_z}{\partial z}\right)=0$$

Radial velocity is very small and hence can be neglected $(v_r = 0)$

$$\rho\left(\frac{1}{r}\frac{\partial(rv_r)}{\partial r}+\frac{1}{r}\frac{\partial v_\theta}{\partial \theta}+\frac{\partial v_z}{\partial z}\right)=0$$

Angular velocity can be neglected $(v_\theta = 0)$

Radial and angular velocities are very small and hence are cancelled $(v_r = v_\theta = 0)$; as a result, the change of velocity v_z in the z direction is zero. Thus, the continuity equation is:

$$\frac{dv_z}{dz}=0$$

The z-component of the Navier-Stocks equation in cylindrical coordinate after applying the assumptions is as follows:

Velocity in the r-direction equals to zero $(v_r = 0)$ From continuity $\frac{\partial v_z}{\partial z}=0$ From continuity $\frac{\partial v_z}{\partial z}=0$

$$\rho\frac{\partial v_z}{\partial t}+\rho\left(v_r\frac{\partial v_z}{\partial r}+\frac{v_\theta}{r}\frac{\partial v_z}{\partial \theta}+v_z\frac{\partial v_z}{\partial z}\right)=\mu\left(\frac{1}{r}\frac{\partial}{\partial r}\left(r\frac{\partial v_z}{\partial r}\right)+\frac{1}{r^2}\frac{\partial^2 v_z^2}{\partial \theta^2}+\frac{\partial^2 v_z}{\partial z^2}\right)-\frac{\partial p}{\partial z}+\rho g_z$$

Steady state Angular velocity can be neglected $(v_\theta = 0)$ v_z does not change with θ No gravity in the z-direction

After simplifying, we have:

$$\mu\left(\frac{1}{r}\frac{\partial}{\partial r}\left(r\frac{\partial v_z}{\partial r}\right)\right)-\frac{\partial p}{\partial z}=0$$

Specify the pressure gradient in the z direction as a constant: $\Delta p/\Delta z=\varphi$. Substitute G in the simplified partial differential equation:

$$\frac{\partial}{\partial r}\left(r\frac{\partial v_z}{\partial r}\right)=\frac{\varphi r}{\mu}$$

First integration:

$$r\frac{\partial v_z}{\partial r}=\frac{\varphi r^2}{2\mu}+C_1$$

Divide both sides by r:

$$\frac{\partial v_z}{\partial r}=\frac{\varphi r}{2\mu}+\frac{C_1}{r}$$

Second integration:

$$v_z = \frac{\varphi r^2}{4\mu} + C_1 \ln r + C_2$$

Boundary conditions:

$$\text{B.C.1: at } r = kR,\ v_z = V \Rightarrow V = G(kR)^2/4\mu + C_1 \ln(kR) + C_2$$
$$\text{B.C.2: at } r = R,\ v_z = 0 \Rightarrow 0 = G(R)^2/4\mu + C_1 \ln(R) + C_2$$

Subtract the second equation from the first equation:

$$V = \frac{\varphi}{4\mu}(k^2R^2 - R^2) + C_1 \ln\frac{kR}{R}$$

Simplify:

$$V = \frac{\varphi R^2(k^2 - 1)}{4\mu} + C_1 \ln k$$

Rearrange:

$$C_1 = \frac{V - \dfrac{\varphi R^2(k^2-1)}{4\mu}}{\ln k}$$

Substitute C_1 in Equation 4.3 to get C_2.

$$0 = \frac{GR^2}{4\mu} + \frac{V - \dfrac{\varphi R^2(k^2-1)}{4\mu}}{\ln k} \ln R + C_2$$

Rearrange:

$$C_2 = -\frac{GR^2}{4\mu} - \left(\frac{V - \dfrac{\varphi R^2(k^2-1)}{4\mu}}{\ln k} \ln R \right)$$

Substitute C_1 and C_2 in developed velocity profile equation:

$$V_z = \frac{\varphi r^2}{4\mu} + \left[\frac{V - \dfrac{\varphi R^2(k^2-1)}{4\mu}}{\ln k} \ln r \right] - \frac{\varphi R^2}{4\mu} - \left[\frac{V - \dfrac{\varphi R^2(k^2-1)}{4\mu}}{\ln k} \ln R \right]$$

Arrange and simplify:

$$v_z = \frac{\varphi}{4\mu}(r^2 - R^2) + \frac{V - \dfrac{\varphi R^2(k^2-1)}{4\mu}}{\ln k}\ln\frac{r}{R}$$

To find the velocity profile of the same system if the pressure gradient is zero ($\partial p/\partial z = 0$):

$$\mu\left(\frac{1}{r}\frac{\partial}{\partial r}\left(r\frac{\partial v_z}{\partial r}\right)\right) = 0$$

Perform double integration:

$$v = C_1 \ln r + C_2$$

Boundary conditions:

B.C.1: at $r{=}kR,\ v_z = V \Rightarrow V = \ln(kR) + C_2$
B.C.2: at $r{=}R,\ v_z = 0 \Rightarrow 0 = C_1\ln(R) + C_2$

Subtract the second equation from the previous one:

$$V = C_1 \ln\frac{kR}{R}$$

Rearrange for first integration constant, C_1:

$$C_1 = \frac{V}{\ln k}$$

Substitute C_1 in main developed velocity profile equation to get C_2:

$$0 = \frac{V}{\ln k}\ln(R) + C_2$$

Rearrange for C_2:

$$C_2 = -\frac{V}{\ln k}\ln(R)$$

Substitute C_1 and C_2 in the velocity profile equation:

$$v_z = \frac{V}{\ln k}\ln r - \frac{V}{\ln k}\ln(R)$$

After simplifying, the following velocity profile is determined for the dyeing process:

$$v_z = \frac{V}{\ln k}\left(\ln\frac{r}{R}\right)$$

To check if the final equation is correct or not, the boundary conditions will be substituted in the equation. Use the first boundary condition:

B.C.1: at $r = kR$, $v_z = V$

Replace r by kR:

$$v_z = \frac{V}{\ln K}\left(\ln\frac{kR}{R}\right) = \frac{V}{\ln k}\ln k = V$$

Use the second boundary condition:

B.C.2: at $r = R$, $v_z = 0$

Replace r by R:

$$v_z = \frac{V}{\ln k}\left(\ln\frac{R}{R}\right) = \frac{V}{\ln k}\times 0 = 0$$

Both boundary conditions satisfy the equation, so the equation is correct.

Example 4.7: Velocity Profile in Waterfalls

Waterfalls are a popular example of distributed systems. Consider water, with constant density ρ and viscosity μ, flowing from a waterfall cliff. The flow is inclined at an angle of β compared to the vertical (Figure 4.33). The width of the water film in the z direction is w. The flow is laminar. End effects can be neglected considering that this is a very large and highly elevated waterfall. The thickness of the fluid layer is δ. You are

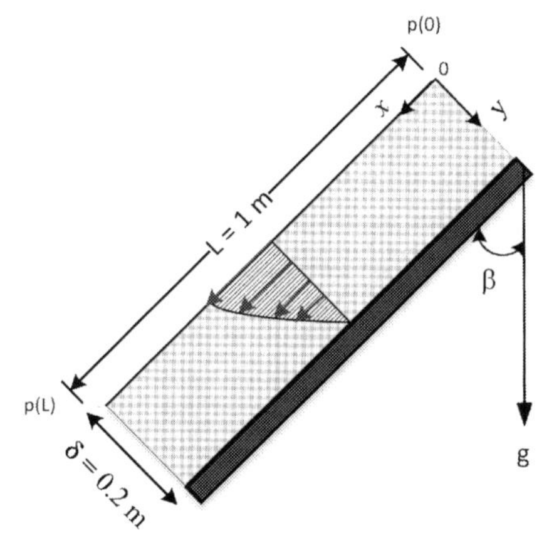

FIGURE 4.33
Flow of Newtonian fluid down an inclined plane.

required as an engineer to determine an expression for the velocity profile of the water flow in a waterfall and then find an expression for the maximum and average velocities.

Solution

There are several assumptions that can be made to simplify the problem in hand. These assumptions include the following:

- Steady state.
- Laminar flow.
- Incompressible fluid (constant density, ρ).
- Negligible pressure gradient.
- Newtonian fluid (constant viscosity, μ).

The laminar flow of a liquid film down an inclined plane can be modeled in the same manner as flow between parallel plates. The flow between parallel plates is symmetric about the centerline, which leads to the velocity gradient being zero along the centerline. For the flow of a liquid film driven by gravity, the shear stress at the free liquid surface is negligible. This means that the velocity gradient is negligible at liquid air interface, $y=0$. The continuity equation for fluid flow in Cartesian coordinates is:

The velocities in the y and z directions are neglected $(v_z = v_y = 0)$

$$\frac{\partial v_x}{\partial x} + \frac{\partial v_y}{\partial y} + \frac{\partial v_z}{\partial z} = 0$$

The velocities in the y and z directions are neglected $(v_y = 0, v_z = 0)$. The momentum equation; the x component of Navier-Stokes equation is:

$$\rho\frac{\partial v_x}{\partial t} + \rho\left(v_x\frac{\partial v_x}{\partial x} + v_y\frac{\partial v_x}{\partial y} + v_z\frac{\partial v_x}{\partial z}\right)$$

$$= \mu\left(\frac{\partial^2 v_x}{\partial x^2} + \frac{\partial^2 v_x}{\partial y^2} + \frac{\partial^2 v_x}{\partial z^2}\right) - \frac{\partial p}{\partial x} + \rho g_x$$

$\partial v_x/\partial t = 0$ (steady state)
$v_y = 0, v_z = 0$ (fluid flows in the x-direction)
$\partial v_x/\partial x = 0$ (from the continuity equation)
$\partial^2 v_x/\partial x^2$ and $\partial^2 v_x/\partial z^2 = 0$ (velocity changes in the y-direction only)

The free surface of the liquid film is exposed to the atmosphere everywhere so that there can be no pressure gradient in the direction of the flow. This means that $p(0) = p(L)$. Therefore, we can write the momentum equation in the simplified form:

Steady state; $(v_z = v_y = 0)$; No external pressure was applied; $\rho g\cos\beta$

$$\rho\frac{\partial v_x}{\partial t} + \rho\left(v_x\frac{\partial v_x}{\partial x} + v_y\frac{\partial v_x}{\partial y} + v_z\frac{\partial v_x}{\partial z}\right) = \mu\left(\frac{\partial^2 v_x}{\partial x^2} + \frac{\partial^2 v_x}{\partial y^2} + \frac{\partial^2 v_x}{\partial z^2}\right) - \frac{\partial P}{\partial x} + \rho g_x$$

From continuity $\frac{\partial v_x}{\partial x} = 0$; From continuity $\frac{\partial v_x}{\partial x} = 0$; No change in the velocity v_x in the z-direction (2D flow)

$$0 = \mu\left(\frac{\partial^2 v_x}{\partial y^2}\right) - 0 + \rho g\cos\beta$$

Rearrange:

$$\mu\left(\frac{\partial^2 v_x}{\partial y^2}\right) = -\rho g\cos\beta$$

Divide by the viscosity:

$$\frac{\partial^2 v_x}{\partial y} = -\frac{\rho g\cos\beta}{\mu}$$

Performing double integration generates two arbitrary constants (C_1, C_2):

$$v_x = -\frac{\rho g\cos\beta}{2\mu}y^2 + C_1 y + C_2$$

To evaluate the two arbitrary constants of integration, we need two boundary conditions. We can use the no-slip conditions at the wall and negligible shear stress at the liquid free surface.

B.C.1: $y = 0,\ dv_x/dy = 0 \Rightarrow C_1 = 0$ (negligible shear stress)
B.C.2: $y = \delta,\ v_x = 0$ (no slip conditions) $\rightarrow C_2 = \rho g\delta\cos\beta/\mu$

Substitute C_2 in the velocity profile equation and rearrange:

$$v_x = -\frac{\rho g\cos\beta}{2\mu}y^2 + \frac{\rho g\delta^2\cos\beta}{2\mu} = \frac{\rho g\delta^2\cos\beta}{2\mu}\left(1 - \left(\frac{y}{\delta}\right)^2\right)$$

The velocity profile in a falling liquid film in laminar flow is represented by one-half of the parabola that is obtained for flow between parallel plates. To find an expression for v_{max}, substitute $y = 0$ (liquid free surface):

$$v_{max} = \frac{\rho g\delta^2\cos\beta}{2\mu}$$

The volumetric flow rate (Q) in the film is:

$$Q = \frac{\rho g \delta^2 \cos\beta}{2\mu} \int_0^\delta \left(1 - \left(\frac{y}{\delta}\right)^2\right) w dy = \frac{\rho g \delta^2 w \cos\beta}{2\mu} \left[\delta - \frac{\delta^3}{3\delta^2}\right]$$

The volumetric flow rate (Q) of liquid film on an inclined plate is:

$$Q = \frac{\rho g \delta^3 w \cos\beta}{3\mu}$$

The equation shows the relationship between the volumetric flow rate Q and the film thickness δ for a given volumetric rate of flow. The film thickness can be determined from the relationship between the volumetric flow rate and the film thickness.

Example 4.8: Fluid Flow in Vertical Falling Film

In a gas absorption experiment, a fluid flows upward through a small circular tube and then fluid flows downward in laminar flow on the outside of the circular tube (Figure 4.34). Determine the velocity distribution in the falling film, and neglect end effects. Assume the following: width, 0.1cm; height, 1cm; inlet fluid velocity, 4.35 cm/s; R = 0.1 cm.

Solution

There are several assumptions that can be made to shorten the problem form:

- Steady state system.
- The flow is only in z-direction.
- The pressure gradient is zero because we have already requested that the end effects be neglected.

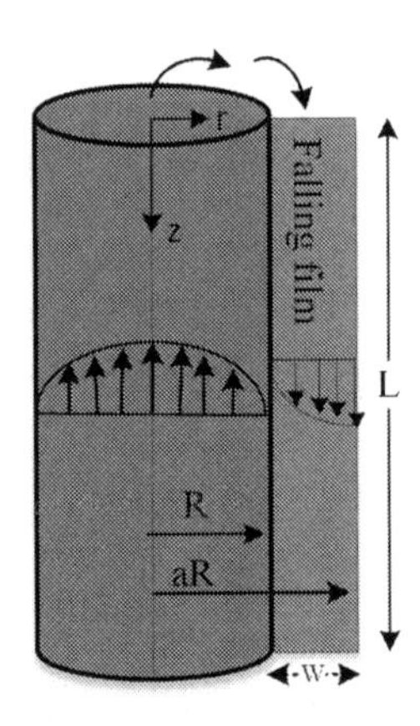

FIGURE 4.34
Gas liquid absorption film.

Cylindrical coordinates will be used to model this system. For flow in cylindrical coordinates, the differential equation for the continuity equation is expressed by:

$$\rho\left(\frac{1}{r}\frac{\partial(rv_r)}{\partial r}+\frac{1}{r}\frac{\partial v_\theta}{\partial\theta}+\frac{\partial v_z}{\partial z}\right)=0$$

Radial velocity is very small and hence can be neglected ($v_r = 0$)

$$\rho\left(\frac{1}{r}\frac{\partial(rv_r)}{\partial r}+\frac{1}{r}\frac{\partial v_\theta}{\partial\theta}+\frac{\partial v_z}{\partial z}\right)=0$$

Angular velocity can be neglected ($v_\theta = 0$)

Radial and angular velocity are very small and hence are neglected ($v_r = v_\theta = 0$); accordingly, the change of the velocity in the z-direction is neglected ($dv_z/dz = 0$); this means the velocity profile is fully developed. The z-component of the momentum equation for fluid flow in the z-direction in a cylindrical coordinate is:

$$\rho\frac{\partial v_z}{\partial t}+\rho\left(v_r\frac{\partial v_z}{\partial r}+\frac{v_\theta}{r}\frac{\partial v_z}{\partial\theta}+v_z\frac{\partial v_z}{\partial z}\right)$$

$$=\mu\left(\frac{1}{r}\frac{\partial}{\partial r}\left(r\frac{\partial v_z}{\partial r}\right)+\frac{1}{r^2}\frac{\partial^2 v_z^2}{\partial\theta^2}+\frac{\partial^2 v_z}{\partial z^2}\right)-\frac{\partial p}{\partial z}+\rho g_z$$

Velocity in the r-direction equals to zero ($v_r = 0$)

From continuity $\frac{\partial v_z}{\partial z} = 0$

From continuity $\frac{\partial v_z}{\partial z} = 0$

$$\rho\frac{\partial v_z}{\partial t}+\rho\left(v_r\frac{\partial v_z}{\partial r}+\frac{v_\theta}{r}\frac{\partial v_z}{\partial\theta}+v_z\frac{\partial v_z}{\partial z}\right)=\mu\left(\frac{1}{r}\frac{\partial}{\partial r}\left(r\frac{\partial v_z}{\partial r}\right)+\frac{1}{r^2}\frac{\partial^2 v_z^2}{\partial\theta^2}+\frac{\partial^2 v_z}{\partial z^2}\right)-\frac{\partial p}{\partial z}+\rho g_z$$

Steady state

Angular velocity can be neglected ($v_\theta = 0$)

v_z does not change with θ

ρg

Neglect the pressure gradient ($\partial p/\partial z$) because the fluid is exposed to atmospheric pressure. The equation is reduced to the following form:

$$0=\mu\left(\frac{1}{r}\frac{\partial}{\partial r}\left(r\frac{\partial v_z}{\partial r}\right)\right)+\rho g$$

Rearange:

$$\frac{\partial}{\partial r}\left(r\frac{\partial vz}{\partial r}\right)=-\frac{\rho g r}{\mu}$$

Perform double integration:

$$v_z(r)=-\frac{\rho g r^2}{4\mu}+C_1\ln r+C_2$$

The arbitrary integration constants are determined from the following boundary conditions:

B.C.1: at $r = aR$, $(\partial v_z)/(\partial r) = 0$ (negligible shear stress), the shear stress between liquid and air is zero.
B.C.2: at $r = R$, $v_z = 0$ (no-slip conditions)

Implement boundary conditions to find the arbitrary integration constants C_1 and C_2:

$$r\frac{dv_z}{dr} = -\frac{\rho g r^2}{2\mu} + C_1$$

From boundary condition 1: $C_1 = \rho g(aR)^2/2\mu$

The second boundary condition will be substituted in the general velocity distribution profile equation to find C_1 as follows:

$$0 = -\frac{\rho g R^2}{4\mu} + \frac{\rho g\ (aR)^2}{2\mu}\ln R + C_2$$

Rearrange:

$$C_2 = \frac{\rho g R^2}{4\mu} - \frac{\rho g (aR)^2}{2\mu}\ln R$$

Substitute C_1 and C_2 to determine the velocity distribution:

$$v_z = -\frac{\rho g r^2}{4\mu} + \frac{\rho g (aR)^2}{2\mu}\ln r + \frac{\rho g R^2}{4\mu} - \frac{\rho g (aR)^2}{2\mu}\ln R$$

Reorganize the equation to obtain its better final velocity profile shape:

$$v_z(r) = \frac{\rho g R^2}{4\mu}\left[1 - \left(\frac{r}{R}\right)^2 + 2a^2 \ln\left(\frac{r}{R}\right)\right]$$

COMSOL Simulation

Repeat the above example using the flowing portion of the falling film (width, 0.1 cm; height, 1 cm) and proceed as follows:

1. Start COMOL, model wizard, select 2D space dimension, add laminar flow physics, add study, stationary.
2. Right-click on geometry and select rectangle (0.1, 1 cm).
3. Right-click on Laminar flow and select "Inlet." Set the velocity to 0.1 m/s (top boundary).
4. Right-click Laminar flow and select "outlet" (bottom boundary).
5. Right-click on Laminar flow and select "wall" in the wall condition. Select "Slip" (right boundary). Specify "wall" for

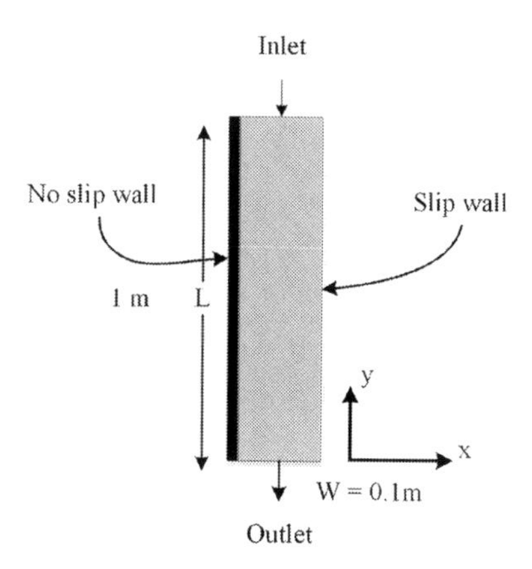

FIGURE 4.35
Schematic of water falling film.

the left boundary; the "wall" is under "No slip" conditions. The wall boundary conditions are shown in Figure 4.35. The water is falling downward under the effect of gravitational force. The right-hand side boundary in COMSOL is shown in Figure 4.36. Figure 4.37 shows the surface plot velocity profile. The diagram depicts the velocity across the film width. The COMSOL "Line figure" makes it clear that the velocity is zero for the fluid close to the wall and increases in the x-direction toward the liquid-air interface where shear stress is almost zero (Figure 4.38).

Comparison between analytical solution and model predictions show that the results are consistent (Figure 4.38).

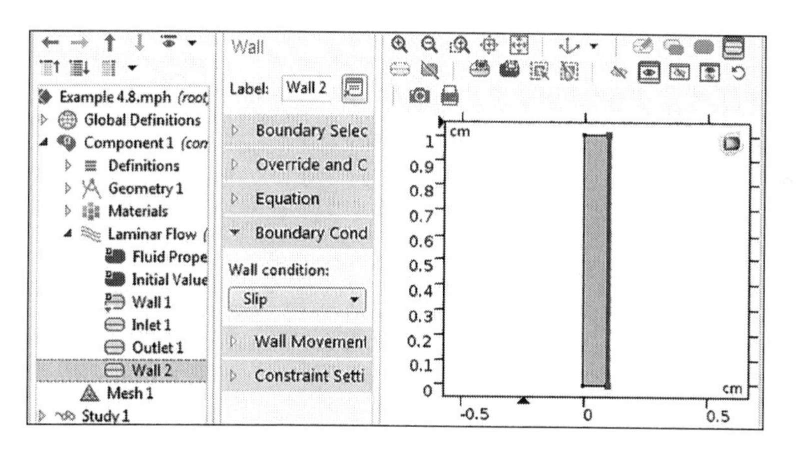

FIGURE 4.36
Right-side boundary slip wall.

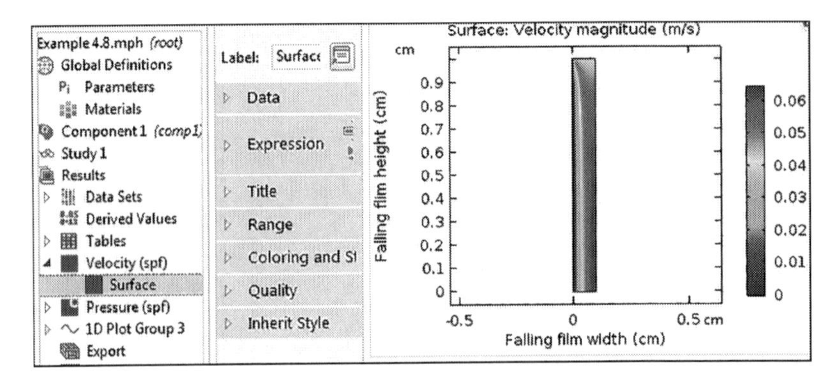

FIGURE 4.37
Surface profile of velocity distribution in a water-falling film.

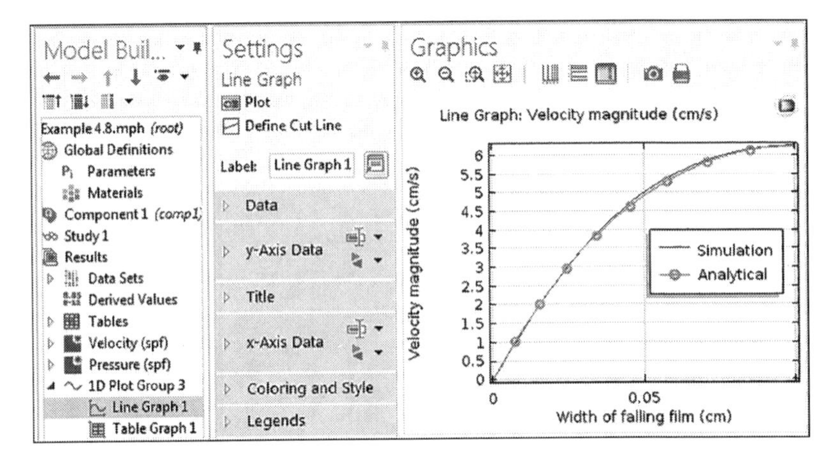

FIGURE 4.38
Velocity profile across the falling film at length = 0.5 cm.

Example 4.9: Cross Flow Heat Exchanger

A cross-flow heat exchanger exchanges thermal energy from one airstream to another in an air handling unit. Crossflow heat exchangers are commonly used in gas heating or cooling. A cross-flow heat exchanger is made of thin metal panels, generally aluminum. A traditional cross-flow heat exchanger has a square cross-section. In this example, two large porous plates are separated by a distance L (Figure 4.39). An incompressible constant-property fluid fills the channel formed by the fixed plates. An axial pressure gradient dp/dx is applied to the fluid to set it in motion. The same fluid is also injected through the lower plate with a normal velocity V extracted from the upper plate at the same velocity. There is no gravity force in the x and y direction. Using continuity equation find velocity profile in x-direction (v_x)

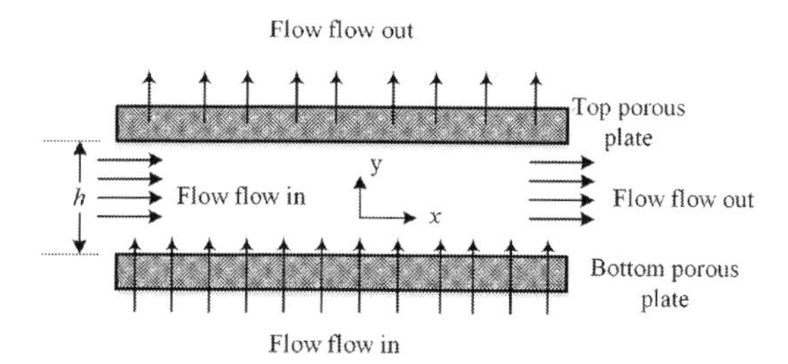

FIGURE 4.39
Schematic of fluid flow between two porous plates.

Solution

The following assumptions are expected in the solution

1. Steady state ($\partial v_x / \partial t = 0$).
2. Incompressible flow (constant density).
3. 2D flow ($v_z = 0, \partial v_z / \partial z = 0$).
4. Fully developed ($\partial v_x / \partial x = 0$).
5. Constant pressure gradient ($\partial p / \partial x =$ constant).
6. No gravity force in the x and y direction ($g_x = g_y = 0$).

The continuity equation in Cartesian coordinates is simplified as follows:

The velocity in z direction is neglected

$$\frac{\partial v_x}{\partial x} + \frac{\partial v_y}{\partial y} + \frac{\partial v_z}{\partial z} = 0$$

Fully developed flow

Apply the assumption of fully developed and 2D flow:

$$\frac{\partial v_y}{\partial y} = 0 \rightarrow v_y = V$$

The x-momentum of the Navier Stocks equation in Cartesian coordinates:

$$\rho \frac{\partial v_x}{\partial t} + \rho \left(v_x \frac{\partial v_x}{\partial x} + v_y \frac{\partial v_x}{\partial y} + v_z \frac{\partial v_x}{\partial z} \right)$$

$$= \mu \left(\frac{\partial^2 v_x}{\partial x^2} + \frac{\partial^2 v_x}{\partial y^2} + \frac{\partial^2 v_x}{\partial z^2} \right) - \frac{\partial p}{\partial x} + \rho g_x$$

Applying preceding assumption, the momentum equation is reduced to:

Steady state | The velocity in z direction is neglected | No gravity in the x direction

$$\rho\frac{\partial v_x}{\partial t} + \rho\left(v_x\frac{\partial v_x}{\partial x} + v_y\frac{\partial v_x}{\partial y} + v_z\frac{\partial v_x}{\partial z}\right) = \mu\left(\frac{\partial^2 v_x}{\partial x^2} + \frac{\partial^2 v_x}{\partial y^2} + \frac{\partial^2 v_x}{\partial z^2}\right) - \frac{\partial P}{\partial x} + \rho g_x$$

Fully developed flow | Fully developed flow | No change in the velocity v_x in the z-direction (2D flow)

To solve the differential equation analytically, assume that the solution is:

$$v_x = C_1 e^{\gamma y} - \left(\frac{\partial p}{\partial x}\right)\frac{1}{\rho V} y + C_2$$

Derivative of the first term in the differential equation:

$$\frac{\partial v_x}{\partial y} = C_1 \gamma e^{\gamma y} - \left(\frac{\partial p}{\partial x}\right)\frac{1}{\rho V}$$

Derivative of the second term in the differential equation:

$$\frac{\partial^2 v_x}{\partial y^2} = C_1 \gamma^2 e^{\gamma y}$$

Replace in the differential equation:

$$\rho V\left[C_1 \gamma e^{\gamma y} - \left(\frac{\partial p}{\partial x}\right)\frac{1}{\rho V}\right] = \mu(C_1\gamma^2 e^{\gamma y}) - \frac{\partial p}{\partial x}$$

$$\gamma = \frac{\rho V}{\mu}$$

Therefore:

$$v_x = C_1 e^{\frac{\rho V}{\mu} y} - \left(\frac{\partial p}{\partial x}\right)\frac{1}{\rho V} y + C_2$$

Boundary conditions

B.C.1: $v_x = 0$ at $y = 0$ and $y = h \Rightarrow 0 = C_1 + C_2 \rightarrow C_2 = -C_1$

B.C.2: at $y = h,\ v_x = 0$: $\quad 0 = C_1 e^{\frac{\rho V}{\mu} h} - \left(\frac{\partial p}{\partial x}\right)\frac{1}{\rho V} h + C_2$

Replace $C_2 = -C_1$ and solve for C_1:

$$0 = C_1 e^{\frac{\rho V}{\mu} h} - \left(\frac{\partial p}{\partial x}\right)\frac{1}{\rho V} - C_1$$

Rearrange to find C_1 and C_2 as follows:

$$C_1 = \frac{-\left(\frac{\partial p}{\partial x}\right)\frac{1}{\rho V}}{1 - e^{\frac{\rho V h}{\mu}}}$$

$$C_2 = \frac{\left(\frac{\partial p}{\partial x}\right)\frac{1}{\rho V}}{1 - e^{\frac{\rho V h}{\mu}}}$$

Substitute C_1 and C_2:

$$v_x = \frac{\left(\frac{\partial p}{\partial x}\right)\frac{1}{\rho V}}{1 - e^{\frac{\rho V h}{\mu}}} e^{\frac{\rho V}{\mu} y} - \left(\frac{\partial p}{\partial x}\right)\frac{1}{\rho V} y - \frac{\left(\frac{\partial p}{\partial x}\right)\frac{1}{\rho V}}{1 - e^{\frac{\rho V h}{\mu}}}$$

Rearrange:

$$v_x = \left(\frac{\partial p}{\partial x}\right)\frac{h}{\rho V}\left(\frac{1 - e^{\frac{\rho V}{\mu} y}}{1 - e^{\frac{\rho V h}{\mu}}} - \frac{y}{h}\right)$$

Example 4.10: Newtonian Fluid Flow in a Plane Narrow Slit

Consider an incompressible fluid (of constant density ρ), moving upward in laminar flow at steady state in a plane with a narrow slit of length *L* and width *W* formed by two flat vertical parallel walls that are a distance 2*B* apart (Figure 4.40). The fluid is flowing upward (against the gravitational force). End effects may be neglected because

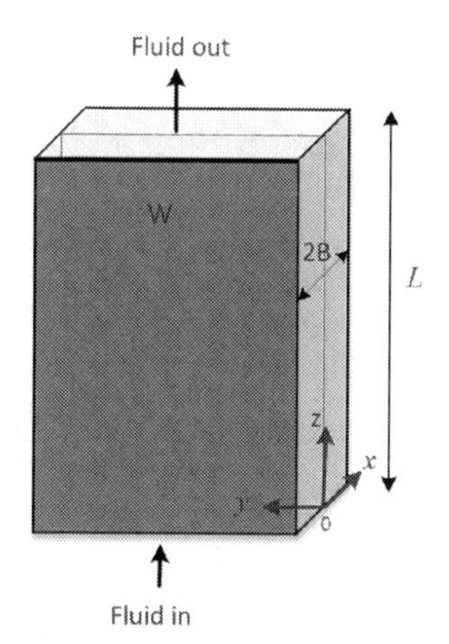

FIGURE 4.40
Fluid flow in two vertical parallel plates.

$B << W << L$. The fluid flows under the influence of both a pressure difference Δp and gravity.

1. Determine the velocity profile for a Newtonian fluid (of constant viscosity μ).
2. Obtain expressions for the maximum velocity.

Solution

Consider the following assumptions:

- Steady state.
- Flow in z-direction.
- Fluid flow is under the influence of pressure difference Δp and gravity.

For a narrow-slit plane, the natural choice is rectangular Cartesian coordinates. Since the fluid flow is in the z-direction, $v_x = 0$, $v_y = 0$, and only v_z exists:

The velocities in x and y direction are neglected

$$\frac{\partial v_x}{\partial x} + \frac{\partial v_y}{\partial y} + \frac{\partial v_z}{\partial z} = 0$$

The fluid velocity in the z-direction is considered. Velocities in the x and y directions are negligible ($v_x = v_y = 0$); consequently, $(\partial v_z / \partial z = 0)$ for a fully developed velocity profile.

Since the fluid flows under the influence of both a pressure difference Δp and gravity, fluid flow is in the opposite direction of gravitational force $(-\rho g)$. The z-momentum balance in a Cartesian coordinate:

$$\rho\frac{\partial v_z}{\partial t} + \rho\left(v_x\frac{\partial v_z}{\partial x} + v_y\frac{\partial v_z}{\partial y} + v_z\frac{\partial v_z}{\partial z}\right)$$

$$= \mu\left(\frac{\partial^2 v_z}{\partial x^2} + \frac{\partial^2 v_z}{\partial y^2} + \frac{\partial^2 v_z}{\partial z^2}\right) - \frac{\partial p}{\partial z} + \rho g_z$$

Simplify after applying the preceding assumptions:

Steady state; The velocities in x and y direction are neglected; From continuity $\frac{\partial v_z}{\partial z} = 0$

$$\rho\frac{\partial v_z}{\partial t} + \rho\left(v_x\frac{\partial v_z}{\partial x} + v_y\frac{\partial v_z}{\partial y} + v_z\frac{\partial v_z}{\partial z}\right) = \mu\left(\frac{\partial^2 v_z}{\partial x^2} + \frac{\partial^2 v_z}{\partial y^2} + \frac{\partial^2 v_z}{\partial z^2}\right) - \frac{\partial p}{\partial z} + \rho g_z$$

From continuity $\frac{\partial v_z}{\partial z} = 0$; No change in the velocity in the y direction; $-\rho g$

Reorganize:

$$\mu\left(\frac{\partial^2 v_z}{\partial x^2}\right) = \frac{\partial p}{\partial z} + \rho g$$

Assume the pressure gradient in the z direction is constant, $\partial P/\partial z = (p_L - p_0)/L$.

Let $P = p + \rho g h$, where h is the distance upward from any chosen reference plane, and h is in the direction opposite to gravity; hence $P_0 = p_0 + \rho g(0) = p_0$, $P_L = p_L + \rho g L$, and:

$$\frac{\partial p}{\partial z} + \rho g = \frac{p_L - p_0}{L} + \rho g = \frac{p_L + \rho g L - p_0}{L}$$

$$= \frac{(p_L + \rho g L) - (p_0 + 0)}{L} = \frac{P_L - P_0}{L} = \frac{\Delta P}{L}$$

$$\frac{\partial^2 v_z}{\partial x^2} = \frac{\Delta P}{\mu L}$$

The result of the double integration of the above equation:

$$v_z = \frac{\Delta P}{2\mu L} x^2 + C_1 x + C_2$$

The arbitrary integration constants C_1 and C_2 are evaluated from the following axial symmetry and no-slip boundary conditions:

B.C.1: at $x = 0$, $\frac{\partial v_z}{\partial x} = 0$ (axial symmetry)

B.C.2: at $x = B$, $v_z = 0$ (no slip conditions)

Substitute B.C.1 in the velocity profile equation to get C_1:

$$\frac{dv_z}{dx} = \frac{\Delta P}{\mu L} x + C_1$$

$$0 = \frac{\Delta P}{\mu L}(0) + C_1 \rightarrow C_1 = 0$$

Substitute the B.C.2 and C_1 in the velocity profile equation and get C_2:

$$v_z = \frac{\Delta P}{2\mu L} x^2 + C_1 x + C_2$$

$$0 = \frac{\Delta P}{2\mu L} B^2 + 0(B) + C_2$$

$$C_2 = -\frac{\Delta P}{2\mu L} B^2$$

Substitute C_1 and C_2 in the velocity profile to obtain the following expression for velocity, v_z:

$$v_z = \frac{\Delta P}{2\mu L}x^2 - \frac{\Delta P}{2\mu L}B^2$$

Rearrange and take B^2 as common factor:

$$v_z = \frac{P_0 - P_L}{2\mu L}B^2\left(1-\left(\frac{x}{B}\right)^2\right)$$

Obtain expressions for the maximum velocity (v_{max}). When $x=0$, the velocity is at its maximum value and the velocity is a function of maximum velocity:

$$v_{max} = \frac{P_0 - P_L}{2\mu L}B^2$$

$$v_z = v_{max}\left(1-\left(\frac{x}{B}\right)^2\right)$$

Example 4.11: Coaxial Cylinder Rotating with Angular Velocities

An incompressible isothermal fluid is in laminar flow between two coaxial cylinders (Figure 4.41). The inner and outer sides have radii of kR and R, respectively. The inner and outer cylinders are rotating at angular velocities v_i and v_o, respectively. The end effects may be neglected. Determine the steady-state velocity distribution in the fluid.

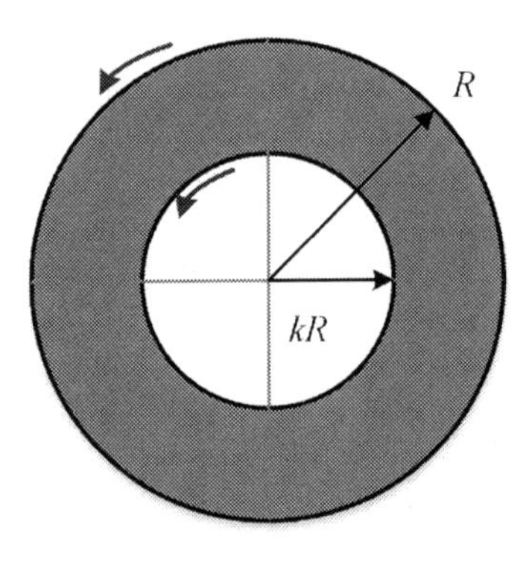

FIGURE 4.41
Coaxial cylinder rotating in the same direction.

Solution

The following assumptions are considered in the solution:

- Incompressible fluid between two coaxial rotating cylinders.
- Isothermal operation.
- Steady state.
- No gravitation force as the cylinders are horizontals.
- No pressure drops.

Employ the balance equations in cylindrical coordinates. Cylindrical coordinates will be used to model this system. For flow in cylindrical coordinate, the differential equation for the continuity equation is simplified and expressed by:

$$\rho\left(\frac{1}{r}\frac{\partial(rv_r)}{\partial r}+\frac{1}{r}\frac{\partial v_\theta}{\partial \theta}+\frac{\partial v_z}{\partial z}\right)=0$$

$(v_r = 0)$ $(v_z = 0)$

Since the cylinders are rotating in the angular direction:

$$v_r=0,\ v_z=0\rightarrow\frac{\partial v_\theta}{\partial\theta}=0$$

The following is the equation of motion for the θ-component of total momentum:

$$\rho\frac{\partial v_\theta}{\partial t}+\rho\left(v_r\frac{\partial v_\theta}{\partial r}+\frac{v_\theta}{r}\frac{\partial v_\theta}{\partial\theta}+\frac{v_r v_\theta}{r}+v_z\frac{\partial v_\theta}{\partial z}\right)$$

$$=\mu\left(\frac{\partial}{\partial r}\left(\frac{1}{r}\frac{\partial r v_\theta}{\partial r}\right)+\frac{1}{r^2}\frac{\partial^2 v_\theta}{\partial\theta^2}+\frac{2}{r^2}\frac{\partial v_r}{\partial\theta}+\frac{\partial^2 v_\theta}{\partial z^2}\right)-\frac{\partial p}{\partial\theta}+\rho g_\theta$$

After applying the simplifications:

Steady state; $(v_r = 0)$; From continuity; $(v_r = 0)$; $(v_z = 0)$

$$\rho\frac{\partial v_\theta}{\partial t}+\rho\left(v_r\frac{\partial v_\theta}{\partial r}+\frac{v_\theta}{r}\frac{\partial v_\theta}{\partial\theta}+\frac{v_r v_\theta}{r}+v_z\frac{\partial v_\theta}{\partial z}\right)$$

No external pressure was applied

$$=\mu\left(\frac{\partial}{\partial r}\left(\frac{1}{r}\frac{\partial r v_\theta}{\partial r}\right)+\frac{1}{r^2}\frac{\partial^2 v_\theta}{\partial\theta^2}+\frac{2}{r^2}\frac{\partial v_r}{\partial\theta}+\frac{\partial^2 v_\theta}{\partial z^2}\right)-\frac{\partial p}{\partial\theta}+\rho g_\theta$$

From continuity; $(v_r = 0)$; v_z does not change with θ; No gravity in the angular direction

To solve the simplified equation, double integration is required. The first integration:

$$\frac{\partial(rv_\theta)}{\partial r} = C_1 r$$

Second integration:

$$rv_\theta = \frac{1}{2}C_1 r^2 + C_2$$

Divide each term by r and rearrange. The integration constant is to be determined through appropriate boundary conditions:

$$v_\theta = \frac{C_1 r}{2} + \frac{C_2}{r}$$

The angular speed is equal to radius multiplied by the angular velocity:

B.C.1: at $r = kR$, $v_\theta = v_i kR$
B.C.2: at $r = R$, $v_\theta = v_o R$

Accordingly, apply the first boundary condition:

$$v_o R = \frac{C_1 R}{2} + \frac{C_2}{R}$$

Substitute the second boundary condition:

$$v_i kR = \frac{C_1 kR}{2} + \frac{C_2}{kR}$$

Multiply the equation resulting from the substitution of BC1 ($v_o R = C_1 R/2 + C_2/R$) by k:

$$v_o kR = \frac{C_1 kR}{2} + \frac{C_2 k}{R}$$

Subtract one from the other:

$$kR(v_i - v_o) = \frac{C_1 kR}{2} - \frac{C_1 kR}{2} + \frac{C_2}{kR} - \frac{C_2 k}{R}$$

Rearrange:

$$kR(v_i - v_o) = C_2\left(\frac{1}{kR} - \frac{k}{R}\right)$$

The integration constant result from the second integration, C_2:

$$C_2 = \frac{kR(v_i - v_o)}{\left(\frac{1}{kR} - \frac{k}{R}\right)}$$

Divide the terms in the above equation by k:

$$\frac{v_o R}{k} = \frac{C_1 R}{2k} + \frac{C_2}{kR}$$

Subtract the two equations:

$$v_i kR - \frac{v_o R}{k} = \frac{C_1 kR}{2} - \frac{C_1 R}{2k} + \frac{C_2}{kR} - \frac{C_2}{kR}$$

Rearrange:

$$v_i kR - \frac{v_o R}{k} = \frac{C_1}{2}\left(kR - \frac{R}{k}\right)$$

The integration constant resulting from the substitution of boundary condition 1:

$$C_1 = \frac{2\left(v_i kR - \frac{v_o R}{k}\right)}{\left(kR - \frac{R}{k}\right)}$$

After substituting the integration constants, the angular velocity profile is:

$$v_\theta = \frac{\left(v_i kR - \frac{v_o R}{k}\right)}{\left(kR - \frac{R}{k}\right)} r + \frac{1}{r}\frac{kR(v_i - v_o)}{\left(\frac{1}{kR} - \frac{k}{R}\right)}$$

Divide each term in the equation by kR:

$$v_\theta = \frac{\left(v_i - \frac{v_o}{k^2}\right)}{\left(1 - \frac{1}{k^2}\right)} r + \frac{1}{r}\frac{(v_i - v_o)}{\left(\frac{1}{k^2 R^2} - \frac{1}{R^2}\right)}$$

Reorganize:

$$v_\theta = \frac{\left(v_i - \frac{v_o}{k^2}\right)}{\left(1 - \frac{1}{k^2}\right)} r + \frac{(v_i - v_o)}{\left(\frac{1}{k^2} - 1\right)}\frac{R^2}{r}$$

Multiply by k^2:

$$v_\theta = \frac{(k^2 v_i - v_o)}{(k^2 - 1)} r + \frac{(v_i - v_o)}{(1 - k^2)}\frac{k^2 R^2}{r}$$

Or it can be of the following form (multiply and divide the first term by −1):

$$v_\theta = \frac{(v_o - k^2 v_i)}{(1-k^2)} r + \frac{(v_i - v_o)}{(1-k^2)} \frac{k^2 R^2}{r}$$

Take as a common factor: $kR/(1-k^2)$.

Finally, the fluid velocity is described by the following equation:

$$v_\theta = \frac{kR}{(1-k^2)} \left\{ (v_o - k^2 v_i) \frac{r}{kR} + (v_i - v_o) \frac{kR}{r} \right\}$$

Example 4.12: Liquid Coated Film

Liquid coated film is used to provide surface enhancements for printability and heat-seal coatings. In this example, a thin film of thickness δ connects a liquid coating film that is being pulled up from a processing bath by rollers with a steady-state velocity V_1 at an angle θ to the horizontal (Figure 4.42). As the film leaves the bath, it entrains some liquid; the velocity of liquid in contact with the film is V_0. There is no net flow of liquid because as much is being pulled up by the film as is falling back by gravity. Find the velocity distribution of the liquid coating film.

Solution

Note the following assumptions:

- Steady state.
- The flow is in the x-direction only.
- Incompressible fluid.

The continuity equation in Cartesian coordinates

The velocities in z and y direction are neglected

$$\frac{\partial v_x}{\partial x} + \frac{\partial v_y}{\partial y} + \frac{\partial v_z}{\partial z} = 0$$

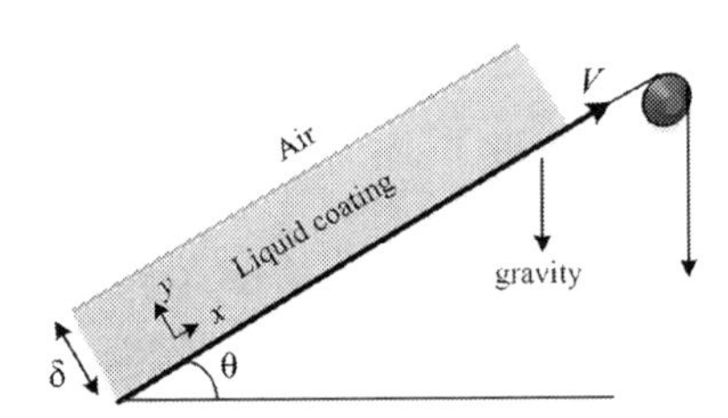

FIGURE 4.42
Inclined liquid coating film.

Since the fluid moving with the sliding slip is in the x-direction, $v_y = 0, v_z = 0$, and only v_x exists; thus, $dv_x/dx = 0$.

The x-component of velocity in a steady, incompressible flow in Cartesian coordinates is:

$$\rho\frac{\partial v_x}{\partial t} + \rho\left(v_x\frac{\partial v_x}{\partial x} + v_y\frac{\partial v_x}{\partial y} + v_z\frac{\partial v_x}{\partial z}\right)$$

$$= \mu\left(\frac{\partial^2 v_x}{\partial x^2} + \frac{\partial^2 v_x}{\partial y^2} + \frac{\partial^2 v_x}{\partial z^2}\right) - \frac{\partial p}{\partial x} + \rho g_z$$

Assume steady state, and neglect velocity in the x and y direction. The equation is simplified to:

Steady state

The velocity in z direction is neglected

$$\rho\cancel{\frac{\partial v_x}{\partial t}} + \rho\left(v_x\cancel{\frac{\partial v_x}{\partial x}} + v_y\cancel{\frac{\partial v_x}{\partial y}} + v_z\cancel{\frac{\partial v_x}{\partial z}}\right) = \mu\left(\cancel{\frac{\partial^2 v_x}{\partial x^2}} + \frac{\partial^2 v_x}{\partial y^2} + \cancel{\frac{\partial^2 v_x}{\partial z^2}}\right) - \frac{\partial P}{\partial x} + \rho g_x$$

From continuity

$v_y = 0$

From continuity

No change in the velocity v_x in the z-direction (2D flow)

$$0 = \mu\left(\frac{d^2 v_x}{dy^2}\right) - \frac{dp}{dx} + \rho g_x$$

The gravity is in the opposite direction of the fluid movement; accordingly:

$$g_x = -g\sin\theta$$

Replace g_x in the generated velocity profile equation:

$$0 = \mu\left(\frac{d^2 v_x}{dy^2}\right) - \frac{dp}{dx} + \rho(-g\sin\theta)$$

and adjust it to get:

$$\frac{d^2 v_x}{dy^2} = \frac{1}{\mu}\left(\frac{dp}{dx} + \rho(g\sin\theta)\right)$$

Symbolize:

$$\frac{dp}{dx} + \rho(g\sin\theta) = \varphi$$

Substitute φ:

$$\frac{d^2 v_x}{dy^2} = \frac{\varphi}{\mu}$$

The first integration:

$$\frac{dv_x}{dy} = \frac{\varphi}{\mu} y + C_1$$

The second integration:

$$v_x = \frac{\varphi}{\mu} \frac{y^2}{2} + C_1 y + C_2$$

Boundary conditions

B.C.1. $y = \delta, \dfrac{dv_x}{dy} = 0$ (shear stress at a free liquid surface)

Substituting boundary condition 1:

$$0 = \frac{\varphi}{\mu} \delta + C_1$$

Rearrange to obtain the following:

$$C_1 = -\frac{\varphi}{\mu} \delta$$

B.C.2 at $y = 0, v_x = V_1$

Before replacing, take the second integration of:

$$\frac{dv_x}{dy} = \frac{\varphi}{\mu} y + C_1$$

Substitute C_1:

$$v_x = \frac{\varphi}{\mu} \frac{y^2}{2} - \frac{\varphi}{\mu} \delta y + C_2$$

Use the second B.C.2 to find C_2:

$$V_1 = \frac{\varphi}{\mu} \left(\frac{(0)^2}{2} - \delta(0) \right) + C_2$$

Accordingly:

$$V_1 = C_2$$

Substitute C_1 and C_2 and rearrange to get:

$$v_x = \frac{\varphi}{\mu} \left(\frac{y^2}{2} - \delta y \right) + V_1$$

Simplify by taking y common factor:

$$V_x = y \frac{\varphi}{\mu} \left(\frac{y}{2} - \delta \right) + V_1$$

COMSOL Simulation

Let's solve the above example using COMSOL. The sliding wall is very slow; hence, make the velocity of the sliding wall 0.001 m/s, the width of the film $\delta = 0.1$ m, and the length of the film 1 m (Figure 4.43).

The boundary condition of the moving wall is shown in Figure 4.44.

Fluid is under constant atmospheric pressure and the only effects are due to gravitational force and the movement of the film. The inlet and exit boundary conditions are set for the inlet and exit pressure as $p = 0$ (gauge pressure). The results are shown in Figure 4.45.

The line velocity profile across the film is shown in Figure 4.46.

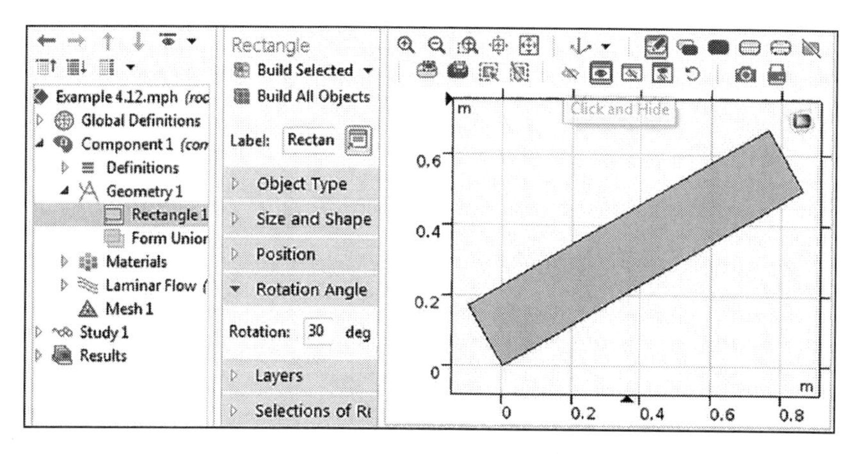

FIGURE 4.43
Dimensions of rectangular geometry; rotation is 30°.

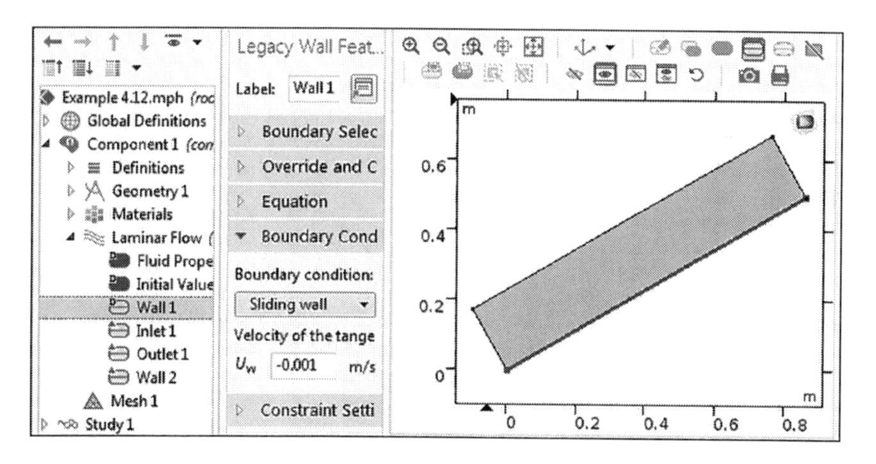

FIGURE 4.44
Boundary condition of moving wall.

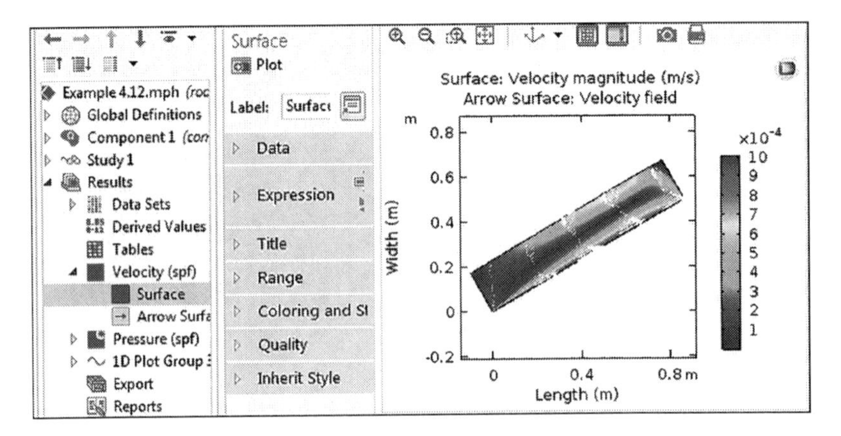

FIGURE 4.45
COMSOL simulation results.

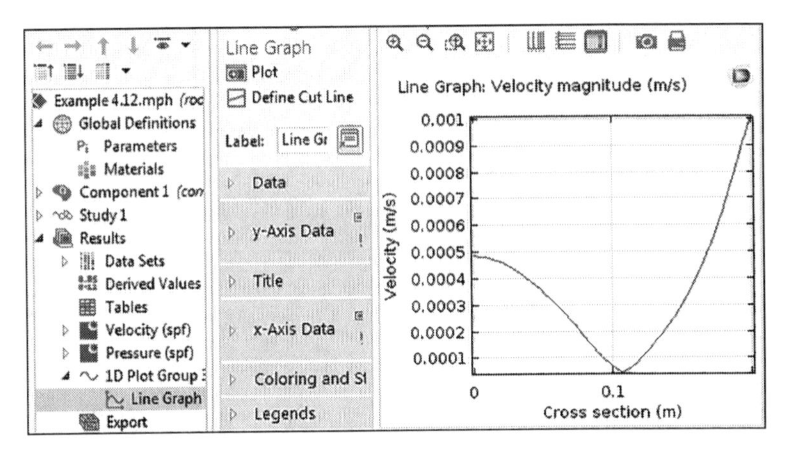

FIGURE 4.46
Velocity profile across the film at distance mid length of the coated film.

Example 4.13: Polymer Extruder

Study a polymer extrusion nuzzle that forms a tube of polymer. The dye of length L is shown in Figure 4.47. A pressure difference $p_1 - p_2$ causes a liquid of viscosity μ and density ρ to flow steadily from left to right in the annular area between two fixed concentric cylinders. Note that p_1 is selected to be the inlet pressure and p_2 to correspond to the extruder exit pressure. The inner cylinder is solid, whereas the outer one is hollow; their radii are r_1 and r_2, respectively. The problem, which could occur in extrusion of plastic r_2 tubes, is to find the velocity profile in the annular space.

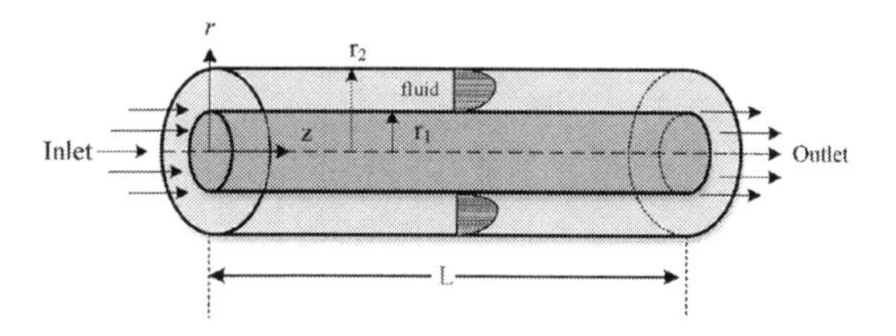

FIGURE 4.47
Schematic of polymer extrusion nuzzle.

Solution

Apply the following assumptions:

- The velocity flow is only in the z-direction, so $v_r = v_\theta = 0$.
- Gravity acts vertically downward so that $g_z = 0$.
- The axial velocity is independent to the angular location, so $(dv_z/d\theta = 0)$.

The cylindrical coordinates will be used to model this system. For flow in cylindrical coordinates, the differential equation for the continuity equation is simplified and expressed by:

Radial velocity is very small and hence can be neglected $(v_r = 0)$

$$\rho\left(\frac{1}{r}\frac{\partial(rv_r)}{\partial r} + \frac{1}{r}\frac{\partial v_\theta}{\partial \theta} + \frac{\partial v_z}{\partial z}\right) = 0$$

Angular velocity can be neglected $(v_\theta = 0)$

Radial and angular velocities are very small and hence are eliminated. Thus, the change of the velocity in the z-direction is zero $(\partial v_z/\partial z = 0)$, and the change of v_z in the z-direction is zero (fully developed velocity profile). The momentum equation for fluid flow in the z-direction in a cylindrical coordinate is:

$$\rho\frac{\partial v_z}{\partial t} + \rho\left(v_r\frac{\partial v_z}{\partial r} + \frac{v_\theta}{r}\frac{\partial v_z}{\partial \theta} + v_z\frac{\partial v_z}{\partial z}\right)$$

$$= \mu\left(\frac{1}{r}\frac{\partial}{\partial r}\left(r\frac{\partial v_z}{\partial r}\right) + \frac{1}{r^2}\frac{\partial^2 v_z^2}{\partial \theta^2} + \frac{\partial^2 v_z}{\partial z^2}\right) - \frac{\partial p}{\partial z} + \rho g_z$$

Applying the preceding assumptions, reduce the momentum equation to the following simplified form:

Velocity in the r-direction equals to zero $(v_r = 0)$ — From continuity $\frac{\partial v_z}{\partial z} = 0$ — From continuity $\frac{\partial v_z}{\partial z} = 0$

$$\rho\frac{\partial v_z}{\partial t} + \rho\left(v_r\frac{\partial v_z}{\partial r} + \frac{v_\theta}{r}\frac{\partial v_z}{\partial \theta} + v_z\frac{\partial v_z}{\partial z}\right) = \mu\left(\frac{1}{r}\frac{\partial}{\partial r}\left(r\frac{\partial v_z}{\partial r}\right) + \frac{1}{r^2}\frac{\partial^2 v_z^2}{\partial \theta^2} + \frac{\partial^2 v_z}{\partial z^2}\right) - \frac{\partial p}{\partial z} + \rho g_z$$

Angular velocity can be neglected $(v_\theta = 0)$ — v_z does not change with θ — No gravity in the z-direction

$$\mu\left(\frac{1}{r}\frac{d}{dr}\left(r\frac{dv_z}{dr}\right)\right) = \frac{dp}{dz}$$

Multiply both sides of the equation by r/μ:

$$\frac{d}{dr}\left(r\frac{dv_z}{dr}\right) = \frac{dp}{dz}\frac{1}{\mu}r$$

Integrate both sides:

$$r\frac{dv_z}{dr} = \frac{1}{2\mu}\frac{dp}{dz}r^2 + C_1$$

Divide both sides by r:

$$\frac{dv_z}{dr} = \frac{1}{2\mu}\frac{dp}{dz}r + \frac{C_1}{r}$$

Integrate both sides for the second time:

$$v_z = \frac{1}{4\mu}\frac{dp}{dz}r^2 + C_1 \ln r + C_2$$

The arbitrary integration constants (C_1, C_2) are obtained using the following boundary conditions:

B.C.1: at $r = r_1$, $v_z = 0$ (no-slip condition)
B.C.2: at $r = r_2$, $v_z = 0$ (no-slip condition)

Note that in fluid dynamics, the no-slip condition for viscous fluids assumes that, at a solid boundary, the fluid will have zero velocity relative to the boundary. The fluid velocity at all fluid–solid boundaries is equal to that of the solid boundary. Substitute the first boundary condition:

$$0 = \frac{1}{4\mu}\frac{dp}{dz}r_1^2 + C_1 \ln r_1 + C_2$$

Substitute B.C.2:

$$0 = \frac{1}{4\mu}\frac{dp}{dz}r_2^2 + C_1 \ln r_2 + C_2$$

Subtract equations to obtain C_1 and C_2:

$$0 = \frac{1}{4\mu}\frac{dp}{dz}(r_2^2 - r_1^2) + C_1\left(\ln\frac{r_2}{r_1}\right)$$

$$C_1 = -\frac{1}{4\mu}\frac{(r_2^2 - r_1^2)}{\ln(r_2/r_1)}$$

$$C_2 = -\frac{1}{4\mu}\frac{dp}{dz}r_2^2 + \frac{k}{4\mu}\frac{(r_2^2 - r_1^2)}{\ln(r_2/r_1)}\ln(r_2)$$

Substitute C_1 and C_2 in the velocity profile equation and rearrange to produce the following equation:

$$v_z = -\frac{1}{4\mu}\frac{\ln(r/r_1)}{\ln(r_2/r_1)}\frac{dp}{dz}\left((r_2^2 - r_1^2) - (r^2 - r_1^2)\right)$$

Note that the maximum velocity occurs slightly before the halfway point in the progressing from the inner cylinder to the outer cylinder. Prove that the maximum velocity is at $r = r_{\max}$ where:

$$r_{\max} = \sqrt{\frac{1}{2}\frac{(r_2^2 - r_1^2)}{\ln(r_2/r_1)}}$$

COMSOL Solution

Assume the length of the extruder is 1 m and the diameter is 0.1 m. These can be simulated in Cartesian coordinates (Figure 4.48).

The surface velocity distribution profile is shown in Figure 4.49.

The line velocity distribution across the diameter of the width is shown in Figure 4.50.

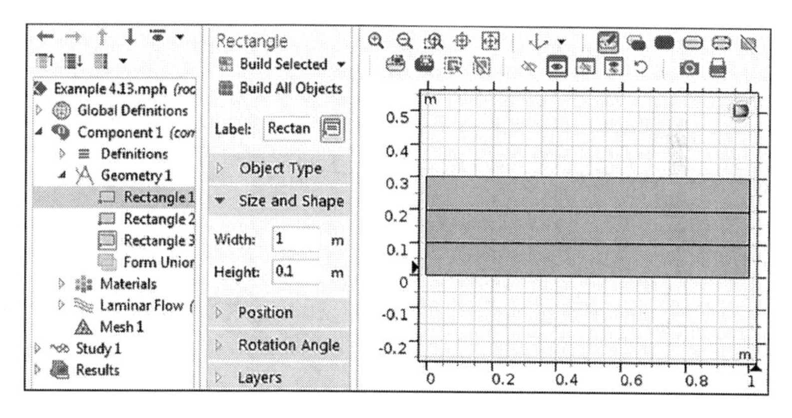

FIGURE 4.48
COMSOL geometrical section.

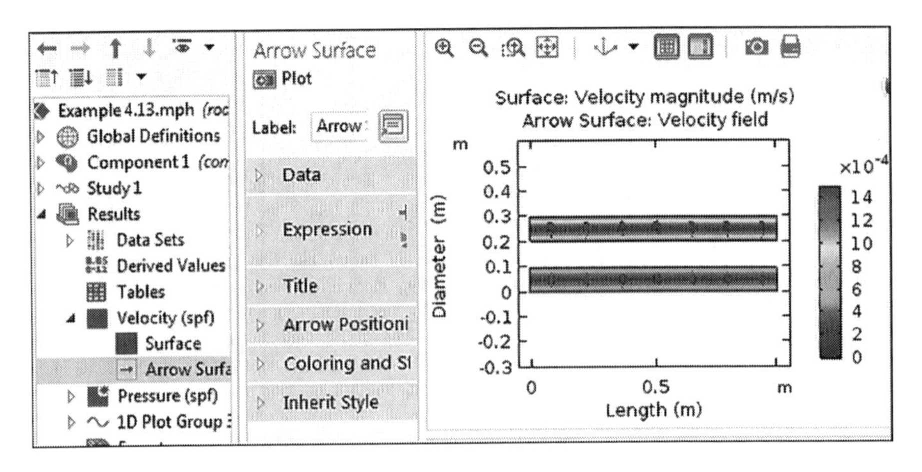

FIGURE 4.49
Physics and boundary conditions used in the simulation.

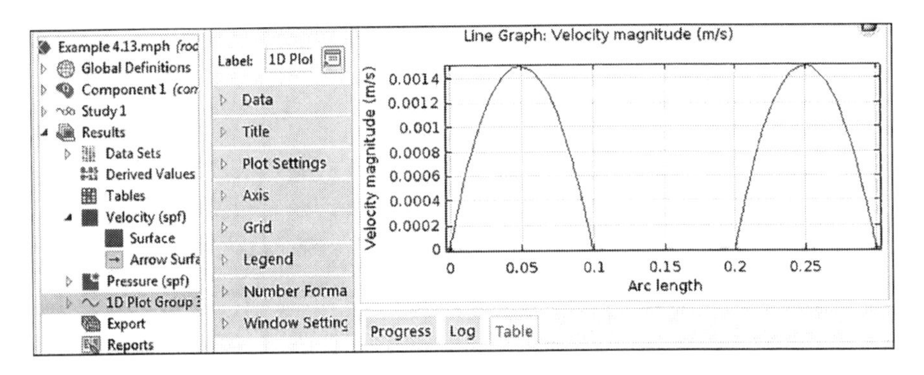

FIGURE 4.50
Velocity distribution across the diameter of the polymer flowing section.

PROBLEMS

4.1 Laminar Flow Between Two Parallel Plates

Imagine laminar flow between two parallel plates. Starting with the equation of change, obtain the velocity profile which describes the incompressible flow between two parallel vertical plates. The distance between the two plates is L. The right plate is at rest, while the left is moving upward at a constant velocity, V. Prove that the velocity profile is:

$$v_y = \frac{1}{2\mu}\left[\rho g + \frac{dp}{dy}\right](x^2 - Lx) + V\left(1 - \frac{x}{L}\right)$$

4.2 Laminar Flow Through a Circular Pipe

An incompressible viscous fluid undergoes steady laminar flow through a circular pipe of radius R. Assume one-dimensional flow and a fully developed velocity profile. Prove that the velocity at any point in the pipe to the maximum velocity is:

$$v_z = v_{z,\max}\left[1-\left(\frac{r}{R}\right)^2\right], \quad v_{z,\max} = -\frac{R^2}{4\mu}\frac{dp}{dz}$$

4.3 Flow Between Two Vertical Concentric Cylinders

Fluid at constant properties positioned between two vertical concentric cylinders. The outer cylinder is rotating with an angular velocity, w, and the inner cylinder is at rest. The inner radius of the concentric cylinders is r_1 and the outer radius is r_2. Show that the velocity distribution profile for the given problem is:

$$v_\theta = wr_2\frac{\left(\frac{r_1}{r}-\frac{r}{r_1}\right)}{\left(\frac{r_1}{r_2}-\frac{r_2}{r_1}\right)}$$

4.4 Laminar Flow Outside a Vertical Flat Sheet

Fluid film flows at constant density and viscosity downward in laminar flow outside a vertical flat sheet. The thickness of the film is δ. Determine the velocity distribution in the falling film (neglect end effects).

4.5 Fluid Flows Down an Inclined Infinite Plane

Consider a liquid that flows steadily down an inclined infinite plane in a uniform film of thickness h. The plane is inclined at an angle θ to the horizontal, and the free surface of the liquid is in contact with the atmosphere. The flow is fully developed. Determine the velocity distribution of the liquid in the film. The pressure distribution across the thickness of the liquid film is the hydrostatic pressure distribution (this implies that $\partial p/\partial x = 0$).

$$v_x = \frac{\rho g \sin\theta}{\mu}\left(hy - \frac{y^2}{2}\right)$$

4.6 Fluid Flow Between Two Concentric Tubes

Consider the case of two concentric tubes for which the radius of the inside wall of the outer tube is R and the radius of the outer wall of the inner tube is kR, where $k < 1$ and length L. A viscous fluid with viscosity p flows in a laminar fashion in the annulus between the two tubes. (a) Show that, for fully developed flow in the annulus, the velocity profile is given by:

$$v_z = \frac{\Delta p}{4\mu L} R^2 \left[1 - \left(\frac{r}{R} \right)^2 - \frac{1-k^2}{\ln(1/k)} \ln \left(\frac{R}{r} \right) \right]$$

4.7 Fluid Flowing Between Two Flat and Horizontal Plates

Find the velocity profile of liquid water flowing between two flat and horizontal plates. The vertical distance between the two plates is 0.01 m and the length is 0.5 m. The pressure difference between the inlet and the outlet is 1 Pa. The fluid is driven by pressure gradient. Neglect the effect of gravitational force. Use the physical properties of water that is built into COMSOL.

4.8 Flow of Air in a Vertical Pipe

Simulate the laminar flow of air in a vertical pipe of length 6.5 cm and radius 0.625 cm. The air density is $1.2\ \text{kg/m}^3$, and the air viscosity is 1.82×10^{-5} Pa.s. The fluid inlet average velocity is 1 m/s. Use axial symmetry for simulation purpose.

4.9 Laminar Flow of a Film of Water

For the steady laminar flow of a film of water thickness, $H = 0.01$ m down an inclined plane of angle $\beta = 30°$ with respect to vertical. The length of the is $L = 1$ m; the width of the film is $W = 0.1$ m. The fluid velocity varies with position in the film. Show that the velocity, as a function of position in rectangular coordinates, is:

$$v_z = \frac{H^2 \rho g (\cos\beta)}{2\mu} \left(1 - \frac{x^2}{H^2} \right)$$

4.10 Flow of Incompressible Fluid Between Two Parallel Plates

The steady flow of incompressible fluid occurs between two parallel plates of width W (z-direction) and length L (x-direction). The distance between the plats is H (y-direction). The top plate moves in the x-direction at speed V; the flow is also assisted by a driving pressure in the x-direction, which is slightly higher at the entrance P_o than at the exit P_L. Neglect the effect of gravity. Neglect the velocity variation in the z-direction. Assume wide flow away from edges. Prove that the velocity profile is:

$$v_x(y) = \frac{(P_L - P_o)}{2\mu L} \left((y^2 - yH) + \frac{V}{H} y \right)$$

References

1. Geankoplis, C. J., 2003. *Transport Processes and Separation Process Principles (includes unit operations)*, 4th ed., New York: Prentice Hall.
2. Deen, W. M., 2012. *Analysis of Transport Phenomena*, 2nd ed., New York: Oxford University Press.
3. Bird, R. B., W. E. Stewart, E. N. Lightfoot, 2007. *Transport Phenomena*, 2nd ed., Hoboken, NJ: John Wiley & Sons.
4. Slattery, J. C., 1999. *Advanced Transport Phenomena*, New York: Cambridge University Press.
5. Plawsky, J. L., 2014. *Transport Phenomena Fundamentals*, 3rd ed., Boca Raton, FL: CRC Press.
6. Cengel, Y., J. Cimbala, 2013. *Fluid Mechanics Fundamentals and Applications*, 3rd ed., New York: McGraw-Hill Science.

5

Mass Transport of Distributed Systems

Mass transfer defines the transport of mass from one point to another, and it is one of the main topics of transport phenomena. The theory of mass transfer allows for the computation of mass flux in a system and the distribution of the mass of distinct species over time and space in such a system. The most precise method to express these conservation laws and constitutive relations is to use differential equations to describe a system. Solving the equations that describe transport phenomena and interpreting the results is an effective technique to know the systems being considered. This chapter covers modeling of various distributed parameter systems that involve mass transfer. The developed models were solved using COMSOL Multiphysics 5.3a.

LEARNING OBJECTIVES

- Develop mathematical models for distributed parameters systems (mass transport).
- Describe the system by partial differential equations.
- Simplify and solve the system equations manually and by COMSOL software.

5.1 Introduction

Mass transfer is the transport of a substance in liquid and gaseous media. When the medium is at rest, the mass transfer driving force is the difference of concentrations in adjacent regions. In such cases, the mechanism is molecular diffusion. This mass transfer is described by Fick's law. The diffusion coefficient is the proportionality factor D in Fick's law. The molar flux due to diffusion is proportional to the concentration gradient. In case of movement of the entire system, there is convective flux term. The sum of all mass fluxes, including the convective term, results in the continuity equation for the mixture. An example is the diffusion of gas in membrane tubes, which is of high importance in gas separation techniques. Diffusivity or diffusion coefficient is defined as the proportionality constant between the molar flux due to molecular diffusion and the gradient in the concentration of the species, and it is the driving force for diffusion. Diffusivity can be found in Fick's law and numerous other equations of physical chemistry. If the diffusivity of one substance is higher with respect to another, they diffuse into each other at a faster rate. Usually a compound's diffusion coefficient is approximately 10,000 times greater in gas (air) than that in liquid (water). For example, the diffusion coefficient of carbon dioxide in air is $1.6\times10^{-5}\,m^2/s$, and in water the carbon dioxide diffusion coefficient is $1.6\times10^{-9}\,m^2/s$ [1]. The diffusion of gas molecules in porous media involves molecular interactions between gas molecules as well as collisions between gas molecules and the porous media. The structure of the solid and the interaction with the solutes are important for the rate of diffusion. The diffusion of solutes through the porous solids plays an important role in many processes. The conservation laws and constitutive relations use differential equations to describe mass transfer systems. The differential equations of mass transfer are general equations describing mass transfer in all directions and in all conditions. The differential equation for mass transfer is obtained by applying the law of conservation of mass (mass balance) to a differential control volume representing the system; the resultant equation is called the continuity equation and takes two forms [2–4]:

- Total continuity equation: the law of conservation of mass on the total mass of the system [in – out = accumulation].
- Component continuity equation: the law of conservation of mass to an individual component [in – out + generation – consumption = accumulation].

The following sections describe various mass transfer systems. Mathematical models were developed for those systems and solved manually when possible and numerically using COMSOL Multiphysics 5.3a.

5.2 Diffusion of Gas through a Membrane Tube

Membrane technology covers the transport of substances between two sectors with the help of permeable membranes. In general, mechanical separation processes for separating gaseous or liquid streams use membrane technology. Membrane separation processes operate without heating and consequently use less energy than conventional thermal separation processes such as distillation, sublimation, or crystallization. The cross section of a hollow fiber membrane tube is shown in Figure 5.1. Suppose a gas mixture is flowing in the lumen side of a membrane tube. Obtain an expression for the concentration profile at which gas will penetrate out in terms of gas diffusivity and the membrane tube.

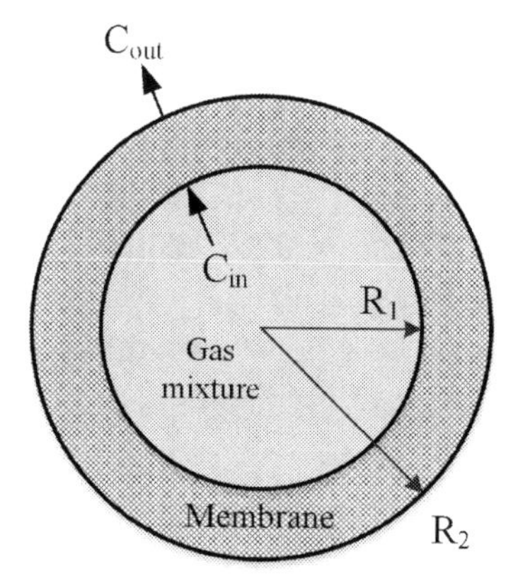

FIGURE 5.1
Diffusion of gas through membrane section.

Consider the following assumptions:

- Steady state.
- Transportation of gas takes place by diffusion only.
- Diffusion in z and θ directions is negligible.
- Diffusivity of gas in membrane pores D_{AB} is a constant.
- No reaction takes place in the system.

The component mole balance equation for gas mass transport in membrane contactor is described by the differential component mole Equation 5.1 in cylindrical coordinates:

$$\frac{\partial C_A}{\partial t}+\left(v_r\frac{\partial C_A}{\partial r}+v_\theta\frac{1}{r}\frac{\partial C_A}{\partial \theta}+v_z\frac{\partial C_A}{\partial z}\right)=D_A\left(\frac{1}{r}\frac{\partial}{\partial r}\left(r\frac{\partial C_A}{\partial r}\right)+\frac{1}{r^2}\frac{\partial^2 C_A}{\partial \theta^2}+\frac{\partial^2 C_A}{\partial z^2}\right)+R_A \quad (5.1)$$

The preceding assumptions state that there is no convection term $(v_r=v_\theta=v_z=0)$, you can neglect diffusion in the z and θ directions $\left(D_A(1/r^2)(\partial^2 C_A/\partial\theta^2)=0,\ D_A(\partial^2 C_A/\partial z^2)=0\right)$, and no reaction is taking place in the membrane ($R_A=0$); hence, the gas component mole balance equation is simplified to the following form after apply the assumptions:

$$\frac{\partial C_A}{\partial t} + \left(v_r \frac{\partial C_A}{\partial r} + v_\theta \frac{1}{r}\frac{\partial C_A}{\partial \theta} + v_z \frac{\partial C_A}{\partial z}\right) = D_A\left(\frac{1}{r}\frac{\partial}{\partial r}\left(r\frac{\partial C_A}{\partial r}\right) + \frac{1}{r^2}\frac{\partial^2 C_A}{\partial \theta^2} + \frac{\partial^2 C_A}{\partial z^2}\right) + R_A$$

Steady state; $(v_r = 0)$; $(v_\theta = 0)$; $(v_z = 0)$; Neglect diffusion in θ and z; No reaction

The continuity component mole balance is reduced to the following form:

$$\frac{1}{r}\frac{d}{dr}\left(r\left(-D_{AB}\frac{dC_A}{dr}\right)\right) = 0 \tag{5.2}$$

Equation 5.2 is solved via double integration:

$$C_A = \frac{-C_1}{D_{AB}}\ln r + C_2 \tag{5.3}$$

where C_1 and C_2 are the arbitrary constants of integration. The integration constants are determined by applying the following boundary conditions (B.C.):

B.C.1: at $r = R_1$, $C_A = C_{\text{in}}$

B.C.2: at $r = R_2$, $C_A = C_{\text{out}}$

Substitution the two boundary conditions:

$$C_{\text{in}} = \frac{-C_1}{D_{AB}}\ln R_1 + C_2 \tag{5.4}$$

$$C_{\text{out}} = \frac{-C_1}{D_{AB}}\ln R_2 + C_2 \tag{5.5}$$

Subtract Equation 5.4 from Equation 5.5:

$$C_{\text{out}} - C_{\text{in}} = \frac{-C_1}{D_{AB}}\ln R_2 - \frac{-C_1}{D_{AB}}\ln R_1 = \frac{C_1}{D_{AB}}\ln R_1/R_2 \tag{5.6}$$

Solve for C_1:

$$C_1 = \left(C_{\text{out}} - C_{\text{in}}\right)\frac{D_{AB}}{\ln R_1/R_2} \tag{5.7}$$

Substitute C_1 in Equation 5.4 to obtain C_2:

$$C_2 = C_{in} + \frac{\left(C_{out} - C_{in}\right)\frac{D_{AB}}{\ln R_1/R_2}}{D_{AB}} \ln R_1 \tag{5.8}$$

Substitute C_1 and C_2 in the general concentration profile Equation 5.3:

$$C_A = \frac{-\left(C_{out} - C_{in}\right)\frac{D_{AB}}{\ln R_1/R_2}}{D_{AB}} \ln r + C_{in} + \frac{\left(C_{out} - C_{in}\right)\frac{D_{AB}}{\ln R_1/R}}{D_{AB}} \ln R_1 \tag{5.9}$$

Rearrange:

$$C_A = C_{in} + \frac{\left(C_{out} - C_{in}\right)\frac{D_{AB}}{\ln R_1/R_2}}{D_{AB}} \ln R_1/r \tag{5.10}$$

The equation can be arranged as follows:

$$\frac{C_A - C_{in}}{\left(C_{out} - C_{in}\right)} = \frac{\ln R_1/r}{\ln R_1/R_2} \tag{5.11}$$

5.3 Mass Transfer with Chemical Reaction

Mass transfer with chemical reaction requires simultaneous consideration of molecular diffusion, fluid mechanics, and chemical reaction kinetics. Chemical processes involving diffusion usually contain chemical reactions. Often diffusion and reaction occur in the same district, and the two rate phenomena are coupled so closely that they must be treated simultaneously. Mass transport with chemical reaction systems is described by mathematical models that correspond to several physical phenomena; the most common are the change in space and time of the concentration of one or more chemical substances, the local chemical reactions in which the substances are transformed into each other, and diffusion that causes the substances to spread out over a surface in space. Mathematically, reaction–diffusion systems take the form of partial differential equations. For example, consider the diffusion of component A in a slab of solid catalyst coupled with the following chemical reaction: $A \rightarrow B$. The reaction takes place in the slab of catalyst shown in Figure 5.2. The concentration inside the slab varies with both the position z and time t. The rate of reaction is $r = kC_A$. The system is assumed at steady state, which

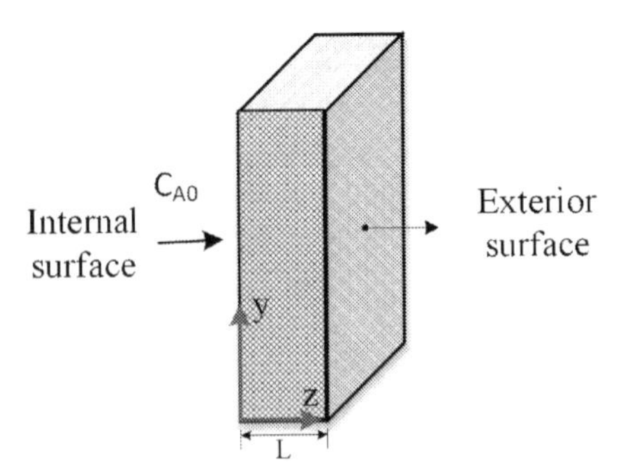

FIGURE 5.2
Diffusion with chemical reaction inside a slab catalyst.

means that there is no accumulation term. Determine the concentration profile of species A, considering the following assumptions:

- Steady state.
- Transportation of gas takes place by diffusion only.
- Diffusion in x and y directions are negligible.
- Diffusivity of gas in the membrane tube is constant.

The component mole balance equation of component A diffuses through the catalyst slab and is described by the differential equation of mass transfer in Cartesian coordinates:

$$\frac{\partial C_A}{\partial t}=-\left(v_x\frac{\partial C_A}{\partial x}+v_y\frac{\partial C_A}{\partial y}+v_z\frac{\partial C_A}{\partial z}\right)+D_{AB}\left(\frac{\partial^2 C_A}{\partial x^2}+\frac{\partial^2 C_A}{\partial y^2}+\frac{\partial^2 C_A}{\partial z^2}\right)+R_A \tag{5.12}$$

Equation 5.12 is the differential molar balance on species A. It states that the rate of accumulation of moles of A at a point in space (left-hand side) equals the rate at which moles of A are convicted into that point plus the rate at which moles of A diffuse and are produced at that point by chemical reactions. D_A represents the diffusion coefficient of (A) inside the catalyst particle. Starting with the equation of change for mass transfer in Cartesian coordinates, assume steady state, no bulk motion, and diffusion only in the z-direction.

Steady state $\quad (v_x = 0) \quad (v_y = 0) \quad (v_z = 0) \quad$ Neglect diffusion in x and y

$$\frac{\partial C_A}{\partial t}=-\left(v_x\frac{\partial C_A}{\partial x}+v_y\frac{\partial C_A}{\partial y}+v_z\frac{\partial C_A}{\partial z}\right)+D_{AB}\left(\frac{\partial^2 C_A}{\partial x^2}+\frac{\partial^2 C_A}{\partial y^2}+\frac{\partial^2 C_A}{\partial z^2}\right)+R_A$$

Equation 5.12 is simplified to Equation 5.13:

$$D_A \frac{d^2C_A}{dz^2} + R_A = 0 \tag{5.13}$$

The reaction rate (R_A):

$$R_A = -kC_A \tag{5.14}$$

The following boundary conditions could be used to solve this second-order ODE:

B.C.1: at $z = 0,\ C_A = C_{Ao}$

B.C.2: at $z = L,\ \frac{dC_A}{dz} = 0$

The bulk flow concentration C_{Ao} at the interior surface of the slab is enforced by the first boundary condition, while the second boundary condition signifies that the concentration is finite at the exterior surface of the catalyst slab. The equation can be rearranged to the following form:

$$\frac{d^2C_A}{dz^2} - \alpha^2 C_A = 0 \tag{5.15}$$

where $\alpha^2 = k/D_A$.

The analytical solution of the equation is:

$$C_A = C_1 e^{\alpha z} - C_2 e^{-\alpha z} \tag{5.16}$$

The arbitrary constants of integration C_1 and C_2 can be obtained from the boundary conditions:

From B.C.1:

$$C_{Ao} = C_1 - C_2 \tag{5.17}$$

From B.C.2:

$$0 = C_1 \alpha e^{\alpha L} + C_2 \alpha e^{-\alpha L} \tag{5.18}$$

Rearrange:

$$C_2 = -C_1 e^{2\alpha L} \tag{5.19}$$

Substitute C_2 in Equation 5.17:

$$C_1 = C_{Ao} + C_2 = C_{Ao} - C_1 e^{2\alpha L}, \qquad C_1 = \frac{C_{Ao}}{1 + e^{2\alpha L}} \tag{5.20}$$

$$C_2 = -\frac{C_{Ao}}{1+e^{2\alpha L}} e^{2\alpha L}$$

Substitute the arbitrary constants of integration (C_1 and C_2) in Equation 5.16:

$$C_A = \frac{C_{Ao}}{1+e^{2\alpha L}} e^{\alpha z} + \frac{C_{Ao}\, e^{2\alpha L}}{1+e^{2\alpha L}} e^{-\alpha z} \tag{5.21}$$

Rearrange by taking $C_{Ao}/1+e^{2\alpha L}$ as a common factor:

$$C_A = \frac{C_{Ao}}{1+e^{2\alpha L}} \left\{ e^{\alpha z} + e^{2\alpha L} e^{-\alpha z} \right\} \tag{5.22}$$

5.4 Plug Flow Reactor

The plug flow reactor (PFR) model is used to describe chemical reactions in continuous flowing systems of cylindrical geometry. The chemical reaction in plug flow reactor proceeds as the chemicals travel through the PFR. In this type of reactor, the changing reaction rate creates a gradient with respect to distance crossed; at the inlet to the PFR, the rate is very high, but as the concentrations of the reagents decrease and the concentration of the products increase, the reaction rate slows. The ideal PFR model assumes no axial mixing: any element of fluid traveling through the reactor does not mix with fluid upstream or downstream from it, as implied by the term *plug flow*. Consider the case of a first-order reaction taking place in an isothermal tubular reactor. Assume plug flow conditions, and develop a model for the reaction process in the plug flow tubular reactor shown in Figure 5.3.

The plug flow conditions mean that the density, concentration, and velocity change with the axial direction only $\left(v_r = v_\theta = 0, \partial C_A/\partial r = \partial C_A/\partial \theta = 0\right)$. Two mechanisms are used to ensure that mass transfer is taking place: convection and diffusion. The component mole balance of species A considering cylindrical coordinates is:

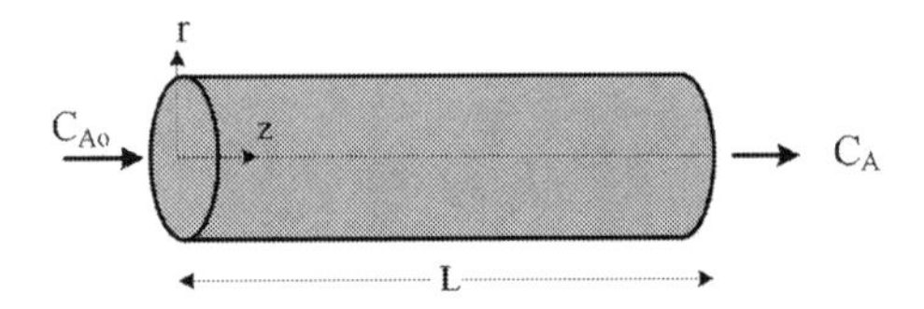

FIGURE 5.3
Isothermal plug flow reactor.

$$\frac{\partial C_A}{\partial t} + \left(v_r \frac{\partial C_A}{\partial r} + v_\theta \frac{1}{r}\frac{\partial C_A}{\partial \theta} + v_z \frac{\partial C_A}{\partial z} \right) = D_{AB}\left(\frac{1}{r}\frac{\partial}{\partial r}\left(r \frac{\partial C_A}{\partial r} \right) + \frac{1}{r^2}\frac{\partial^2 C_A}{\partial \theta^2} + \frac{\partial^2 C_A}{\partial z^2} \right) + R_A \tag{5.23}$$

where:

D_{AB} is the binary diffusion coefficient

R_A is the reaction molar rate per unit volume

The general transient component balance equation in the cylindrical coordinates can be simplified after applying the plug flow assumption to the following form:

$(v_r = 0)$ $(v_\theta = 0)$ Neglect diffusion in r and θ

$$\frac{\partial C_A}{\partial t} + \left(v_r \frac{\partial C_A}{\partial r} + v_\theta \frac{1}{r}\frac{\partial C_A}{\partial \theta} + v_z \frac{\partial C_A}{\partial z} \right) = D_{AB}\left(\frac{1}{r}\frac{\partial}{\partial r}\left(r \frac{\partial C_A}{\partial r} \right) + \frac{1}{r^2}\frac{\partial^2 C_A}{\partial \theta^2} + \frac{\partial^2 C_A}{\partial z^2} \right) + R_A$$

Equation 5.23 is simplified to Equation 5.24:

$$\frac{\partial C_A}{\partial t} + \left(v_z \frac{\partial C_A}{\partial z} \right) = D_{AB}\left(\frac{\partial^2 C_A}{\partial z^2} \right) + R_A \tag{5.24}$$

where:

$$R_A = -kC_A \tag{5.25}$$

Substitute the reaction rate:

$$\frac{\partial C_A}{\partial t} + \left(v_z \frac{\partial C_A}{\partial z} \right) = D_{AB}\left(\frac{\partial^2 C_A}{\partial z^2} \right) - kC_A \tag{5.26}$$

In this partial differential equation (PDE), the state variable (C_A) depends on both t and z.

The PDE is reduced at steady state to the following second order ODE:

$$0 = -\left(v_z \frac{dC_A}{dz} \right) + D_{AB}\left(\frac{d^2 C_A}{dz^2} \right) - kC_A \tag{5.27}$$

with the following boundary conditions (B.C.):

B.C.1: at $z = 0$, $C_A = C_{A0}$ (inlet concentration of A)

B.C.2: at $z = L$, $dC_A/dz = 0$ (convective flux at the exit of the reactor)

The first boundary condition presents the concentration at the entrance of the reactor, while the second boundary condition expresses that there is no flux at the exit length of the reactor. The further simplified design equation of plug flow reactor can be obtained by assuming that the convective term $\left(v_z\,(dC_A/dz)\right)$ is much larger than the diffusion term $\left(D_{AB}\,(d^2C_A/dz^2)\right)$. Accordingly, the diffusion term can be eliminated from Equation 5.27, and hence second order ODE is reduced to the following simple first order equation:

$$v_z \frac{dC_A}{dz} = -kC_A \tag{5.28}$$

The velocity $v_z\,(\text{m/s})$ is defined as the volumetric flow rate $\text{V}\,(\text{m}^3/\text{s})$ divided by the cross-section area $A(\text{m}^2)$ of the cylindrical plug flow reactor (i.e., $v_z = \text{V}/A$). The molar flow rate used in reactor design course is $F_A\,(\text{mol/s})$ equals to $C_A \times \text{V}$, and the differential volume of the PFR, $dV = Adz$, accordingly, after substation of v_z and dV in Equation 5.28, the equation is reduced to the design equation of plug flow reactor studied in the reactor design course as:

$$\frac{dF_A}{dV} = R_A \tag{5.29}$$

5.5 Diffusion of Gas in Solid

Diffusion is the ability of one substance to move into another substance. Diffusion in a solid is very slow. The diffusion of gas molecules in porous media involves molecular interactions between gas molecules as well as collisions between gas molecules and the porous media. As gas molecules travel through the porous media, one of three mechanisms can occur, depending on the characteristic of the diffusing gas species and the intrinsic microstructure of the porous media. The three mechanisms are molecular diffusion, viscous diffusion, and Knudsen diffusion. The following case demonstrated the diffusion of gas in solid substrate.

Consider the case of pure oxygen (O_2) gas as it flows through a tube and diffuses at the tube-substrate interface into a substrate. The concentration of the dissolved oxygen at the tube-substrate interface is $C_A^* = P_A/H$, where P_A is the partial pressure of oxygen in the tube, and H is the Henry's constant. The dissolved oxygen is consumed by the material per a zero-order reaction with $R_A = -m$, where m is the metabolic O_2 consumption rate for the matter. In this case, molecular diffusion controls the mass transfer process, and the

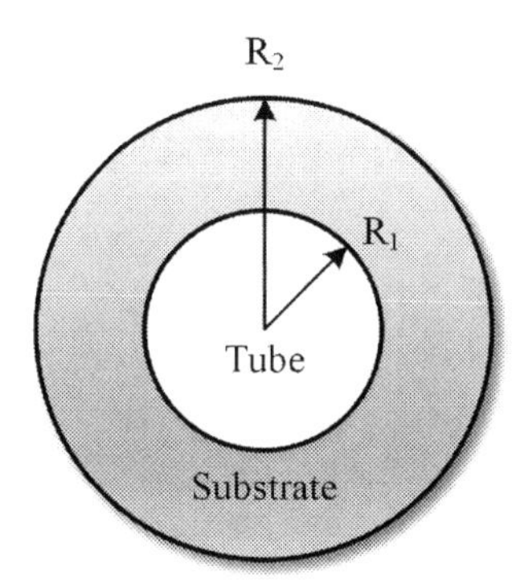

FIGURE 5.4
Dissolved oxygen at the tube-substrate interface.

metabolic consumption of oxygen within the skin serves as the sink for oxygen mass transfer (Figure 5.4). Develop a mathematical model to predict the steady-state concentration profile of the dissolved oxygen in the substrate material, $C_A(r)$.

5.5.1 Assumptions

Consider the following assumptions:

- Steady state.
- Flux of oxygen in the r direction only.
- Isothermal operation.
- Molecular diffusion controls the process (no bulk flow).
- Consumption of dissolved O_2 is a zero-order process, $R_A = -m$.

The continuity equation of the total mass balance:

$$\rho\left(\frac{1}{r}\frac{\partial(rv_r)}{\partial r}+\frac{1}{r}\frac{\partial v_\theta}{\partial \theta}+\frac{\partial v_z}{\partial z}\right)=0 \tag{5.30}$$

Component mole balance equation in cylindrical coordinates:

$$\frac{\partial C_A}{\partial t}+\left(v_r\frac{\partial C_A}{\partial r}+v_\theta\frac{1}{r}\frac{\partial C_A}{\partial \theta}+v_z\frac{\partial C_A}{\partial z}\right) = D_{AB}\left(\frac{1}{r}\frac{\partial}{\partial r}\left(r\frac{\partial C_A}{\partial r}\right)+\frac{1}{r^2}\frac{\partial^2 C_A}{\partial \theta^2}+\frac{\partial^2 C_A}{\partial z^2}\right)+R_A \tag{5.31}$$

Since there is no bulk flow, $v_r = v_z = v_\theta = 0$, then the component mole balance equation after applying the preceding assumptions is reduced to Equation 5.32:

$$\underbrace{\frac{\partial C_A}{\partial t}}_{\text{Steady state}} + \left(\underbrace{v_r \frac{\partial C_A}{\partial r}}_{(v_r = 0)} + \underbrace{v_\theta \frac{1}{r}\frac{\partial C_A}{\partial \theta}}_{(v_\theta = 0)} + \underbrace{v_z \frac{\partial C_A}{\partial z}}_{(v_z = 0)}\right) = D_{AB}\left(\frac{1}{r}\frac{\partial}{\partial r}\left(r\frac{\partial C_A}{\partial r}\right) + \underbrace{\frac{1}{r^2}\frac{\partial^2 C_A}{\partial \theta^2} + \frac{\partial^2 C_A}{\partial z^2}}_{\text{Neglect diffusion in } \theta \text{ and } z}\right) + \underbrace{R_A}_{-m}$$

Equation 5.31 is reduced to Equation 5.32:

$$0 = D_{AB}\left(\frac{1}{r}\frac{\partial}{\partial r}\left(r\frac{\partial C_A}{\partial r}\right)\right) - m \tag{5.32}$$

Reorganize and divide by D_{AB}:

$$\frac{\partial}{\partial r}\left(r\frac{\partial C_A}{\partial r}\right) = \frac{m}{D_{AB}} r \tag{5.33}$$

Integration of Equation 5.33 generates the first arbitrary integration constant C_1:

$$r\frac{\partial C_A}{\partial r} = \frac{m}{2D_{AB}} r^2 + C_1 \tag{5.34}$$

Divide both sides of Equation 5.34 by r to get Equation 5.35:

$$\frac{\partial C_A}{\partial r} = \frac{m}{2D_{AB}} r + \frac{C_1}{r} \tag{5.35}$$

Performing the second integration creates the second arbitrary integration constant C_2:

$$C_A = \frac{m}{4D_{AB}} r^2 + C_1 \ln r + C_2 \tag{5.36}$$

The following boundary conditions are used to get the expressions of C_1 and C_2:

B.C.1: at $r = R_1$, $C_A = C_A^* = \dfrac{P_A}{H}$

B.C.2: at $r = R_2$, $\dfrac{\partial C_A}{\partial r} = 0$

Substitute B.C.2 in Equation 5.35:

$$\frac{\partial C_A}{\partial r} = \frac{m}{2D_{AB}} r + \frac{C_1}{r} \tag{5.37}$$

$$0 = \frac{m}{2D_{AB}} R_2 + \frac{C_1}{R_2} \tag{5.38}$$

Rearrange to obtain an expression of C_1:

$$C_1 = \frac{-m}{2D_{AB}} R_2^2 \tag{5.39}$$

Substitute C_1 in Equation 5.36:

$$C_A = \frac{m}{4D_{AB}} r^2 + \left(\frac{-m}{2D_{AB}}\right) R_2^2 \ln r + C_2 \tag{5.40}$$

Substitute boundary conditions 1 in Equation 5.40:

$$C_A^* = \frac{m}{4D_{AB}} R_1^2 - \frac{m}{2D_{AB}} R_2^2 \ln R_1 + C_2 \tag{5.41}$$

The arbitrary integration constant C_2 is found in Equation 5.42:

$$C_2 = C_A^* - \frac{m}{2D_{AB}}\left(\frac{R_1^2}{2} - R_2^2 \ln R_1\right) \tag{5.42}$$

Finally, after replacing the obtained integration constants into Equation 5.36, the concentration profile is described by Equation 5.43:

$$C_A(r) = C_A^* + \frac{m}{4D_{AB}}\left(r^2 - R_1^2\right) - \frac{mR_2^2}{2D_{AB}} \ln \frac{r}{R_1} \tag{5.43}$$

5.6 Diffusion with Chemical Reaction

Diffusion control in the gas phase is infrequent because rates of diffusion of molecules are generally very high. Diffusion control is more likely in solution where diffusion of reactants is slower due to the greater number of collisions with solvent molecules. Reactions that occur so quickly are considered diffusion-controlled reactions. Reactants in distinct phases (heterogeneous reactions) are candidates for diffusion control such as reactions involving catalysis and enzymatic reactions. If the reaction rate is affected by stirring or agitation, then under those conditions the reaction is diffusion controlled. Consider the case of diffusion from a solid sphere with chemical

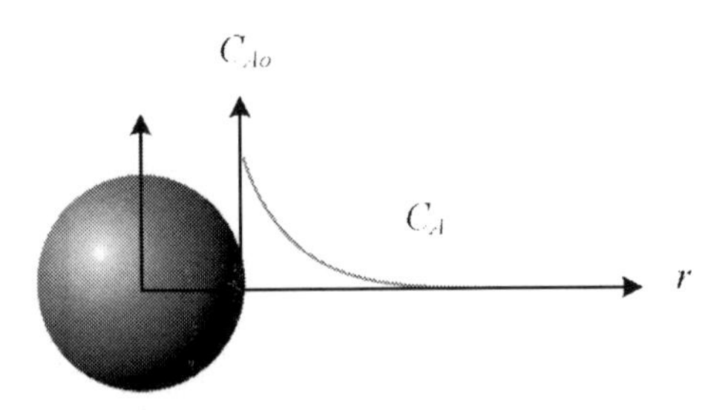

FIGURE 5.5
Diffusion with chemical reaction from spherical particle into liquid solution.

reaction; in this situation a solid sphere made of substance A with radius R and density ρ is suspended in a liquid B. Solid A experiences a first-order homogeneous chemical reaction, with rate constant k_A being slightly soluble in liquid B. Let C_{A0} be the molar solubility of A in B, and D_{AB} the diffusivity of A in B (Figure 5.5). Develop an expression for the concentration profile of A in the solution B.

Consider the following assumption to find the concentration profile of component A in the liquid solution B:

- Steady state $\left(\partial C_A/\partial t = 0\right)$.
- Ignoring convection effect $\left(\left(v_r\,(\partial C_A/\partial r) + (v_\theta/r)(\partial C_A/\partial\theta) + (v_\varnothing/r\sin(\theta))(\partial C_A/\partial\varnothing)\right) = 0\right)$.
- Diffusion is in r direction only $\left(D_{AB}\dfrac{1}{r^2\sin(\theta)}\dfrac{\partial}{\partial\theta}\left(\sin(\theta)\dfrac{\partial C_A}{\partial\theta}\right) = D_{AB}\dfrac{1}{r^2\sin^2(\theta)}\dfrac{\partial^2 C_A}{\partial\varnothing^2} = 0\right)$.

Start with the overall total material balance in spherical coordinate:

$$\rho\left(\frac{1}{r^2}\frac{\partial\left(r^2 v_r\right)}{\partial r} + \frac{1}{r\sin(\theta)}\frac{\partial\left(v_\theta\sin(\theta)\right)}{\partial\theta} + \frac{1}{r\sin(\theta)}\frac{\partial v_\varnothing}{\partial\varnothing}\right) = 0 \tag{5.44}$$

The differential component mole balance equation for species A in spherical coordinates:

$$\begin{aligned}\frac{\partial C_A}{\partial t} + \left(v_r\frac{\partial C_A}{\partial r} + \frac{v_\theta}{r}\frac{\partial C_A}{\partial\theta} + \frac{v_\varnothing}{r\sin(\theta)}\frac{\partial C_A}{\partial\varnothing}\right) &= D_{AB}\left(\frac{1}{r^2}\frac{\partial}{\partial r}\left(r^2\frac{\partial C_A}{\partial r}\right)\right.\\ &\left.+\frac{1}{r^2\sin(\theta)}\frac{\partial}{\partial\theta}\left(\sin(\theta)\frac{\partial C_A}{\partial\theta}\right) + \frac{1}{r^2\sin^2(\theta)}\frac{\partial^2 C_A}{\partial\varnothing^2}\right) + R_A = 0\end{aligned} \tag{5.45}$$

Since the convective term is neglected $(v_r = v_\theta = v_z = 0)$ and assuming steady state, neglect convective mass transfer and the concentration gradient in the angular component coordinate. The general differential component male balance equation is simplified to the following form:

$$D_{AB}\left(\frac{1}{r^2}\frac{\partial}{\partial r}\left(r^2\frac{\partial C_A}{\partial C_A}\right)\right) + R_A = 0 \tag{5.46}$$

The reaction rate of component A:

$$R_A = -k_A C_A \tag{5.47}$$

Substitute reaction rate in Equation 5.46:

$$D_{AB}\left(\frac{1}{r^2}\frac{\partial}{\partial r}\left(r^2\frac{\partial C_A}{\partial C_A}\right)\right) - k_A C_A = 0 \tag{5.48}$$

Rearrange the equation by dividing each term by D_{AB}:

$$\frac{1}{r^2}\frac{\partial}{\partial r}\left(r^2\frac{\partial C_A}{\partial C_A}\right) - \frac{k_A}{D_{AB}}C_A = 0 \tag{5.49}$$

Let: $a^2 = k_A/D_{AB}$. Rearrange:

$$\frac{1}{r^2}\frac{\partial}{\partial r}\left(r^2\frac{\partial C_A}{\partial C_A}\right) - a^2 C_A = 0 \tag{5.50}$$

The analytical solution of Equation 5.50 is exposed in Equation 5.51 [3]:

$$C_A = \frac{C_1}{r}e^{ar} + \frac{C_2}{r}e^{-ar} \tag{5.51}$$

The arbitrary constants of integration are found using the following boundary conditions:

$$\text{At } r = \infty,\ C_A = 0 \rightarrow C_1 = 0$$

$$\text{At } r = R,\ C_A = C_{Ao} \rightarrow C_2 = C_{Ao}R/e^{-aR}$$

Substitute expressions of the arbitrary constants of integration C_1 and C_2 in Equation 5.51 to attain the steady-state concentration profile:

$$C_A = C_{Ao}\frac{R}{r}\frac{e^{-ar}}{e^{-aR}} = C_{Ao}\frac{R}{r}e^{a(R-r)} \tag{5.52}$$

5.7 Leaching of Solute from Solid Particles

The separation process of solute being extracted from the solid is called liquid-solid leaching or simply leaching. Leaching is the process of extracting substances from a solid by dissolving them in a liquid. In the chemical processing industry, leaching has a variety of commercial applications, including separation of metal from ore using acid, and sugar from beets using hot water. In an ideal leaching process, all the solute is dissolved by the solvent and none of the carrier is dissolved. The solid mixture consists of particles, inert insoluble carrier *A*, and solute *B*. In a typical leaching process, the solvent, *C*, is added to the solid mixture to be separated to selectively dissolve *B*. The overflow from the stage is free of solids and consists of only solvent *C* and dissolved *B*. The underflow consists of a slurry of liquid of similar composition in the liquid overflow and solid carrier *A*.

Consider the case of a solid sheet containing compound *A* that is to be leached out by a solvent *B* for solid surface. The concentration of *A* in the bulk stream is $C_{A\delta}$ and at the surface of the solid sheet is C_{As}, which is the saturation concentration of A in B. Assume that the solvent B is flowing over the flat sheet solid under turbulent flow conditions. Assume a stagnant film of thickness δ around the solid through which *A* is leached out into the bulk stream (Figure 5.6). Determine the concentration profile of component *A* in the liquid film.

Consider the following assumptions to determine the concentration profile of component *A* in the liquid film:

- Steady state $\left(\partial C_A/\partial t = 0\right)$.
- Mass transfer is only by diffusion $\left(v_x = v_y = v_z = 0\right)$.
- Component *B* is stagnant.
- Diffusivity of *A* in *B*, D_{AB} is a constant.
- Mass transport in the *x* and *y* directions is negligible: $\left(D_{AB}\left(\partial^2 C_A/\partial x^2 + \partial^2 C_A/\partial y^2\right) = 0\right)$.
- No reaction is taking place, $R_A = 0$.

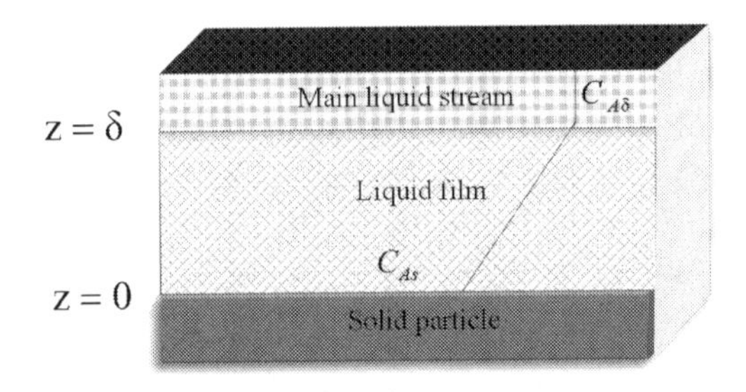

FIGURE 5.6
Mass transfer through a stagnant liquid film.

Apply the component mole balance equation in Cartesian coordinates:

$$\frac{\partial C_A}{\partial t} + \left(v_x \frac{\partial C_A}{\partial x} + v_y \frac{\partial C_A}{\partial y} + v_z \frac{\partial C_A}{\partial z} \right)$$

$$= D_{AB} \left(\frac{\partial^2 C_A}{\partial x^2} + \frac{\partial^2 C_A}{\partial y^2} + \frac{\partial^2 C_A}{\partial z^2} \right) + R_A \quad (5.53)$$

After executing the proceeding assumptions, Equation 5.53 is simplified to Equation 5.54:

$$\frac{d^2 c_A}{dz^2} = 0 \quad (5.54)$$

Double integration of Equation 5.54 generates the two arbitrary constants of integration:

$$C_A = C_1 z + C_2 \quad (5.55)$$

The following boundary conditions can be used to determine expressions of the two integration constants C_1 and C_2.

B.C.1: at $z = 0,\ C_A = C_{As}$

B.C.2: at $z = \delta,\ C_A = C_{A\delta}$

Using the former boundary conditions, we obtained an expression of C_1 and C_2:

$$C_1 = \frac{(C_{A\delta} - C_{As})}{\delta} \quad (5.56)$$

and hence, $C_2 = C_{As}$.

Finally, the concentration profile along the z direction is described by the linear algebraic Equation 5.57:

$$C_A = \frac{C_{A\delta} - C_{As}}{\delta} z + C_{As} \quad (5.57)$$

5.8 Applied Examples

This section contains a few examples to give the reader a better understanding of the topic of mass transfer in distributed systems. In most of the examples, model equations for mass transfer in distributed systems were developed and solved manually and by using COMSOL Multiphysics.

Example 5.1: Ventilation Duct

Ventilation refers to the process of removal of inside air from a building, room, or a confined space and replacing it with natural fresh air. In this example we will study the progress of ethanol concentration in a rectangular duct when the air in the duct is stagnant. The aeration duct consists of one 1 m long straight fragment with a quadratic side 0.4 m. The concentration of ethanol at the inlet is homogenous and is 1 mol/m^3; the outlet concentration can be assumed to be zero. The diffusivity for ethanol in air is 1.35×10^{-5} m^2/s. Develop an expression for the ethanol concentration profile in the aeration duct.

Solution

Consider the following assumption in developing the model equations:

- Steady state.
- Exit concentration is zero.
- Diffusion is in the x-direction only.

The schematic diagram of the ventilation duct is presented in Figure 5.7. Ethanol inlet concentration is 1 mol/m^3.

Ethanol component mole balance through stagnant air is represented by the following general differential equation in Cartesian coordinates:

$$\frac{\partial C_A}{\partial t}+\left(v_x\frac{\partial C_A}{\partial x}+v_y\frac{\partial C_A}{\partial y}+v_z\frac{\partial C_A}{\partial z}\right)=D_{AB}\left(\frac{\partial^2 C_A}{\partial x^2}+\frac{\partial^2 C_A}{\partial y^2}+\frac{\partial^2 C_A}{\partial z^2}\right)+R_A$$

Based on the former assumptions, steady state $(\partial C_A/\partial t=0)$, the convective term is neglected $(v_x=v_y=v_z=0)$. The diffusion is in the x-direction only $(\partial^2 C_A/\partial y^2=\partial^2 C_A/\partial z^2=0)$, no reaction is taking place in the duct $(R_A=0)$, and the driving force is the difference in the inlet and exit concentration of ethanol in the duct. Accordingly, the component mole balance equation is reduced to the following second order ordinary differential equation:

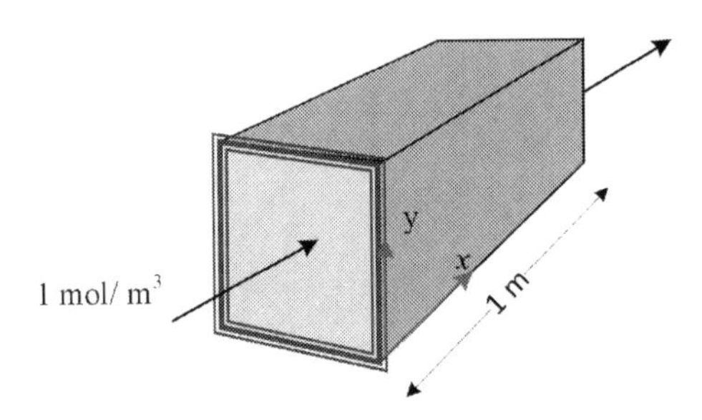

FIGURE 5.7
Schematic of ventilation duct.

$$0 = D_{AB}\left(\frac{\partial^2 C_A}{\partial x^2}\right)$$

The first integration generates the first arbitrary integration constant C_1:

$$\frac{dC_A}{dx} = C_1$$

The second integration generates the second arbitrary integration constant, C_2:

$$C_A = C_1 x + C_2$$

The expressions of the two-arbitrary integration constant are obtained after implementing the following boundary conditions:

B.C.1: at $x = 0$, $C_A = C_{Ao} = 1\ \text{mol m}^{-3}$ from the first boundary conditions, $C_2 = C_{Ao}$
B.C.2: at $x = L$, $C_A = 0$ from this second boundary condition $C_1 = -C_{Ao}/L$.

Substitute the obtained expressions of the integration constants C_1 and C_2, Hereafter, the expression for the concentration profile along the duct is defined by the following linear equation:

$$C_A = -\left(C_{A0}/L\right)x + C_{A0}$$

The analytical solution of the concentration profile is shown in Figure 5.8. The figure shows the linear drop of ethanol concentration versus distance

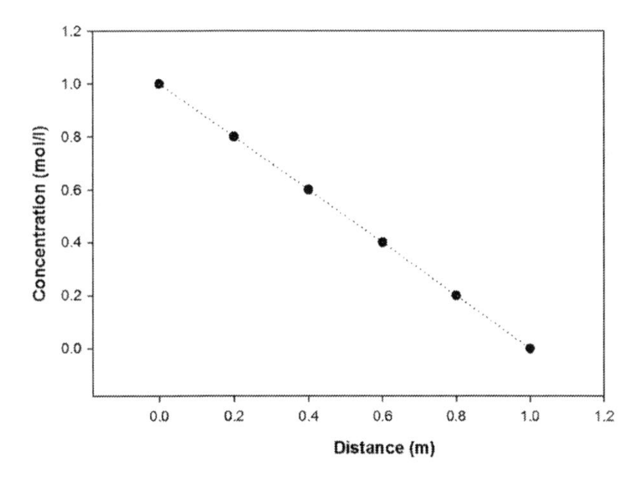

FIGURE 5.8
Concentration profile of ethanol in the duct.

through the length of the duct. The diagram revealed that the concentration decreases linearly with the length of the duct.

COMSOL Solution

Start COMSOL Multiphysics and click Model Wizard, then select 2D coordinates, for the physics select transport of dilute species, and stationary for the study case. Specify the boundary condition, right-click "Transport of dilute species," select "concentration," and name it as inlet concentration. Right-click again, select "concentration," and name it as outlet concentration (Figure 5.9). For left-side boundary conditions, enter $1\,mol/m^3$ for the inlet concentration boundary conditions and zero for the right-side outlet concentration. Click on "Transport properties," and enter the diffusion coefficient $1\times10^{-9}m^2/s$. Specify "No Flux" for the top and bottom boundary conditions. The model and the boundary conditions should look like those in Figure 5.9. The surface concentration crosses the duct as shown in Figure 5.10. The arrows represent the diffusive flux. The ethanol concentration decreases along the length of the duct.

The concentration profile at the centerline of the duct is generated as follows: Right-click "Data Sets" and select "Cut Line 2D," then right click on "Results" and select "1D Plot Group 1." Right-click "1D Plot Group 1" and select "Line Graph 1." The result should look like that shown in Figure 5.11. Manual calculation and numerical simulation predictions are identical.

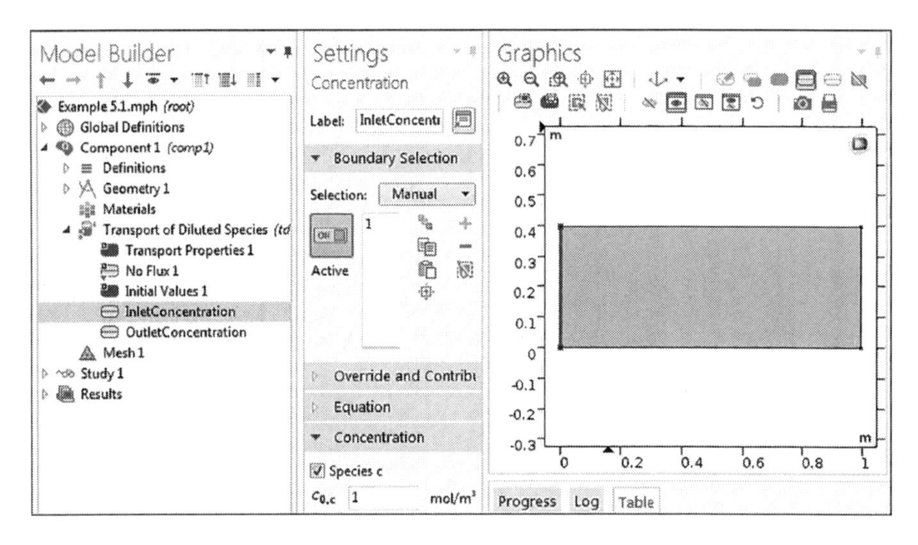

FIGURE 5.9
Physics and boundary conditions.

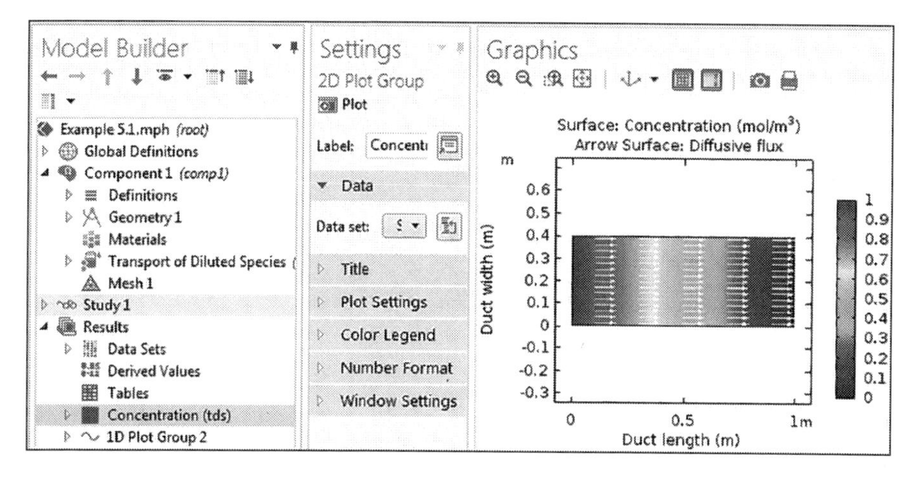

FIGURE 5.10
Ethanol concentration profile throughout the duct.

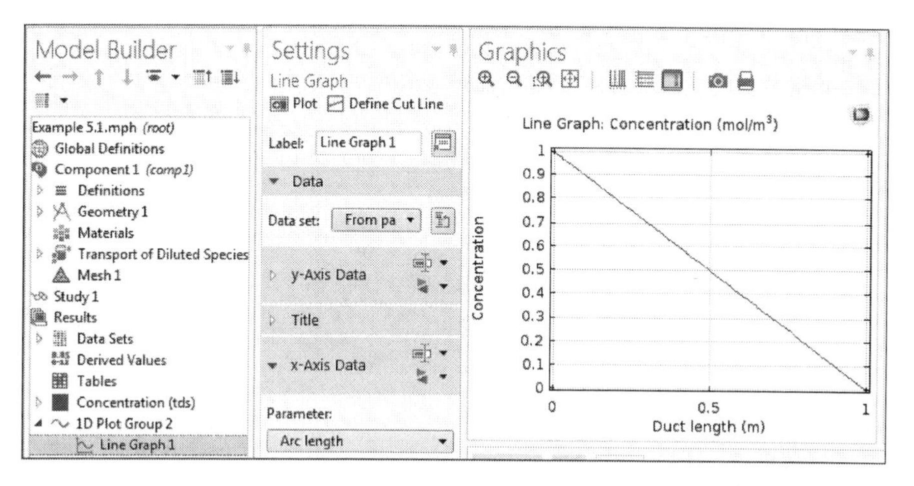

FIGURE 5.11
Concentration profile in the middle of the duct at $w = 0.2\,\text{m}$.

Example 5.2: Diffusion in Porous Solid

Flow and diffusion through porous media represent a field of study with several scientific and engineering applications, including catalysis, water purification, soil mechanics, and petroleum engineering. In this example, a block of porous silica is completely saturated with water where potassium chloride KCl is diffusing in the porous silica. The silica has a void fraction ε of 0.3, and the tortuosity τ is 4.0. The pores are filled with water at 298 K. At one face the concentration of KCl is held at $0.1\,\text{mol/m}^3$, and at the other face 2 mm apart, fresh water flows rapidly to maintain a negligible concentration of KCl at the other side. Neglect any other resistance, but consider that, in the porous solid, the diffusivity of KCl in water is $D_{AB} = 1.87 \times 10^{-9} \text{m}^2/\text{s}$. Calculate the diffusion flux of KCl at steady state, and, using COMSOL Multiphysics, compute the concentration profile through the porous solid.

Solution

The simple analytical solution of the diffusion flux of KCl in porous solid is determined as follows:

The effective diffusivity, $D_{\text{eff}} = D_{AB}\varepsilon/\tau$

$$N_A = \frac{D_{\text{eff}}(C_{A0} - C_{AL})}{(z_2 - z_1)} = \frac{0.3 \times 1.87 \times 10^{-9}(0.1 - 0)}{4(0.002 - 0)} = 7.0 \times 10^{-9} \frac{\text{mol KCl}}{\text{s.m}^2}$$

COMSOL Simulation

Start a new case in COMSOL and select the physics and boundary conditions as shown in Figure 5.12. Type in the space near the left-side boundary conditions "Concentration 1," $0.1 \,\text{mol/m}^3$, and for the right-side boundary condition, typ in "Concentration 2," zero. Click on "Transport Properties 1" and enter the value of $(\varepsilon/\tau)D_{AB}$ for the diffusion coefficient $\left((0.3/4)1.87 \times 10^{-9} \text{m}^2/\text{s}\right)$. Set "No Flux" boundary conditions for the top and bottom boundary conditions. The surface concentration of the simulation results is depicted in Figure 5.13. The diagram reveals that the concentration is maximum at the surface of the porous silica solid surface and decreases through the porous solid particle.

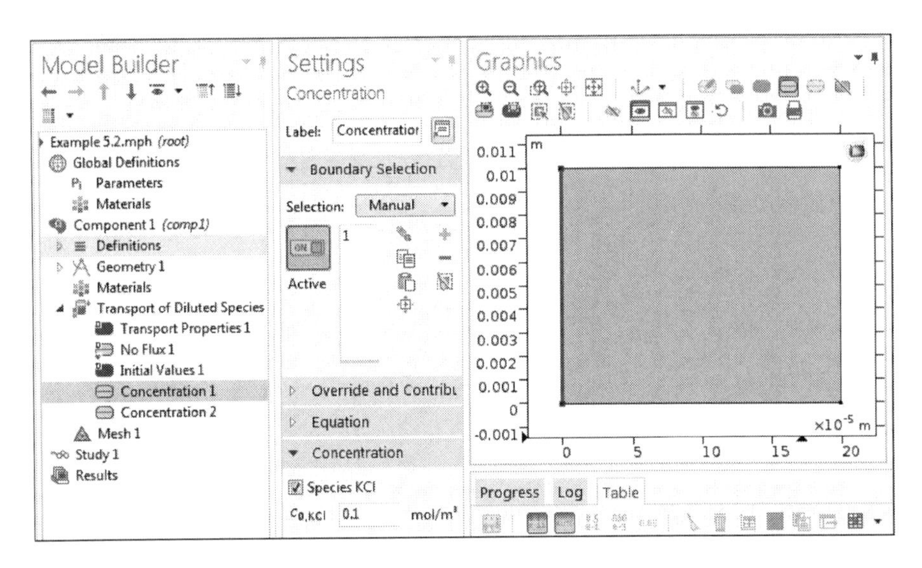

FIGURE 5.12
Physics and boundary conditions used in COMSOL.

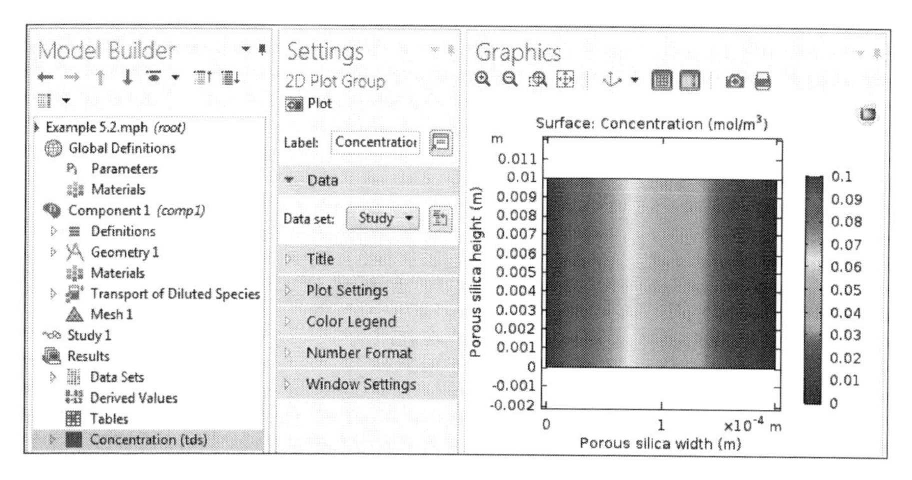

FIGURE 5.13
Simulation results, surface concentration of KCL in porous silica.

Example 5.3: Drug Diffusion in Membrane Shield

Nicotine cover is made of a reservoir of drug sealed onto the surface of a polymeric membrane. The device is constructed by inserting a drug solution into the reservoir (labeled "drug reservoir" in Figure 5.14a) and then locking the reservoir. Initially, the polymer membrane is free from any of the drug material. When in use, the bottom of the polymer membrane is adhered to the surface of the skin and drug diffuses from the reservoir down through the membrane into the skin (Figure 5.14b). Essentially, the drug remains at a constant concentration C_{A0} in the reservoir while its being used. The length of the reservoir is L = 1 mm, $C_{A0} = 100$ mg/mL, diffusion coefficient of the drug in the membrane is $D = 5 \times 10^{-8}$ cm^2/s; the concentration of the drug in the membrane/concentration of drug in reservoir fluid is determined for the solubility: S = 0.01. The nicotine molecular weight is MW = 162.23 g/mole. Determine how long it will take after the device is fixed to the skin for the drug to penetrate to the exterior (bottom) surface of the polymer membrane.

Solution

Consider the following assumptions in solving the nicotine diffusion in the drug reservoir through the membrane layer:

- Mass transfer is only by diffusion $\left(v_x \dfrac{\partial C_A}{\partial x} = v_y \dfrac{\partial C_A}{\partial y} = v_z \dfrac{\partial C_A}{\partial z} = 0\right)$.
- Diffusivity of A in membrane is a constant $(D_{AB} = \text{constant})$.
- Mass transport is in x-direction only $\left(D_{AB}\left(\dfrac{\partial^2 C_A}{\partial y^2} + \dfrac{\partial^2 C_A}{\partial z^2}\right) = 0\right)$.

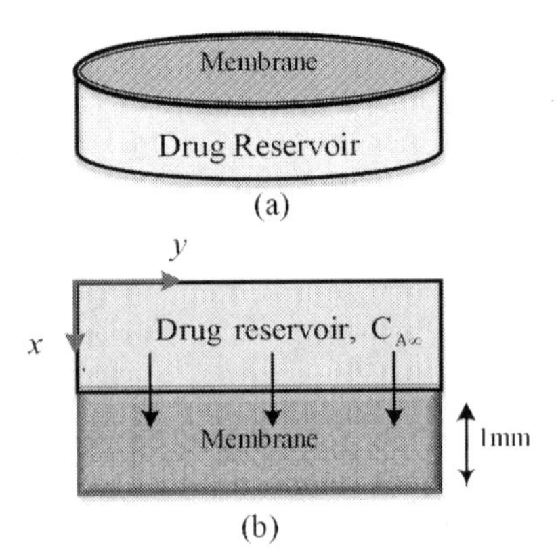

FIGURE 5.14
(a) Drug reservoir before use, (b) drug reservoir when in use.

Nicotine component mole balance through the membrane layer is solved by engaging the general differential component mole equation of change in Cartesian coordinates:

$$\frac{\partial C_A}{\partial t}+\left(v_x\frac{\partial C_A}{\partial x}+v_y\frac{\partial C_A}{\partial y}+v_z\frac{\partial C_A}{\partial z}\right)=D_{AB}\left(\frac{\partial^2 C_A}{\partial x^2}+\frac{\partial^2 C_A}{\partial y^2}+\frac{\partial^2 C_A}{\partial z^2}\right)+R_A$$

No reaction is taking place ($R_A=0$) and the convective term is neglected ($v_x=v_y=v_z=0$). The diffusion is in the x-direction is ($\partial^2 C_A/\partial y^2=0$, $\partial^2 C_A/\partial z^2=0$). Accordingly, the component mole balance equation for the drug concentration across the membrane is described by the as follows, with the concentration of nicotine (C_A) in mol/m^3:

$$\frac{\partial C_A}{\partial t}=D_{AB}\left(\frac{\partial^2 C_A}{\partial x^2}\right)$$

The initial and boundary conditions:
Initial conditions: at $t=0$, $C_A=C_{A0}=100$ mg/mL

B.C.1: for $t>0$, $x=0$, $C_A=C_{A0}=100$ mg/mL
B.C.2: for $t>0$, $x=\infty$, $C_A=0$

The analytical solution:

$$C_A=C_{A0}+\left(C_{As}-C_{A0}\right)*\operatorname{erfc}\left(\frac{z}{2\sqrt{D_{AB}t}}\right)$$

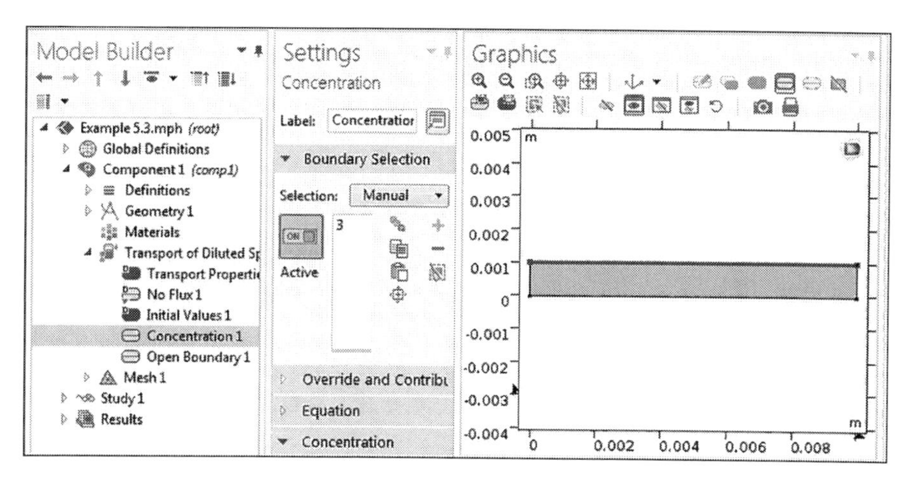

FIGURE 5.15
Setting of the top boundary condition.

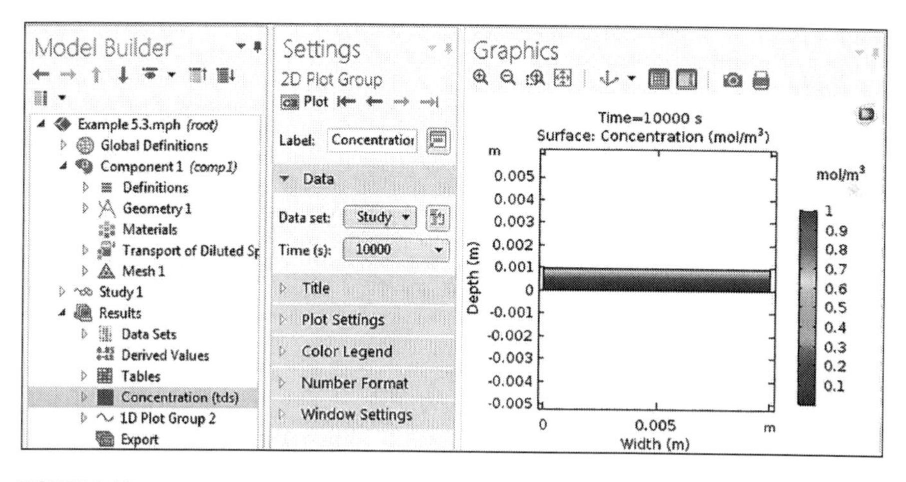

FIGURE 5.16
COMSOL simulation results.

COMSOL Simulation

The COMSOL model builder with the boundary conditions is shown in Figure 5.15.

The software simulation prediction is shown in Figure 5.16.

For a two-dimensional cut line at width of (0.005,0) and (0.005, 0.001), the change in the concentration of drug with time is shown in Figure 5.17.

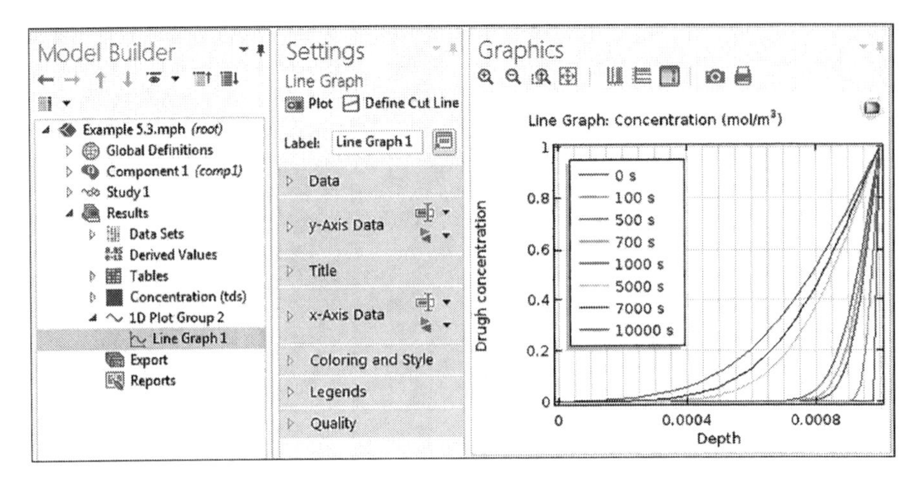

FIGURE 5.17
Drug concentration versus time.

Example 5.4: Absorption of CO_2 in a Wetted Column

Global warming is caused by the emission of greenhouse gases. Most of the emitted greenhouse gases is carbon dioxide (CO_2). In this example, carbon dioxide is being absorbed from air via a wetted wall column. The column is provided with fresh water at the top of the column, and air containing 5 mole percent CO_2 is supplied at the bottom, as described in Figure 5.18. Water is flowing downward at a very low velocity of 7.96×10^{-6} m/s. The change in gas composition may be neglected due to the state of the air. The gas phase resistance to mass transfer is negligible. It is acceptable to neglect axial diffusion. Develop an expression for the carbon dioxide concentration in water.

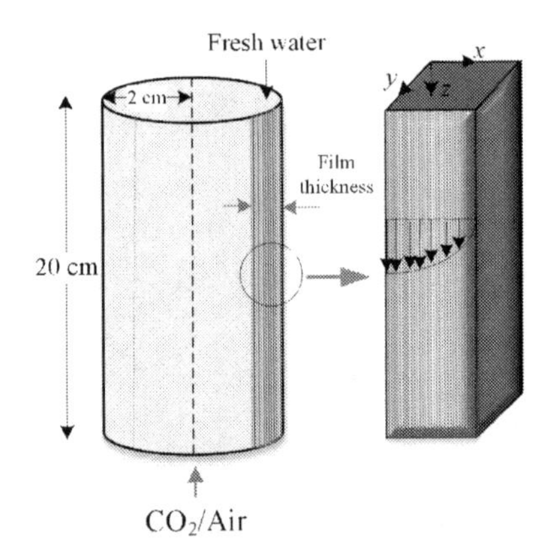

FIGURE 5.18
Schematic diagram of wetted wall absorption column.

Data

Diffusivity of carbon dioxide in water: $D_{CO_2/H_2O} = 2\times10^{-9}\,m^2/s$.

Solubility of CO_2 in water $= 0.15 g\ CO_2/100\ g\ H_2O$ at CO_2 partial pressure of 2 atm.

Density of water is $1 g/cm^3$.

Solution

The concentration profile is supposed to change in the x and z directions; hence, the differential mass balance equation in Cartesian coordinates is:

$$\frac{\partial C_A}{\partial t} + \left(v_x \frac{\partial C_A}{\partial x} + v_y \frac{\partial C_A}{\partial y} + v_z \frac{\partial C_A}{\partial z} \right)$$

$$= D_{AB}\left(\frac{\partial^2 C_A}{\partial x^2} + \frac{\partial^2 C_A}{\partial y^2} + \frac{\partial^2 C_A}{\partial z^2} \right) + R_A$$

There is no reaction that is taking place, and the v_x and v_y components of velocity are neglected. The diffusion is in the x and z directions. The steady-state component mole balance equation of carbon dioxide through the wetted shell is described by the following equation:

$$\left(v_z \frac{\partial C_A}{\partial z} \right) = D_{AB}\left(\frac{\partial^2 C_A}{\partial x^2} \right)$$

The diffusion is mostly in the x-direction, and the transport of A in the z-direction will be primarily by convection $\left(v_z\left(\partial C_A/\partial z\right)\right)$, not diffusion $D_{AB}\left(\partial^2 C_A/\partial z^2 = 0\right)$. The velocity profile in the thin film is described by the following velocity profile equation:

$$v_z = \frac{1}{2}\frac{\rho g \delta^2}{\mu}\left[1 - \frac{x^2}{\delta^2}\right]$$

The boundary conditions are:

B.C.1: at $z=0$, $C_A = 0$
B.C.2: at $x=0$, $C_A = C_{A0}$
B.C.3: at $x = \delta$, $\partial C_A/\partial x = 0$

Assume that the flow of air is at its maximum velocity:

$$v_z = v_{z,\max} = \frac{1}{2}\frac{\rho g \delta^2}{\mu}$$

The partial differential equation is reduced to:

$$v_{z,\max}\frac{\partial C_A}{\partial z} = D_{AB}\frac{\partial^2 C_A}{\partial x^2}$$

where the analytical solution will be:

$$C_A = C_{A0}\left[1-\operatorname{erf}\left(\frac{x}{\sqrt{4D_{AB}\,z / v_{z,\max}}}\right)\right]$$

COMSOL Simulation

The physical solubility of CO_2 in water is $m = 0.8314$. The COMSOL screen snapshot for the laminar flow of fresh water is shown in Figure 5.19. The other borders are the "no-slip" wall, and on the right side, these are "slip wall" in the middle and the "outlet" from the bottom.

Figure 5.20 shows the solubility (m) of CO_2 in water at the gas-wetted wall interface. The inlet air concentration at 1 atm total pressure and 5 moles% CO_2, the balance is air is shown in Figure 5.21.

The surface plot of the carbon dioxide concentration is depicted in Figure 5.22. The figure makes clear that concentration of CO_2 in air is almost constant, as stated in the problem statement. However, the water starts fresh with zero CO_2 concentration and then the concentration of CO_2 increases downward. Simulation predictions reveal that the CO_2 concentration decreases along the wetted wall (Figure 5.23).

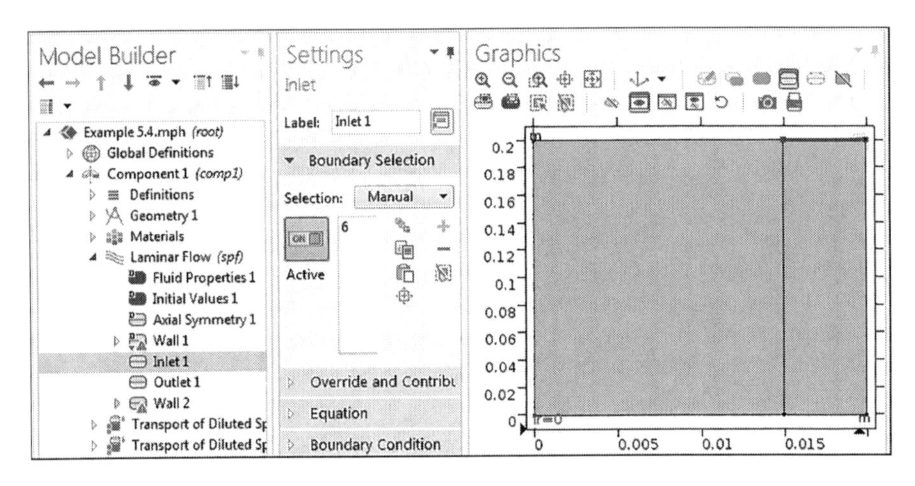

FIGURE 5.19
Boundary conditions and the inlet of the fresh water.

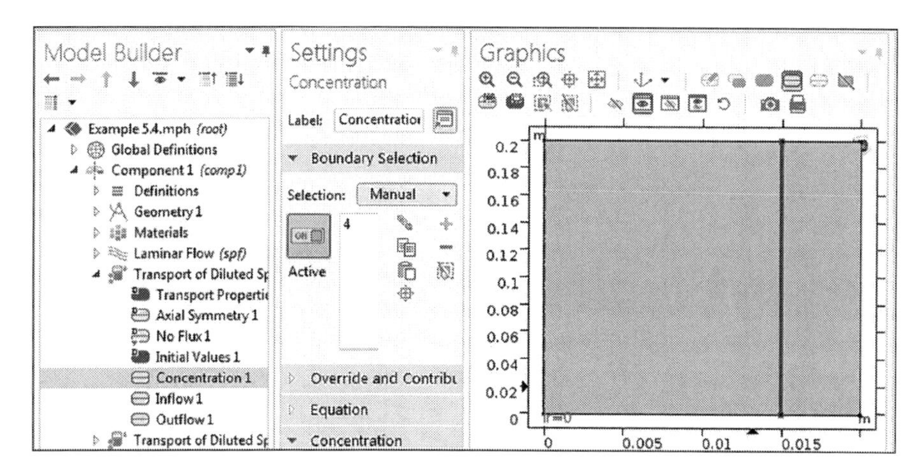

FIGURE 5.20
Air-water interface boundary condition.

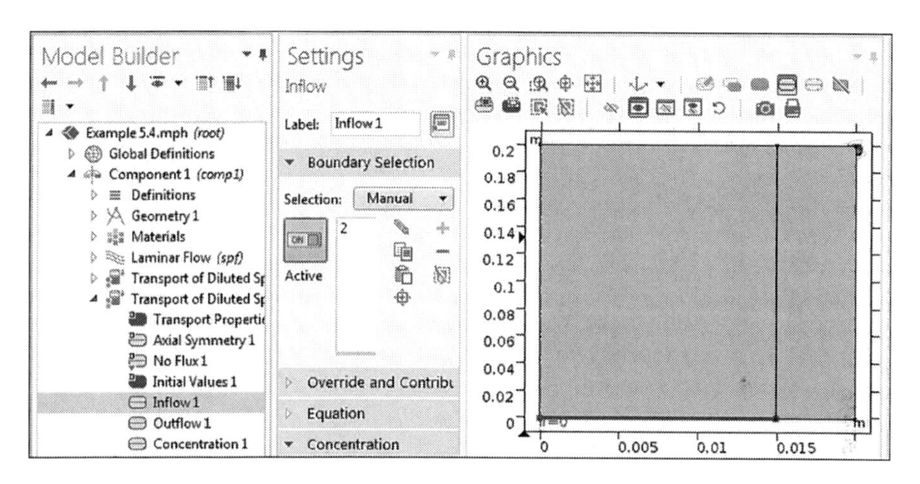

FIGURE 5.21
Gas inlet conditions, 1 atm pressure, 5 moles% CO_2.

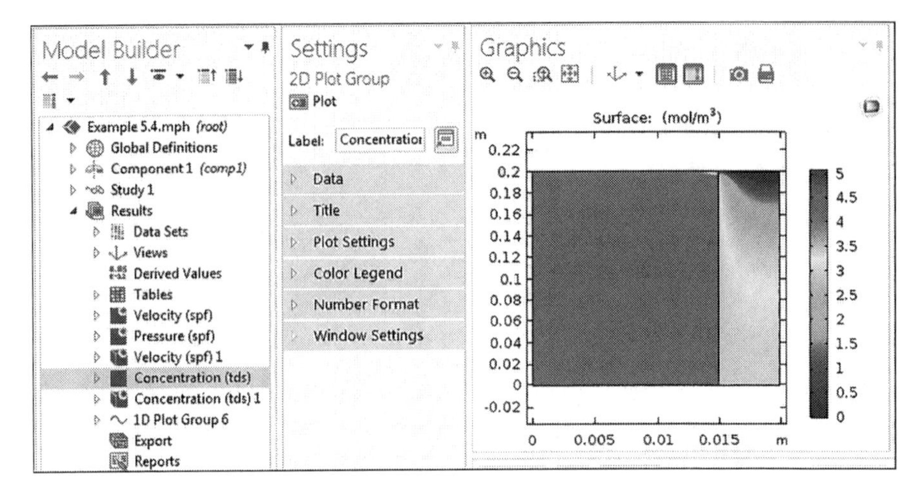

FIGURE 5.22
Surface plot concentration of CO_2.

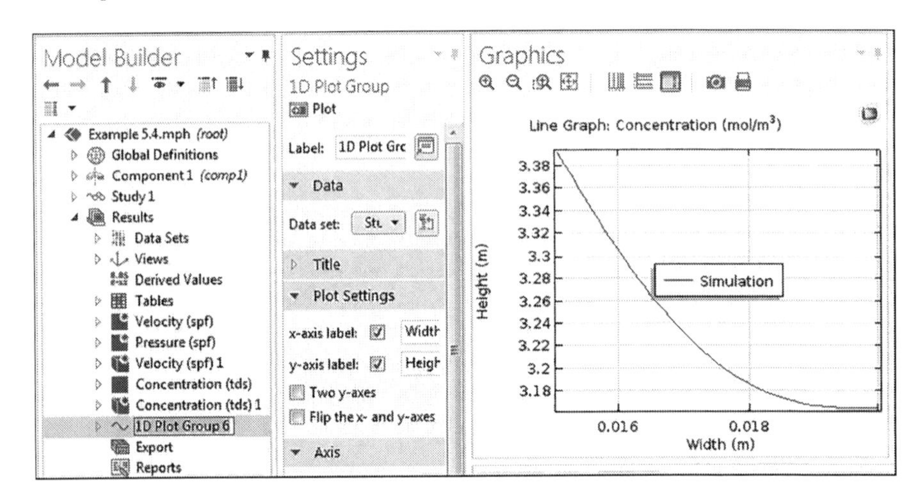

FIGURE 5.23
Simulation results for concentration versus width.

Example 5.5: Separation of Helium from Natural Gas

Diffusivity of helium in Pyrex glass is very high in comparison to the other gases. Because of this, Pyrex glass tubes can be used for separation of helium from natural gas. Suppose a natural gas mixture is flowing in the Pyrex tube with dimensions as shown in Figure 5.24. Obtain an expression for the concentration profile at which helium will diffuse out in terms of the diffusivity of helium through the wall of Pyrex tube, the interfacial concentration of helium in the Pyrex tube wall, and dimensions of the tube. At 500°C, the diffusivity of helium in the Pyrex

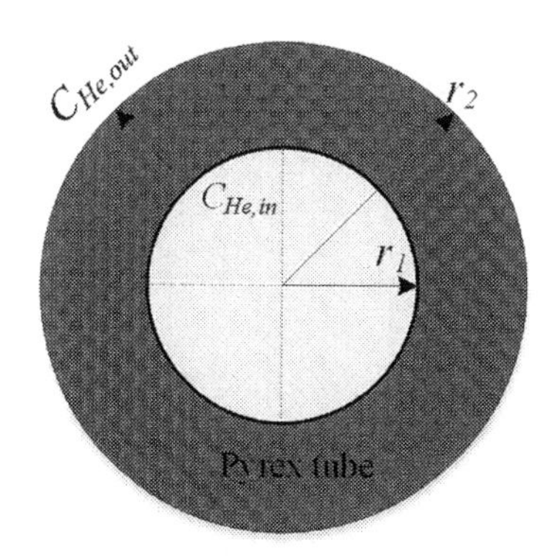

FIGURE 5.24
Separation of He from natural gas by Pyrex tube.

tube is $D_{He-pyrex} = 2\times10^{-12}\,m^2/s$; for some steady gas/helium mixtures $C_{in} = 10^{-5}\,g/cm^3$. The Pyrex thickness is $\delta = 0.5$ mm; the inner radius is 1.5 mm.

Solution

Consider the following assumptions to obtain the expression for the concentration profile in the Pyrex glass.

- Steady state $(\partial C_A/\partial t = 0)$.
- Transportation of helium takes place by diffusion only $(v_r = v_\theta = v_z = 0)$.
- Diffusion in z and θ, directions are negligible $((1/r^2)(\partial^2 C_A/\partial_\theta{}^2) = \partial^2 C_A/\partial z^2 = 0)$.
- Diffusivity of helium in Pyrex D_{He} is a constant.
- No reaction takes place in the system ($R_A = 0$).

The component mole balance equation in cylindrical coordinates:

$$\frac{\partial C_A}{\partial t} + \left(v_r \frac{\partial C_A}{\partial r} + v_\theta \frac{1}{r}\frac{\partial C_A}{\partial \theta} + v_z \frac{\partial C_A}{\partial z}\right)$$

$$= D_{AB}\left(\frac{1}{r}\frac{\partial}{\partial r}\left(r\frac{\partial C_A}{\partial r}\right) + \frac{1}{r^2}\frac{\partial^2 C_A}{\partial \theta^2} + \frac{\partial^2 C_A}{\partial z^2}\right) + R_A$$

Apply former assumptions, and the equation of change in cylindrical coordinate is reduced to the following equation:

$$D_{AB}\frac{1}{r}\frac{d}{dr}\left(r\frac{dC_A}{dr}\right) = 0$$

Perform double integration to obtain the following equation:

$$C_A = \frac{-C_1}{D_{AB}} \ln r + C_2$$

where C_1 and C_2 are the arbitrary constants of integration. The boundary conditions are as follows:

B.C.1: at $r = r_1$, $C_A = C_{He,1}$
B.C.2: at $r = r_2$, $C_A = C_{He,2}$

Substitution of boundary condition 1:

$$C_{He,1} = \frac{-C_1}{D_{AB}} \ln r_1 + C_2$$

Substitution of the second boundary condition:

$$C_{He,2} = \frac{-C_1}{D_{AB}} \ln r_2 + C_2$$

Subtract equations to get the following equation:

$$C_{He,2} - C_{He,1} = \frac{-C_1}{D_{AB}} \ln r_2 - \frac{-C_1}{D_{AB}} \ln r_1 = \frac{C_1}{D_{AB}} \ln \frac{r_1}{r_2}$$

The first integration constant, C_1:

$$C_1 = \left(C_{He,2} - C_{He,1}\right) \frac{D_{AB}}{\ln r_1 / r_2}$$

The second integration constant, C_2:

$$C_2 = C_{He,1} + \frac{\left(C_{He,2} - C_{He,1}\right) \frac{D_{AB}}{\ln r_1 / r_2}}{D_{AB}} \ln r_1$$

Substitute C_1 and C_2 to obtain the following equation:

$$C_A = \frac{-\left(C_{He,2} - C_{He,1}\right) \frac{D_{AB}}{\ln r_1/r_2}}{D_{AB}} \ln r + C_{He,1} + \frac{\left(C_{He,2} - C_{He,1}\right) \frac{D_{AB}}{\ln r_1/r_2}}{D_{AB}} \ln r_1$$

Rearrange:

$$C_A = C_{He} = C_{He,1} + \frac{(C_{He,2} - C_{He,1}) \frac{D_{AB}}{\ln r_1/r_2}}{D_{AB}} \ln \frac{r_1}{r}$$

The equation can be arranged such that:

$$C_{He} - C_{He,1} = (C_{He,2} - C_{He,1}) \frac{\ln r_1/r}{\ln r_1/r_2}$$

The concentration of helium inside the Pyrex tube:

$$C_{He}, \text{in} = \frac{1 \times 10^{-5}\,\text{g}}{\text{cm}^3} \times \frac{10^6\,\text{cm}^3}{1\,\text{m}^3} \times \frac{\text{mol}}{4\,\text{g}} = 2.5\,\text{mol/m}^3$$

COMSOL Simulation

Start a new case in COMSOL, then click on "Model Wizard," select 2D, select the physics "Transport of Diluted Species," and click "Add." Click "Study" and select "Stationary," then click "Done" to enter the model builder.

Geometry: right-click on "Geometry" and select two circles of radius 1.5 mm and 2.0 mm. The zone of interest is the Pyrex thickness (d = 0.5 mm); accordingly, take the difference between the two circles. To do so, right-click "Geometry 1," and select "Booleans and Partitions"/ Difference.

Material: for material properties, right-click on "Material," then click "add Material from library." Search for Helium, and after finding it, click "Add to component 1." The model with the boundary conditions is shown in Figure 5.25.

Boundary conditions: The system is bounded by internal and external surface concentrations of helium gas. To get the concentration

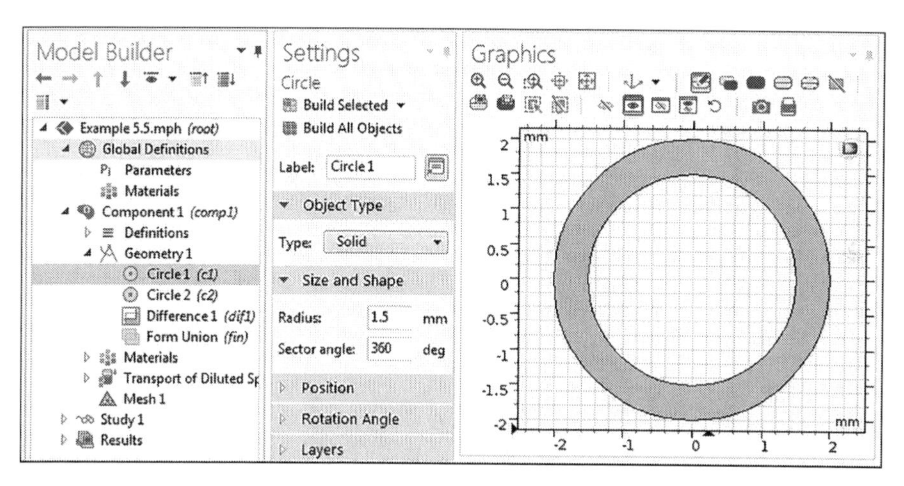

FIGURE 5.25
Model builder and boundary conditions.

boundaries, right-click on "Transport of Diluted species" and click on "Concentration." Repeat for the second concentration boundary conditions. For the internal concentration, set it to $C_{He,in} = 2.5\ mol/m^3$, and to zero for the external surface concentration. Default mesh is sufficient for this type of problem and it is ready. Click on "Study," then click "Compute." The final model should look like Figure 5.26.

A comparison of line graph concentration between analytical and simulation results is shown in Figure 5.27. The results are in excellent agreement.

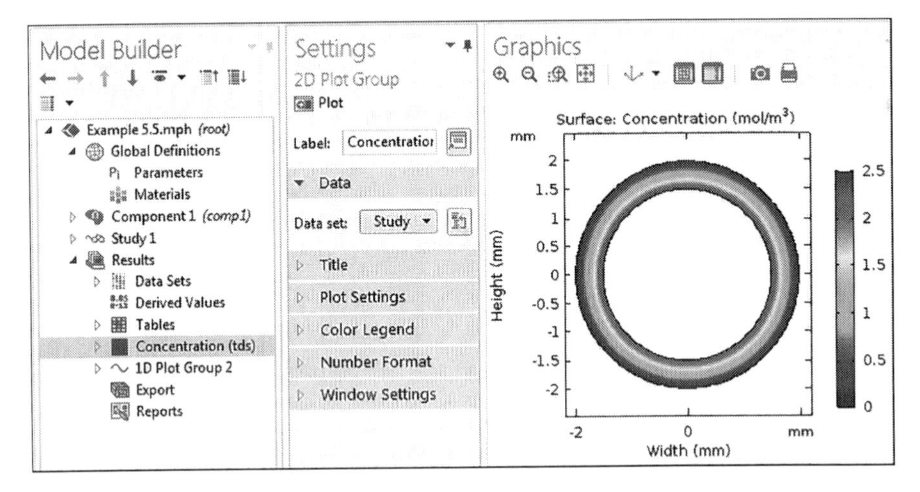

FIGURE 5.26
Model builder and simulation results.

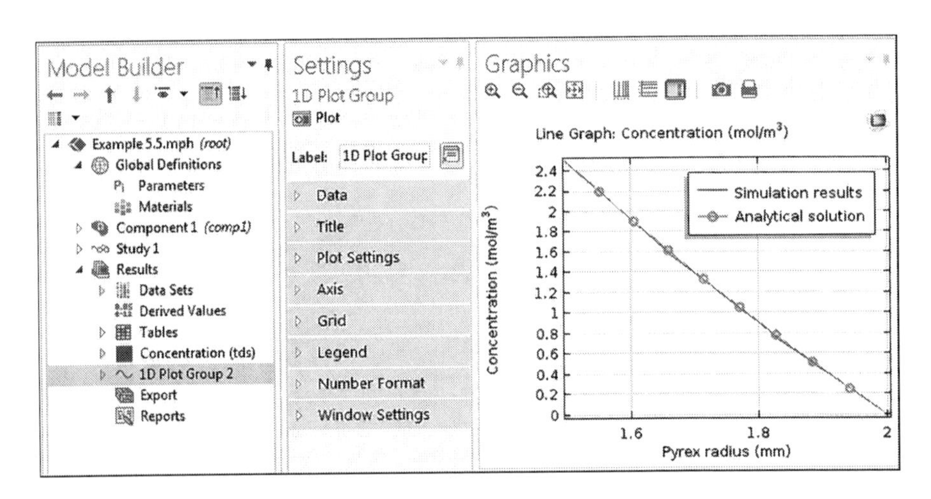

FIGURE 5.27
Comparison between simulation and analytical results.

Example 5.6: Extraction of Soybean Oil

Soybean oil, one of the most widely consumed cooking oils, is a vegetable oil extracted from the seeds of the soybean. Processed soybean oil is used as a base for printing inks and oil paints. In this example, a spherical soybean particle having an average diameter 2.00 mm and density $\rho = 577\,\text{kg/m}^3$ is considered as a sample of an extraction process (Figure 5.28). The soybean flakes contain 20% soybean oil that is to be leaked out by pure hexane solvent. The hexane solvent is flowing over the spherical particles under turbulent flow conditions. Assume a stagnant film of thickness $\delta = 0.5\,\text{mm}$ around the spherical particles through which soybean oil is leached out into the bulk stream. The soybean oil density is $917\,\text{kg/m}^3$, and viscosity is $\mu = 0.05$ Pa.s. The average molecular weight of soybean oil methyl esters is 292.2. The effective diffusivity is $D_{\text{eff}} = 1.0 \times 10^{-11}\,\text{m}^2/\text{s}$. Determine the concentration profile.

Assumptions

Consider the following assumptions in determining the concentration profile of soybean oil in the stagnant liquid film of hexane solvent:

- Steady state $\left(\partial C_A/\partial t = 0\right)$.
- Mass transfer is only by diffusion $\left(v_x = v_y = v_z = 0\right)$.
- B is stagnant, and diffusivity of A in B $\left(D_{AB}\right)$ is a constant.
- Diffusion is in the z-direction only $\left(\partial^2 C_A/\partial x^2 = \partial^2 C_A/\partial y^2 = 0\right)$.
- No reaction is taking place $\left(R_A = 0\right)$.

The differential equation of mass transport of the oil component (A) in the stagnant liquid layer of the hexane solvent states that:

$$\frac{\partial C_A}{\partial t} + \left(v_x \frac{\partial C_A}{\partial x} + v_y \frac{\partial C_A}{\partial y} + v_z \frac{\partial C_A}{\partial z}\right) = D_{AB}\left(\frac{\partial^2 C_A}{\partial x^2} + \frac{\partial^2 C_A}{\partial y^2} + \frac{\partial^2 C_A}{\partial z^2}\right) + R_A$$

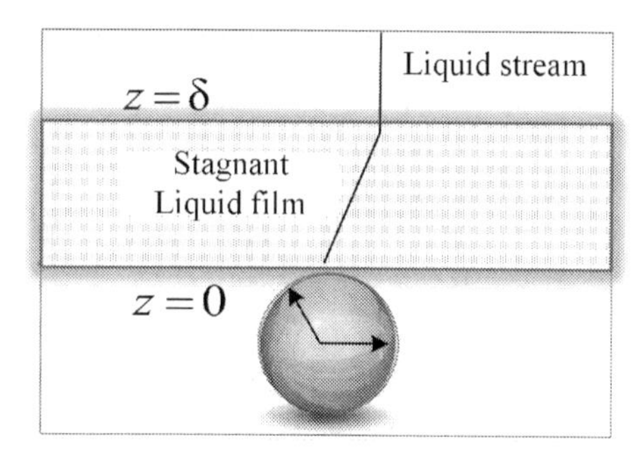

FIGURE 5.28
Mass transfer through a stagnant liquid film.

After applying the earlier assumption, the equation may be simplified as:

$$\frac{d^2 c_A}{dz^2} = 0$$

Double integration to get the following equation with two integration constants (C_1,C_2):

$$C_A = C_1 z + C_2$$

The following boundary conditions can be used to find the arbitrary constants of integration, C_1 and C_2:

B.C.1: at $z = 0$, $C_A = C_{As}$
B.C.2: at $z = \delta$, $C_A = C_{A\delta}$

Using the above boundary conditions, we obtain:

$$C_1 = \frac{C_{A\delta} - C_{As}}{\delta} \text{ and } C_2 = C_{As}$$

Finally, the concentration profile of oil along the z direction is given by:

$$C_A = \frac{C_{A\delta} - C_{As}}{\delta} z + C_{As}$$

The mass of the soybean flakes:

$$\text{Mass} = V \times \rho = \frac{4\pi r^3}{3} \times \rho = \frac{4\pi(0.001)^3}{3} \times \frac{577 \text{ kg}}{\text{m}^3} = 2.42 \times 10^{-6} \text{kg}$$

The mass of soybean oil is 20% of the total soybean particle:

$$m_{\text{oil}} = 0.2 \times 2.42 \times 10^{-6} \text{kg} = 0.484 \times 10^{-6} \text{kg}$$

Concentration of soybean oil in the soybean particle:

$$C_{\text{oil}} = \frac{m_{\text{oil}}}{V} = \frac{0.484 \times 10^{-6} \text{kg}}{4.19 \times 10^{-9} \text{m}^3} \times \frac{1000 \text{ g}}{1 \text{ kg}} \times \frac{\text{mol}}{292.2 \text{ g}} = 395 \text{mol/m}^3$$

COMSOL Simulation

The model builder and the boundary conditions are shown in Figure 5.29. The COMSOL simulation result is depicted in Figure 5.30. The concentration is maximum around the solid particle and decreases as it departures from the solid particle.

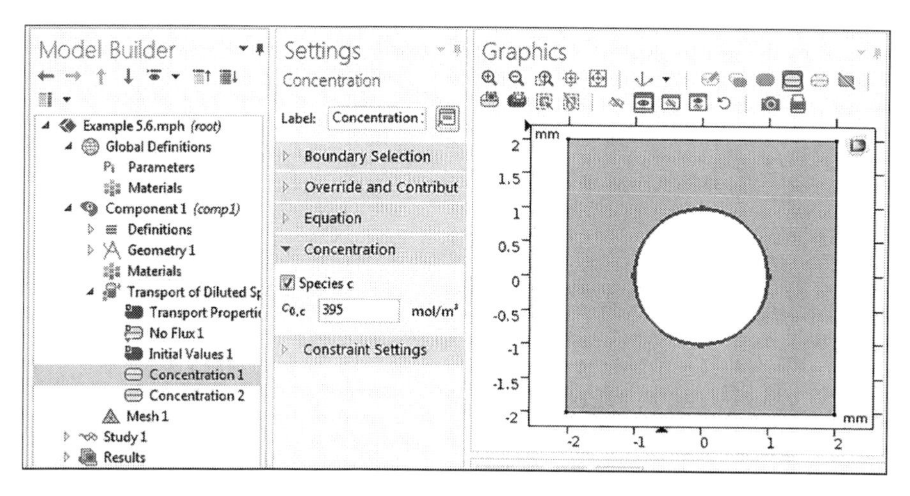

FIGURE 5.29
Model builder with boundary conditions.

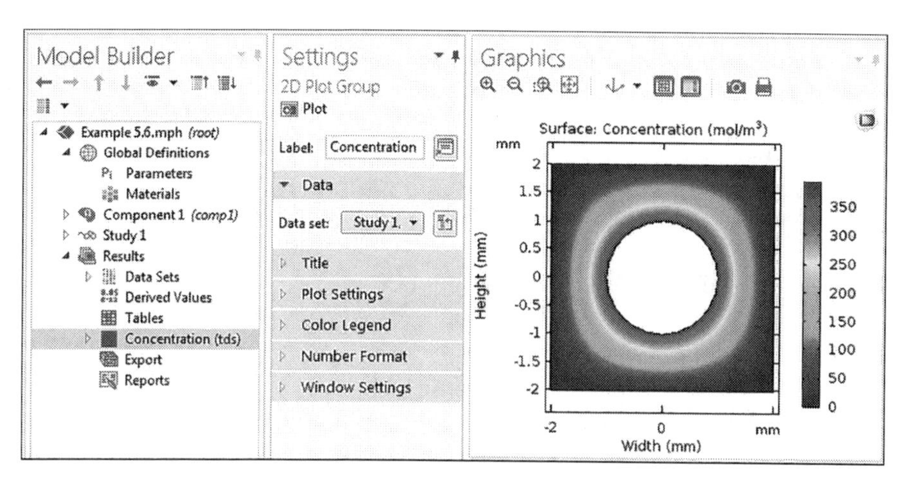

FIGURE 5.30
Simulation predictions.

Example 5.7: Diffusion of Water Droplet in Stagnant Air

Consider a small drop of water at the bottom of a tube; it is diffusing in stagnant air and filling the tube, as shown in Figure 5.31. To accelerate the diffusion process, air is blown over the top of the tube such that the concentration of water at the top can be assumed negligible. Naturally this case is transient; however, we can assume it is steady state to determine the concentration profile of water along the channel. Let us consider a spherical shape of the drop of water with a 1 mm radius. To simplify the analysis, assume a perfectly cylindrical channel of 7.5 mm height and 2.5 mm diameter. The diffusion coefficient of water in air is $1\times10^{-6}\ \text{m}^2/\text{s}$.

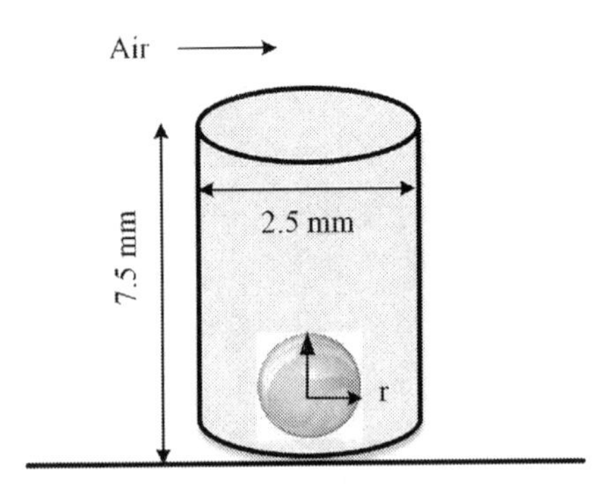

FIGURE 5.31
Diffusion of water droplet in stagnant air.

The surface concentration of water in the vapor phase at that boundary of the water drop can be determined using the vapor pressure of water. At 20°C, the vapor pressure of water is approximately 2.33 kPa. Assume the ideal gas law and that the vapor pressure of water is equivalent to a concentration of 0.95 mol/m^3.

Solution

Consider the following assumptions in determining the water vapor concentration along the tube length:

- Steady state, $\partial C_A/\partial t = 0$.
- Transportation of water vapor takes place by diffusion only ($v_r = v_\theta = v_z = 0$).
- Diffusion in the z and θ directions are negligible: $\left((1/r^2)(\partial^2 C_A/\partial\theta^2) = \partial^2 C_A/\partial z^2 = 0\right)$.
- Diffusivity of water vapor in air is constant: $(D_{AB} = \text{Constant})$.
- No reaction takes place in the system $(R_A = 0)$.

The mass transfer occurs by diffusion (neglect the convective term). The flow molar concentration of species of water vapor into the stagnant layer considering the cylindrical coordinate system is determined engaging the equation of change:

$$\frac{\partial C_A}{\partial t} + \left(v_r \frac{\partial C_A}{\partial r} + v_\theta \frac{1}{r}\frac{\partial C_A}{\partial \theta} + v_z \frac{\partial C_A}{\partial z}\right)$$

$$= D_{AB}\left(\frac{1}{r}\frac{\partial}{\partial r}\left(r\frac{\partial C_A}{\partial r}\right) + \frac{1}{r^2}\frac{\partial^2 C_A}{\partial \theta^2} + \frac{\partial^2 C_A}{\partial z^2}\right) + R_A$$

where D_{AB} is the binary diffusion coefficient of water vapor in air. Appling the former assumptions, the general steady-state component mole balance equation in the cylindrical coordinate can be reduced to the following equation:

$$0 = D_{AB}\left(\frac{1}{r}\frac{\partial}{\partial r}\left(r\frac{\partial C_A}{\partial r}\right)\right)$$

Performing double integration generates the two arbitrary constants of integration (C_1 and C_2):

$$C_A = C_1 \ln r + C_2$$

The integration constants are obtained from the following boundary conditions:

$$\text{B.C.1: at } r = r_1,\ C_A = C_{As} \Rightarrow \quad C_{As} = C_1 \ln r_1 + C_2$$
$$\text{B.C.2: at } r = r_2,\ C_A = 0 \Rightarrow \quad 0 = C_1 \ln r_2 + C_2$$

Subtract the equation of boundary conditions 2 from 1 and rearrange to obtain the expression of integration constant C_1:

$$C_1 = \frac{C_{As}}{\ln r_1 / r_2}$$

The expression of the second arbitrary integration constant C_2 is:

$$C_2 = -\frac{C_{As}}{\ln r_1 / r_2} \ln r_2$$

Substitute C_1 and C_2 into the concentration profile equation: $C_A = C_1 \ln r + C_2$ gives the concentration profile of water vapor along the tube length:

$$C_A = \frac{C_{As}}{\ln(r_1 / r_2)} \ln\left(\frac{r}{r_2}\right)$$

COMSOL Simulation

Double-click on COMSOL and open a new case using model wizard. The model wizard will appear in the interface; this will guide you through selecting the appropriate geometry, physics, and study type (stationary or time dependent). This problem can be described in just two dimensions because of the angular symmetry. The following steps are used in depicting the concentration profile via COMSOL software:

1. Click on 2D.
2. Select the appropriate physics, in this case "transport of diluted species," then click "Add."
3. Click on the green arrow "Study."
4. Choose the study type. Because we are assuming steady state (even though this problem is inherently transient), the appropriate study type will be "stationary." To finish click Done.

Right-click on geometry and select both a rectangle and a circle. The region of interest is the region between the water droplet surface and the top of the cylindrical tunnel where water concentration is zero. Subtract the circular water region from the rectangular air region using "Boolean" operation; to do this:

1. Right-click "geometry" and add a rectangle. Set height = 7.5 [mm] and width = 2.5 [mm]. The center of the rectangle should be at the origin.
2. Right-click "geometry" to add a circle. Set the radius = 0.5 [mm] and set the center position to $x = 0$, $y = -1.25$ [mm]. After these two steps your geometry should look like the one shown in Figure 5.32.
3. Right-click on "geometry," select "Booleans and Partition," then select "Difference" for the menu of different possible Boolean

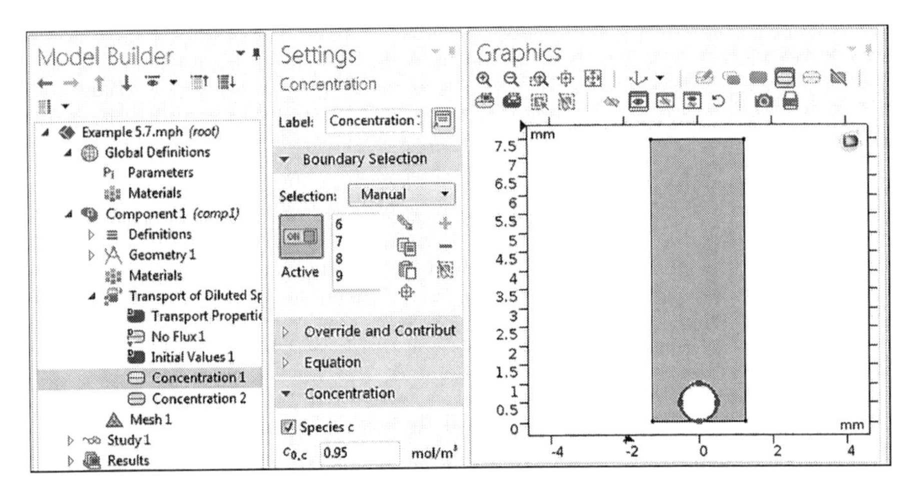

FIGURE 5.32
COMSOL model builder boundary conditions.

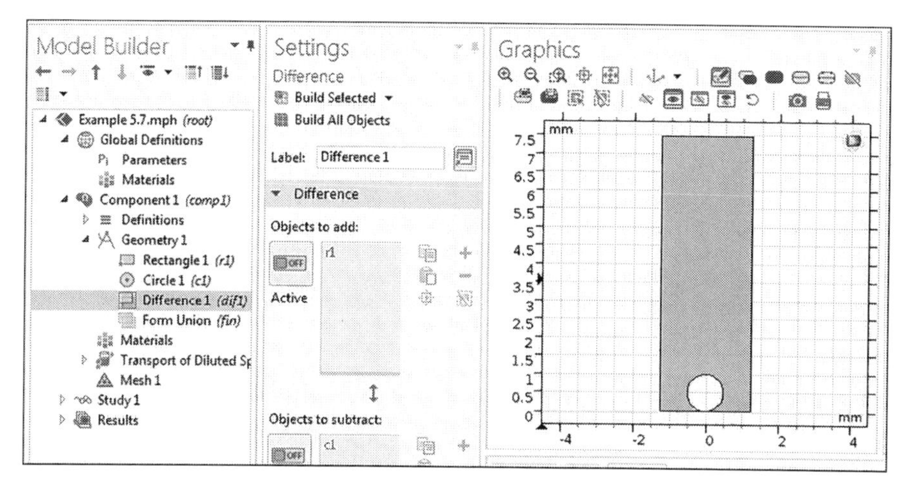

FIGURE 5.33
Booleans and partitions for the difference between a rectangle and a circle.

operations. The area of interest is the difference between rectangle and circle as shown in Figure 5.33. Click "Build all Objects."

4. Under the "Transport of Diluted Species" tab, you will see the "Transport properties" tab. Left-click it to open the interface. Under "Diffusion coefficient," set D_c equal to the appropriate value of 1×10^{-6} m^2/s.
5. Input the appropriate boundary conditions to the problem. Since our two conditions are values of concentration at two surfaces, we will need to add two "concentration" boundary conditions. The concentration boundary condition is found by right-clicking the "Transport of Diluted Species" tab and left-clicking "concentration" since we have two boundaries.
6. Boundary "Concentration 1" is the boundary at the surface of the water droplet. This surface will have the concentration of pure water vapor at the droplet surface at a temperature of 20°C, a partial pressure of 2.34 kPa, approximately, 0.95 mol/m^3.
7. For boundary condition "Concentration 2," add the negligible concentration of water condition to the top surface. At this surface, we can assume $C_A = 0$.
8. The problem is now ready to be solved as the meshing by default will be fine for a problem like this. Right-click on "study" and left-click "compute" in the menu it brings up. The surface plot for the water vapor concentration profile is depicted in Figure 5.34. The concentration is maximum at the surface of the droplet and decreases upward until it reaches zero at the surface of the tube due to the continuous blowing of fresh air that makes the concentration of water at the surface insignificant.

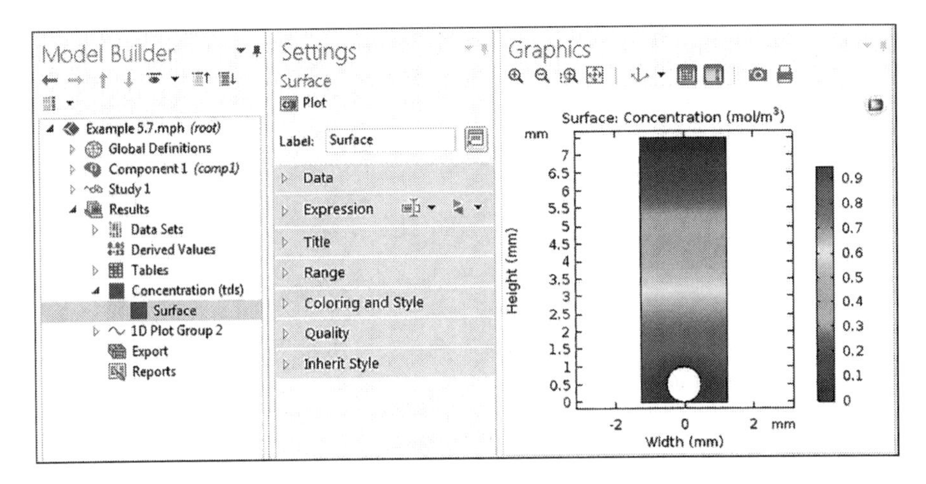

FIGURE 5.34
Surface plot for water vapor concentration in air.

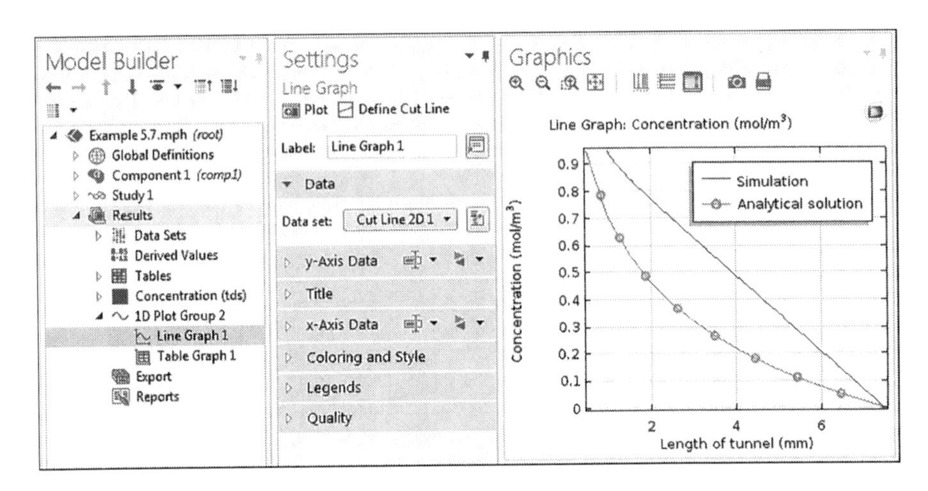

FIGURE 5.35
Comparison of analytical and simulation predictions.

A comparison between analytical solution and simulation predictions is shown in Figure 5.35. We now need to add a one-dimensional plot group to display this cut line data. Right-click on "Results" and left-click "1D Plot Group"; this will add the one-dimensional plot group where the data can be displayed. Right-click the "1D plot group" tab and select "Line Graph." In the "Line Graph" interface choose "Cut line 2D" in the data set box and click the paint brush (graph button at top right of the interface).

The simulation results revealed that the concentration is maximum at the surface of the solid particle and decreases along the length of the tube.

Example 5.8: Diffusion of Oil Spill

Oil spill from vehicles occurs on the surface of the ground. Determine the concentration profile of the contaminating oil concentration profile inside the soil. Assume the spill is large in x and y directions relative to the z direction, and the diffusivity of oil in the soil is $D_{AB} = 1\times10^{-12}\,\text{m}^2/\text{s}$. The soil density is $2000\,\text{kg/m}^3$; oil density is $876.5\,\text{kg/m}^3$. The oil spill concentration on the surface of the ground is $1\,\text{mol/m}^3$; the maximum allowable concentration of oil in the soil is $8\times10^{-6}\,\text{nmol/m}^3$.

Solution

Consider the following assumptions to determine the concentration of oil in the soil:

- Transient state $(\partial C_A/\partial t \neq 0)$.
- Transportation of oil takes place by diffusion only $(v_x = v_y = v_z = 0)$.
- Diffusion in x and y directions are negligible $(D_{AB}(\partial^2 C_A/\partial x^2) = D_{AB}(\partial^2 C_A/\partial y^2) = 0)$.
- No reaction takes place in the benzene spill area $(R_A = 0)$.

The mass transfer occurs by diffusion in the z direction only. The diffusion and molar flow rate of benzene from the spill into the ground is described by Cartesian coordinates. The differential mass balance of mass transport of oil (A) in the soil states that:

$$\frac{\partial C_A}{\partial t} + \left(v_x \frac{\partial C_A}{\partial x} + v_y \frac{\partial C_A}{\partial y} + v_z \frac{\partial C_A}{\partial z} \right) = D_{AB}\left(\frac{\partial^2 C_A}{\partial x^2} + \frac{\partial^2 C_A}{\partial y^2} + \frac{\partial^2 C_A}{\partial z^2} \right) + R_A$$

Based on the considered assumptions, the component mole balance equation of oil concentration through the ground is described by the following simplified equation:

$$\frac{\partial C_A}{\partial t} = D_{AB}\left(\frac{\partial^2 C_A}{\partial z^2} \right)$$

where D_{AB} is the binary diffusion coefficient of oil spill in earth. The system initial and boundary conditions are:
Initial conditions: at $t = 0$, $C_A = 0$ (initial conditions, benzene concentration in soil)

B.C.1: at $t > 0$, $z = 0$, $C_A = C_{As} = 1\ \text{mol/m}^3$
B.C.2: at $t > 0$, $z = \infty$, $C_A = 0$

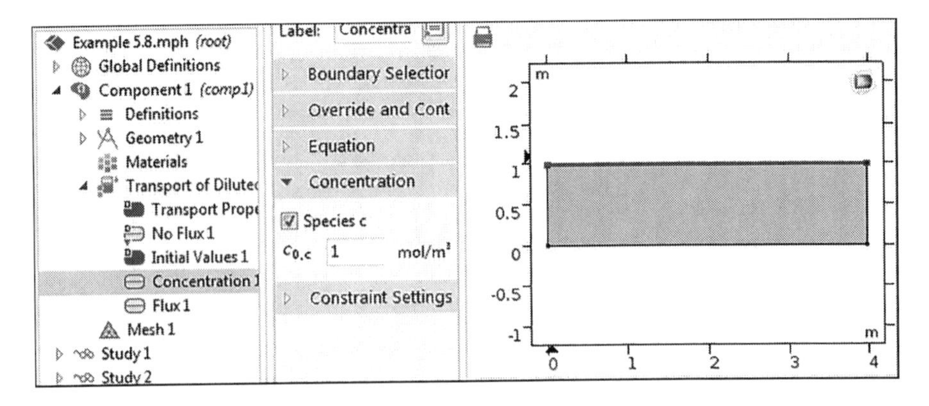

FIGURE 5.36
Boundary conditions of top and bottom borders, the other boundaries are set to no flux.

The simplified unsteady-state partial differential equation that represent the concentration of benzene into the ground:

$$\frac{\partial C_A}{\partial t} = D_{AB}\frac{\partial^2 C_A}{\partial z^2}$$

The analytical solution:

$$C_A = C_{A0} + \left(C_{As} - C_{A0}\right) * \operatorname{erfc}\left(\frac{z}{2\sqrt{D_{AB}t}}\right)$$

The concentration oil at certain distance in the soil is estimated from the following:

$$t = \frac{z^2}{4D_{AB}}\left\{\operatorname{erfc}^{-1}\left(\frac{C_A - C_{A0}}{C_{As} - C_{A0}}\right)\right\}^{-2}$$

COMSOL Solution

The transient system partial differential equation is solved using COMSOL Multiphysics 5.3a. The boundary conditions are shown in Figure 5.36.

The concentration profile of oil spill in the soil is shown in Figure 5.37.

Example 5.9: Diffusion of Benzoic Acid in Pure Water

Pure water at 25°C flows through a smooth metal pipe of 6 cm internal diameter with an average velocity of $1.5 \times 10^{-3}\,\text{m/s}$. A fully developed velocity profile is established in the pipe. The metal pipe is cast from benzoic acid. The length of the pipe made of a benzoic acid is 2 m. Calculate the concentration of benzoic acid in water at the exit of the pipe. For water at 25°C, the density is $\rho = 1000\,\text{kg/m}^3$, viscosity is $\mu = 8.92 \times 10^{-4}\,\text{kg/m}\cdot\text{s}$, and diffusivity of benzoic acid in water is $D_{AB} = 1.21 \times 10^{-9}\,\text{m}^2/\text{s}$. The Schmidt number is $\text{Sc} = 737$, the saturation

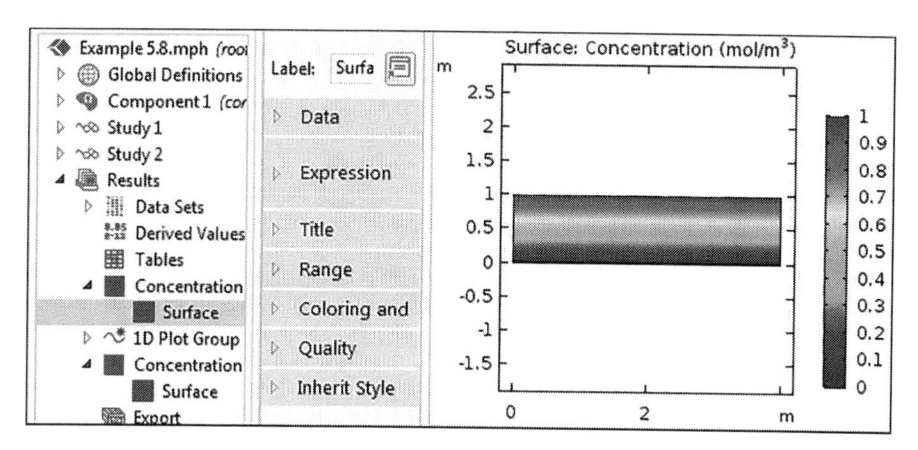

FIGURE 5.37
Steady-state surface concentration of oil in the soil.

solubility of benzoic acid (A) in water (B) = $3.412\ \text{kg/m}^3$, and the molecular weight of benzoic acid is $MW_A = 122.12\ \text{g/mol}$.

Solution

Consider the following assumptions in determining an expression for benzoic acid concentration in pure water:

- Steady state $\left(\partial C_A/\partial t = 0\right)$.
- Neglect convective diffusion $(v_r(\partial C_A/\partial r)+v_\theta(1/r)(\partial C_A/\partial\theta)+v_z(\partial C_A/\partial z))=0$.
- Diffusion in θ and z directions are negligible: $D_{AB}\left((1/r)\left(\partial/\partial r\left(r(\partial C_A/\partial r)\right)+(1/r^2)(\partial^2 C_A/\partial\theta^2)\right)=0\right)$.
- No reaction takes place in the system ($R_A = 0$).

The mass transfer occurs by diffusion in the r direction and convection in the z direction. The diffusion and molar flow rate of benzoic acid from the pipe skin into the running water through the pipe is described by the cylindrical component coordinate:

$$\frac{\partial C_A}{\partial t}+\left(v_r\frac{\partial C_A}{\partial r}+v_\theta\frac{1}{r}\frac{\partial C_A}{\partial\theta}+v_z\frac{\partial C_A}{\partial z}\right)=D_{AB}\left(\frac{1}{r}\frac{\partial}{\partial r}\left(r\frac{\partial C_A}{\partial r}\right)+\frac{1}{r^2}\frac{\partial^2 C_A}{\partial\theta^2}+\frac{\partial^2 C_A}{\partial z^2}\right)+R_A$$

where D_{AB} is the binary diffusion coefficient of water vapor in air. The general steady-state component mole balance equation in the cylindrical coordinate is reduced after implementing the assumption as follows:

$$0 = D_{AB}\left(\frac{1}{r}\frac{\partial}{\partial r}\left(r\frac{\partial C_A}{\partial r}\right)\right)$$

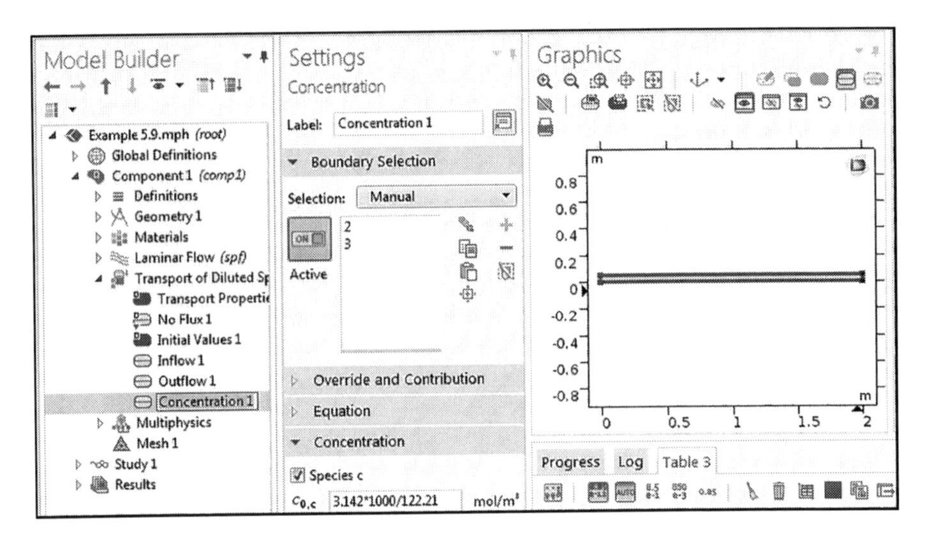

FIGURE 5.38
The average concentration of benzoic acid in water at the exit of the pipe.

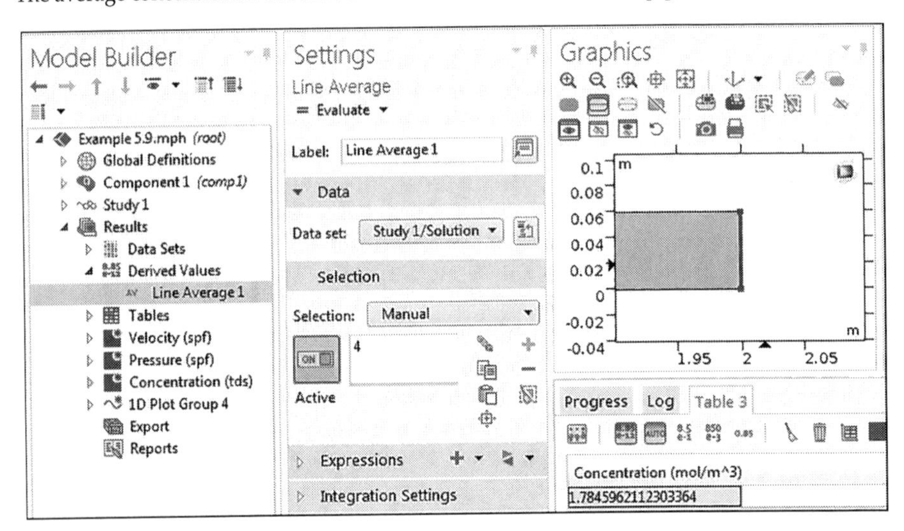

FIGURE 5.39
Average concentration of benzoic acid in water at the exit of the pipe.

The problem is solved using COMSOL Multiphysics as shown in Figure 5.38.

The COMSOL predictions of the average concentration at the exit of the pipe is 1.78 mol/m^3 as shown in Figure 5.39.

$$C_b, \text{exit} = 1.78 \frac{\text{mol}}{\text{m}^3} \times 122.21 \frac{\text{g}}{\text{mol}} \times \frac{1\,\text{kg}}{1000\,\text{g}} = 0.22\,\text{kg/m}^3$$

$$C_b, \text{out} = 0.22\,\text{kg/m}^3$$

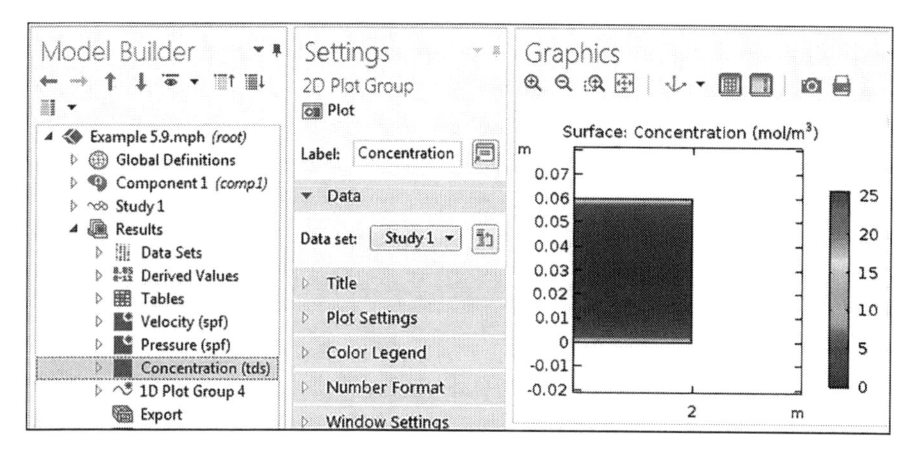

FIGURE 5.40
Surface concentration profile at the exit portion of the pipe.

The concentration profile of benzoic acid at the exit region of the pipe is shown in Figure 5.40. The predictions reveal that the diffusion of benzoic acid through the water is very slow.

PROBLEMS

5.1 Diffusion of Nitrogen into Steel

A sheet of steel 1.5 mm thick has nitrogen atmospheres on both sides at 1200°C and is permitted to achieve a steady-state diffusion condition. The diffusion coefficient for nitrogen in steel at this temperature is $6\times10^{-11}\ m^2/s$, and the diffusion flux is found to be $1.2\times10^{-7}\ kg/m^2s$. The concentration of nitrogen at the surface of the steel at the high-pressure surface is $143\ mol/m^3$. Calculate the concentration of nitrogen at a distance 1 mm from the high-pressure side.

Answer: $71.57\ mol/m^3$

5.2 Diffusion of Nitrogen into Pure Iron

Nitrogen gas from the gaseous phase is diffused into pure iron at 675°C. If the surface concentration of the iron is maintained at $0.2\ mol/m^3 N_2$, the diffusion coefficient for nitrogen in iron bar at 675°C is $2.5\times10^{-11}\ m^2/s$. What will be the concentration 0.3 mm from the surface after 5 h?

Answer: $0.15\ mol/m^3$

5.3 Diffusion of Carbon in Iron–Carbon Alloy

A carbon concentration of 0.45 wt% at is required at a position 2 mm into an iron–carbon alloy that initially contains 0.20 wt% C. The surface concentration is to be maintained at 1.30 wt% C, and the treatment is to be conducted at 1000°C. The diffusion coefficient of C in an iron–carbon alloy at 1000°C is $1.93 \times 10^{-11} m^2/s$. Determine the time necessary to achieve the desired percent C (0.45 wt%).

Answer: 19.7 h

5.4 Hardening of Steel Surface

The wear resistance of a steel shaft is to be improved by hardening its surface. This is to be accomplished by increasing the nitrogen content within an outer surface layer because of nitrogen diffusion into the steel; the nitrogen is to be supplied from an external nitrogen-rich gas at an elevated and constant temperature. The initial nitrogen content of the steel is 0.0025 wt%, whereas the surface concentration is to be maintained at 0.45 wt%. For this treatment to be effective, a nitrogen content of 0.12 wt% must be established at a position 0.45 mm below the surface. The diffusion of coefficient of nitrogen in iron at 625°C is $1.12 \times 10^{-11} m^2/s$. Specify an appropriate heat treatment in terms of temperature and time for a temperature of 625°C. For this treatment to be effective, calculate a nitrogen content at a position 0.45 mm below the surface after 2 hours.

Answer: 0.12 wt%

5.5 Diffusion of CO_2 Through Rubber

A flat plug 30 mm thick having an area of $4.0 \times 10^{-4} m^2$ and made of vulcanized rubber is used for closing an opening in a container. The gas CO_2 at 25°C and 2.0 atm pressure is inside the container. The solubility of CO_2 gas is 0.90 m^3 gas (STP of 0°C and 1 atm) per m^3 rubber per atm pressure of CO_2. The diffusivity is 1.1×10^{-10} m^2/s. Calculate the total leakage or diffusion of CO_2 through the plug; assume that the partial pressure of CO^2 outside is zero.

Answer: 1.178 x 10^{13} kmol/s

5.6 Carburization of Steel Using Methane

Carbon steel containing 0.25 wt% is carburized using methane. Methane is maintained at the surface of the steel at 1.2 wt% carbon. The diffusivity of carbon in carbon steel is $1.6 \times 10^{-11} m^2/s$. Determine the time it takes to achieve a carbon content of 0.80% carbon at a position 0.5 mm below the surface.

Answer: 7.1 h

5.7 Unsteady-State Diffusion in a Slab of Three Insulated Surfaces

A solid slab 0.01 thick has an initial surface concentration of solute A of 1.00 mol/m^3. The diffusivity of *A* in the solid is $D_{AB} = 1.0 \times 10^{-10} m^2/s$.

All surfaces of the slab are insulated except the top surface. The surface concentration is suddenly dropped to zero concentration and held there. Unsteady-state diffusion occurs in the one x direction with the rear surface insulated. Determine the concentration profile using COMSOL after 1 hour.

5.8 Solute Diffusion in a Slab of Two Insulated Phases

A solid slab 0.01 thick has an initial uniform concentration of solute A of 1.00 mol/m^3. The diffusivity of A in the solid is $D_{AB} = 1.0 \times 10^{-10} \text{m}^2/\text{s}$. All surfaces of the slab are insulated except the top rear surfaces. The surface concentration is suddenly dropped to zero concentration and held there. Unsteady-state diffusion occurs in the one x direction with the rear surface insulated. Determine the concentration profile using COMSOL Multiphysics.

5.9 Solute Diffusion in a Solid Slab

A solid slab 0.01 thick has an initial uniform concentration of solute A of 1.00 mol/m^3. The diffusivity of A in the solid is $D_{AB} = 1.0 \times 10^{-10} \text{m}^2/\text{s}$. The bottom surface is the only insulated surface. The surface concentration is suddenly dropped to zero concentration and held there. Steady-state diffusion occurs in the one x direction with the rear surface insulated. Determine the concentration profile using COMSOL Multiphysics.

5.10 Mass Transfer from a Pipe Wall

Pure water at 25°C is flowing at a velocity of 0.0305 m/s in a tube having an inside diameter of 6.35 mm. The tube is 1.22 m, and the walls are coated with benzoic acid. The solubility of benzoic acid in water is 29.48 mol/m^3. Calculate the average concentration of benzoic acid at the outlet of the tube. The diffusivity of benzoic acid in water is $1.245 \times 10^{-9} \text{m}^2/\text{s}$. The Schmidt number benzoic acid in water at 25°C is approximately 702 and the Reynold number is 221.4. Use the correlation

$$\frac{W}{D_{AB}\rho L} = N_{\text{Re}} N_{\text{sc}} \left(\frac{D}{L}\right)\left(\frac{\pi}{2}\right)$$

$$\frac{C_A - C_{A0}}{C_{As} - C_{A0}} = 5.5\left(\frac{W}{D_{AB}\rho L}\right)^{-\frac{2}{3}}$$

Answer: 2.906 mol/m^3

References

1. Fogler, H. S., 2014. *Elements of Chemical Reaction Engineering*, 4th ed., New York: Pearson.
2. Incropera, F. P., D. P. DeWitt, 1996. *Fundamentals of Heat and Mass Transfer*, 4th ed., New York: John Wiley & Sons.
3. Bird, R. B., W. E. Stewart, E. N. Lightfoot, 2007. *Transport Phenomena*, 2nd ed., New York: John Wiley & Sons.
4. Geankoplis, C. J., 2009. *Transport Processes and Separation Process Principals, Include Unit Operations*, 4th ed., Upper Saddle River, NJ: Prentice Hall.

6

Heat Transfer Distributed Parameter Systems

Heat transfer of distributed parameter systems is distinguished by the fact that the states depend on spatial position. The heat transfer equation of the distributed system is a partial differential equation that describes the distribution of heat or variation in temperature in a specific region over time. In some cases, exact solutions of the simplified equations are possible; in other cases, the equations must be solved numerically using computational methods. In this chapter, COMSOL Multiphysics 5.3a is used to solve the cases requiring numerical computation.

LEARNING OBJECTIVES

- Develop mathematical models for distributed parameters systems (energy balance).
- Represent systems by partial differential equations (PDEs).
- Solve the system model equations using COMSOL Multiphysics software.
- Compare COMSOL predictions with manual calculations.

6.1 Introduction

The exchange of thermal energy between physical systems is referred to as heat transfer. The rate of heat transfer depends on factors such as the temperature of the system and the properties of the dominant medium through which the heat is transferred. The direction of heat transfer moves from a region of higher temperature to another region of lesser temperature and is governed by the second law of thermodynamics. Heat can be transferred from one place to another by three methods: conduction in solids, convection of fluids (liquids or gases), and radiation through anything that will allow radiation to pass. The method used to transfer heat is usually the one that is the most efficient. Conduction occurs when two objects at different temperatures are in contact with each other. Heat flows from the warmer to the cooler object until both objects are at the same temperature. Convection in liquids and gases is usually the most efficient way to transfer heat. Convection occurs when warmer areas of a liquid or gas rise to cooler areas in the liquid or gas. As this happens, cooler liquid or gas takes the place of the warmer areas, which have risen higher. Both conduction and convection require matter to transfer heat. Radiation is a method of heat transfer that does not rely upon any contact between the heat source and the heated object. For example, we feel heat from the sun even though we are not touching it.

Heat is a form of energy that can be transferred from one object to another; the usual symbol for heat is Q. Common units for measuring heat are the Joule and Calorie in the SI system. Temperature is a measure of the ability of a substance to transfer heat energy to another physical system; it is a measure of the amount of energy controlled by the molecules of a substance. Temperature is a relative measure of how hot or cold a substance is and can be used to forecast the trend of heat transfer. The traditional symbol for temperature is T. The scales for measuring temperature in SI units are the Celsius and Kelvin temperature scales [1–3].

6.1.1 Equations of Energy

The temperature distribution in various geometrical units for a Newtonian fluid with constant thermal conductivity is [4–6]:

1. Equation of change for heat transfer in Cartesian coordinates

$$\rho C_p \frac{\partial T}{\partial t} + \rho C_p \left(v_x \frac{\partial T}{\partial x} + v_y \frac{\partial T}{\partial y} + v_z \frac{\partial T}{\partial z} \right) = k \left(\frac{\partial^2 T}{\partial x^2} + \frac{\partial^2 T}{\partial y^2} + \frac{\partial^2 T}{\partial z^2} \right) + \Phi_H \tag{6.1}$$

2. Equation of change for heat transfer in cylindrical coordinates

$$\rho C_p \frac{\partial T}{\partial t} + \rho C_p \left(v_r \frac{\partial T}{\partial r} + \frac{v_\theta}{r} \frac{\partial T}{\partial \theta} + v_z \frac{\partial T}{\partial z} \right) = k \left(\frac{1}{r} \frac{\partial}{\partial r} \left(r \frac{\partial T}{\partial r} \right) + \frac{1}{r^2} \frac{\partial^2 T}{\partial \theta^2} + \frac{\partial^2 T}{\partial z^2} \right) + \Phi_H \tag{6.2}$$

3. Equation of change for heat transfer in spherical coordinates

$$\rho C_p \frac{\partial T}{\partial t} = \frac{1}{r^2} \frac{\partial}{\partial r} \left(k r^2 \frac{\partial T}{\partial r} \right) + \frac{1}{r^2 sin\theta} \frac{\partial}{\partial \theta} \left(k sin\theta \frac{\partial T}{\partial \theta} \right) + \frac{1}{r^2 \sin^2 \theta} \frac{\partial}{\partial \phi} \left(k \frac{\partial T}{\partial \phi} \right) + \Phi_H \tag{6.3}$$

The method of separation of variables is useful in the determination of solutions to heat transfer problems in cylindrical, Cartesian, and spherical coordinates. The following are selected cases to illustrate the applications of equations of energy to obtain temperature distribution for various unit processes.

6.2 Heat Transfer from a Fin

Heat transfer from surfaces is increased for many applications using fins. Typically, the material of the fin possesses a high thermal conductivity. The fin's high thermal conductivity will allow high amounts of heat to be conducted from the wall through the fin. The design of cooling fins is required in many situations and thus we examine heat transfer in a fin as a way of defining some criteria for design. In this case, a study of the temperature distribution in a cylindrical metallic fin of radius R and length L will be demonstrated (Figure 6.1). The fin made of copper is originally at a uniform temperature of T_0. The ratio of the fin diameter to fin length is very small. Suppose that one end of the fin is brought into contact with a hot fluid of temperature T_h, and the surface area of the fin is exposed to ambient temperature T_a.

Copper has high thermal conductivity that makes the heat transfer by conduction significant. If the radius of the rod is very small compared to rod length, then radial temperature gradient can be neglected. If the rod diameter is large enough, the thermal distribution in the radial direction cannot be neglected. The following transport equation for heat transfer in cylindrical

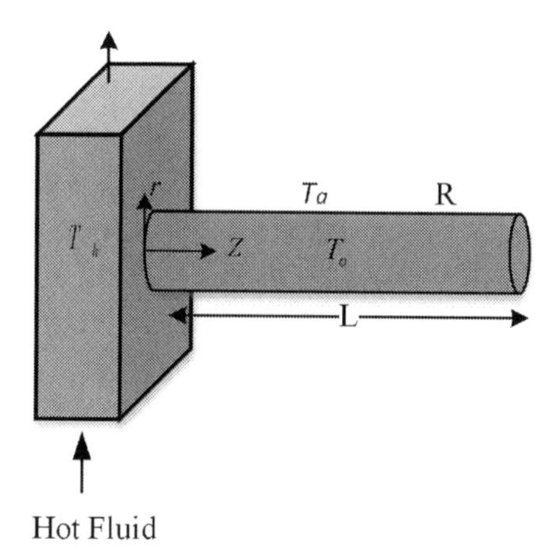

FIGURE 6.1
Temperature profile in cylindrical fin.

coordinates can be written to develop a mathematical model to describe the fin radial and axial temperature change:

$$\rho C_p \frac{\partial T}{\partial t} + \rho C_p \left(v_r \frac{\partial T}{\partial r} + v_\theta \frac{1}{r} \frac{\partial T}{\partial \theta} + v_z \frac{\partial T}{\partial z} \right) = k \left(\frac{1}{r} \frac{\partial}{\partial r} \left(r \frac{\partial T}{\partial r} \right) + \frac{1}{r^2} \frac{\partial^2 T}{\partial \theta^2} + \frac{\partial^2 T}{\partial z^2} \right) + \Phi_H \quad (6.4)$$

The fin is stationary; accordingly, the terms containing the velocity may be discarded. Neglecting convective terms and heat transfer in the angular direction gives the following simplified partial different equation:

Temperature gradient in the radial direction is neglected

$$\rho C_p \frac{\partial T}{\partial t} + \rho C_p \left(v_r \frac{\partial T}{\partial r} + v_\theta \frac{1}{r} \frac{\partial T}{\partial \theta} + v_z \frac{\partial T}{\partial z} \right) = k \left(\frac{1}{r} \frac{\partial}{\partial r} (r \frac{\partial T}{\partial r}) + \frac{1}{r^2} \frac{\partial^2 T}{\partial \theta^2} + \frac{\partial^2 T}{\partial z^2} \right) + Q$$

Convection heat transfer is neglected

$$\rho C_p \frac{\partial T}{\partial t} = k \left(\frac{1}{r} \frac{\partial}{\partial r} \left(r \frac{\partial T}{\partial r} \right) + \frac{\partial^2 T}{\partial z^2} \right) + Q \quad (6.5)$$

If we assume steady-state conditions and the ratio of fin diameter to length is very small, then the radial effect is neglected and the partial differential equation is reduced to Equation 6.6:

$$0 = k \left(0 + \frac{\partial^2 T}{\partial z^2} \right) + Q \quad (6.6)$$

The heat transferred per unit volume (Q) from the fin walls is described by Equation 6.7:

$$Q = h \frac{A_s}{V_b} (T_a - T) = h \frac{2\pi RL}{\pi R^2 L} (T_a - T) \quad (6.7)$$

where:
- A_s is the fin surface area
- V_b is the fin volume

The fin perimeter (P) and fin cross-sectional area (A_c) can be substituted in the energy balance equation:

$$P = 2\pi R, \; A_c = \pi R^2 \quad (6.8)$$

If the fin has a constant cross section along its length, the area and perimeter are constant and the differential equation for temperature is simplified to:

$$0 = \frac{d^2T}{dz^2} - \frac{hP}{kA_c}(T - T_a) \tag{6.9}$$

Let $m^2 = hP/(kA_c)$ and $\theta(z) = (T - T_a)$:

$$0 = \frac{d^2\theta}{dz^2} - m^2\theta \tag{6.10}$$

The solution to the shortened equation is:

$$\theta(z) = C_1e^{mz} + C_2e^{-mz}$$

To find the temperature distribution, the previous equation can be solved for the constants C_1 and C_2 as follows:

B.C.1: at $z = 0$, $\theta = \theta_h$, wall of the fin

B.C.2: at $z = L$, $d\theta/dz = 0$, thermal insulation, no heat loss from the fin axial end

The temperature distribution:

$$\frac{\theta}{\theta_h} = \frac{coshm(L - x)}{coshmL} \tag{6.11}$$

6.3 Radial Temperature Gradients in an Annular Chemical Reactor

An annular chemical reactor is a packed bed with catalyst particles positioned between two coaxial cylinders (Figure 6.2). The catalytic reaction is being carried out at constant pressure in a packed bed region. The inner and

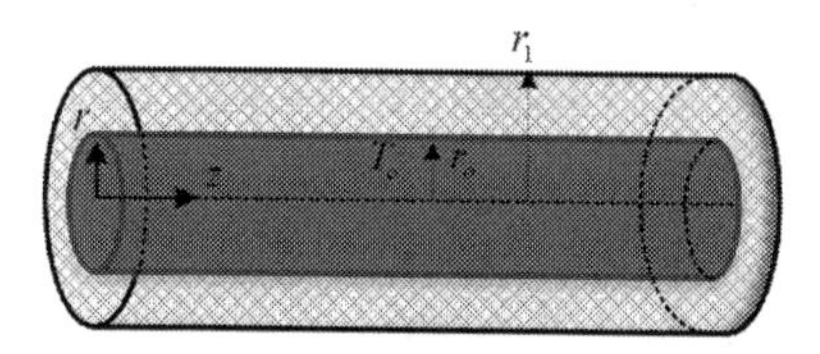

FIGURE 6.2
An annular chemical reactor consists of a packed bed of catalyst.

outer cylinders have radii of r_o and r_1, respectively. Assume that there is no heat transfer through the surface of the inner cylinder, which is at a constant temperature T_0. The chemical catalytic reaction is exothermic and releases heat at a constant uniform volumetric rate $Q(\mathrm{W/m^3})$ throughout the reactor. The reactor thermal conductivity k is constant. Derive the radial temperature distribution in the annular catalytic reactor.

The following assumptions are considered in simplifying the model equations:

- Steady state
- Neglect of convective heat
- Neglect of the temperature gradients in the axial and angular direction

The heat equation is a parabolic partial differential equation that describes the distribution of heat (or variation in temperature) in each region over time. Use the general differential equation that describes the heat transfer in cylindrical coordinate:

$$\rho C_p \frac{\partial T}{\partial t} + \rho C_p \left(v_r \frac{\partial T}{\partial r} + v_\theta \frac{1}{r} \frac{\partial T}{\partial \theta} + v_z \frac{\partial T}{\partial z} \right) = k \left(\frac{1}{r} \frac{\partial}{\partial r} \left(r \frac{\partial T}{\partial r} \right) + \frac{1}{r^2} \frac{\partial^2 T}{\partial \theta^2} + \frac{\partial^2 T}{\partial z^2} \right) + \Phi_H \quad (6.12)$$

This partial differential equation depends on time and three more variables: z, θ, and r. For catalytic reactor, convective heat can be neglected; accordingly:

No convective heat

$v_r = v_\theta = v_z = 0$

$$\rho C_p \frac{\partial T}{\partial t} + \rho C_p \left(\cancel{v_r \frac{\partial T}{\partial r}} + \cancel{v_\theta \frac{1}{r} \frac{\partial T}{\partial \theta}} + \cancel{v_z \frac{\partial T}{\partial z}} \right) = k \left(\frac{1}{r} \frac{\partial}{\partial r} \left(r \frac{\partial T}{\partial r} \right) + \frac{1}{r^2} \frac{\partial^2 T}{\partial \theta^2} + \frac{\partial^2 T}{\partial z^2} \right) + \Phi_H$$

The steady-state assumption implies that $\partial T/\partial t = 0$.

Steady state

No convective heat

$v_r = v_\theta = v_z = 0$

$$\rho C_p \cancel{\frac{\partial T}{\partial t}} + \rho C_p \left(\cancel{v_r \frac{\partial T}{\partial r}} + \cancel{v_\theta \frac{1}{r} \frac{\partial T}{\partial \theta}} + \cancel{v_z \frac{\partial T}{\partial z}} \right) = k \left(\frac{1}{r} \frac{\partial}{\partial r} \left(r \frac{\partial T}{\partial r} \right) + \frac{1}{r^2} \frac{\partial^2 T}{\partial \theta^2} + \frac{\partial^2 T}{\partial z^2} \right) + \Phi_H$$

Temperature gradients in the axial and angular direction are neglected:

$$\frac{\partial^2 T}{\partial z^2} = \frac{\partial^2 T}{\partial \theta^2} = 0$$

Steady state — No convective heat $v_r = v_\theta = v_z = 0$ — Temperature gradients in the axial and angular direction is neglected

$$\rho C_p \frac{\partial T}{\partial t} + \rho C_p \left(v_r \frac{\partial T}{\partial r} + v_\theta \frac{1}{r}\frac{\partial T}{\partial \theta} + v_z \frac{\partial T}{\partial z} \right) = k\left(\frac{1}{r}\frac{\partial}{\partial r}\left(r\frac{\partial T}{\partial r} \right) + \frac{1}{r^2}\frac{\partial^2 T}{\partial \theta^2} + \frac{\partial^2 T}{\partial z^2} \right) + \Phi_H$$

where Φ_H represents the rate of heat generation by chemical reaction per unit volume. If we assume steady-state conditions, then the partial differential equation is reduced to:

$$0 = k\left(\frac{1}{r}\frac{\partial}{\partial r}\left(r\frac{\partial T}{\partial r} \right) \right) + Q \tag{6.13}$$

The partial differential equation can be further rearranged:

$$\frac{d}{dr}\left(r\frac{dT}{dr} \right) = -\frac{Q}{k} r \tag{6.14}$$

On the first integral:

$$r\frac{dT}{dr} = -\frac{Q}{2k} r^2 + C_1 \tag{6.15}$$

The second integral:

$$T = -\frac{Q}{4k} r^2 + C_1 ln r + C_2 \tag{6.16}$$

Using the boundary conditions, the arbitrary constant of integrations C_1 and C_2 can be determined from boundary conditions. The first boundary condition recommends no heat transfer through the inner cylindrical wall of the annulus.

B.C.1 : $r = r_o$, $dT/dr = 0$, the first arbitrary constant of integration C_1 is:

$$C_1 = -\frac{Q}{2k} r_o^2$$

The second boundary condition suggests that the heat is constant at the inner cylinder wall of the annulus.

B.C.2: $r = r_o$, $T = T_o$, and the second arbitrary constant of integration C_2 is:

$$C_2 = T_0 + \frac{Q}{4k} r_o^2 - \frac{Q}{2k} r_o^2 ln r_o$$

After substituting the integration constants, the temperature profile is expressed by:

$$T = \frac{Qr_o^2}{4k}\left\{1-\left(\frac{r}{r_o}\right)^2 + 2ln\left(\frac{r}{r_o}\right)\right\} + T_0 \tag{6.17}$$

6.4 Heat Transfer in a Nonisothermal Plug-Flow Reactor

The gas phase reaction takes place under nonisothermal conditions in a plug-flow reactor. Control over the temperature in the reactor is essential to achieve reasonable conversion. The reaction is exothermic, and hence the heat of reaction is removed via a cooling jacket surrounding the reactor, as shown in Figure 6.3. The average temperature of the cooling water in the jacket is assumed to be at T_w. The reactants enter the tubular reactor at temperature T_o and inlet concentration C_{Ao}. The reactor length is L. Develop the model equations for the temperature profile along the axial length of the tube. In this example neglect the heat transfer in the radial and angular directions.

The following assumptions are made for the energy balance:

- Steady state.
- Energy flow will be due to bulk flow (convection) and conduction.
- No radial variation in velocity, concentration, temperature, or reaction rate.

The system is characterized by the general differential equation for the change of temperature in cylindrical coordinates:

$$\rho C_p \frac{\partial T}{\partial t} + \rho C_p\left(v_r \frac{\partial T}{\partial r} + v_\theta \frac{1}{r}\frac{\partial T}{\partial \theta} + v_z \frac{\partial T}{\partial z}\right) = k\left(\frac{1}{r}\frac{\partial}{\partial r}\left(r\frac{\partial T}{\partial r}\right) + \frac{1}{r^2}\frac{\partial^2 T}{\partial \theta^2} + \frac{\partial^2 T}{\partial z^2}\right) + Q \tag{6.18}$$

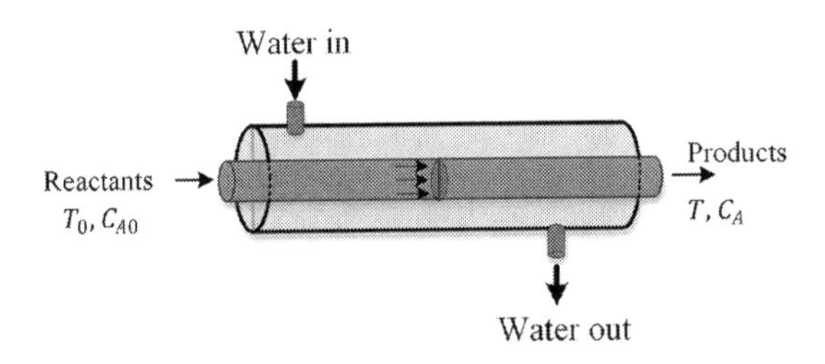

FIGURE 6.3
Nonisothermal plug flow reactor.

The heat is generated due to chemical reaction and through the wall in terms of the cooling jacket:

$$Q = -\Delta H_r k_o e^{-\frac{E}{RT}} C_A - h\left(\frac{\pi DL}{A_c L}\right)(T - T_w) \tag{6.19}$$

where A_c is the cross-sectional area of the reactor. Since the fluid is incompressible, it satisfies the equation of continuity. Substituting these expressions and expanding gives:

$$\rho C_p \frac{\partial T}{\partial t} = -\rho C_p\left(v_z \frac{\partial T}{\partial z}\right) + k\left(\frac{\partial^2 T}{\partial z^2}\right) - \Delta H_r k_o e^{-\frac{E}{RT}} C_A - h\left(\frac{\pi D}{A_c}\right)(T - T_w) \tag{6.20}$$

At steady state ($\partial T/\partial t = 0$), the partial differential equation is reduced to:

$$0 = -\rho C_p\left(v_z \frac{dT}{dz}\right) + k\left(\frac{d^2 T}{dz^2}\right) - \Delta H_r k_o e^{-\frac{E}{RT}} C_A - h\left(\frac{\pi D}{A}\right)(T - T_w) \tag{6.21}$$

The following boundary conditions are considered:

$$\text{at } z = 0 \; T(z) = T_o$$
$$\text{at } z = L, dT/dz = 0$$

The first boundary condition states that the temperature at the entrance of the reactor and the second boundary condition indicates that there is no heat flux at the exit length of the reactor.

6.5 Temperature Profile across a Composite Plane Wall

A composite material is made from two or more constituent materials with significantly different physical or chemical properties. Composite material produces a material with characteristics different from the individual ones when combined. In this example, the composite wall is constructed from different materials on opposite sides of the wall that are bonded together, one forming the facing of the wall and the other is the backup. The plane wall consists of two materials (Figure 6.4). The wall is exposed to convection on the right side. The wall of first material has uniform heat generation q, and thermal conductivity k_1, and thickness L_1. Material 2 has no heat generation with thermal conductivity k_2, and thickness L_2. The inner surface of material 1is well insulated, while the outer surface of material 2 is cooled using a water stream with T_a and heat transfer coefficient h. Develop the steady-state temperature distribution profile for both materials and determine the temperature of the insulated surface and the temperature of the cooled surface.

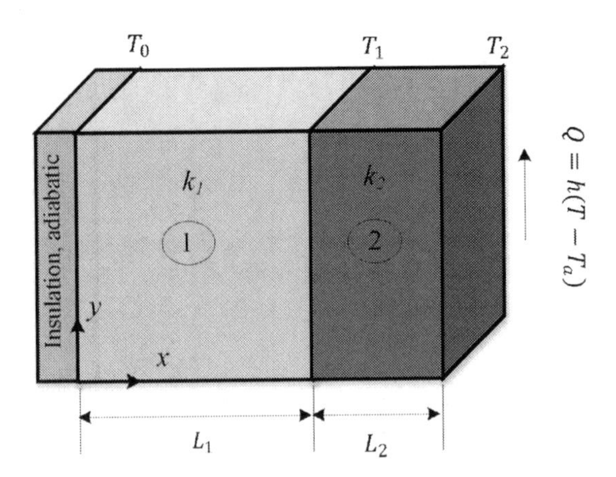

FIGURE 6.4
Schematic of composite walls.

Consider the following assumptions in developing the steady-state temperature distribution profile for both materials:

- Heat transfer is steady since there is no indication that it changes with time.
- Heat transfer is one dimensional.
- Thermal conductivities are constant.
- The thermal contact resistance at the interface is negligible.
- The heat transfer coefficient incorporates the radiation effects.

Based on the steady state and no bulk motion assumptions, heat is transferred in the x-direction by conduction, and heat is lost to air by convective heat flux. The problem is solved using the general differential equation for heat transfer in Cartesian coordinates:

$$\rho C_p \frac{\partial T}{\partial t} + \rho C_p \left(v_x \frac{\partial T}{\partial x} + v_y \frac{\partial T}{\partial y} + v_z \frac{\partial T}{\partial z} \right) = k \left(\frac{\partial^2 T}{\partial x^2} + \frac{\partial^2 T}{\partial y^2} + \frac{\partial^2 T}{\partial z^2} \right) + \Phi_H \quad (6.22)$$

Based on the preceding assumptions, the general differential equation is simplified as follows:

No convective heat flux

Steady state $\quad v_r = v_\theta = v_z = 0$

$$\rho C_p \frac{\partial T}{\partial t} + \rho C_p \left(v_x \frac{\partial T}{\partial x} + v_y \frac{\partial T}{\partial y} + v_z \frac{\partial T}{\partial z} \right) = k \left(\frac{\partial^2 T}{\partial x^2} + \frac{\partial^2 T}{\partial y^2} + \frac{\partial^2 T}{\partial z^2} \right) + \Phi_H$$

Thermal symmetry about the center line and no variation in the axial direction

6.5.1 Energy Balance on Wall 1

The first wall has a heat generation term is bounded with insulated surface, and the interface of material two; consequently, the model partial differential equation is reduced to:

$$0 = k_1\left(\frac{\partial^2 T}{\partial x^2}\right) + \dot{q} \tag{6.23}$$

The first integral generates the first arbitrary constant of integration C_1:

$$\frac{dT}{dx} = -\frac{\dot{q}}{k_1}x + C_1 \tag{6.24}$$

The second integral generates the second arbitrary constant of integration C_2:

$$T = -\frac{\dot{q}}{2k_1}x^2 + C_1 x + C_2 \tag{6.25}$$

The integration constants are found using the following boundary conditions: Left side of plane 1 is insulated, insulated wall implies that:

B.C.1: at $x = 0$, $dT/dx = 0$, this boundary conditions leads to: $C_1 = 0$

At the interface between the two planes, the wall temperature is maintained at temperature T_i.

B.C.2: at $x = L_1$, $T_1 = T_i$

Substitute boundary condition 2 and rearrange:

$$T_i = -\frac{\dot{q}}{2k_1}L_1^2 + C_2 \tag{6.26}$$

Rearrange to get the second arbitrary constant of integration C_2:

$$C_2 = T_i + \frac{\dot{q}}{2k_1}L_1^2 \tag{6.27}$$

Substitute C_1 and C_2 in the general temperature distribution of plane one:

$$T = -\frac{\dot{q}}{2k_1}x^2 + T_i + \frac{\dot{q}}{2k_1}L_1^2 \tag{6.28}$$

Rearrange:

$$T = \frac{\dot{q}}{2k_1}(L_1^2 - x^2) + T_i \tag{6.29}$$

where the interface temperature between the two planes T_i is:

$$T_i = T - \frac{\dot{q}}{2k_1}(L_1^2 - x^2) \tag{6.30}$$

6.5.2 Energy Balance on Wall 2

Under steady-state assumptions, no convective, and no heat generation in wall 2, the model partial differential equation is reduced to:

$$0 = k_2\left(\frac{\partial^2 T}{\partial x^2}\right) \tag{6.31}$$

Perform the first integral:

$$\frac{dT}{dx} = C_{21} \tag{6.32}$$

The first integral generates the first arbitrary constant of integration C_{21}, and the second integral generates C_{22}; hence the general temperature profile in wall 2 is:

$$T = C_{21}x + C_{22} \tag{6.33}$$

The boundary conditions are given by:

At the right side of wall 1, convective heat transfer from the wall implies that at

B.C.1: at $x = L_1 + L_2,\ q = -k_2\, dT/dx = h(T_2 - T_a)$

This boundary condition is obtained by taking the first derivative of the general temperature profile equation:

$$\frac{dT}{dx} = C_{21} \tag{6.34}$$

Multiply both sides by $-k_2$, replace $-k_2\, dT/dx$ by $h(T_2 - T_a)$, and divide both sides of the equation by k_2, as follows:

$$C_{21} = -\frac{h}{k_2}(T_2 - T_a) \tag{6.35}$$

At the plane's interface, the wall temperature is maintained at constant temperature T_i:

B.C.2: at $x = L_1,\ T = T_i$

Substitute boundary condition 2 in the temperature distribution of partition 2:

$$T_i = -\frac{h}{k_2}(T_2 - T_a)(L_1) + C_{22} \tag{6.36}$$

Rearrange to obtain the second arbitrary constant of integration C_{22}:

$$T_i + \frac{h}{k_2}(T_2 - T_a)L_1 = C_{22} \tag{6.37}$$

Substitute the integration constants in the general energy balance equation of plane 2:

$$T = -\frac{h}{k_2}(T_2 - T_a)x + T_i + \frac{h}{k_2}(T_2 - T_a)L_1 \tag{6.38}$$

Reorganize to obtain the temperature distribution profile of the walls:

$$T = T_i + \frac{h}{k_2}(T_2 - T_a)(L_1 - x) \tag{6.39}$$

6.6 Applied Examples

The following examples are solved analytically when possible and numerically by employing the efficient software package COMSOL Multiphysics 5.3a. Manual and software results are compared; in most cases the analytical results and simulation predictions are consistent.

Example 6.1: Heat Transfer in a Shield Electrical Heater

Consider the cylindrical electrical heating rod shown in Figure 6.5. The rod is surrounded by a copper material 0.05 m thick, which also starts 0.05 m away from the center of the rod. The entire assembly is immersed in a fluid maintained at constant temperature. Assume the temperature of the heater is constant at 400 K at R_1. Initially, the temperature of fluid

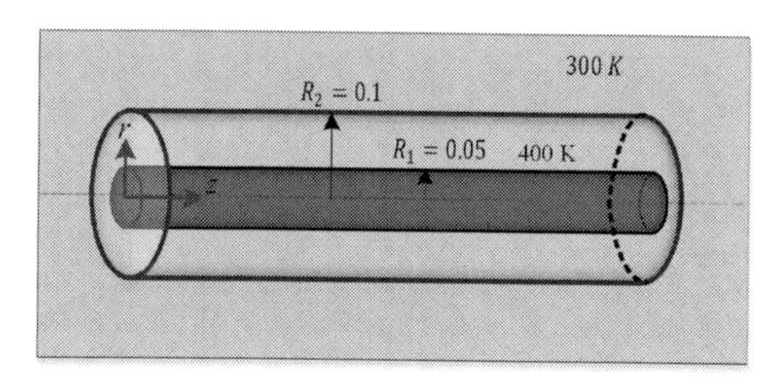

FIGURE 6.5
Heat transfer within heart shield.

surrounding the shield is 300 K, the same temperature of the surrounding cover at R_2. Develop an expression for the temperature distribution $T(r)$ in the shield. Determine the steady-state temperature profile within the copper casing in the case without and with heat generation source per unit volume (100,000 kW/m^3).

Solution

Consider the following assumptions in determining the temperature profile in the heater shield:

- Steady-state conditions.
- One-dimensional radial conduction.
- Constant properties.
- Uniform volumetric heat generation within the rod.
- Outer surface is adiabatic and at constant temperature.

The general differential energy balance equation in cylindrical coordinate:

$$\rho C_p \frac{\partial T}{\partial t} + \rho C_p \left(v_r \frac{\partial T}{\partial r} + v_\theta \frac{1}{r}\frac{\partial T}{\partial \theta} + v_z \frac{\partial T}{\partial z} \right)$$

$$= k\left(\frac{1}{r}\frac{\partial}{\partial r}\left(r\frac{\partial T}{\partial r} \right) + \frac{1}{r^2}\frac{\partial^2 T}{\partial \theta^2} + \frac{\partial^2 T}{\partial z^2} \right) + \Phi_H$$

Based on the steady-state operation, no heat is transferred by convection within the rod wall. The convective term is lost. Neglect end effects by taking the energy balance far from both ends of the rod. To determine $T(r)$ and for the prescribed conditions, we have the appropriate and reduced form of the heat equation:

No convective heat

Steady state $\quad v_r = v_\theta = v_z = 0 \quad$ One dimensional radial conduction

$$\rho C_p \frac{\partial T}{\partial t} + \rho C_p \left(v_r \frac{\partial T}{\partial r} + v_\theta \frac{1}{r}\frac{\partial T}{\partial \theta} + v_z \frac{\partial T}{\partial z} \right) = k\left(\frac{1}{r}\frac{\partial}{\partial r}\left(r\frac{\partial T}{\partial r} \right) + \frac{1}{r^2}\frac{\partial^2 T}{\partial \theta^2} + \frac{\partial^2 T}{\partial z^2} \right) + \Phi_H$$

no internal heat generation

The general differential energy balance equation is reduced to the following form:

$$\frac{1}{r}\frac{\partial}{\partial r}\left(r\frac{\partial T}{\partial r} \right) = 0$$

Performing double integration generates two arbitrary constants of integrations, C_1 and C_2:

$$T(r) = C_1 \ln r + C_2$$

To evaluate C_1 and C_2, the following two boundary conditions are required:

B.C.1: at $r = R_1 = 0.05\ m,\ T = T_1 = 400\,\text{K}$
B.C.2: at $r = R_2 = 0.1\ m,\ T = T_2 = 300\,\text{K}$

Consequently, from the first boundary condition:

$$T_1 = C_1 ln\, R_1 + C_2,$$

Substitute the second boundary condition:

$$T_2 = C_1 ln\, R_2 + C_2$$

Solve for C_1 and C_2 by subtracting the former equations to obtain:

$$C_1 = \frac{(T_1 - T_2)}{(Ln\, R_1/R_2)}$$

Substitute C_1 in any of the previous equations to get:

$$C_2 = T_1 - \frac{T_1 - T_2}{Ln\dfrac{R_1}{R_2}} ln\, R_1$$

Substitute C_1 and C_2 to obtain the general temperature distribution profile of the fluid located between the two coaxial cylinders:

$$T(r) = \frac{T_1 - T_2}{Ln\dfrac{R_1}{R_2}} \ln r + T_1 - \frac{T_1 - T_2}{Ln\dfrac{R_1}{R_2}} ln\, R_1$$

Rearrange by taking the common factor: $T_1 - T_2/Ln(R_1/R_2)$

$$T(r) = \frac{T_1 - T_2}{Ln\dfrac{R_1}{R_2}}\left(\ln\frac{r}{R_1}\right) + T_1$$

The analytical solution is depicted in Figure 6.6. The diagram reveals that there is almost a linear decay in the temperature along the radius of the insulation shield.

COMSOL Simulation

COMSOL Multiphysics is a finite element analysis (FEA) solver and simulation software package for various physics and engineering applications, especially coupled phenomena. In addition, COMSOL Multiphysics allows for entering coupled systems of PDEs. The following is a

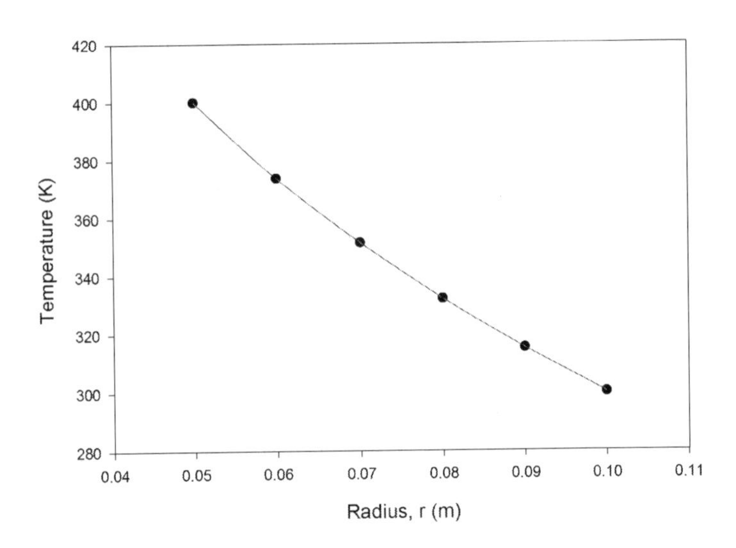

FIGURE 6.6
Temperature profile across the radius of the cylinders.

step-by-step procedure on how to solve energy balance problems using COMSOL Multiphysics 5.3a:

1. Start omsol Multiphysics 5.3a.
2. COMSOL will start with the Model Wizard, which will ask you to choose the coordinate system for the model, the relevant physics to the problem, and the category of study you wish to perform (time dependent or stationary).
3. Define the equations, parameters, and variables relevant to the model.
4. Construe the geometry of the model (Geometry).
5. Select the materials you wish to use in your model (Materials).
6. Select the initial boundary conditions and bulk for your system for each physics you are using. This will be entered separately for each different physics you are using (e.g., you will need to enter these for laminar flow and again for heat transfer if you are using both).
7. Choose the appropriate element size (Mesh).
8. Adjust the solver parameters.
9. Click compute (Study).
10. Display the desired results by selecting the Results icon. Note that these steps are always necessary when you are building a model. The order is also variable depending on the complexity of the model.

Procedure

It is rational to define this problem in two dimensions (2D) as follows:

1. The COMSOL can be started by clicking on the COMSOL Multiphysics 5.3a icon.
2. When COMSOL starts, click on the Model Wizard. Define the space dimension for this problem by selecting 2D.

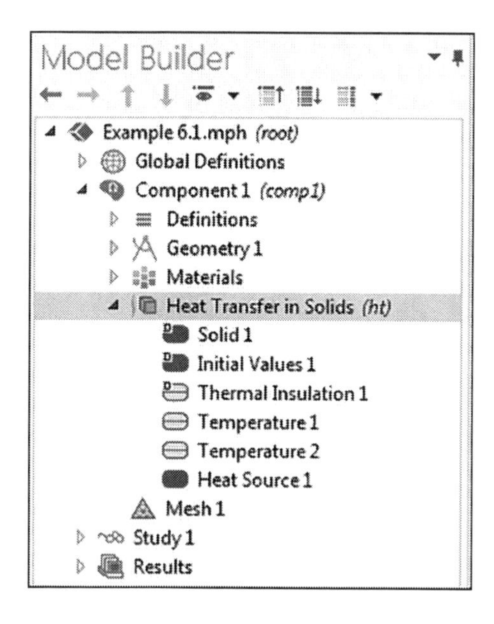

FIGURE 6.7
Selection of physics.

3. The next step is to select the applicable physics for the model. In this case, heat transfer in solids will be selected. This can be found under the Heat Transfer module. Click Add. The figure should look that in Figure 6.7.
4. Click "Study" at the bottom right corner. The last step in the Model Wizard is to select the type of study you would like to perform on the model. In our case, stationary will be sufficient to find the steady-state solution to this problem. As with the physics, add the stationary study by left-clicking on "Stationary" below the preset studies icon.
5. Click "Done" at the bottom right of the Model Wizard to finish.

Model Builder and Saving

Saving the model can be achieved by clicking "File" at the top left corner of the screen and then selecting "Save As," as is the case with most programs.

Geometry

The geometry consists of only one rectangle (the shield):

1. The first step in creating the rectangle is to click the "Geometry 1" icon in the model builder menus and select "Rectangle." See the appearance of the menu shown in Figure 6.8.
2. After adding the rectangle, the dimensions of this rectangle need to be adjusted to fit the dimensions in the problem. This

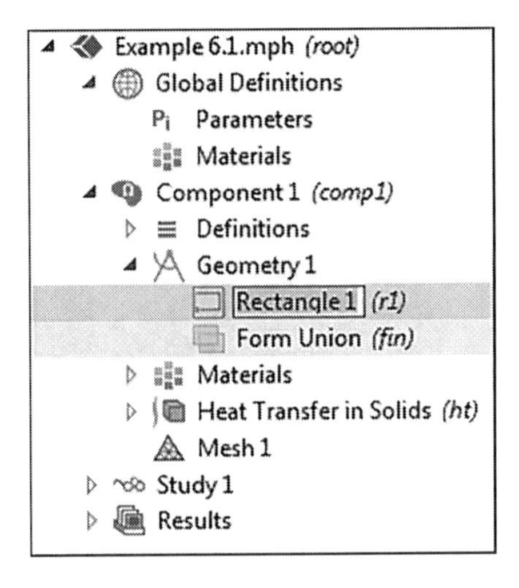

FIGURE 6.8
Selection of model geometry.

can be achieved by clicking the white rectangle located just to the left of the geometry icon. The expanded geometry tab will show all the subtabs contained within geometry. If the rectangle was added correctly, the tab called Rectangle 1 will be visible. This tab contains all the information regarding this object and the tools to adjust the dimensions and position of this rectangular object by left-clicking the tab labeled Rectangle 1.

3. If all the aforementioned steps have been successfully completed, your screen should resemble the one shown in Figure 6.9. Notice that the corner of the rectangle has been placed at the origin (position $x = 0$, $y = 0$) by default and has an equal width and height of 1 m. For this problem, the height needs to be adjusted

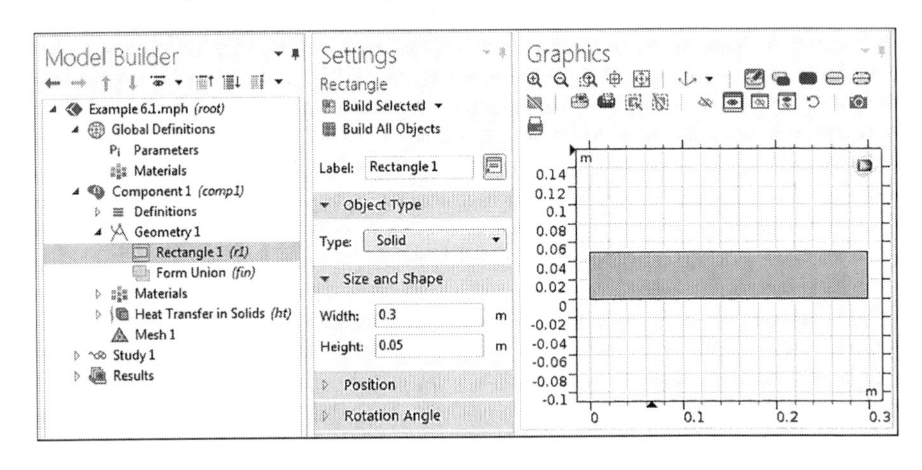

FIGURE 6.9
Specifying size and shape of the geometry.

FIGURE 6.10
Material selection.

to 5 cm (0.05 m), and the width needs to be adjusted to 30 cm (0.3 m). These values are then inserted into the designated fields, after which the blue building icon at the top right corner of the rectangle menus is pressed. Adding your rectangle to the model will be achieved by clicking the "Build All Objects" button.

4. The "Zoom Extents" button allows the user to get the graphical interface of COMSOL to center on the rectangle and adjust the axis bounds.

Materials

There are two ways to give the rectangle thermal properties, such as heat capacity and thermal conductivity. We can either add these directly under the "Heat Transfer" tab or we can select a material to build the rectangle from. In this problem, the rectangle is made of copper, and we will do this using the "Materials" tab.

1. When you right-click on "Materials" tab and then left-click "Add Materials," your screen should look like the screen shown in Figure 6.10.
2. From the above illustrated "Material Browser," a search bar can be seen that allows you to enter the name of the material in question. COMSOL will use its built-in database to find any matches for that material. Enter "copper" into the search bar and click "search."
3. Right-click on the "Copper" icon in the "Built-in" tab. Your screen should now look like the one shown in Figure 6.11. Once you have left-clicked on "Add Material to Component 1," you would have now added copper to all domains by default, which means the rectangle now has the properties of solid copper.

Heat Transfer

The boundary, bulk, and initial conditions for the equations of heat conduction can be inserted under the "Heat Transfer" tab.

Boundary Conditions

In this case, we only have four boundary conditions. Time-dependent studies are used in conjunction with initial conditions. Bulk conditions apply to the entire domain, not just a boundary. In our case, we have

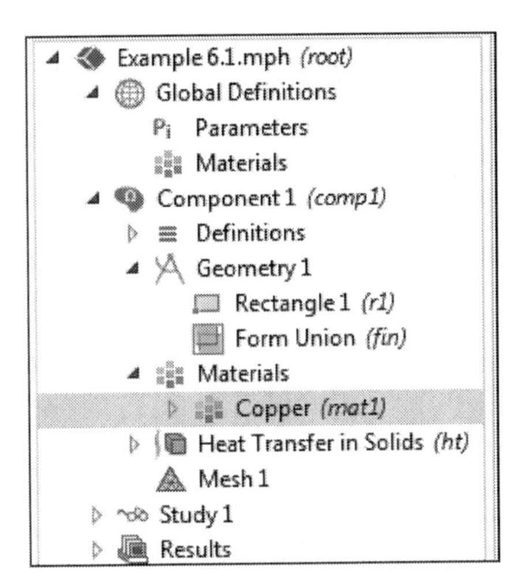

FIGURE 6.11
Copper is the material used for the study.

one boundary in contact with the heated rod, which has a temperature of 400 k. All other boundaries in contact with the thermostat bath are at 300 k temperature.

1. To insert these boundary conditions, open the "Heat Transfer" tab first by left-clicking the white triangle located to the left of the "Heat Transfer" icon. Your screen should take the appearance shown in Figure 6.12.
2. To open a menu containing the various types of bulk and boundary conditions, right-click the "Heat Transfer" icon. Go through this menu and select "Temperature." A new icon that says "Temperature" will appear under the initial values. This is where we will insert one of our two temperature conditions.
3. By repeating step 2 you can add another temperature boundary condition. After adding the two boundary conditions, your screen should take the appearance shown in Figure 6.13.
4. The temperature boundary conditions need a specific value and location. Let's start with the warm surface. Start by left-clicking the icon labeled "Temperature 1." We need to conduct two steps here: The first step is to add the surface to which we wish to

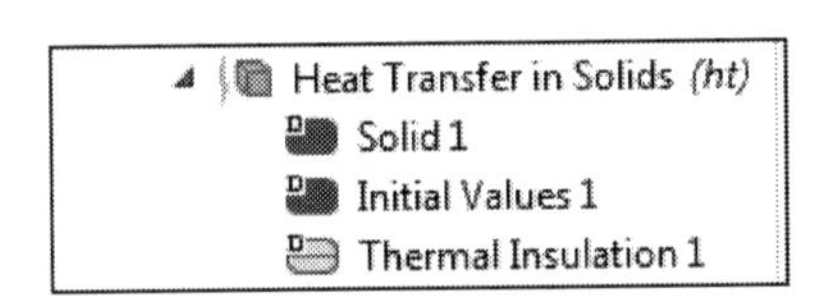

FIGURE 6.12
Boundary conditions.

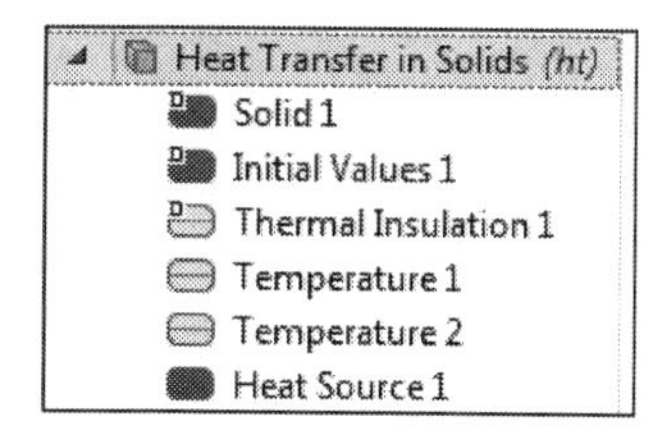

FIGURE 6.13
Adding temperature as a boundary condition.

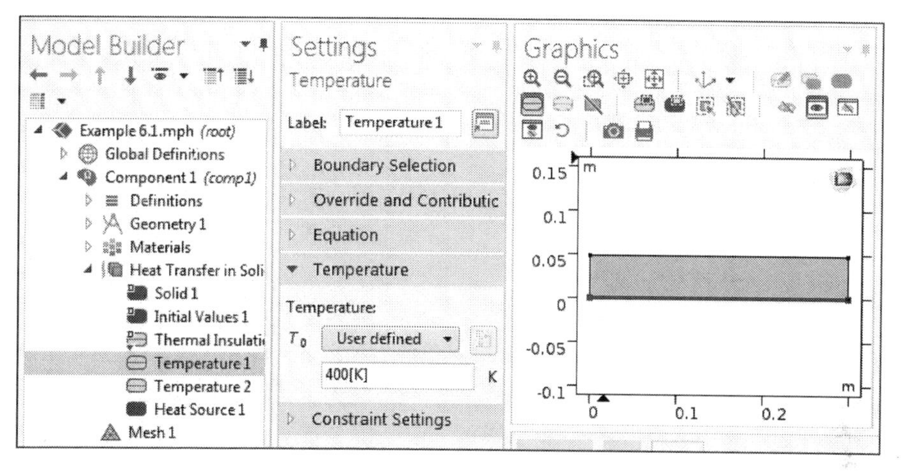

FIGURE 6.14
Specifying boundary conditions linked with temperature 2.

apply this boundary condition, and the second step is to give a value to this temperature. Now set the temperature to 400 K by typing 400 into the T_0 field (Figure 6.14). We now need to apply the cooler boundary condition. This can be done by clicking the icon labeled "Temperature 2," which opens the interface. Select the top and side boundaries to apply the boundary condition. Then enter 300 into the T_0 field.

Click on "Mesh 1" and under setting, select normal under "Element size." See Figure 6.15.

Since this concludes our activities within the "Heat Transfer" tab, we can now proceed to calculate the solution. To determine the solution to our PDE, we simply right-click on the "Study" tab and click on the green equals sign. After the solution is converged, the profile of the surface temperature will be displayed as shown in Figure 6.16.

Results

To examine the temperature at any given point, left-click the point you wish to probe and the result will be displayed under the results tab.

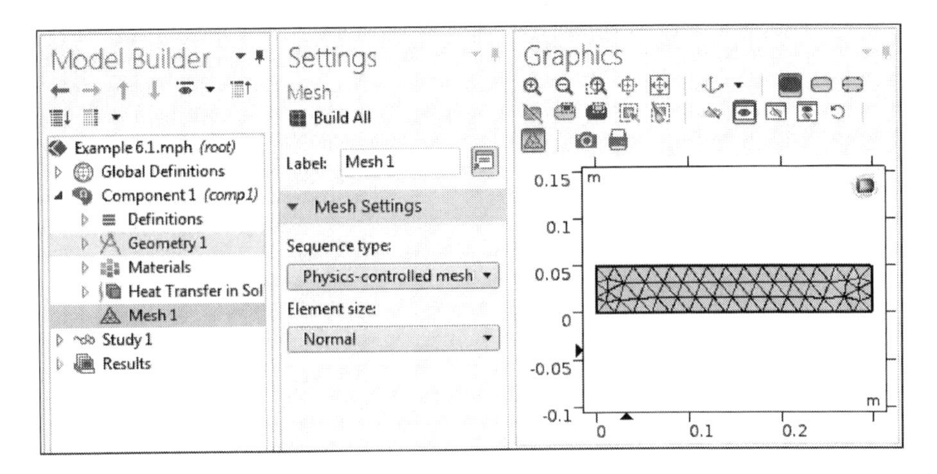

FIGURE 6.15
Mesh settings.

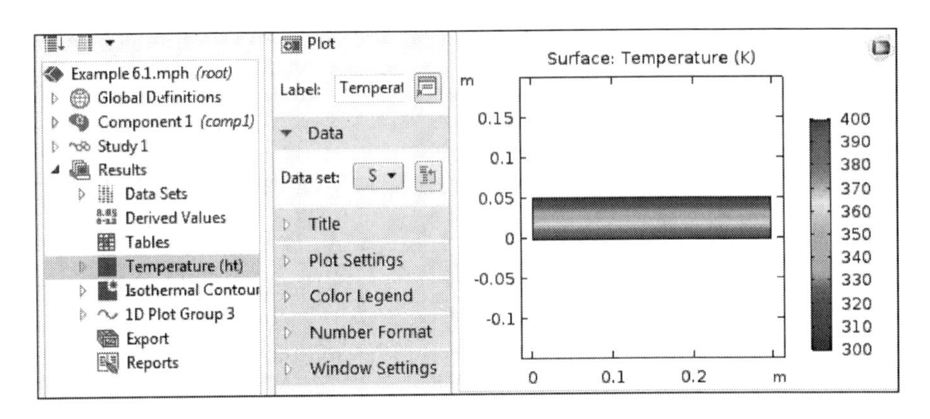

FIGURE 6.16
Surface temperature diagram.

Line Graph

To produce a graph that displays the temperature profile along a line, we will need to add a "cut line" to the solution and display the temperature along it. This can be achieved as follows:

1. Right-click on the "Data sets" button under the "results" tab and select "2D cut line" from the popup menu.
2. The two points defining the "cut line" need to be selected. In this case, we will have our "cut line" start at point (0.15,0) and end at point (0.15,0.05). To do this, enter these coordinates into the "Cut line 2D1" screen that will come up after left-clicking on the "Cut line 2D1" icon under the data sets tab.

3. Press the Plot button in the top left corner of the "Cut line 2D" screen to have the cut line displayed. Your cut line should look like Figure 6.17.
4. We now need to add a "1D plot group" to the results. It's important to note that COMSOL uses a right-click interface for addition of most options. So right-click on "Results" and left-click the on "1D plot group."
5. We also need to add a line graph to our "1D plot group." To achieve this, right-click on "1D Plot Group" and choose "Line Graph" from the popup menu. This will add a line graph under the "1D plot group."
6. Finally, left-click on "Line graph," and for "data set," select "Cut Line 2D 1." This will show the temperature everywhere along the created cut line. To create the graph, left-click on the "Plot" button. Figure 6.18 shows the temperature profile in the absence of a heat source. As can be observed, the temperature decreases linearly from the heated surface to the cooled surface.

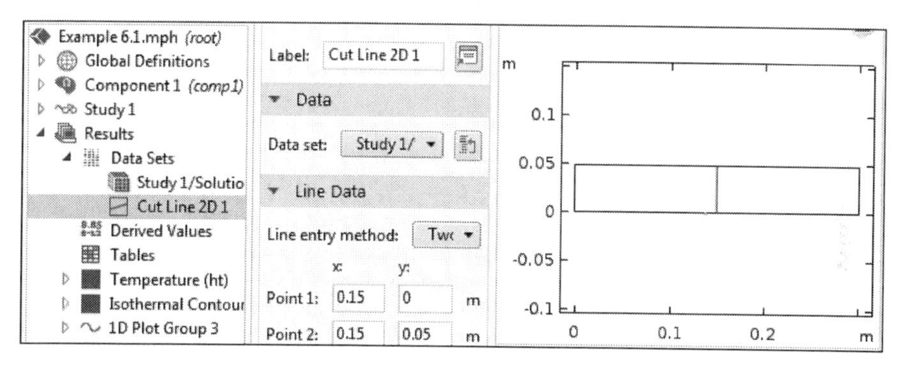

FIGURE 6.17
Selection of cutline 2D.

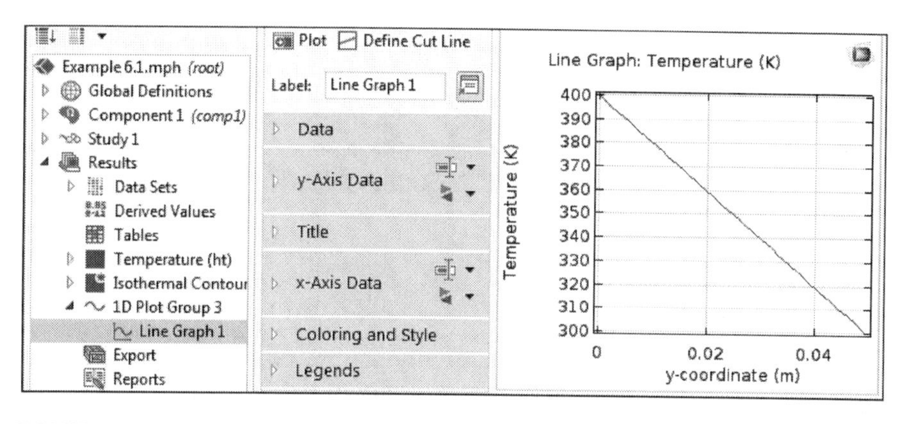

FIGURE 6.18
Line graph generated from data of the 2D cutline (without heat source).

Now it is possible to go back and change some of our boundary or bulk conditions. For example, let's change the lateral surfaces to perfect the insulators. This can be done as follows:

7. Go back to the "Heat Transfer" icon and left-click the arrow located just to the right of this icon to open all the options.
8. Go to the boundary condition by clinking on the "Temperature 2" icon, and de-select the lateral surfaces. This results in the upper surface being the only constant at a temperature of 300 K. You can deselect a subdomain by left-clicking it and then pressing the "minus" button. Now the lateral surfaces will be insulated by default.
9. Right-click on "Study," then press compute. The resulting screen should be like the one shown in Figure 6.19. Notice how only the region of the rectangle close to the lateral surfaces has changed from before. If you analyze the temperature profile along the cut line, you shouldn't see a lot of change because this cut line was exactly in the middle of our rectangle where the side effects were minimal.

   We will now add a heat generation term. This is a bulk condition and can be added using the same procedure as the temperature boundary conditions.

10. Go back up to the "Heat Transfer" icon and right-click to open the list of possible boundary and bulk conditions. Left-click on "Heat Source." This will add a "Heat Source 1" icon within "Heat Transfer" menu. Then left-click this to open the interface.
11. We need to add the domain over which this condition applies, and as a bulk condition, it will apply over the entire geometry. So left-click the rectangle and then left-click the plus sign, as done previously.

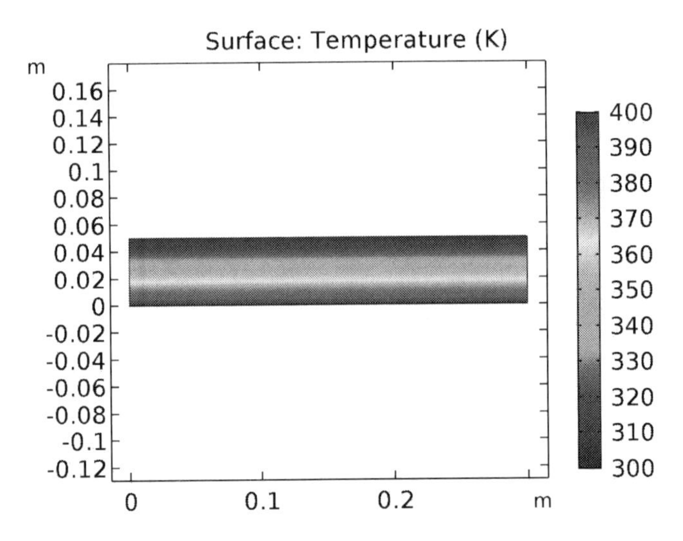

FIGURE 6.19
Temperature surface plot diagram in the absence of heat source.

12. The value of heat generation per volume term needs to be added. We will use 100,000,000 W/m^3, as shown in Figure 6.20.
13. Again, after adjusting any boundary or bulk condition(s), a new solution must be found. Right-click on the "Study" icon and press compute (Figure 6.21). To examine the temperature profile for this solution, click on your previously made line graph displaying the temperature across the cut line. This should look like the one in Figure 6.21. Note how this differs from the solution without heat generation. The maximum temperature is no longer at the heated surface but is instead near the center of the rectangle because of the large amount of heat being produced throughout the entire volume. The temperature profile is taken across the plane at a length of 0.15 m. The temperature increased due to the entered heat source, then decreased until it reached the top boundary conditions, $T = 300$ K (Figure 6.22).

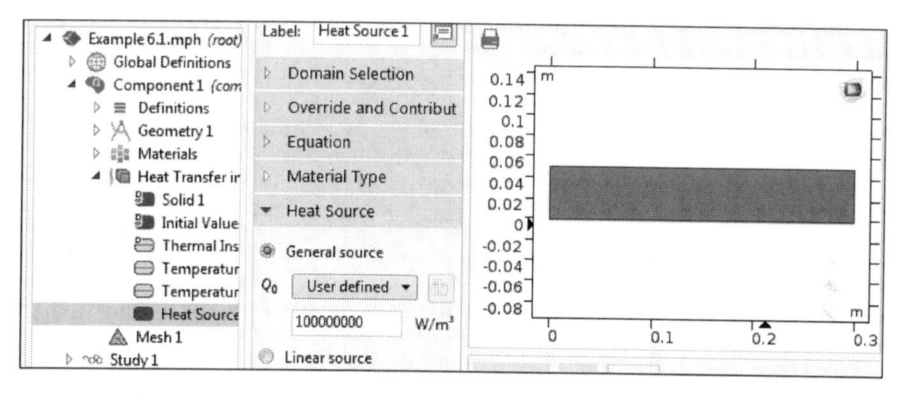

FIGURE 6.20
Heat source.

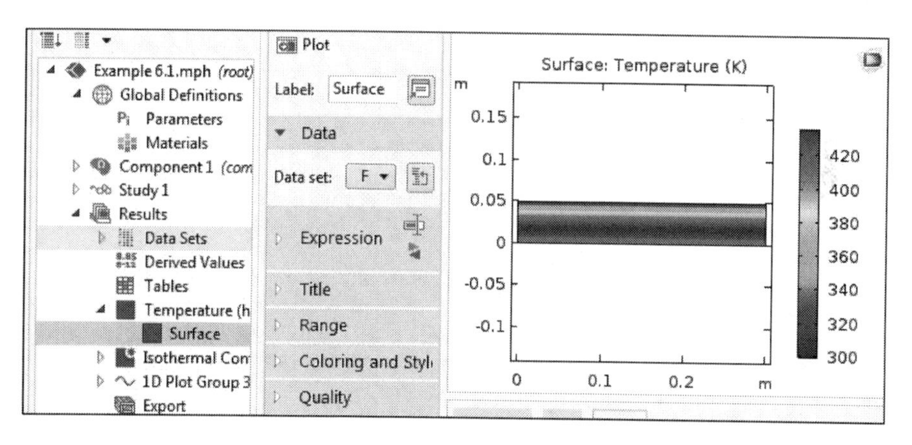

FIGURE 6.21
Temperature surface plot diagram in the presence of heat source.

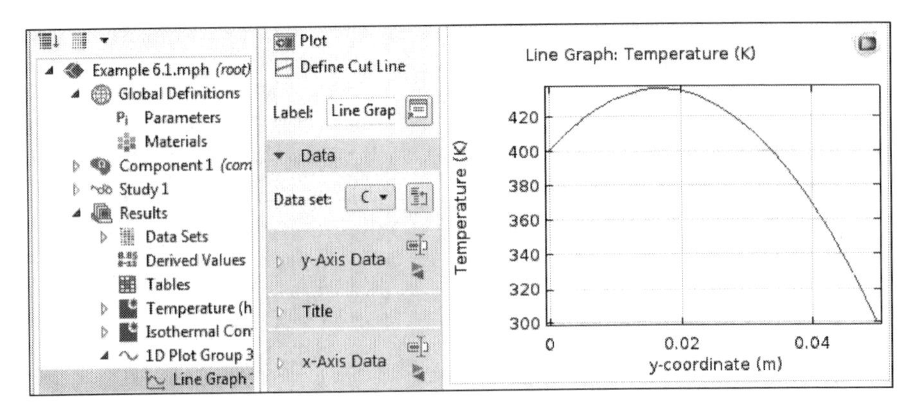

FIGURE 6.22
Line plot in the presence of heat source across the plane at $x = 0.15$ m.

2D Axisymmetric Heat Transfer

Now we will solve the same problem but without the reduction of the problem into rectangular coordinates. To avoid redundancy, only the significant steps which differ from those in Example 6.1 will be explained in detail.

Startup

1. The first thing that must be done is to start a new model, either by restarting COMSOL or by selecting the "New" option in the "File" menu.
2. After starting a new model, you will now select the "2D Axisymmetric" option instead of simply 2D. This option is ideal for problems with symmetry about an axis because it will take whatever geometry you create and rotate it around an axis.
3. Finally, select the "Heat Transfer" option as your physics and the "Stationary" option as your study.

Geometry

Now our geometry will be created; it is important to note that this is where the biggest differences exist between this model and the previous one.

1. Right-click on the geometry icon and add a rectangle.
2. Place the corner at $z = 0$ m and $r = 0.05$ m. Notice that our geometry will be spun around the line $r = 0.05$. Click the "Build All" icon to obtain the geometry shown in Figure 6.23.

Materials

Select copper as the material and apply this to the geometry, as done previously.

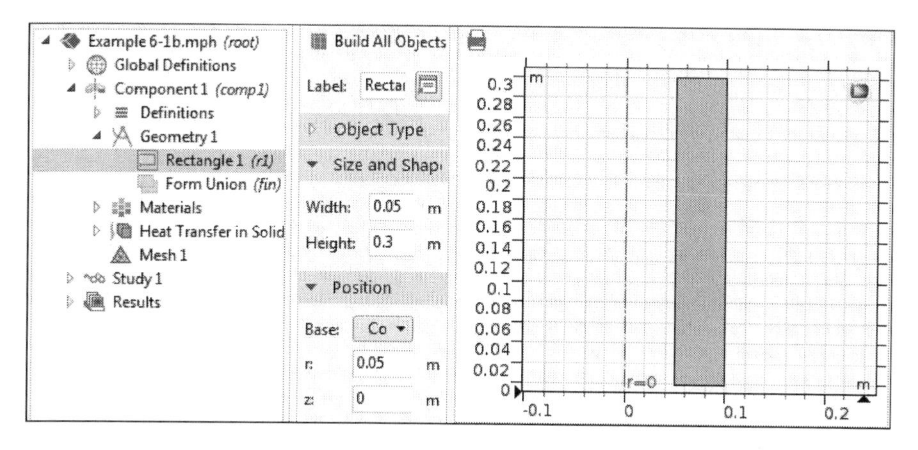

FIGURE 6.23
2D axisymmetric.

Heat Transfer

The same boundary conditions used previously will be used again. Namely, at $r = R_1$, $T = 400$ K; at $r = R_2$, $T = 300$ K; at $z = 0$, $T = 300$ K; and at $z = 0.3$ m, $T = 300$ K. This means two different temperature conditions must be added. This can be achieved by right-clicking on "heat transfer" and selecting "temperature." Enter the relevant temperatures in the temperature field, and select the appropriate surfaces to apply these boundaries.

Study

With the completion of the model, we move to examine the solution. Right-click "Study" and left-click "Compute." The 3D surface temperature profile result should be obtained as shown in Figure 6.24. The 3D diagram does not reveal much about the actual solution.

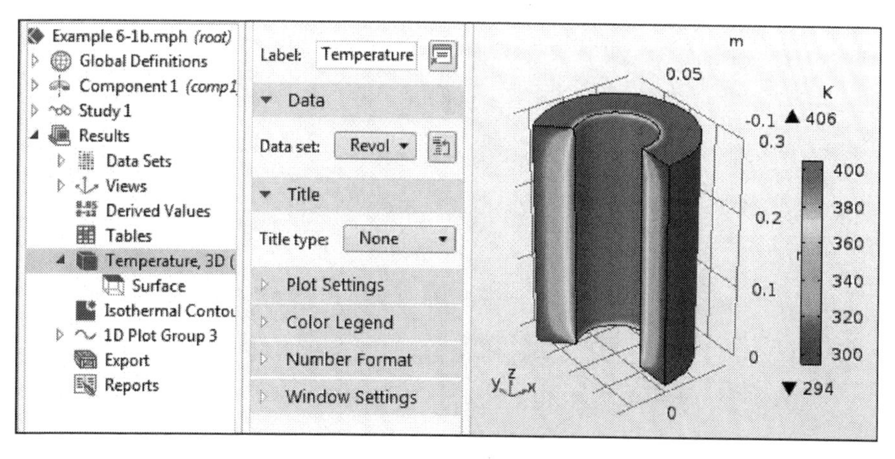

FIGURE 6.24
3D-axial symmetry temperature profile.

To gain a better understanding of the temperature profile, we will add a "Cut Line" as before.

1. Right-click on "Data Sets" under the "Results" tab. Click "Cut Line 2D."
2. The two points for the cut line should be ($r = 0.05$ m, $z = 0.15$ m) and ($r = 0.10$, $z = 0.15$ m).
3. Right-click on "Results" and add a 1D plot group.
4. Right-click on "1D plot group 1" and add a "Line Graph."
5. Select "Cut line 2D" as the data source in the line graph interface and click the plot icon to generate the graph. The analytical and simulation prediction match very well, as shown in Figure 6.25. Compare the predicted solution with the analytical solution. Right-click on "Definition," then select Function/Analytic. Type the analytical equation in the Expression space: ((400–300)/log (0.05/0.1))*log(r/0.05)+400.

Type r for the Argument, as shown in Figure 6.26. Click on "Plot." There is no need to run the program (no need to click "Compute").

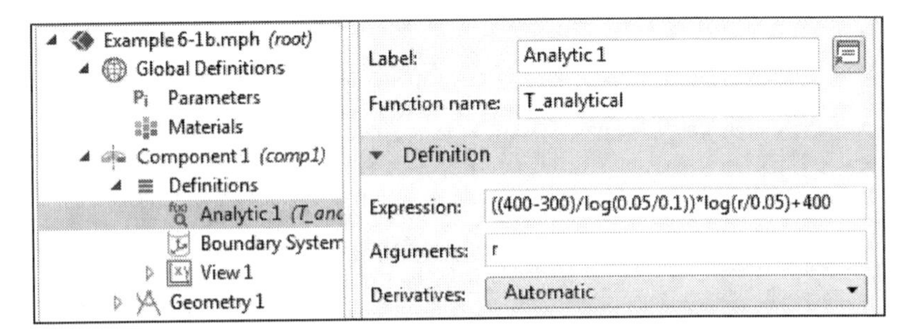

FIGURE 6.25
Axial symmetry temperature profile at $z = 0.15$.

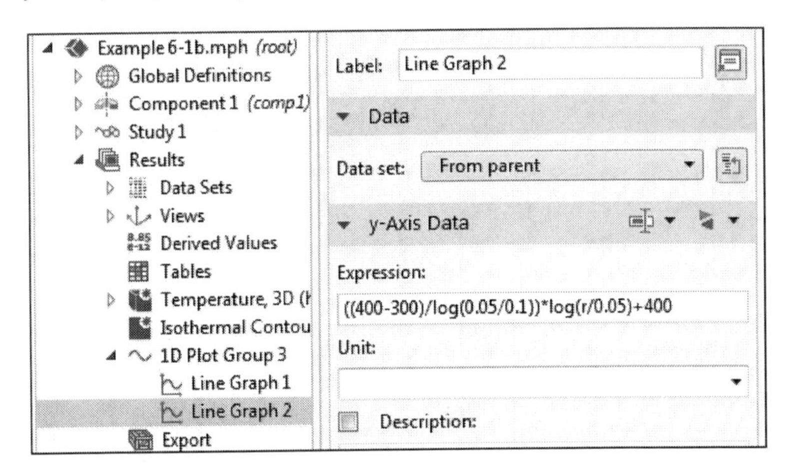

FIGURE 6.26
The analytical expression.

Comparison of COMSOL predictions and the analytical solution with the assumption of neglecting the tube end affect were in good agreement.

Example 6.2: Packed Bed Annular Chemical Reactor

An annular chemical reactor consists of two coaxial cylinders with a packed bed of catalyst between them. The inner cylinder has a radius of r_o, while the outer cylinder has a radius r_1. We can assume that there is no heat transfer through the surface of the inner cylinder, which is maintained at a constant temperature T_o. The heat is released at a uniform volumetric rate Q throughout the reactor by the catalytic reaction. Assume that the reactor's effective thermal conductivity k is constant, and neglect the temperature gradients in the axial direction.

1. Derive a second-order differential equation from the system to describe the radial temperature distribution in the annular reactor, starting with a shell thermal energy balance.
2. Solve the differential equation to establish the radial temperature distribution.

Solution

The schematic diagram of the packed bed reactor is shown in Figure 6.27. Consider the following assumptions in developing the expression for temperature profile in the catalytic section:

- Cylindrical coordinates.
- Steady state.

Given information:

- The temperature of the inner cylinder is constant T_0.
- Uniform volumetric heat rate Q is released throughout the reactor.
- The effective thermal conductivity k is constant.
- Temperature gradients in the axial direction are neglected.

To derive a second-order differential equation that describes the radial temperature distribution in the annular reactor, start with the equation of continuity in cylindrical coordinate:

$$\rho\left(\frac{1}{r}\frac{\partial r v_r}{\partial r}+\frac{1}{r}\frac{\partial v_\theta}{\partial \theta}+\frac{1}{r}\frac{\partial v_z}{\partial z}\right)=0$$

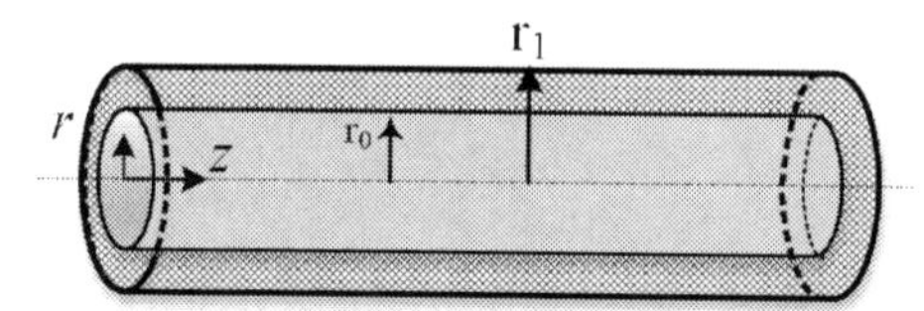

FIGURE 6.27
Heat transfer in packed bed chemical reactor.

No flow: $v_r = v_\theta = v_z = 0$.

No convective heat flux
$v_r = v_\theta = v_z = 0$

$$\rho\left(\frac{1}{r}\frac{\partial r v_r}{\partial r} + \frac{1}{r}\frac{\partial v_\theta}{\partial \theta} + \frac{1}{r}\frac{\partial v_z}{\partial z}\right) = 0$$

The general differential energy balance equation of cylindrical coordinate is:

$$\rho C_p \frac{\partial T}{\partial t} + \rho C_p\left(v_r \frac{\partial T}{\partial r} + v_\theta \frac{1}{r}\frac{\partial T}{\partial \theta} + v_z \frac{\partial T}{\partial z}\right)$$
$$= k\left(\frac{1}{r}\frac{\partial}{\partial r}\left(r\frac{\partial T}{\partial r}\right) + \frac{1}{r^2}\frac{\partial^2 T}{\partial \theta^2} + \frac{\partial^2 T}{\partial z^2}\right) + \Phi_H$$

Simplify the energy balance equation utilizing the previous assumptions:

Steady state; No convective heat $v_r = v_\theta = v_z = 0$; Temperature gradients in the axial and angular direction is neglected

$$\rho C_p \frac{\partial T}{\partial t} + \rho C_p\left(v_r \frac{\partial T}{\partial r} + v_\theta \frac{1}{r}\frac{\partial T}{\partial \theta} + v_z \frac{\partial T}{\partial z}\right) = k\left(\frac{1}{r}\frac{\partial}{\partial r}\left(r\frac{\partial T}{\partial r}\right) + \frac{1}{r^2}\frac{\partial^2 T}{\partial \theta^2} + \frac{\partial^2 T}{\partial z^2}\right) + \Phi_H$$

The heat generation heat term per unit volume, $\Phi_H = q$:

$$0 = k\left(\frac{1}{r}\frac{\partial}{\partial r}\left(r\frac{\partial T}{\partial r}\right)\right) + q$$

Establish the radial temperature distribution by solving the differential equation:

$$\frac{\partial}{\partial r}\left(r\frac{\partial T}{\partial r}\right) = \frac{-q}{k} r$$

Perform the first integration generates the first arbitrary constant of integration C_1:

$$r\frac{dT}{dr} = \frac{-q}{k}\frac{r^2}{2} + C_1$$

Divide each term by r:

$$\frac{dT}{dr} = \frac{q}{k}\frac{r}{2} + \frac{C_1}{r}$$

The second integration results in the second arbitrary constant of integration C_2, and the following temperature distribution profile in the radial direction is obtained:

$$T = \frac{-q}{k}\frac{r^2}{4} + C_1 \ln(r) + C_2$$

The arbitrary constant of integrations is determined by the following two boundary conditions:

B.C.1: at $r = r_o$, $\partial T/\partial r = 0$ (no heat flux at the inner wall of the cylinder)
B.C.2: at $r = r_o$, $T = T_0$ (the inner wall temperature is kept constant)

By substituting B.C.1:

$$0 = \frac{-q}{k}\frac{r_0^2}{2} + C_1$$

Rearrange to obtain: $C_1 = q r_0^2/2k$

By substituting boundary condition 2 and C_1 in the temperature distribution profile, we have:

$$T_0 = \frac{-q}{k}\frac{r_0^2}{4} + \frac{q}{k}\frac{r_0^2}{2}\ln(r) + C_2$$

Rearrange to find C_2:

$$C_2 = T_0 + \frac{q}{k}\frac{r_0^2}{4} - \frac{q}{k}\frac{r_0^2}{2}\ln(r_0)$$

By substituting C_1 and C_2 in the integrated temperature distribution profile, we have:

$$T = \frac{-q}{k}\frac{r^2}{4} + \frac{q}{k}\frac{r_0^2}{2}\ln(r) + \frac{q}{k}\frac{r_0^2}{4} - \frac{q}{k}\frac{r_0^2}{2}\ln(r_0) + T_0$$

The final temperature distribution profile is:

$$T = T_0 + \frac{q}{k}\frac{r_0^2}{4}\left(1 - \left(\frac{r}{r_0}\right)^2 + 2ln\left(\frac{r}{r_0}\right)\right)$$

Example 6.3: Heat Generation Within a Cylindrical Rod

A long solid tube is shielded with an insulator at the outer radius R_2 and cooled at the inner radius R_1. There is a uniform heat generation per unit volume $\dot{q}(\mathrm{W/m^3})$ within the solid. The outer radius R_2 is maintained at constant temperature T_2. If the coolant is available at a temperature T_c, the inner surface is kept at prescribed values of T_2. Obtain the general expression for the temperature distribution in the tube.

Solution

It is known that the solid tube with uniform heat generation is insulated at the outer surface and cooled at the inner surface. Find the following:

- A general solution for the temperature distribution $T(r)$.
- Appropriate boundary conditions and the corresponding form of the temperature distribution.
- Heat removal rate.
- Convection coefficient at the inner surface.

The schematic diagram of the long solid rod is shown in Figure 6.28.

The temperature distribution profile is obtained after considering the following assumptions:

- The system is under steady-state conditions.
- One-dimensional radial conduction.
- Constant properties.
- Uniform volumetric heat generation within wall thickness.
- Outer surface is adiabatic and at constant temperature.

The general differential energy equation in cylindrical coordinate is:

$$\rho C_p \frac{\partial T}{\partial t} + \rho C_p \left(v_r \frac{\partial T}{\partial r} + v_\theta \frac{1}{r}\frac{\partial T}{\partial \theta} + v_z \frac{\partial T}{\partial z} \right)$$
$$= k\left(\frac{1}{r}\frac{\partial}{\partial r}\left(r\frac{\partial T}{\partial r} \right) + \frac{1}{r^2}\frac{\partial^2 T}{\partial \theta^2} + \frac{\partial^2 T}{\partial z^2} \right) + \Phi_H$$

Based on the prior assumptions, steady-state operation and no heat is transferred by convective heat flux within the rod wall; therefore, the convection term is eliminated. To determine $T(r)$ for the prescribed conditions, the appropriate and reduced form of the heat equation:

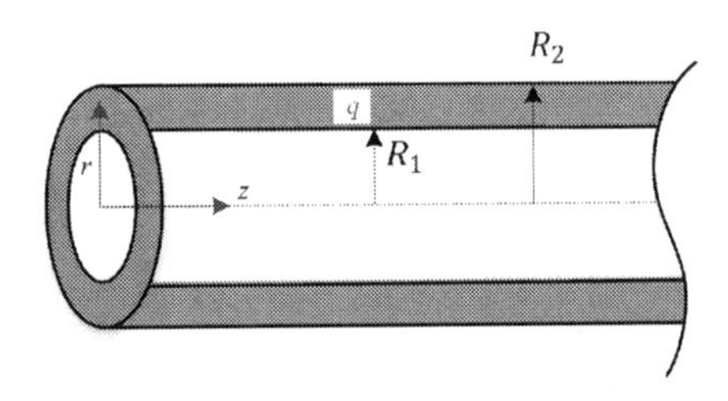

FIGURE 6.28
Schematic of solid rod.

Steady state — No convective heat $v_r = v_\theta = v_z = 0$ — Temperature gradients in the axial and angular direction is neglected

$$\rho C_p \frac{\partial T}{\partial t} + \rho C_p \left(v_r \frac{\partial T}{\partial r} + v_\theta \frac{1}{r}\frac{\partial T}{\partial \theta} + v_z \frac{\partial T}{\partial z} \right) = k \left(\frac{1}{r}\frac{\partial}{\partial r}\left(r \frac{\partial T}{\partial r} \right) + \frac{1}{r^2}\frac{\partial^2 T}{\partial \theta^2} + \frac{\partial^2 T}{\partial z^2} \right) + \Phi_H$$

$$\left(\frac{1}{r}\frac{\partial}{\partial r}\left(r \frac{\partial T}{\partial r} \right) \right) + \frac{q}{k} = 0$$

After performing double integration, two arbitrary constants of integrations are generated (C_1 and C_2):

$$T(r) = -\frac{\dot{q}}{4k} r^2 + C_1 \ln r + C_2$$

To evaluate C_1 and C_2, two boundary conditions must be added. It is suitable to specify both conditions at R_2.

B.C.1: at $r = R_2$, $T = T_2$

Apply Fourier's law at the adiabatic outer surface.

B.C.2: at $r = R_2$, $dT/dr = 0$

Consequently, it follows that (using boundary condition 1):

$$T_2 = -\frac{\dot{q}}{4k} R_2^2 + C_1 ln R_2 + C_2$$

$$C_2 = T_2 + \frac{\dot{q}}{4k} R_2^2 - C_1 \ln R_2$$

Then similarly for boundary conditions 2:

$$0 = -\frac{\dot{q}}{2k} R_2^2 + C_1$$

Hence, from this boundary condition we get:

$$C_1 = \frac{\dot{q}}{2k} R_2^2$$

The general temperature profile in the rod is:

$$T(r) = -\frac{\dot{q}}{4k} r^2 + C_1 \ln r + C_2$$

Substitute arbitrary constant of integrations C_1 and C_2:

$$T(r) = -\frac{\dot{q}}{4k} r^2 + \frac{\dot{q}}{2k} R_2^2 \ln r + T_2 + \frac{\dot{q}}{4k} R_2^2 - \frac{\dot{q}}{2k} R_2^2 \ln R_2$$

After simplifying, the following radial temperature profile equation is obtained:

$$T(r) = T_2 + \frac{\dot{q}}{4k}(R_2^2 - r^2) + \frac{\dot{q}}{2k}R_2^2 ln\frac{r}{R_2}$$

Example 6.4: Heat Transfer in a Thin Slab

A thin slab of construction insulation material is subjected to a heat source that causes volumetric heating to vary along the length of it; see Figure 6.29.

$$q = q_o\left[1 - \frac{x}{L}\right]$$

where:

q_o has a constant value of 1.8×10^5 W/m^3
slab length L is 0.06 m

The thermal conductivity of the slab material is 0.6 W/m.K. The boundary at $x = L$ is perfectly insulated, while the surface at $x=0$ is maintained at a constant temperature of 320 K. The slab density is 1900 kg/m^3, and the specific heat capacity is 840 J/(kg . K). Develop an expression for the temperature distribution $T(x)$ in the slab. Determine an expression for the temperature profile $T(r)$, within the thin slab. What is the value of the temperate near the adiabatic wall [3]?

Solution

Consider the following assumptions in determining the temperature distribution in the thin slab:

- Steady-state system.
- No bulk motion implies that $v_x = v_y = v_z = 0$.
- Heat transfer in the x-direction.

Symbols used in the model:

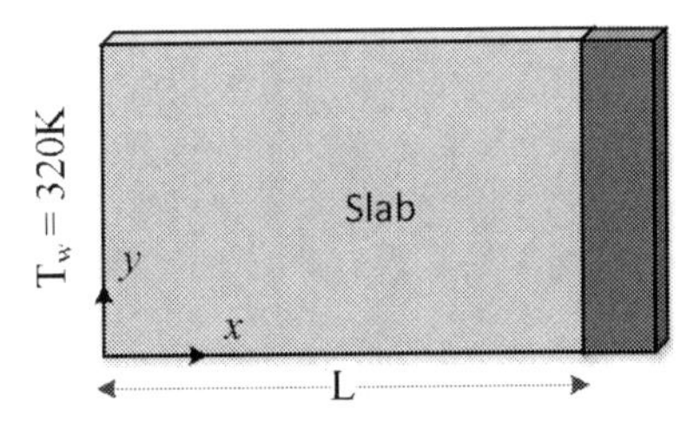

FIGURE 6.29
Temperature profile in a thin slab.

$q^{\cdot}(x)$: the heat flux in the x-direction (kW/m^2)
T: temperature at any point (K)
L: the length of the slab (m)
x: the distance from the surface of the slab (m)
k: thermal conductivity of the slab (W/m.K)

Total mass balance (continuity equation):

$$\rho\left(\frac{\partial v_x}{\partial x}+\frac{\partial v_y}{\partial y}+\frac{\partial v_z}{\partial z}\right)=0$$

There is no flow of fluid in the slab, and assume constant density; thus:

$$v_x = v_y = v_z = 0$$

$$v_x = v_y = v_z = 0$$

$$\rho\left(\frac{\partial v_x}{\partial x}+\frac{\partial v_y}{\partial y}+\frac{\partial v_z}{\partial z}\right)=0$$

Accordingly, all terms of the continuity equation will be eliminated. The general differential energy balance equation in Cartesian coordinates is:

$$\rho C_p\frac{\partial T}{\partial t}+\rho C_p\left(v_x\frac{\partial T}{\partial x}+v_y\frac{\partial T}{\partial y}+v_z\frac{\partial T}{\partial z}\right)=k\left(\frac{\partial^2 T}{\partial x^2}+\frac{\partial^2 T}{\partial y^2}+\frac{\partial^2 T}{\partial z^2}\right)+\Phi_H$$

Based on the preceding assumptions, steady state and no bulk motion, the general differential energy balance equation is simplified as follows:

No convective heat flux

Steady state $\quad v_x = v_y = v_z = 0$

$$\rho C_p\frac{\partial T}{\partial t}+\rho C_p\left(v_x\frac{\partial T}{\partial x}+v_y\frac{\partial T}{\partial y}+v_z\frac{\partial T}{\partial z}\right)=k\left(\frac{\partial^2 T}{\partial x^2}+\frac{\partial^2 T}{\partial y^2}+\frac{\partial^2 T}{\partial z^2}\right)+\Phi_H$$

Heat flux in the x-direction

The result simplifies the general differential equation when $\Phi = \dot{q}$:

$$\frac{\partial^2 T}{\partial x^2}+\frac{\dot{q}}{k}=0$$

Substitute the heat flux, q, as a function of x:

$$\dot{q}=\dot{q}_o\left[1-\frac{x}{L}\right]$$

Substitute the heat flux in the simplified energy balance equation:

$$\frac{d^2T}{dx^2}=-\frac{\dot{q}_o}{k}\left[1-\frac{x}{L}\right]$$

Integrate with respect to x:

$$\frac{dT}{dx}=-\frac{\dot{q}_o}{k}\left[x-\frac{x^2}{2L}\right]+C_1$$

Integrate further:

$$T=-\frac{\dot{q}_o}{k}\left[\frac{x^2}{2}-\frac{x^3}{6L}\right]+C_1x+C_2$$

Boundary conditions:
The heat flux is equal to zero at ($x = L$):
B.C.1: at $x = L$, $\partial T/\partial x = 0$
The temperature at the wall is constant at $x = 0$:
B.C.2: at $x = 0$, $T = T_w$
Apply boundary condition 1:

$$C_1=\frac{\dot{q}_o}{k}\left[\frac{L}{2}\right]$$

Apply boundary condition 2. Substitute $x = 0$ and $T = T_w$ in the flowing equation to obtain the expression for constant C_2:

$$T_w=-\frac{\dot{q}_o}{k}\left[\frac{x^2}{2}-\frac{x^3}{6L}\right]+\frac{1}{k}\times\dot{q}_o\left[\frac{L}{2}\right]+C_2$$

The expression for the second arbitrary integration consent C_2 is $T_w = C_2$. The final form of the temperature profile in the slab is:

$$T=-\frac{\dot{q}_o}{k}\left[\frac{x^2}{2}-\frac{x^3}{6L}\right]+\frac{\dot{q}_o}{k}\left[\frac{L}{2}\right]x+T_w$$

Rearranges to obtain the temperature distribution profile:

$$T=\frac{\dot{q}_oL}{2k}\left[x-\frac{x^2}{L}+\frac{x^3}{3L^2}\right]+T_w$$

The value of the temperature (T_L) at the adiabatic wall ($x = L$) is obtained by replacing x by L:

$$T_L=\frac{\dot{q}_oL}{2k}\left[L-\frac{L^2}{L}+\frac{L^3}{3L^2}\right]+T_w=\frac{\dot{q}_oL}{2k}\left[L-L+\frac{L}{3}\right]+T_w=T_L=\frac{\dot{q}_oL^2}{6k}+T_w$$

Substitute the values of the appropriate parameters to obtain T_L:

$$T_L = 1.8 \times 10^5 \frac{0.06^2}{6 \times 0.6} + 320 = 500\,\text{K}$$

COMSOL Solution

Start COMSOL. Select 2D coordinate system, then select "Heat transfer in solid," steady-state operation "stationary." Right-click "Geometry" and select "rectangle." The width is 0.06 m, and the height is 0.005 m. Right-click "Heat transfer in solid" and then select Temperature (320 K, left side boundary) and symmetry (top and bottom boundaries); the right side of the rod is under thermal insulation boundary. The "Heat Source" is heat generated as a function of the length of the rod (Figure 6.30).

Right-click "Study" and left-click "Compute." The surface plot of temperature across the solid rod fin is shown in Figure 6.31. The temperature

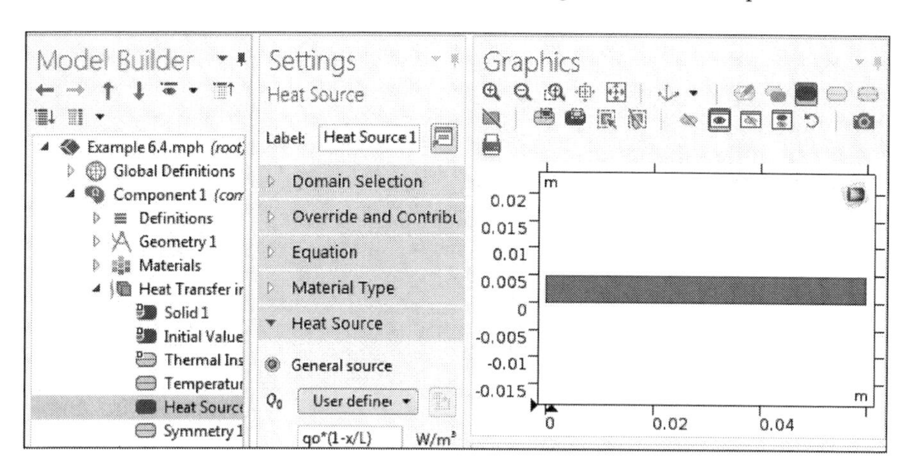

FIGURE 6.30
Setting of heat source as the generation heat per unit volume.

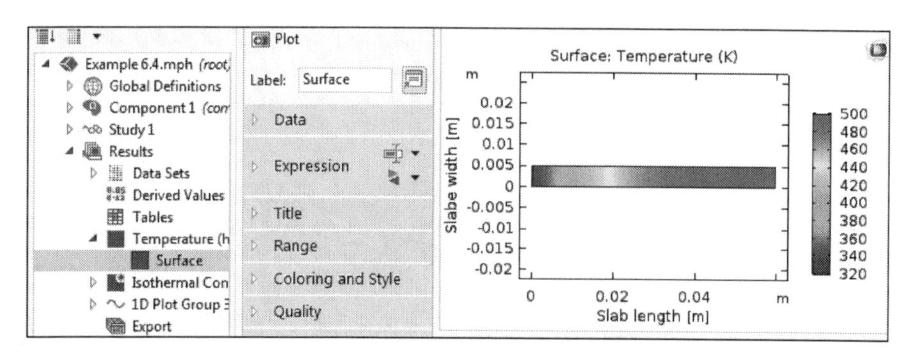

FIGURE 6.31
Surface temperature of slab.

profile shows that the temperature increases along the length of the rod starting from the left-side constant temperature to the adiabatic wall. The increase in temperature is attributed to the presence of the heat source inside the rod. The heat source is a function of the length of the rod; in other words, the heat source increased with rod length.

It can be concluded that simulation predictions are in good agreement with manual calculations.

Example 6.5: Heat Transfer in a Copper Rod

A copper rod is 5 mm in diameter and 100 mm in length; one end of the rod is maintained at 100°C. The surface of the rod is exposed to ambient air at a temperature of $T_a = 25°C$ with a convection heat transfer coefficient of 100 W/m^{2}s. The thermal conductivity of copper is 398 W/m K. Develop an expression for the temperature distribution $T(z)$ in the copper rod and then determine the temperature distribution along the rod (Figure 6.32).

Solution

We have the following assumptions:

- Steady-state operation
- Constant physical properties

The general differential energy balance equation in cylindrical coordinates is:

$$\rho C_p \frac{\partial T}{\partial t} + \rho C_p \left(v_r \frac{\partial T}{\partial r} + v_\theta \frac{1}{r}\frac{\partial T}{\partial \theta} + v_z \frac{\partial T}{\partial z} \right)$$

$$= k\left(\frac{1}{r}\frac{\partial}{\partial r}\left(r\frac{\partial T}{\partial r} \right) + \frac{1}{r^2}\frac{\partial^2 T}{\partial \theta^2} + \frac{\partial^2 T}{\partial z^2} \right) + \Phi_H$$

With steady state and no bulk motion, heat is transferred in the z-direction by conduction, and heat is lost to air by convective heat flux.

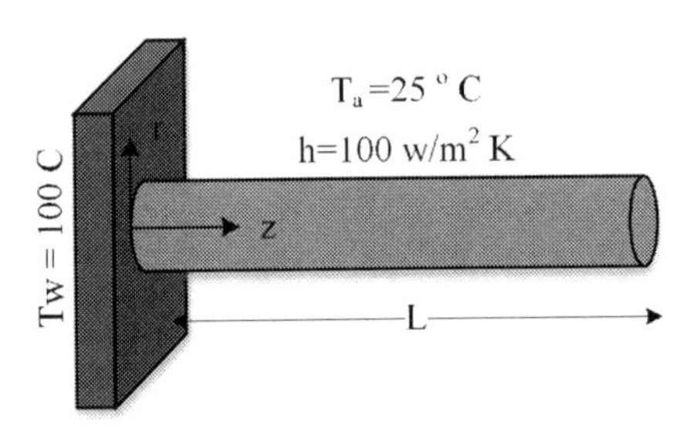

FIGURE 6.32
Heat transfer in a cylindrical rod.

Steady state — No convective heat $v_r = v_\theta = v_z = 0$ — Temperature gradients in the radial and angular direction is neglected

$$\rho C_p \frac{\partial T}{\partial t} + \rho C_p \left(v_r \frac{\partial T}{\partial r} + v_\theta \frac{1}{r}\frac{\partial T}{\partial \theta} + v_z \frac{\partial T}{\partial z} \right) = k \left(\frac{1}{r}\frac{\partial}{\partial r}\left(r \frac{\partial T}{\partial r} \right) + \frac{1}{r^2}\frac{\partial^2 T}{\partial \theta^2} + \frac{\partial^2 T}{\partial z^2} \right) + \Phi_H$$

The general differential equation is simplified to:

$$0 = k\left(\frac{\partial^2 T}{\partial z^2} \right) + \dot{q}, \text{ where, } \dot{q} = \frac{hA_s(T - T_a)}{V_c}$$

Substitute the expression of $\dot{q}$ and rearrange:

$$\frac{d^2T}{dz^2} = -\frac{hA_s(T-T_a)}{kV_c} = -\frac{hPL(T-T_a)}{kA_cL} = -\frac{h\pi DL(T-T_a)}{k(\pi D^2)L} = \frac{h}{kD}(T-T_a)$$

where:

P represents the rod perimeter
A_c represents the fin cross sectional area
A_s represents the face area measured from tube base through x

$$P = \pi D,\ A_c = \pi D^2$$

Finally, the second-order differential equation is:

$$\frac{d^2T}{dz^2} = -\frac{h}{kD}(T - T_a)$$

The analytical solution of the second-order differential equation is:

$$T = T_a + (T_o - T_a) * e^{-\left(\frac{h}{Dk}\right)^{0.5} z}$$

Substitute the given parameter value and plot along the length of the fin rod. The analytical solution is depicted in Figure 6.33. The figure reveals that the temperature decreases with rod length.

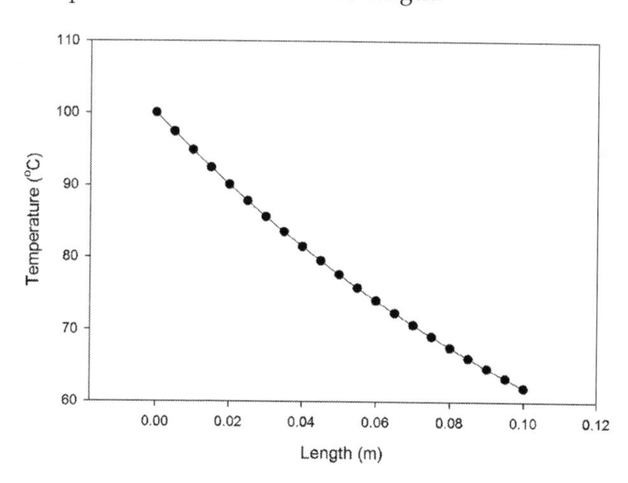

FIGURE 6.33
Analytical solution of temperature versus rod length.

COMSOL Simulation

Start COMSOL. Select 2D, then select "Heat transfer in solid." The system is operating at steady state; hence, select "Stationary." Right-click "geometry" and select "rectangle." The dimensions are shown in Figure 6.34 (height = 5 mm, length = 100 mm).

Right-click on the physics "Heat transfer in solid," and select temperature (inlet temperature = 100°C) and heat flux (surface heat transfer with air, $\dot{q} = h(T_a - T)$), as shown in Figure 6.35.

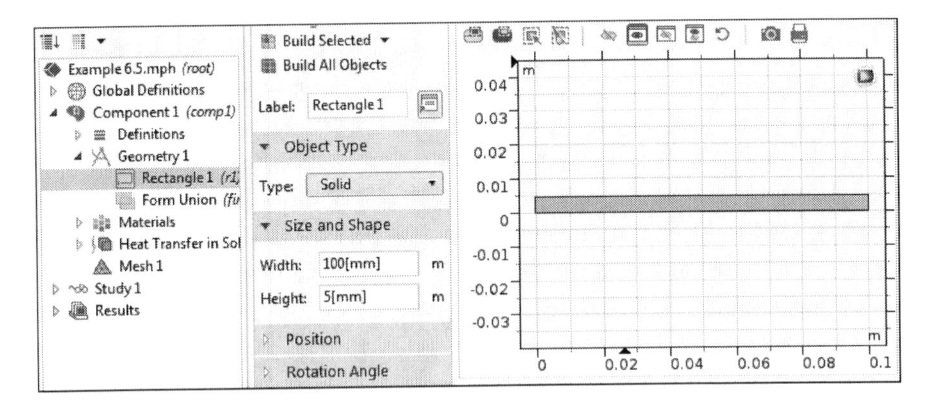

FIGURE 6.34
Specifying model geometry.

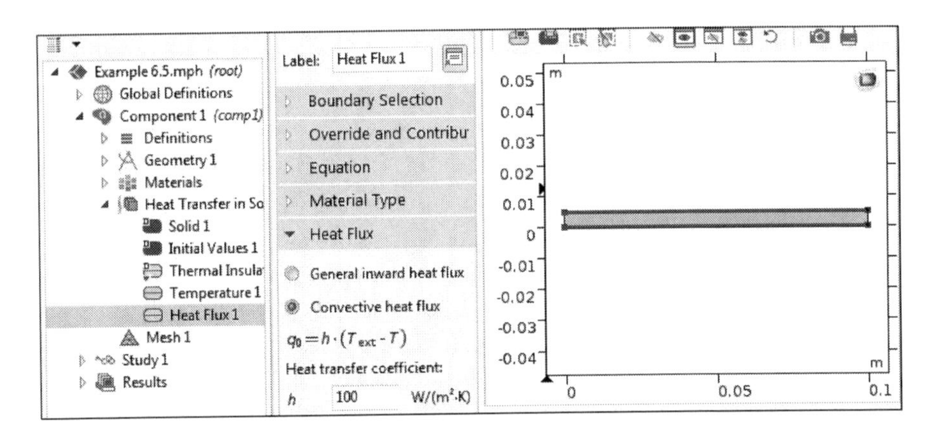

FIGURE 6.35
Specifying heat flux boundary condition.

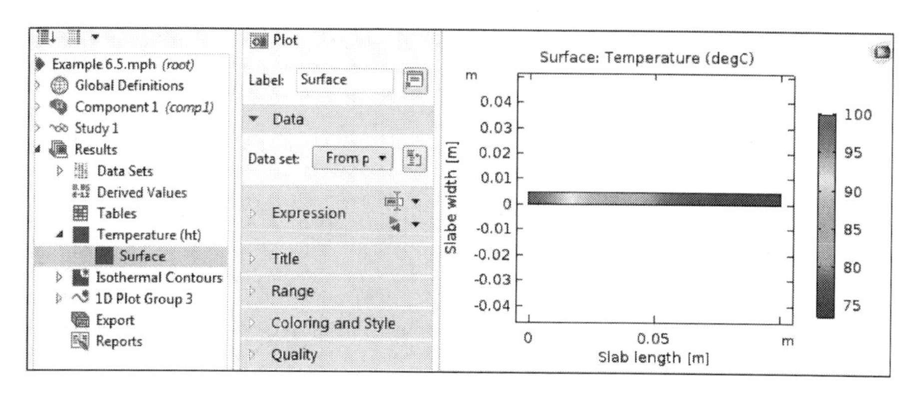

FIGURE 6.36
Surface temperature profile.

After selecting the boundary conditions and entering values for convective heat flux, right-click on "study" and then left-click "compute." The surface temperature profile is shown in Figure 6.36. The surface plot diagram shows the decrease in temperature along the length of the rod. The temperature decreased from 100°C at the left side to approximately around 70°C.

The line graph for the temperature at the center of the fin along the rod length (use cut line by right-clicking on "Data Sets" and selecting "Cut line 2D"). Left-click on the "Cut line 2D" and enter the following two points (0, 2.5 mm) and (100, 2.5 mm) (Figure 6.37).

Right-click on "Results" and select "1D plot group." Then right-click on "1D plot group" and select "Line graph 1." Finally, left-click "Line graph" to see the results shown in Figure 6.38.

Comparison between manual calculations and software predictions show overall a reasonable result within the same trend.

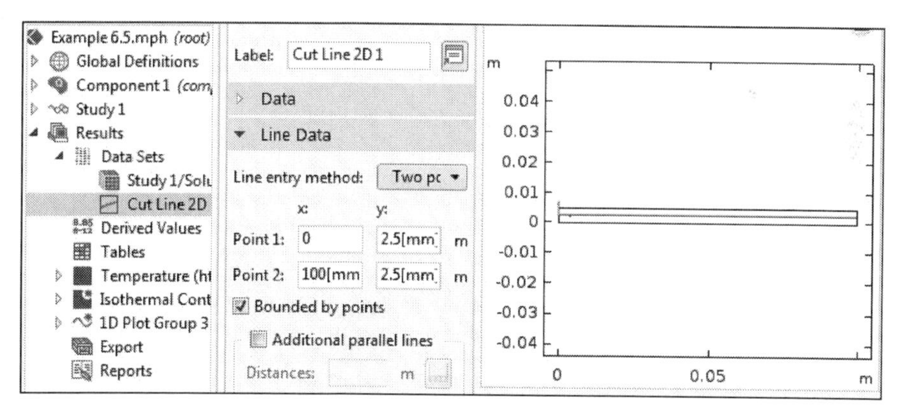

FIGURE 6.37
2D cutline for data generation.

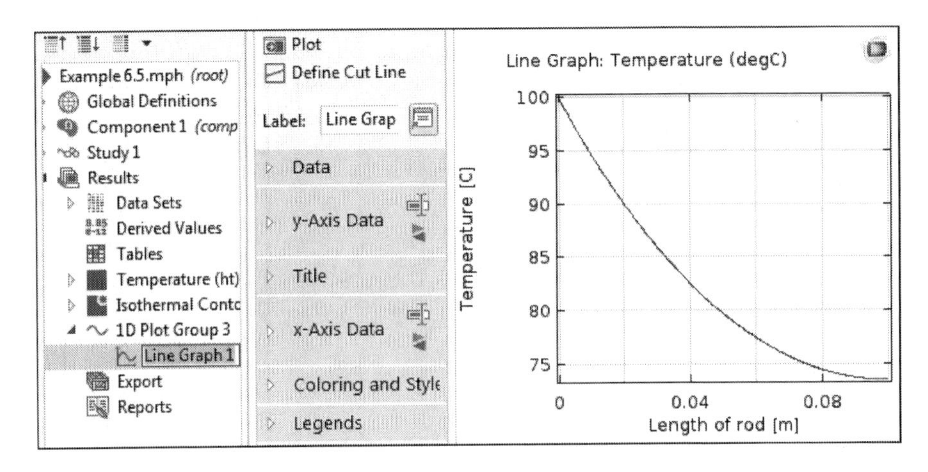

FIGURE 6.38
Temperature along length of rod generated from 2D cutline data.

Example 6.6: Heat Transfer in a Circular Tube

A concentrated slurry flows in a circular tube of radius R as presented in Figure 6.39. The velocity profile is approximately flat over the pipe cross-section because it flows nearly as a solid plug. Accordingly, the velocity in the z-direction is assumed to be constant ($v_z = V$). Heat is added at a uniform constant radial flux q through the tube wall to maintain the temperature of the surface at T_1. You can assume that the thermal conductivity k is constant. Develop an expression for the temperature distribution in the pipe and then calculate the temperature profile at a distance far from the inlet of the pipe.

Solution

Consider the following assumptions in determining the temperature profile in the slurry tank:

- Steady-state operation
- Constant physical properties
- Plug flow

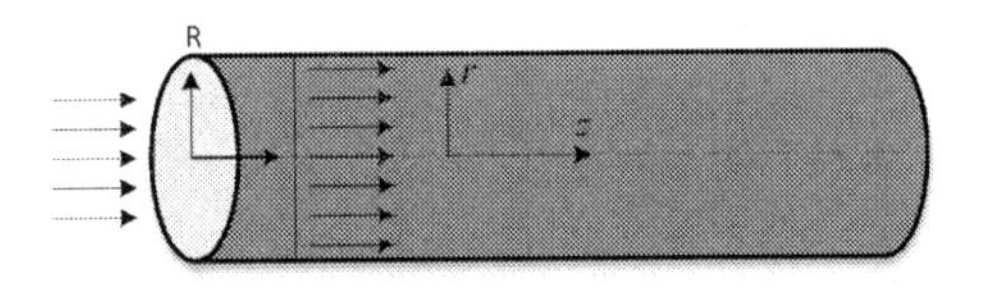

FIGURE 6.39
Plug flow in circular tube being heated by a uniform heat flux.

The general differential energy balance equation in cylindrical coordinates is:

$$\rho C_p \frac{\partial T}{\partial t} + \rho C_p \left(v_r \frac{\partial T}{\partial r} + v_\theta \frac{1}{r} \frac{\partial T}{\partial \theta} + v_z \frac{\partial T}{\partial z} \right)$$

$$= k \left(\frac{1}{r} \frac{\partial}{\partial r} \left(r \frac{\partial T}{\partial r} \right) + \frac{1}{r^2} \frac{\partial^2 T}{\partial \theta^2} + \frac{\partial^2 T}{\partial z^2} \right) + \Phi_H$$

Assume the plug flow to be ($v_r = v_\theta = 0$). The steady-state energy balance equation for constant density fluid is reduced to:

Steady state

No convective heat flux $v_r = v_\theta = 0$

Temperature gradients in the axial and angular direction is neglected

$$\rho C_p \frac{\partial T}{\partial t} + \rho C_p \left(v_r \frac{\partial T}{\partial r} + v_\theta \frac{1}{r} \frac{\partial T}{\partial \theta} + v_z \frac{\partial T}{\partial z} \right) = k \left(\frac{1}{r} \frac{\partial}{\partial r} \left(r \frac{\partial T}{\partial r} \right) + \frac{1}{r^2} \frac{\partial^2 T}{\partial \theta^2} + \frac{\partial^2 T}{\partial z^2} \right) + \Phi_H$$

Temperature gradients in the axial direction is neglected

Accordingly, the energy balance equation is reduced to the following form:

$$\frac{\partial}{\partial r} \left(r \frac{\partial T}{\partial r} \right) = -\frac{q}{k} r$$

Perform the double integration. The first integral generates the first constant C_1:

$$\left(r \frac{\partial T}{\partial r} \right) = -\frac{q}{k} \frac{r^2}{2} + C_1$$

Rearrange by dividing both sides by r:

$$\frac{\partial T}{\partial r} = -\frac{q}{2k} \frac{r^2}{r} + \frac{C_1}{r}$$

The second integration generates the second arbitrary constant of integration C_2; hence, the general temperature profile in the pipe is:

$$T(r) = -\frac{q}{4k} r^2 + C_1 \ln r + C_2$$

The boundary conditions are given by:

At the center of the pipe, axial temperature symmetry is assumed

B.C.1: $r = 0$, $\partial T / \partial r = 0$

This boundary conditions leads to: $C_1 = 0$

At the pipe wall, the temperature is maintained at constant temperature, T_1

B.C.2: at $r = R$, $T = T_1$

Substitute this boundary conditions in the general differential energy equation:

$$T_1 + \frac{q}{4k} R^2 = C_2$$

Substitute C_1 and C_2 in the general temperature profile equation and rearrange:

$$T(r) = T_1 + \frac{q}{4k}\left[R^2 - r^2\right]$$

Example 6.7: Heat Transfer in Composite Planes

A composite plane wall consists of two materials (Figure 6.40). The wall of first material has uniform heat generation per unit volume of $q = 1.5 \times 10^6$ W/m^3, thermal conductivity of $k_1 = 75$ W/mK, and thickness of $L_1 = 0.05$ m. Material 2 has no heat generation with constant thermal conductivity of $k_2 = 150$ W/mK and thickness of $L_2 = 0.02$ m. The inner surface of the first material is well insulated. The second material's outer surface is cooled by a water stream temperature if $T_a = 30$°C and heat transfer coefficient of $h = 1000$ W/m^{2}K. Develop the steady-state temperature distribution profile for both materials, and calculate the temperatures of the insulated surface and the cooled surface. Compare manual calculations with COMSOL simulation results [1].

Solution

The following assumptions are considered:

- There is no indication of any change with time then heat transfer is steady.
- There is thermal symmetry about the center line and no variation in the axial direction, and heat transfer is assumed to be one dimensional.

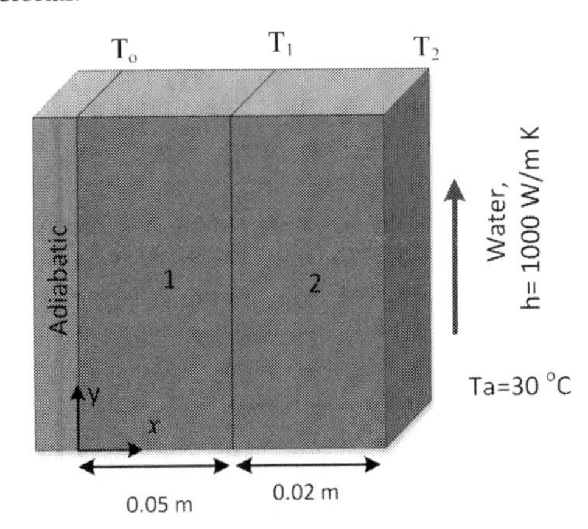

FIGURE 6.40
Schematic diagram of composite walls.

- Thermal conductivities are constant.
- The thermal contact resistance at the interface is assumed to be negligible.
- Heat transfer coefficient is assumed to incorporate the radiation effects, if any.

Based on the steady state and no bulk motion assumptions, heat is transferred in the x-direction by conduction, and heat is lost to air by convective heat flux. The problem is solved using the equation of change for heat transfer in Cartesian coordinates:

$$\rho C_p \frac{\partial T}{\partial t} + \rho C_p \left(v_x \frac{\partial T}{\partial x} + v_y \frac{\partial T}{\partial y} + v_z \frac{\partial T}{\partial z} \right) = k \left(\frac{\partial^2 T}{\partial x^2} + \frac{\partial^2 T}{\partial y^2} + \frac{\partial^2 T}{\partial z^2} \right) + \Phi_H$$

Considering the previous assumptions and no convective flux, we have:

Steady state

No convective heat flux $v_r = v_\theta = v_z = 0$

$$\rho C_p \frac{\partial T}{\partial t} + \rho C_p \left(v_x \frac{\partial T}{\partial x} + v_y \frac{\partial T}{\partial y} + v_z \frac{\partial T}{\partial z} \right) = k \left(\frac{\partial^2 T}{\partial x^2} + \frac{\partial^2 T}{\partial y^2} + \frac{\partial^2 T}{\partial z^2} \right) + \Phi_H$$

Temperature varies in the x-direction only

Energy Balance on Plane 1

Plane one has a heat generation term, and it is bounded with insulation and the interface of material two. Consequently, the model partial differential equation is reduced to:

$$0 = k_1 \left(\frac{\partial^2 T}{\partial x^2} \right) + \dot{q}$$

Preform double integration:

$$T = -\frac{\dot{q}}{2k_1} x^2 + C_1 x + C_2$$

The first integral generates the first arbitrary constant of integration C_1, and the second integral generates C_2. The constants are found using the following boundary conditions:

B.C.1: left side of plane 1 is insulated; an insulated wall implies that:

$$x = 0,\ dT/dx = 0.$$

This boundary condition leads to: $C_1 = 0$

B.C.2: at the interface between the two planes, the wall temperature is maintained at a constant temperature, T_i ($x = L_1, T_1 = T_i$)

Substitute boundary condition 2:

$$T_i = -\frac{\dot{q}}{2k_1} L_1^2 + C_2, \text{ then, } C_2 = T_i + \frac{\dot{q}}{2k_1} L_1^2$$

Substitute C_1 and C_2 in the general temperature distribution of plane and rearrange:

$$T = \frac{\dot{q}}{2k_1}(L_1^2 - x^2) + T_i$$

where the interface temperature between the two planes T_i is:

$$T_i = T - \frac{\dot{q}}{2k_1}(L_1^2 - x^2)$$

Energy Balance on Plane 2

Assume steady state, and no connective and no heat generation in plane 2; thus, the model partial differential equation is reduced to:

$$0 = k_2\left(\frac{\partial^2 T}{\partial x^2}\right)$$

Perform double integration. The first integral generates the first constant C_{21}, and the second integral generates C_{22}. Hence the general temperature profile in the plane 2 is:

$$T = C_{21}x + C_{22}$$

The boundary conditions are given by:

B.C.1: at right side of plane 1, convective heat transfer from the wall implies that at:

$$x = L_1 + L_2,\ q = -k_2\frac{dT}{dx} = h(T_2 - T_a)$$

$$C_{21} = -\frac{h}{k_2}(T_2 - T_a)$$

B.C.2: at the planes interface, the wall temperature is maintained at constant temperature, T_i

$$x = L_1, \quad T = T_i$$

Substitute boundary condition 2 in the temperature distribution of plane 2:

$$T_i = -\frac{h}{k_2}(T_2 - T_a)(L_1) + C_{22}$$

Rearrange:

$$T_i + \frac{h}{k_2}(T_2 - T_a)L_1 = C_{22}$$

Substitute the two integral constants in the general energy balance equation of plane 2:

$$T = -\frac{h}{k_2}(T_2 - T_a)x + T_i + \frac{h}{k_2}(T_2 - T_a)L_1$$

Rearrange:

$$T = T_i + \frac{h}{k_2}(T_2 - T_a)(L_1 - x)$$

Calculate T_2. Since there is no heat generation in material 2, the heat flux into the material at $x = L_1$ equals the heat flux from the material due to convection at $x = L_1 + L_2$:

$$q'' = h(T_2 - T_a)$$

There is an energy balance around the control volume of material 1, where there is no inflow and the energy generation rate must equal the outflow. Accordingly, the heat per unit surface area is:

$$q'' = \dot{q}L_1$$

Combine the two equations, and we have the following for the outer surface temperature:

$$h(T_2 - T_a) = \dot{q}L_1$$

Rearrange:

$$T_2 = T_a + \frac{\dot{q}L_1}{h}$$

Substitute values:

$$T_2 = 30°\text{C} + \frac{1.5\times10^6\ \text{W/m}^3(0.05\ \text{m})}{1000\ \text{W/(m}^2.\text{K)}} = 105°\text{C}$$

The temperature at the interphase surface T_1 is:

$$T_2 = T_i + \frac{h}{k_2}(T_2 - T_a)(L_1 - x)$$

The interface temperature T_i is:

$$105 = T_i + \frac{1000}{150}(105 - 30)(0.05 - 0.07)$$

$$115°\text{C} = T_i$$

The temperature at the insulation side of the plane 1 T_0 is:

$$T_0 = \frac{\dot{q}}{2k}(L_1^2 - x^2) + T_i$$

Substitute given values:

$$T_0 = 115°C + \frac{1.5 \times 10^6 \text{ W/m}^3}{2 \times 75 \text{ W/m.K}}(0.05^2 - 0)$$

The temperature at the insulation-plane 1 interface is:

$$T_0 = 140°C$$

Simulation Using COMSOL

1. Start COMSOL. Click on "Wizard," select "2D," and select "Heat transfer in solid" twice (one for each plane).
2. Right-click on "Geometry" and select rectangle 1. Enter the width and height (Figure 6.41).
3. Right-click on "Geometry" and select rectangle 2. Fill in the width, the height, and the position as shown Figure 6.42.

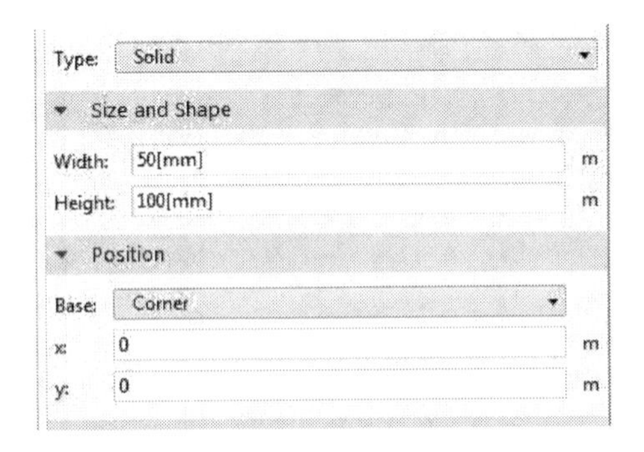

FIGURE 6.41
Specifying rectangle dimensions.

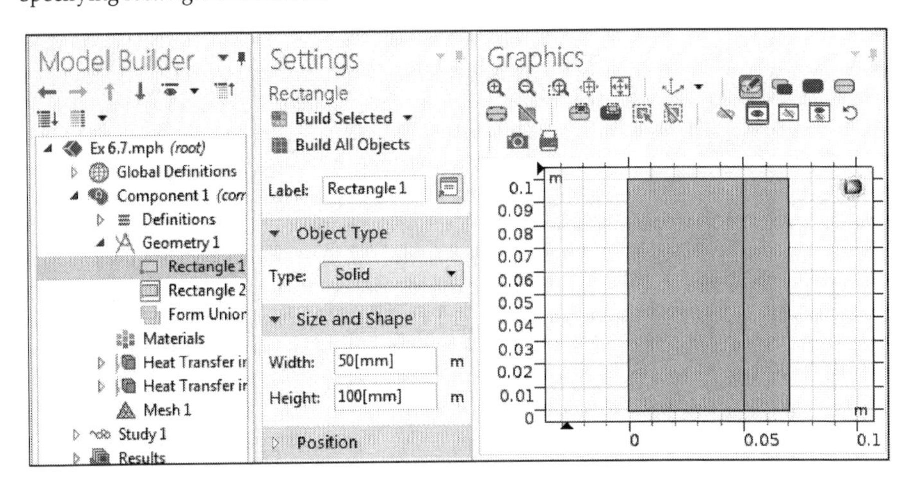

FIGURE 6.42
Model builder settings.

4. Right-click on "heat transfer in solid" and left-click "heat source." Key in the value of the heat source of plane 1 as shown in Figure 6.43.
5. Right-click on "Heat transfer in Solids 2" and left-click "Heat flux." Enter the heat flux as shown in Figure 6.44. The temperature interface between the two planes is shown in Figure 6.45.
6. The surface temperature profile of both planes is shown in Figure 6.46. It can be deduced that the simulation results is in good agreement with manual calculations. The highest temperature is at the insulation-plane 1 interface and is 140°C; the temperature at plane 2-air interface is 105°C , which agrees very well with manual calculations.

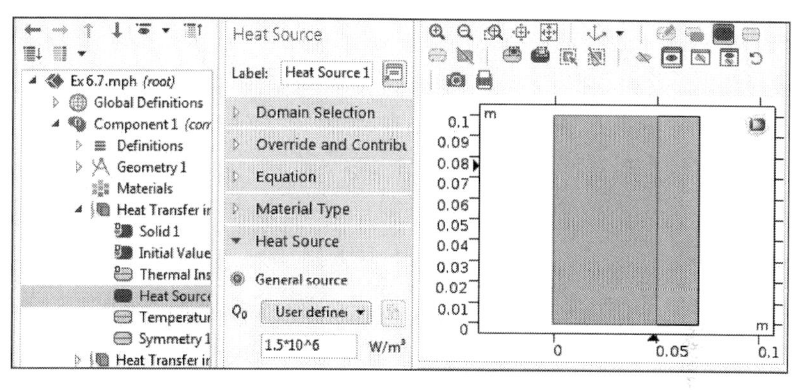

FIGURE 6.43
Specifying heat source per unit volume.

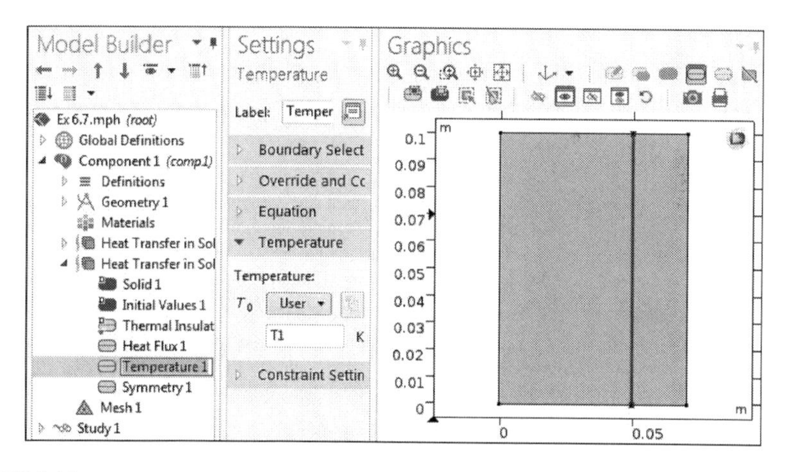

FIGURE 6.44
Specifying heat flux boundary conditions.

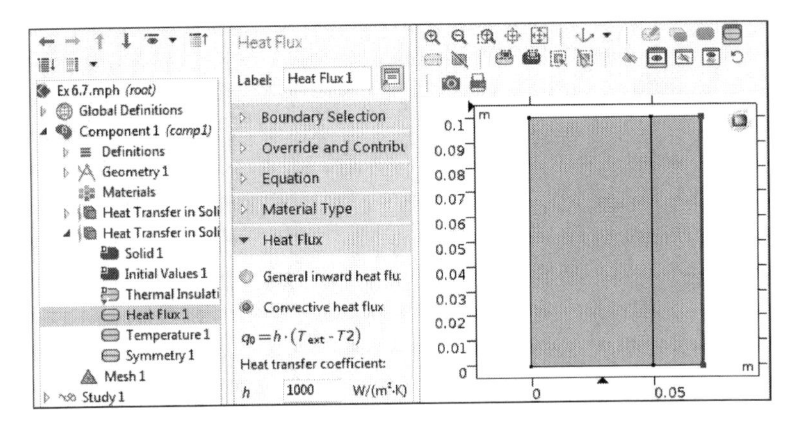

FIGURE 6.45
Interface temperature between two planes.

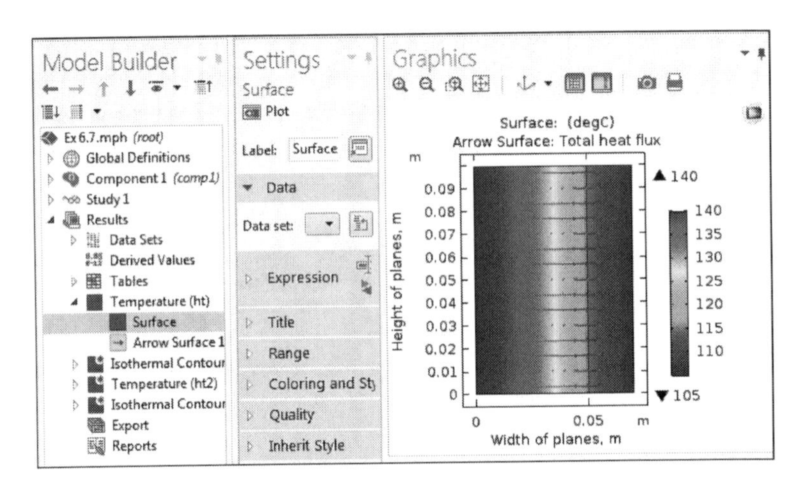

FIGURE 6.46
Temperature distribution across the composite wall.

Example 6.8: Heat Transfer from a Steam Pipe

Steam is the most commonly used heat utility in chemical plants. Understanding how it is used is essential in the study of utility systems. In this example, steam is flowing in a rectangular pipe (5 × 5 cm) made of iron ($k = 58$ W/mK); the pipe thickness is 1 cm and is insulated with 3 cm of rock wool (Figure 6.47). The air surrounding the pipe has a temperature of 20°C. The pipe is filled with steam at 200°C. The heat transfer coefficient to the surrounding air is 15 Wm^{2}K, and the heat transfer coefficient between the steam and the iron is $1 \times 10^5\ \mathrm{W m^{-2} K^{-1}}$. The physical properties of the rock wool are: $k = 0.036\ \mathrm{W\ m^{-1} K^{-1}}$; density $\rho = 90\ \mathrm{kg\ m^{-3}}$; heat capacity $C_p = 840\ \mathrm{J kg^{-1} K^{-1}}$. Simulate the temperature changes in the iron pipe with a rectangular shape using COMSOL Multiphysics.

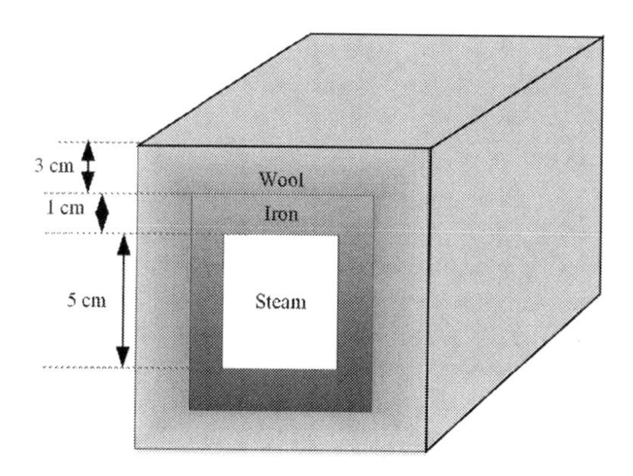

FIGURE 6.47
Temperature changes in an iron pipe with a rectangular transection.

Solution

Consider the following assumptions in determine the temperature distribution in the pipe walls.

- Heat transfer is assumed to be steady since there is no indication of any change with time.
- Heat transfer is assumed to be one dimensional since there is thermal symmetry about the center line and no variation in the axial direction.
- Thermal conductivities are assumed to be constant.
- The thermal contact resistance at the interface is negligible.
- The heat transfer coefficient includes the radiation effects.

The general differential equation of heat transfer in Cartesian coordinates is:

$$\rho C_p \frac{\partial T}{\partial t} + \rho C_p \left(v_x \frac{\partial T}{\partial x} + v_y \frac{\partial T}{\partial y} + v_z \frac{\partial T}{\partial z} \right) = k \left(\frac{\partial^2 T}{\partial x^2} + \frac{\partial^2 T}{\partial y^2} + \frac{\partial^2 T}{\partial z^2} \right) + \Phi_H$$

The heat flux can simply be calculated using the following expression:

$$q = \frac{T_{a1} - T_{a2}}{R_i + R_1 + R_2 + R_o} = \frac{T_{a1} - T_{a2}}{1/h_i + \Delta x_1 / k_1 + \Delta x_2 / k_2 + 1/h_o}$$

Substitute values of known parameters:

$$q = \frac{200 - 20}{\frac{1}{15} + \frac{0.01}{76.2} + \frac{0.03}{0.036} + \frac{1}{10^5}} = 200$$

As the heat flux is constant across the various planes, we have:

$$q = \frac{20 - T_1}{1/15} = 200$$

The temperature interface between the pipe thickness and the insulation layer is, $T_1 = 33°C$.

COMSOL Solution

Figure 6.48 shows the model builder and the heat flux boundary of the inner square pipe.

The heat boundary flux of the outer side of the wooden brick is described in Figure 6.49.

The surface plot of the pipe thickness and insulation layer of rock wool are shown in Figure 6.50.

The temperature across the pipe and rock wool is shown in Figure 6.51. The temperature shows negligible variation in the pipe skin due to high thermal conductivity. By contrast, there is significant change in temperature in the rock wool section of the pipe due to its low thermal conductivity. The figure is built from the generated 2D cutline data shown in Figure 6.52.

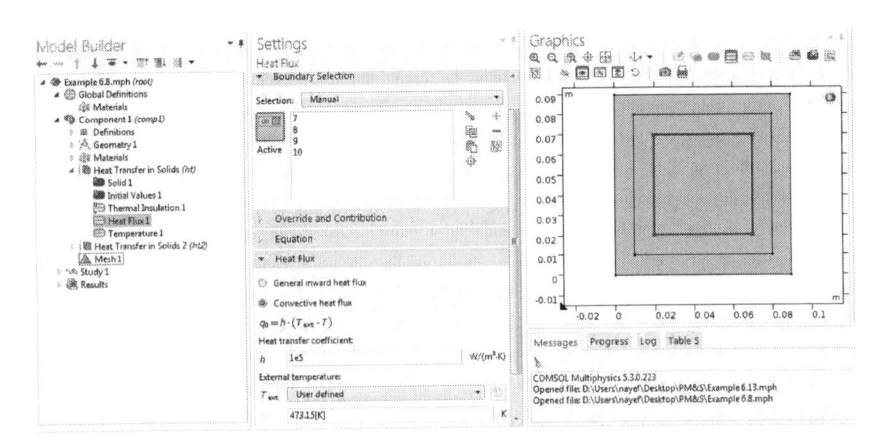

FIGURE 6.48
Heat flux boundary condition for the inner pipe.

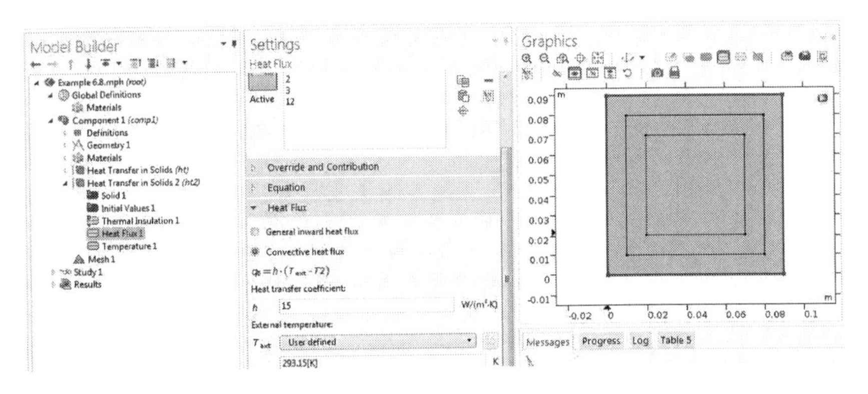

FIGURE 6.49
Heat flux boundary condition for the outer surface of the pipe.

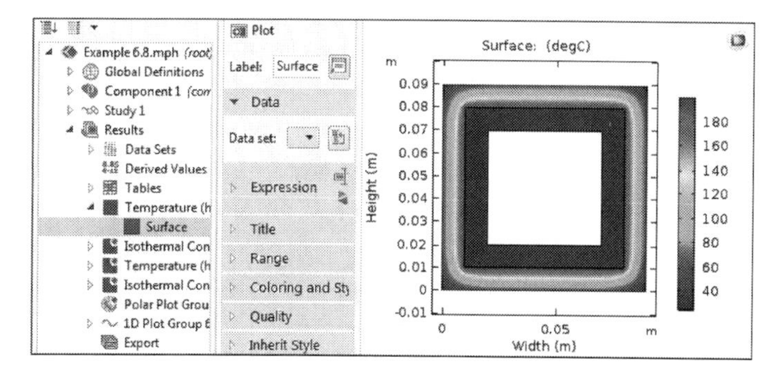

FIGURE 6.50
Temperature surface profile for pipe and rock wool.

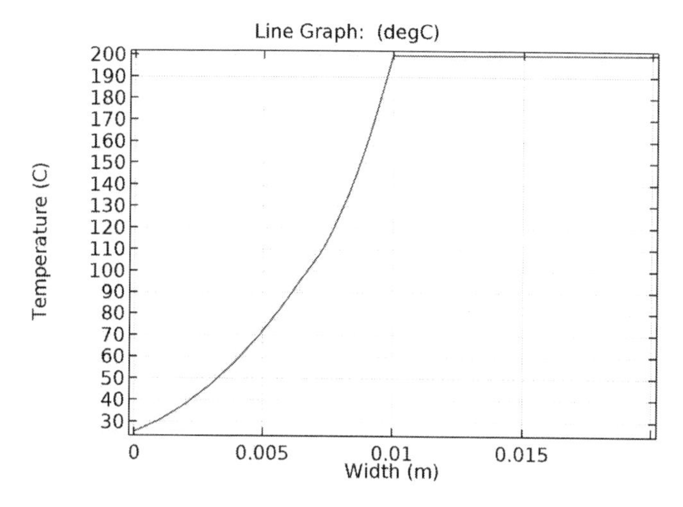

FIGURE 6.51
Temperature profile across the pipe at height = 0.05 m.

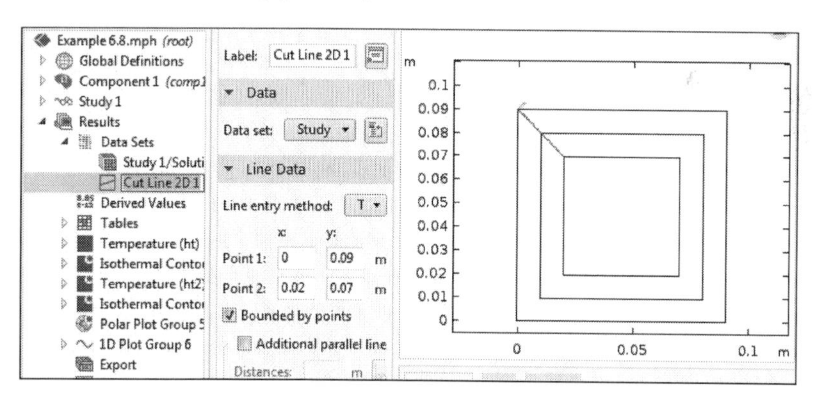

FIGURE 6.52
2D cutline (0,0.09), (0.02, 0.07).

Example 6.9: Heat Transfer from Sphere

Heat energy transferred between a surface and a moving fluid with different temperatures is known as convection. This is a combination of diffusion and bulk motion of molecules. In this example, a heated sphere of diameter $r = 0.1$ m is immersed in a large amount of stagnant fluid. Consider the heat conduction in the fluid surrounding the sphere in the absence of convection. The thermal conductivity of fluid can be taken as constant, $k = 0.7$ W/(m.K). The temperature at the surface of the sphere is $T_R = 70°C$, and the temperature far away (i.e., at $r = \infty$) from the sphere is $T_a = 30°C$. Derive an expression for the temperature T in the surrounding fluid as a function of r for the distance from the center of the sphere (Figure 6.53).

Solution

Consider the following assumptions to determine the temperature distribution in the fluids:

- Steady-state process; heat transfer is assumed steady.
- Thermal symmetry about the center line, then heat transfer is one dimensional.
- Thermal conductivities are constant.
- The thermal contact resistance at the interface is negligible.
- Heat transfer coefficient incorporates the radiation effects.

From the general differential heat balance equation in spherical coordinates:

$$\rho C_p \frac{\partial T}{\partial t} + \rho C_p \left(v_r \frac{\partial T}{\partial r} + \frac{v_\theta}{r} \frac{\partial T}{\partial \theta} + \frac{v_\varnothing}{r\sin(\theta)} \frac{\partial T}{\partial \varnothing} \right)$$

$$= k \left(\frac{1}{r^2} \frac{\partial}{\partial r} \left(r^2 \frac{\partial T}{\partial r} \right) + \frac{1}{r^2 \sin(\theta)} \frac{\partial}{\partial \theta} \left(\sin(\theta) \frac{\partial T}{\partial \theta} \right) + \frac{1}{r^2 \sin^2(\theta)} \frac{\partial^2 T}{\partial \theta^2} \right) + \Phi_H$$

After neglecting convective heat transfer and considering only the heat transfer in the radial direction over a sphere hanged in a surrounding fluid, there is no heat generation because as q is the rate of heat generation per unit volume; in this case, $q = 0$ in the fluid. The thermal conductivity

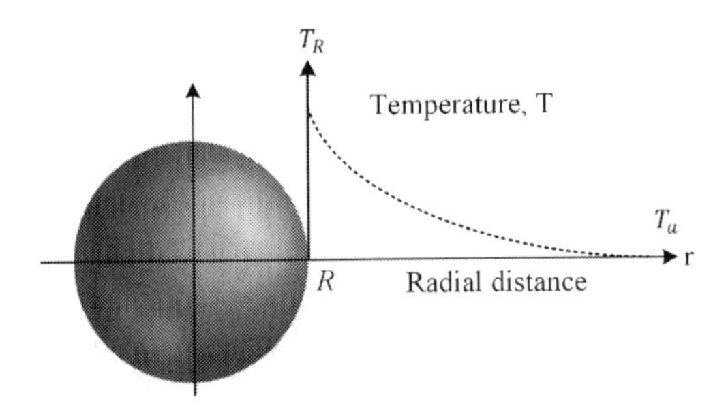

FIGURE 6.53
Heated sphere in a large amount of stagnant fluid.

k for the fluid is constant; on simplifying the energy balance partial differential equations, we have:

$$k\left(\frac{1}{r^2}\frac{\partial}{\partial r}\left(r^2\frac{\partial T}{\partial r}\right)\right)=0$$

Solve the differential equation to get the temperature profile. On the first integration, the first arbitrary constant of integration C_1 is generated:

$$r^2\frac{\partial T}{\partial r}=C_1$$

The second integration generates the second arbitrary constant of integration C_2:

$$T=-\frac{C_1}{r}+C_2$$

The integration constants are determined using the boundary conditions, where R is the radius of the sphere:

B.C.1: $r=\infty, T=T_a$ from this boundary condition $C_2=T_a$
B.C.2: $r=R, T=T_R$, from this boundary condition $C_1=(T_a-T_R)R$

Once the integration constants are substituted, the temperature profile is:

$$T=\left(T_R-T_a\right)\frac{R}{r}+T_a$$

The analytical solution is shown in Figure 6.54. The figure reveals that there is drop in temperature going deeper into the fluid and far away from the sphere surface.

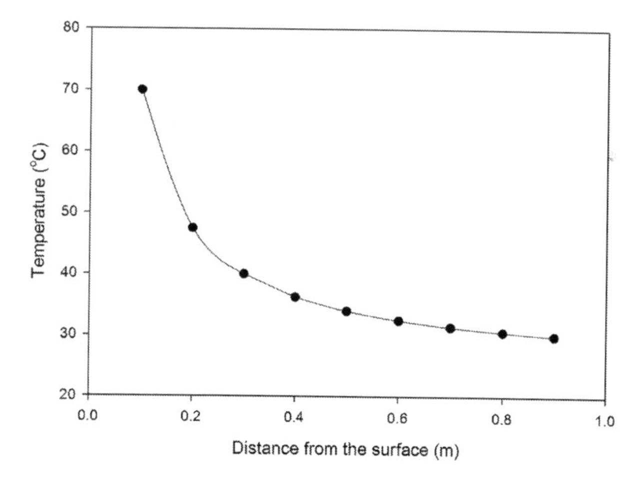

FIGURE 6.54
Analytical solution for temperature versus distance from edge of sphere.

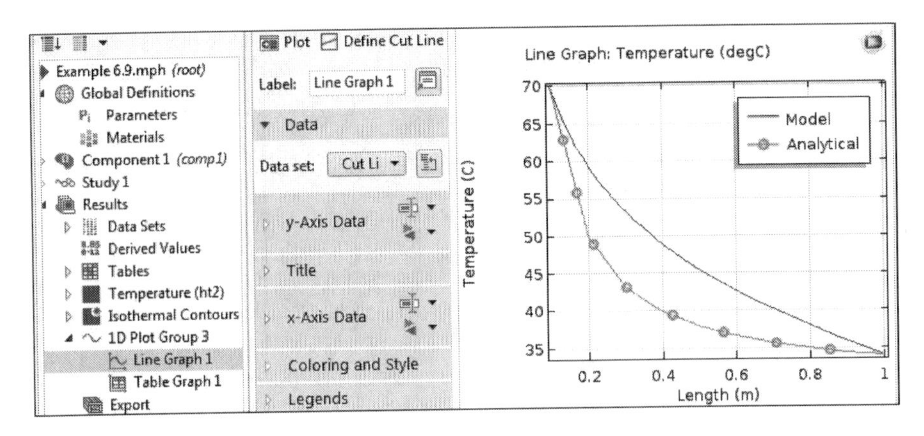

FIGURE 6.55
Temperature distribution from the solid surface (model, analytical).

COMSOL Solution

The model builder in COMSOL with the solution of the model equation is shown in Figure 6.55. Comparison between the analytical solution and COMSOL predictions shows a discrepancy due to the right-side boundary conditions (i.e., $r = \infty$). The temperature at $r = 1$ m is set equal to 34°C, which is obtained from the analytical solution.

The cut line where the data is generated in Figure 6.58 is shown in Figure 6.56. The surface temperature profile is shown in Figure 6.57. The model builder setup with boundary conditions is shown in Figure 6.58.

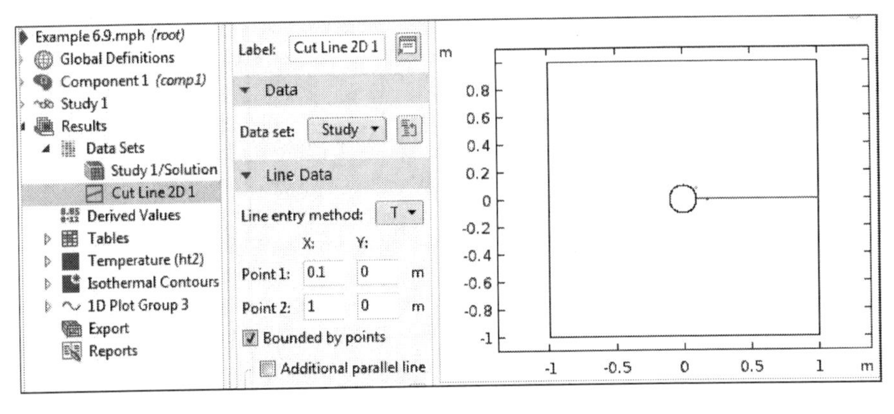

FIGURE 6.56
2D cutline for temperature profile from solid surface through the bulk of fluid.

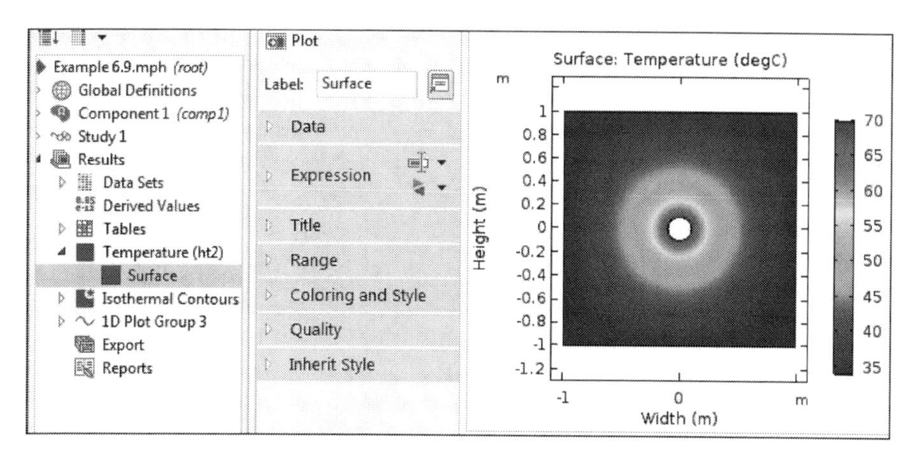

FIGURE 6.57
Surface plot for temperature distribution in the fluid surrounding the sphere.

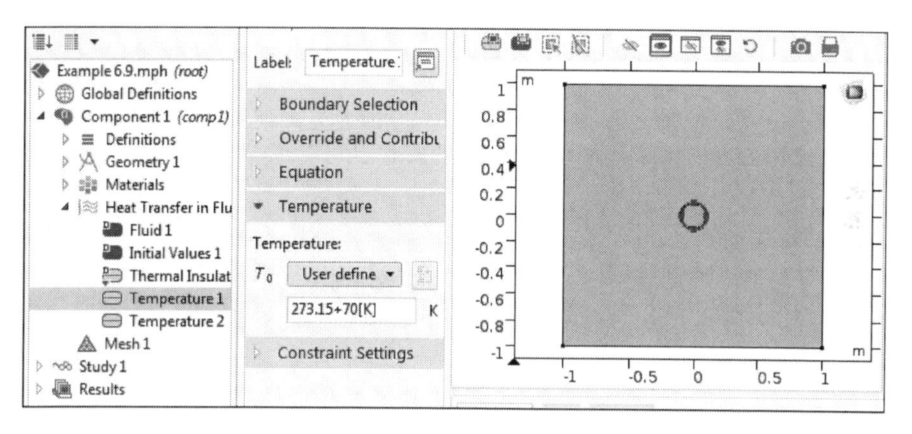

FIGURE 6.58
Model builder setting.

Example 6.10: Transport of Polymeric Solution

A polymeric liquid is transported through a pipe fixed in a wall at a point where the temperature is 650 K (Figure 6.59). A wall constructed of a material has thermal conductivity of $k = 25$ W/mK and a thickness of 1.2 m. The inside surface temperature of the wall is maintained at 925 K. The outside surface of the wall is exposed to air at 300 K with a convective heat transfer coefficient of 23 W/m^{2}K. Develop an expression for the temperature distribution in the wall in terms of $T(x)$. How far from the hot surface should the pipe be located?

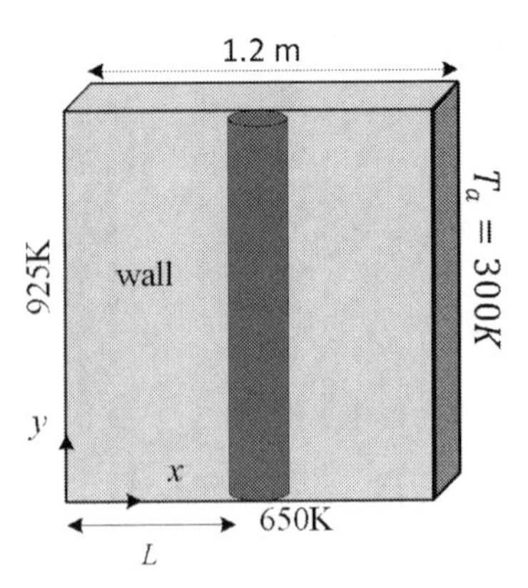

FIGURE 6.59
Schematic of heating a pipe of polymeric solution.

Solution

The equation of change for heat transfer in Cartesian coordinates is:

$$\rho C_p \frac{\partial T}{\partial t} + \rho C_p \left(v_x \frac{\partial T}{\partial x} + v_y \frac{\partial T}{\partial y} + v_z \frac{\partial T}{\partial z} \right) = k \left(\frac{\partial^2 T}{\partial x^2} + \frac{\partial^2 T}{\partial y^2} + \frac{\partial^2 T}{\partial z^2} \right) + \Phi_H$$

Consider the prior assumptions:

Steady state | No convective heat flux $v_r = v_\theta = v_z = 0$ | No heat generation

$$\rho C_p \frac{\partial T}{\partial t} + \rho C_p \left(v_x \frac{\partial T}{\partial x} + v_y \frac{\partial T}{\partial y} + v_z \frac{\partial T}{\partial z} \right) = k \left(\frac{\partial^2 T}{\partial x^2} + \frac{\partial^2 T}{\partial y^2} + \frac{\partial^2 T}{\partial z^2} \right) + \Phi_H$$

Temperature varies in the x-direction only

The general differential energy balance equation is reduced to:

$$0 = k_1 \left(\frac{\partial^2 T}{\partial x^2} \right)$$

Perform double integration, which generates two arbitrary constants of integrations C_1 and C_2:

$$T = C_1 x + C_2$$

The arbitrary constants of integration are found using the following boundary conditions.

B.C.1: at $x = 0$, $T = T_{wall} = 925\text{K}$, from this boundary condition $C_2 = T_{wall}$

B.C.2: at $x = L = 1.2$ m, $q = -k\,dT/dt = -h(T - T_a)$, $T_a = 300$ K

Substitute boundary condition 2:

$$\frac{dT}{dx} = \frac{h}{k}(T - T_a) = C_1$$

Substitute C_1 and C_2 in the resultant temperature profile:

$$T = \frac{h}{k}(T - T_a)x + T_{wall}$$

Find the length where the tube should be located (i.e., at $T = 650$ K):

$$650 = \frac{23}{25}(650 - 300)x + 925$$

Solve for x. The pipe should be fixed at distance, $x = L = 0.996$ m.

COMSOL Simulation

Start COMSOL. Select "heat transfer in solid" as your physics (Figure 6.60).

Add the heat flux as a right-side boundary conditions, as shown in Figure 6.61.

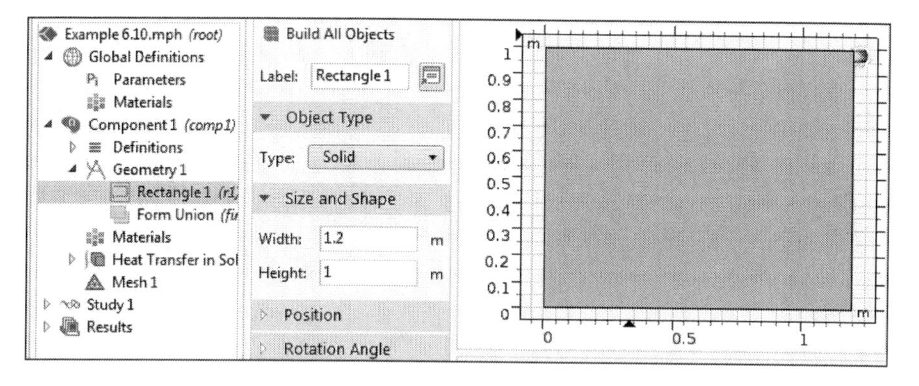

FIGURE 6.60
Selecting rectangular geometry.

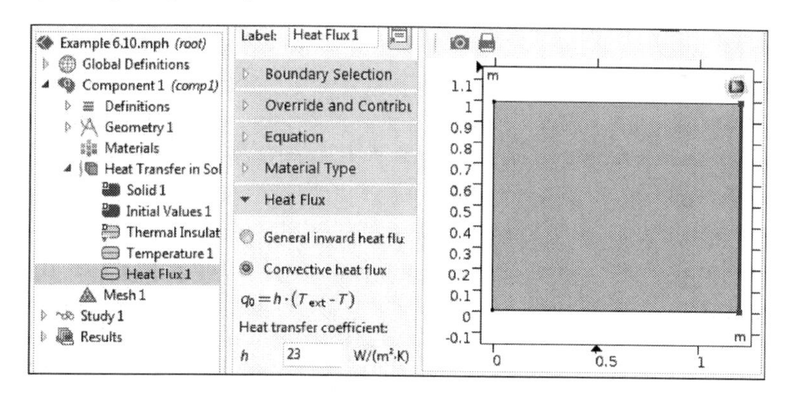

FIGURE 6.61
Setting right-side heat flux boundary condition.

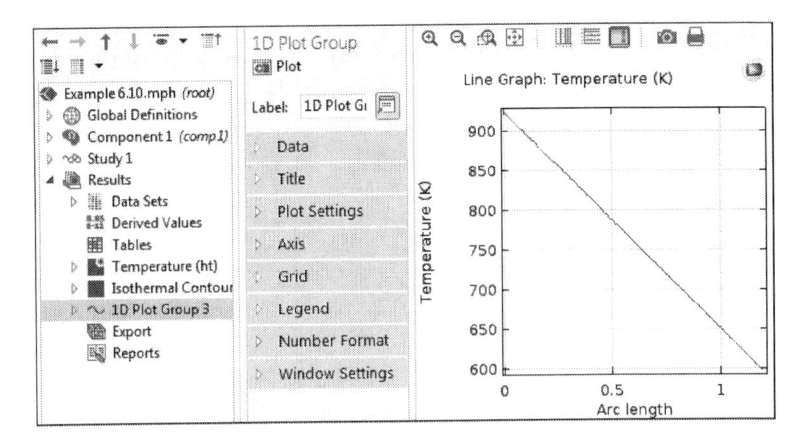

FIGURE 6.62
Temperature profile across the plane at height $H = 0.5$ m.

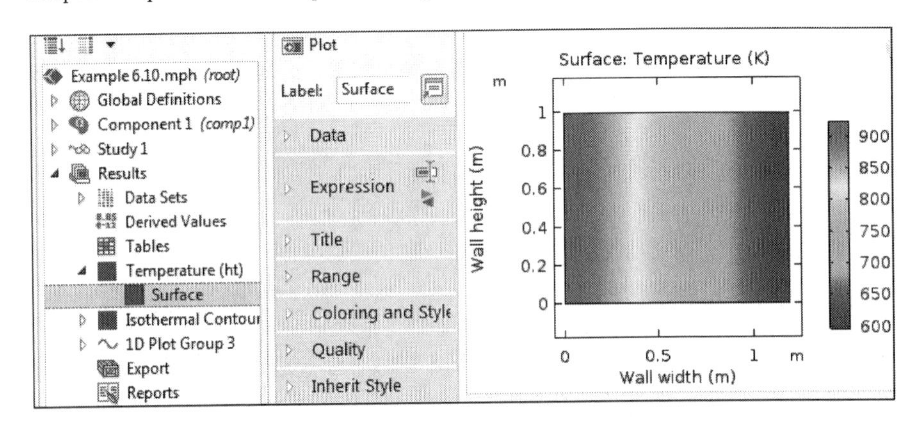

FIGURE 6.63
Surface plot of temperature of the heating wall.

Figure 6.62 shows the line temperature drop across the heating wall. The wall should be placed around 1 m from the hot side of the wall ($T = 925$). The simulation results are in good agreement with manual calculations.

The surface temperature profile across the wall is shown in Figure 6.63.

Example 6.11: Furnace with Composite Walls

A furnace is a device used for high-temperature heating. In this case, the temperature of air inside the furnace is at 1340 K. The wall of the furnace is composed of a 0.10 m layer of fireclay brick ($k_1 = 1.13$ W/m.K) and a 0.05 m thickness of mild steel ($k_2 = 42.9$ W/m.K) on its outside surface (Figure 6.64). Heat transfer coefficients (h) of the inside and outside wall surfaces are 5110 and 45 W/m^2.K, respectively. The outside air is at 295 K. Develop an expression for the temperature distribution in the silicon chip.

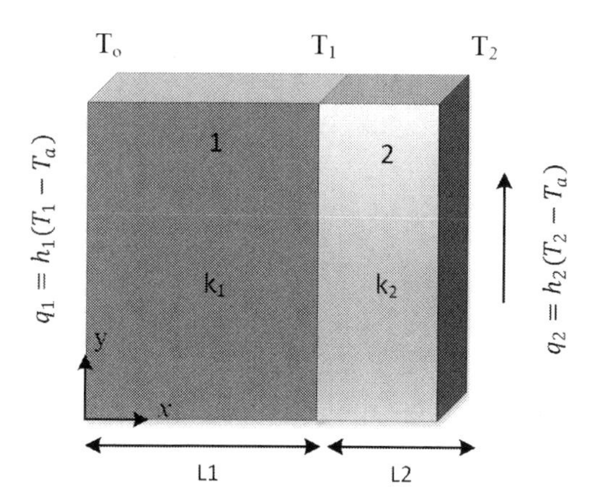

FIGURE 6.64
Composite walls of a furnace.

Determine the temperatures at each surface and at the brick-steel interface. Determine the heat transfer rate per square meter of wall area [3].

Solution

Consider the following assumptions to determine the temperature profile across the walls of the furnace:

- Steady state; no change of temperature with time.
- Assume heat transfer is in one dimension.
- Thermal conductivities are constant.
- The thermal contact resistance at the interface is negligible.

Use the general differential equation of heat transfer in Cartesian coordinates:

$$\rho C_p \frac{\partial T}{\partial t} + \rho C_p \left(v_x \frac{\partial T}{\partial x} + v_y \frac{\partial T}{\partial y} + v_z \frac{\partial T}{\partial z} \right) = k \left(\frac{\partial^2 T}{\partial x^2} + \frac{\partial^2 T}{\partial y^2} + \frac{\partial^2 T}{\partial z^2} \right) + \Phi_H$$

With steady state, no convection heat transfer, and heat in the x-direction, there is no heat generation term. Accordingly, the partial differential equation is reduced to:

Steady state — No convective heat flux $v_r = v_\theta = v_z = 0$ — No heat generation

$$\rho C_p \frac{\partial T}{\partial t} + \rho C_p \left(v_x \frac{\partial T}{\partial x} + v_y \frac{\partial T}{\partial y} + v_z \frac{\partial T}{\partial z} \right) = k \left(\frac{\partial^2 T}{\partial x^2} + \frac{\partial^2 T}{\partial y^2} + \frac{\partial^2 T}{\partial z^2} \right) + \Phi_H$$

Temperature varies in the x-direction only

$$0 = k \left(\frac{\partial^2 T}{\partial x^2} \right)$$

Preform the double integration:

$$T = C_1x + C_2$$

Temperature distribution in wall 1 is:

B.C.1: at $x = 0$, $q = -k_1 dT/dx = -h_1(T_1 - T_{a1})$, $C_1 = h_1/k_1 (T_1 - T_{a1})$
B.C.2: at $x = L_1$, $T = T_i$, $C_2 = T_i - h/k(T_1 - T_{a1})L_1$

According, the temperature distribution in wall 1 is represented by the following equation:

$$T_{w1} = \frac{h_1}{k_1}(T_1 - T_{a1})x + T_i - \frac{h_1}{k_1}L_1(T_1 - T_{a1})$$

Rearrange:

$$T_{w1} = \frac{h_1}{k_1}(T_1 - T_{a1})(x - L_1) + T_i$$

Temperature distribution in wall 2 is:

$$T_2 = C_1x + C_2$$

B.C.1: at $x = L_1, T = T_i$, from this boundary condition, $T_i = C_1L_1 + C_2$
B.C.2: at $x = L_1 + L_2 = L_t$, $q = h_2(T_2 - T_{a2})$, this boundary condition leads to:

$$C_1 = -\left(\frac{h_2}{k_2}\right)(T_2 - T_{a2})$$

Substitute C_1 in the following equation to find C_2:

$$T_i + \left(\frac{h_2}{k_2}\right)(T_2 - T_{a2})L_1 = C_2$$

Substitute C_1 and C_2:

$$T_{w2} = -\left(\frac{h_2}{k_2}\right)(T_2 - T_{a2})x + T_i + \left(\frac{h_2}{k_2}\right)(T_2 - T_{a2})L_1$$

Rearrange:

$$T_{w2} = \left(\frac{h_2}{k_2}\right)(T_2 - T_{a2})(L_1 - x) + T_i$$

The equation contains two unknowns; accordingly, an axillary equation is needed:

$$q = \frac{T_{a1} - T_{a2}}{R_i + R_1 + R_2 + R_o} = \frac{T_{a1} - T_{a2}}{1/h_i + \Delta x_1/k_1 + \Delta x_2/k_2 + 1/h_o}$$

Substitute values:

$$q = \frac{1340 - 295}{\frac{1}{5110} + \frac{0.1}{1.13} + \frac{0.05}{42.9} + \frac{1}{45}} = 9324$$

To find T_1, the heat through each wall is the same:

$$q = \frac{1340 - T_1}{\frac{1}{5110}} = 9324$$

$$T_1 = 1338.18\,\text{K}$$

The temperature distribution of wall 1 is:

$$q = \frac{T_1 - T_i}{\frac{L_1}{k_1}} = \frac{1338.17 - T_i}{\frac{0.1}{1.13}} = 9324$$

$$T_i = -\frac{0.1}{1.13}(9324) + 1338.17$$

Hence, $T_i = 513\,\text{K}$.

The temperature distribution of wall 2 is the temperature of T_2:

$$q = \frac{T_2 - T_i}{\frac{L_2}{k_2}} = \frac{T_2 - 513}{\frac{0.05}{42.9}} = 9324$$

$$T_2 = -\frac{0.05}{42.9}(9324) + 513$$

$$T_2 = 502$$

COMSOL Simulation

Start COMSOL. Select "2D," then right-click "geometry," select "rectangle," and set the dimensions as shown in Figure 6.65.

The heat flux for the boundary material to outside air is set in Figure 6.66. The heat flux for the boundary material to inside air is set in Figure 6.67. The resultant surface plot diagram is shown in Figure 6.68, whereas the line plot curve is shown in Figure 6.69.

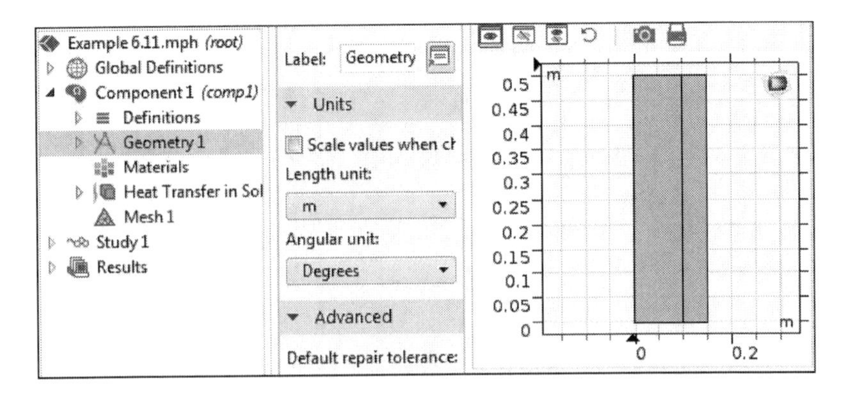

FIGURE 6.65
Selection rectangular coordinate.

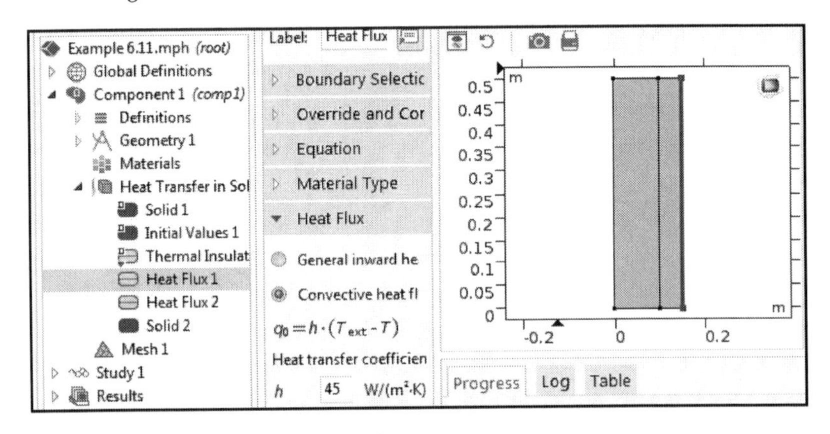

FIGURE 6.66
Setup for heat flux convective term.

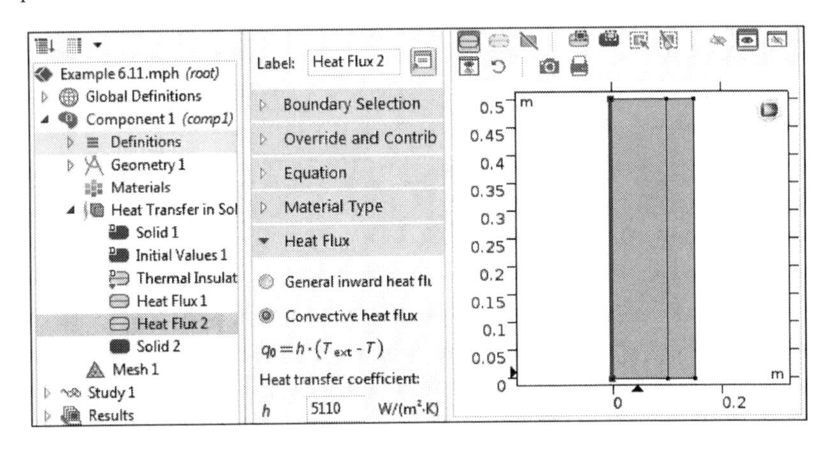

FIGURE 6.67
Heat flux of boundary material to water.

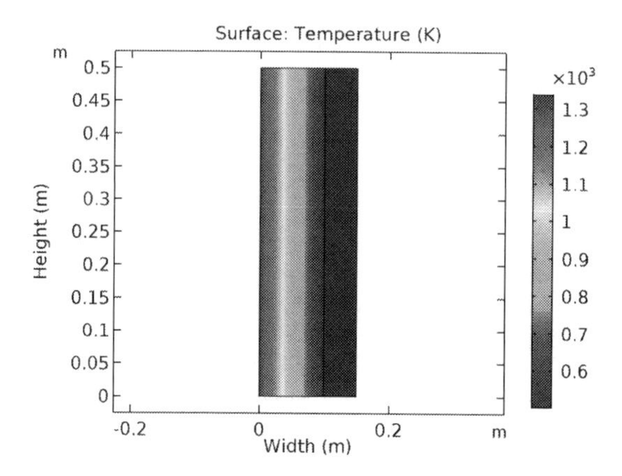

FIGURE 6.68
Surface temperature profile.

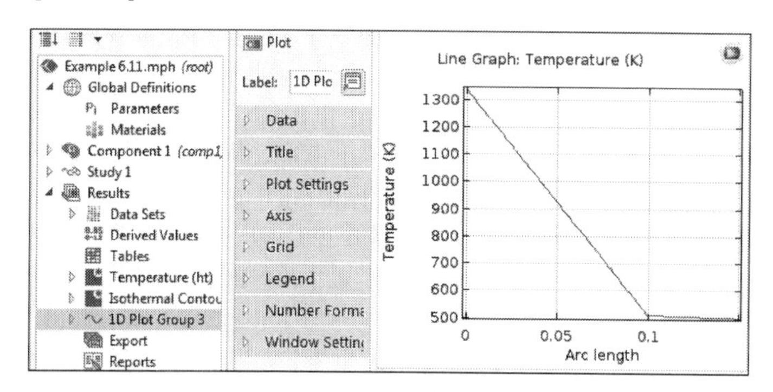

FIGURE 6.69
Line plot of temperature across the tube length.

Example 6.12: Cooling of Silicon Chip

A silicon chip is a very small piece of silicon inside a computer. A square silicon chip (k=150 W/m.K) is of width w = 5 mm on a side and of thickness h = 1 mm (Figure 6.70). The chip is mounted in a substrate such that the front surface is exposed to a coolant, while the sides and back surfaces are insulated. If 4 W are being dissipated in circuits mounted on the back

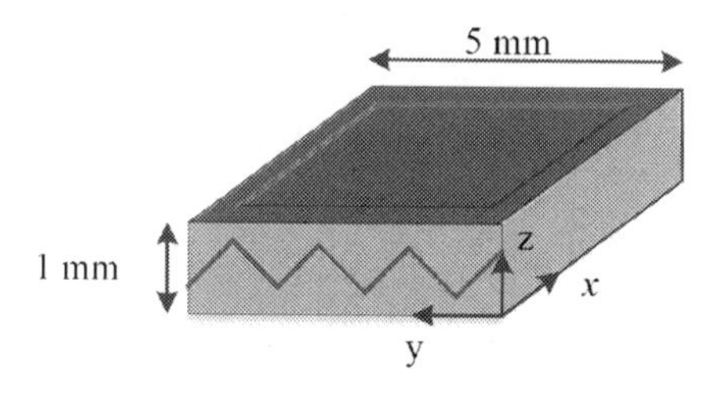

FIGURE 6.70
Cooling of cubic silicon chip.

surface of the chip, develop an expression for the temperature distribution in the chip and determine the steady-state temperature difference between the front and back surfaces.

Solution

Consider the following assumptions in calculating the temperature difference between the front and back surfaces of the chip:

- Assume that the system is in steady-state condition.
- The physical properties are constant.
- Heat dissipation is uniform.
- Negligible heat loss from the back and sides.
- One-dimensional conduction in the chip.

Based on the steady state and no bulk motion assumptions, heat is transferred in the z-direction by conduction, and the temperature of the top surface is fixed by coolant. The problem is solved using the equation of change for heat transfer in Cartesian coordinates:

$$\rho C_p \frac{\partial T}{\partial t} + \rho C_p \left(v_x \frac{\partial T}{\partial x} + v_y \frac{\partial T}{\partial y} + v_z \frac{\partial T}{\partial z} \right) = k \left(\frac{\partial^2 T}{\partial x^2} + \frac{\partial^2 T}{\partial y^2} + \frac{\partial^2 T}{\partial z^2} \right) + \Phi_H$$

No convective heat flux: $v_x = v_y = v_z = 0$

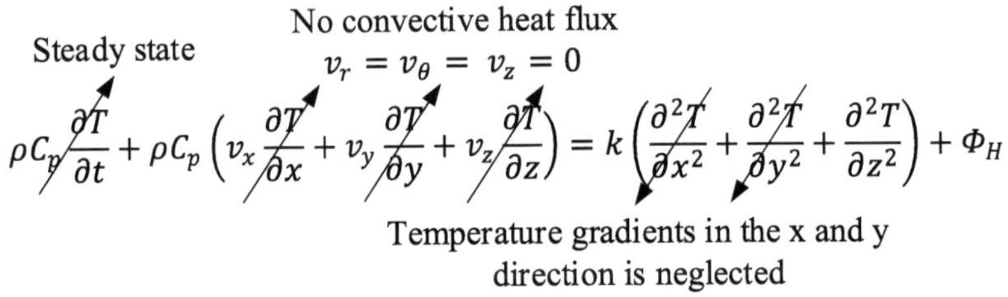

Steady state

No convective heat flux

$v_r = v_\theta = v_z = 0$

$$\rho C_p \frac{\partial T}{\partial t} + \rho C_p \left(v_x \frac{\partial T}{\partial x} + v_y \frac{\partial T}{\partial y} + v_z \frac{\partial T}{\partial z} \right) = k \left(\frac{\partial^2 T}{\partial x^2} + \frac{\partial^2 T}{\partial y^2} + \frac{\partial^2 T}{\partial z^2} \right) + \Phi_H$$

Temperature gradients in the x and y direction is neglected

Energy Balance on Plane 1

Plane 1 has a heat generation term, and it is bounded with insulation and the interface of material two. Consequently, the model partial differential equation is reduced to:

$$0 = k \left(\frac{\partial^2 T}{\partial z^2} \right) + \dot{q}$$

Performing double integrations generates the two arbitrary constants of integration C_1 and C_2:

$$T = -\frac{\dot{q}}{2k} z^2 + C_1 z + C_2$$

Expressions of the arbitrary constants are found using the following boundary conditions:

Left side of plane 1 is insulated, insulated wall implies that:

B.C.1: at $x = 0, dT/dz = 0$

This boundary conditions leads to: $C_1 = 0$.

At the interface between the two planes, the wall temperature is maintained at constant temperature T_i.

B.C.2: at $z = L, T = T_c$

Substitute boundary condition 2 and then reorganize:

$$C_2 = T_c + \frac{\dot{q}}{2k}L^2$$

Substitute C_1 and C_2 in the general temperature distribution of furnace walls:

$$T = -\frac{\dot{q}}{2k}z^2 + T_c + \frac{\dot{q}}{2k}L^2$$

Rearrange:

$$T = \frac{\dot{q}}{2k}(L^2 - z^2) + T_c$$

Rearrange for the temperature difference:

$$T - T_c = \Delta T = \frac{\dot{q}}{2k}(L^2 - z^2)$$

The temperature difference t $z = 1$ mm and temperature at $z = 0$ is:

$$T - T_c = \Delta T = \frac{\dot{q}}{2k}L^2 = \frac{4\ \text{W}/(0.001 \times 0.005 \times 0.005)}{2\left(150\dfrac{\text{W}}{\text{m}}\text{K}\right)}(0.001)^2 = 0.53°\text{C}$$

Note that the q is in the units of W/m^3.

COMSOL Simulation

Start COMSOL. Select "3D simulation," then select "Block" for geometry. Add the block geometry as shown in Figure 6.71.

For physics, select "Heat transfer in solid" as shown in Figure 6.72. The surface plot diagram is shown in Figure 6.73.

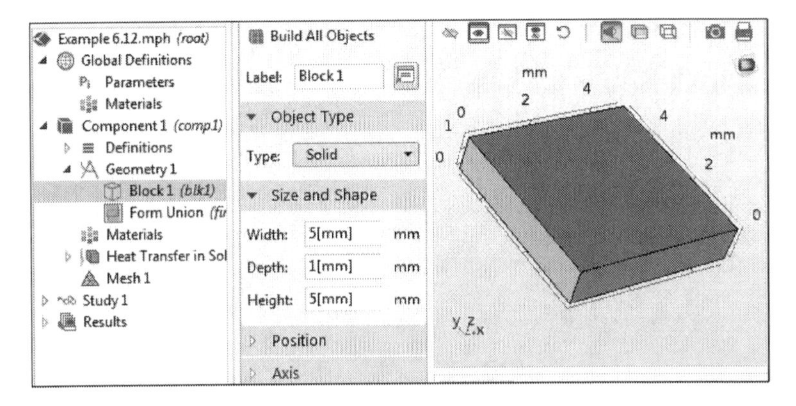

FIGURE 6.71
Specify black geometry.

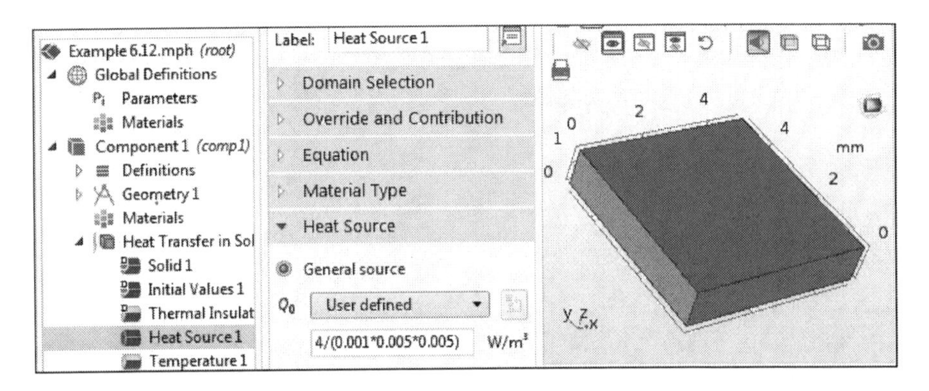

FIGURE 6.72
Heat transfer in solid boundary conditions and heat source.

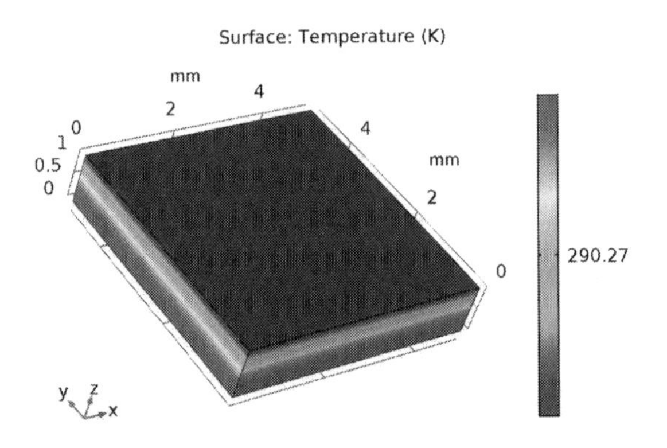

FIGURE 6.73
3D surface plot temperature.

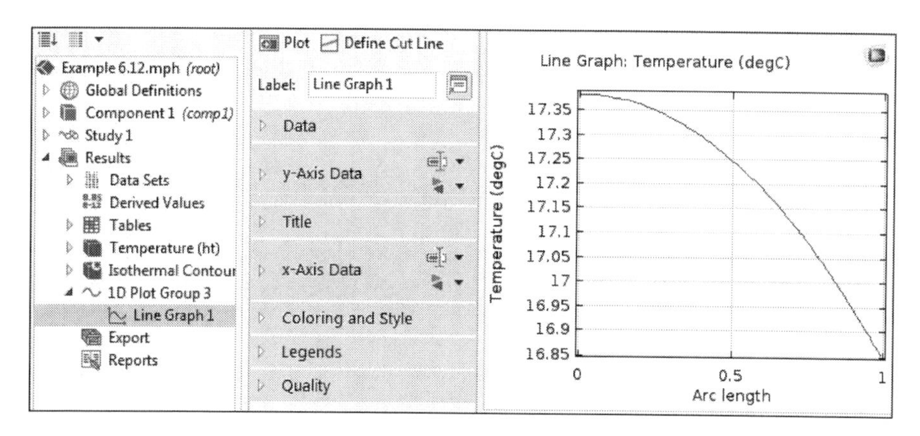

FIGURE 6.74
Temperature distribution curve from bottom to top surface of the chip.

Take a 3D cutline in the middle of the cube. The temperature curve is shown in Figure 6.74. The figure shows the temperature difference between bottom and top surface of the block is almost close to the value of the manual calculation 0.6°C.

Example 6.13: Heat Transfer in the Shield of Electrical Wire

Electrical cables come with various degrees of shielding. The amount of shielding required depends on several factors, including the electrical environment in which the cable is used. Insulated cables are a major type of high-voltage power cable, and their insulation condition is related to conductor temperature. In this example, a long electric copper wire with a radius of 0.003 m and a length of 0.05 m, $k_C = 400$ W/m.°C, is tightly covered with plastic protection that has a thermal conductivity of $k_P = 0.15$ W/m.°C and a thickness of 0.002 m (Figure 6.75). The power dissipated by the cable is 0.2 W. The insulated wire is exposed to a medium at a temperature of $T_a = 27$°C with a heat transfer coefficient of $h = 12$ W/m^2°C. Find an expression for the steady-state temperature profile in the plastic shield. Start with the general differential energy equation and then calculate the steady-state temperature at the interface of the wire and the plastic cover [2].

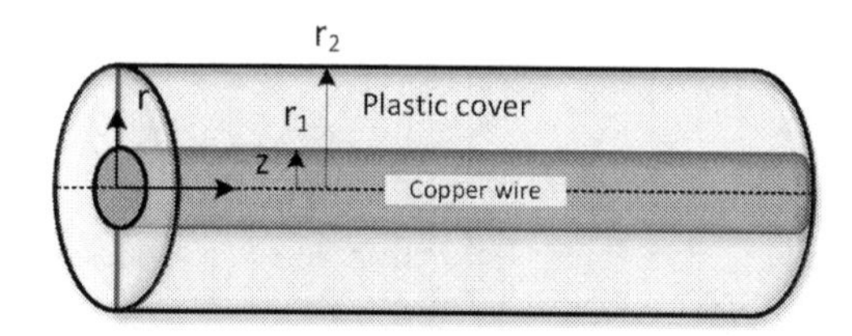

FIGURE 6.75
Electric copper in a thick plastic cover.

Solution

Consider the following assumptions to determine the heat transfer in the plastic cover:

- Heat transfer is steady.
- Heat transfer is a one-dimensional coordinate.
- Thermal conductivities are constant.
- The thermal contact resistance is negligible at the interface.
- The entire heat generated in the resistance wire is transferred to the plastic cover.

At steady conditions, the rate of heat transfer from the wire is equal to the heat generated within the wire, $Q = 0.2\,\text{W}$. The temperature distribution in the plastic shield in cylindrical coordinate is described by the following general partial differential energy equation:

$$\rho C_p \frac{\partial T}{\partial t} + \rho C_p \left(v_r \frac{\partial T}{\partial r} + v_\theta \frac{1}{r}\frac{\partial T}{\partial \theta} + v_z \frac{\partial T}{\partial z} \right)$$

$$= k\left(\frac{1}{r}\frac{\partial}{\partial r}\left(r\frac{\partial T}{\partial r} \right) + \frac{1}{r^2}\frac{\partial^2 T}{\partial \theta^2} + \frac{\partial^2 T}{\partial z^2} \right) + \Phi_H$$

To determine $T(r)$ in the plastic shield of the wire, the appropriate and reduced form of the heat equation for the wire plastic protection layer is:

Steady state

No convective heat flux $v_r = v_\theta = 0$

Temperature gradients in the axial and angular direction is neglected

$$\rho C_p \frac{\partial T}{\partial t} + \rho C_p \left(v_r \frac{\partial T}{\partial r} + v_\theta \frac{1}{r}\frac{\partial T}{\partial \theta} + v_z \frac{\partial T}{\partial z} \right) = k\left(\frac{1}{r}\frac{\partial}{\partial r}\left(r\frac{\partial T}{\partial r} \right) + \frac{1}{r^2}\frac{\partial^2 T}{\partial \theta^2} + \frac{\partial^2 T}{\partial z^2} \right) + \Phi_H$$

Temperature gradients in the axial direction is neglected

No heat generation

The general differential energy balance equation is reduced to:

$$k_p\left(\frac{1}{r}\frac{\partial}{\partial r}\left(r\frac{\partial T}{\partial r} \right) \right) = 0$$

The following two successive integrations are used to obtain the general solution of the differential equation in the wire plastic shield:

$$T = C_1 \ln(r) + C_2$$

The following boundary conditions are used to determine expressions for the two arbitrary constants of integration.

$$\text{B.C.1: at } r = r_1,\ -k\frac{dT}{dr} = \dot{q}_0 = \frac{0.2\ \text{W}}{\text{Area}} = \frac{0.2\ \text{W}}{3.14 * 2 * (0.003) * (0.05)} = 212\ \text{W/m}^2$$

$$-k_p \frac{dT}{dr} = \dot{q}_0 = \frac{C_1}{r_1} \rightarrow C_1 = -\frac{\dot{q}_0}{k_p} r_1$$

$$\text{B.C.2: at } r = r_2,\ q = -k_p \frac{dT}{dt} = h(T - T_a)$$

Substitute boundary condition 2:

$$-k_p \frac{dT}{dt} = \frac{C_1}{r_2} = h(C_1 \ln(r_2) + C_2 - T_a)$$

$$-\frac{C_1}{h k_p r_2} + T_a - C_1 ln(r_2) = C_2$$

Substitute C_1 and rearrange to obtain an expression for C_2:

$$\frac{k_P}{h r_2} \frac{\dot{q}_0}{k_P} r_1 + T_a + \frac{\dot{q}_0}{k_P} r_1 ln(r_2) = C_2$$

Substitute the integration constants to find an expression for the temperature distribution in the wire plastic shield:

$$T = \frac{\dot{q}_0}{k_p} r_1 \left(\frac{k_P}{h r_2} + \ln(r_2) - \ln(r) \right) + T_a$$

The temperature at $r = r_2$:

$$T_2 = \frac{212 \text{W/m}^2}{0.15 \text{ W/(m°C)}} 0.003 \text{ m} \left(\frac{0.15}{12(0.005)} + \ln(0.005) - \ln(0.005) \right) + 27 = 37.6\text{°C}$$

The temperature at $r = r_1$:

$$T_1 = \frac{212 \text{W/m}^2}{0.15 \text{ W/(m°C)}} 0.003 \text{m} \left(\frac{0.15}{12(0.005)} + \ln(0.005) - \ln(0.003) \right) + 27 = 39.76\text{°C}$$

COMSOL Simulation

Start COMSOL. In this case, select "2D axis symmetry" and for the physics, select "Heat transfer in solid." The boundary conditions are heat flux at wire-plastic shield interface. The model builder is shown in Figure 6.76. The 3D temperature surface plot diagram is shown in Figure 6.77. The line plot is shown in Figure 6.78. The line plot diagram shows that the interface temperature is 39.8°C, and the shield-free surface is 37.6°C. The values predicted by the software values are the same as those calculated manually (analytical solution).

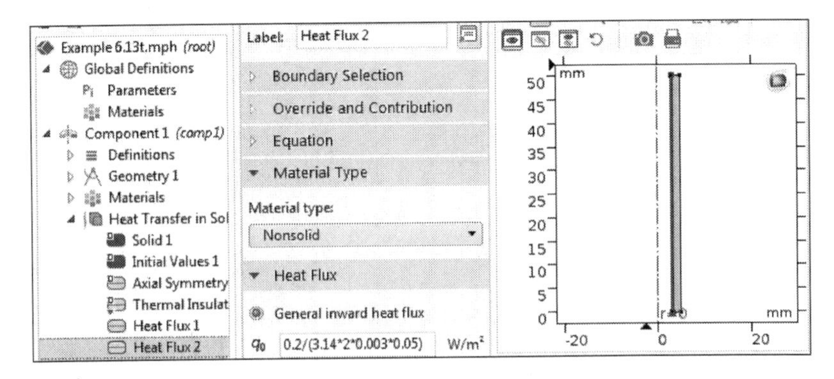

FIGURE 6.76
Model builder and heat flux boundary condition.

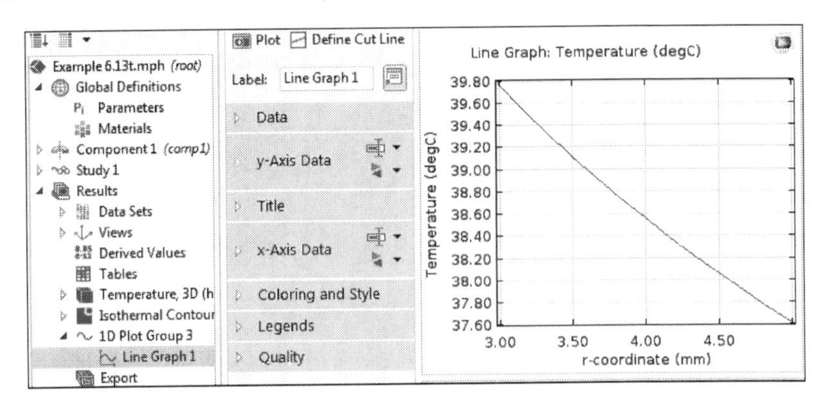

FIGURE 6.77
3D surface temperature of the plastic shield.

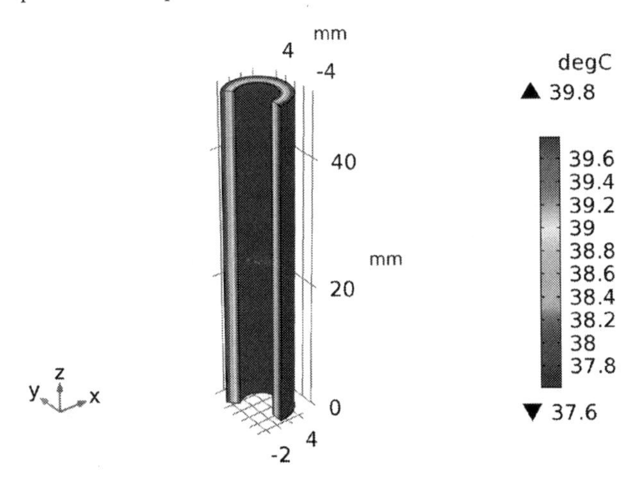

FIGURE 6.78
Temperature profile across the plastic shield.

Example 6.14: Heat Transfer in the Base Plate of an Iron

A clothes iron is generally a triangular surface used to press clothes to remove creases. In this example, a heating source below the base plate of a household iron is 1200 W. The base plate thickness is 0.005 m, and the base area is $A = 0.03\ \text{m}^2$. The thermal conductivity of the base is $k = 15\,\text{W/(m.°C)}$ (Figure 6.79). The inner surface of the base plate is subjected to uniform heat flux generated by the resistance heaters inside, and the outer surface loses heat, by convection, to the surroundings at $T_a = 20°\text{C}$. The convective heat transfer coefficient is $h = 80\,\text{W/(m}^2\text{.°C)}$. Develop an expression for the variation of temperature in the base plate, and calculate the temperatures at the inner and outer surfaces [2].

Solution

The base plate of the iron is a constant heat source. The variation of the temperature in the plate is to be determined. To perform the computation, consider the following assumptions:

- Heat transfer is steady.
- Since the heat transfer area of the plate is large relative to its thickness, the heat transfer is one dimensional.
- Thermal conductivity is constant.
- The heat transfer coefficient incorporates the radiation effects, if any.
- The entire heat generated in the resistance wires is transferred to the plate.

The inner surface of the base plate is subjected to uniform heat flux at a rate of:

$$q_0 = \frac{Q_0}{A_{base}} = \frac{1200\,\text{W}}{0.03\,\text{m}^2} = 40{,}000\ \text{W/m}^2$$

The problem is solved using the equation of change for heat transfer in Cartesian coordinates:

$$\rho C_p \frac{\partial T}{\partial t} + \rho C_p \left(v_x \frac{\partial T}{\partial x} + v_y \frac{\partial T}{\partial y} + v_z \frac{\partial T}{\partial z} \right) = k \left(\frac{\partial^2 T}{\partial x^2} + \frac{\partial^2 T}{\partial y^2} + \frac{\partial^2 T}{\partial z^2} \right) + \Phi_H$$

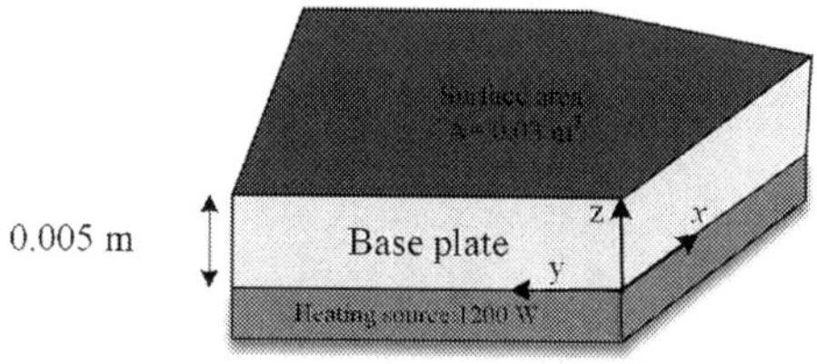

FIGURE 6.79
Household iron 1.2 Kw.

Based on the prescribed assumption:

Steady state — No convective heat flux $v_x = v_y = v_z = 0$ — No heat generation

$$\rho C_p \frac{\partial T}{\partial t} + \rho C_p \left(v_x \frac{\partial T}{\partial x} + v_y \frac{\partial T}{\partial y} + v_z \frac{\partial T}{\partial z} \right) = k \left(\frac{\partial^2 T}{\partial x^2} + \frac{\partial^2 T}{\partial y^2} + \frac{\partial^2 T}{\partial z^2} \right) + \Phi_H$$

Temperature gradients in the x and y direction is neglected

The general differential energy balance equation is reduced to: $0 = k\left(\partial^2 T/\partial z^2\right)$

Two successive integrations are used to obtain the general solution of the differential equation:

$$T = C_1 z + C_2$$

where C_1 and C_2 are arbitrary constant of integrations obtained by applying the following two boundary conditions:

B.C.1: at $x = 0$, $-k\,dT(0)/dz = q_0 = 40{,}000\ \mathrm{W/m^2}$
B.C.2. at $x = L$, $-k\,dT(L)/dz = h(T(L) - T_a)$

From the first boundary condition, we have:

$$-k\frac{dT(0)}{dz} = q_0 = -kC_1, \quad C_1 = -\frac{q_0}{k}$$

Apply the second boundary condition:

$$-k\frac{dT(L)}{dz} = h(T(L) - T_a) \rightarrow -kC_1 = h(C_1 L + C_2 - T_a)$$

Rearrange:

$$-\frac{k}{h}C_1 - C_1 L + T_a = C_2$$

Substitute $C_1 = -q_0/k$ to obtain the expression of C_2:

$$C_2 = \frac{k}{h}\frac{q_0}{k} + \frac{q_0}{k}L + T_a = T_a + \frac{q_0}{k}(k/h + L)$$

After substituting C_1 and C_2 into the general solution, we have:

$$T = T_a + \frac{q_0}{k}\left[L - z + k/h\right]$$

The temperature at $z = 0$:

$$T = 20°\mathrm{C} + \frac{40{,}000\ (\mathrm{W/m^2})}{15\ \mathrm{W/(m\,.°C)}}\left[0.005\ \mathrm{m} - 0 + \frac{15\ \mathrm{W/(m\,.°C)}}{80\ \mathrm{W/(m^2\,.°C)}}\right] = 533°\mathrm{C}$$

The temperature at $z = L$:

$$T = 20°C + \frac{40{,}000\ (W/m^2)}{15\ W/(m\ .°C)}\left[0.005\ m - 0.005 + \frac{15\ W/(m.°C)}{80\ W/(m^2\ .°C)}\right] = 520°C$$

COMSOL Simulation

Considering the base plate of the iron, the model builder setting along with the heat flux boundary condition at the base of the iron are shown in Figure 6.80. "Heat transfer in Solid" is the suitable physics in 2D axis symmetry. The heat flux boundary conditions are used for the top and bottom of the iron base. Convective heat flux is used for the top surface of the iron-free base. The surface plot is shown in Figure 6.81. The software simulation predictions are the same as the analytical solution (Figure 6.82).

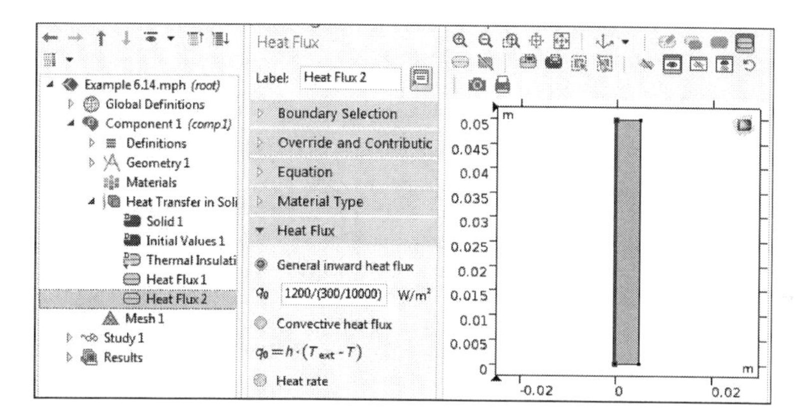

FIGURE 6.80
Model builder and heat source measurement.

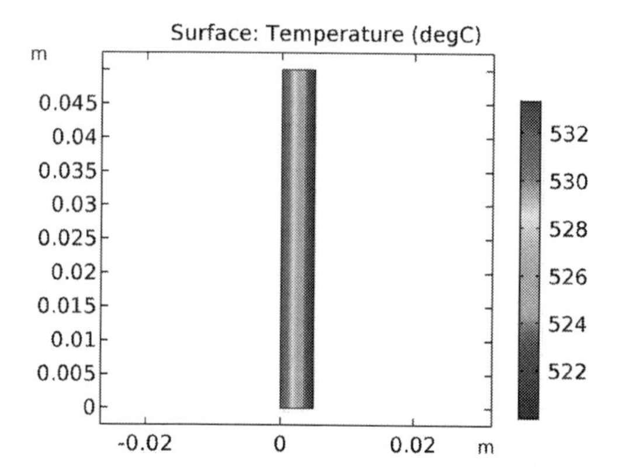

FIGURE 6.81
Surface temperature distribution in the iron plate.

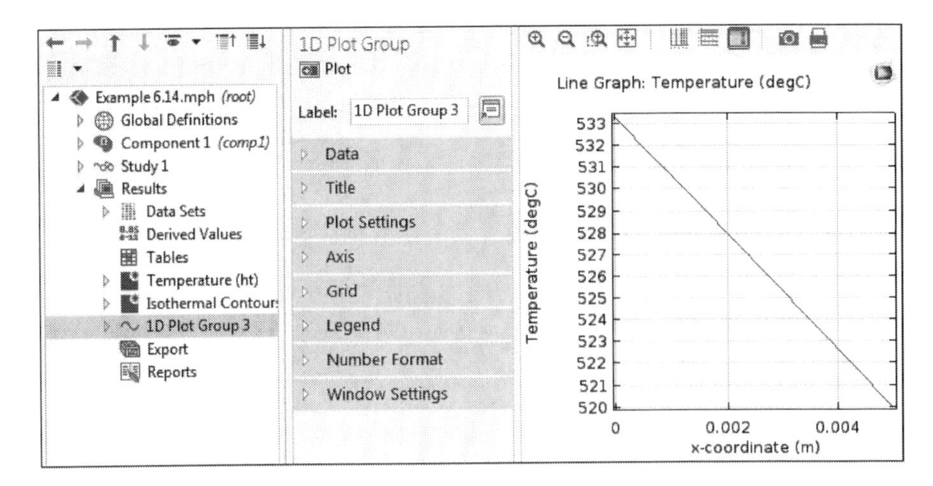

FIGURE 6.82
Line curve for temperature across the iron plate.

Example 6.15: Cooling of Silicon Chip

A silicon chip is a very small piece of silicon inside a computer. A square silicon chip (k=150 W/m.K) has a width of w=5 mm on one side and a thickness of H=1 mm. The chip is mounted in a substrate so that its side surfaces are insulated, while the front top surface is exposed to a coolant (Figure 6.83). A heat of 4 W is dissipated in circuits mounted below the bottom surface of the chip. Develop an expression for the steady-state temperature distribution in the silicon chip starting with equations of change, then find the steady-state temperature difference between the top and back surfaces.

Assumptions

Consider the following assumptions in calculating the steady-state temperature difference between the front and back surfaces:

- The conditions are in a steady state.
- Physical properties are constant.
- Uniform heat dissipation.
- Negligible heat loss from back and sides.
- One-dimensional conduction in the chip.

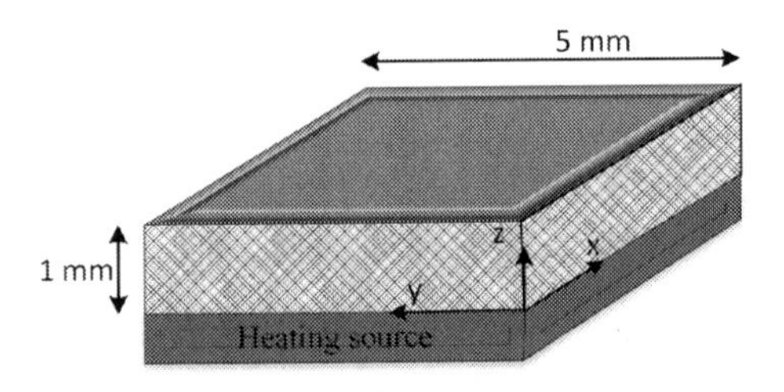

FIGURE 6.83
Schematic of a square silicon chip.

Solution

The problem is solved using the general differential equation for heat transfer in Cartesian coordinates:

$$\rho C_p \frac{\partial T}{\partial t} + \rho C_p \left(v_x \frac{\partial T}{\partial x} + v_y \frac{\partial T}{\partial y} + v_z \frac{\partial T}{\partial z} \right) = k \left(\frac{\partial^2 T}{\partial x^2} + \frac{\partial^2 T}{\partial y^2} + \frac{\partial^2 T}{\partial z^2} \right) + \Phi_H$$

Based on the steady state and no bulk motion assumptions, heat is transferred in the z-direction by conduction, and the temperature of the top surface is fixed by coolant.

Steady state

No convective heat flux
$v_x = v_y = v_z = 0$

No heat generation

$$\rho C_p \frac{\partial T}{\partial t} + \rho C_p \left(v_x \frac{\partial T}{\partial x} + v_y \frac{\partial T}{\partial y} + v_z \frac{\partial T}{\partial z} \right) = k \left(\frac{\partial^2 T}{\partial x^2} + \frac{\partial^2 T}{\partial y^2} + \frac{\partial^2 T}{\partial z^2} \right) + \Phi_H$$

Temperature gradients in the x and y direction is neglected

Plane 1 has no heat generation term, and it is bounded with insulation and the interface of material. Consequently, the model partial differential equation is reduced to:

$$0 = k \left(\frac{\partial^2 T}{\partial z^2} \right)$$

Perform double integration:

$$T = C_1 z + C_2$$

The expressions of the arbitrary constant of integrations C_1 and C_2 are found using the following two boundary conditions.

B.C.1: The bottom side of the chip received inward heat flux, implies that:
$z = 0, -k dT/dz = \dot{q}$, hence: $dT/dz = -\dot{q}/k \rightarrow C_1 = -\dot{q}/k$

B.C.2: The top surface of the chip is kept at constant temperature, at $z = L, T = T_c$.

Substitute boundary condition 2:

$$T_c = -(\dot{q}/k)L + C_2 \rightarrow C_2 = T_c + (\dot{q}/k)L$$

Substitute C_1 and C_2 in the general temperature distribution of the plane and rearrange:

$$T = \frac{\dot{q}}{k}(L - z) + T_c$$

Rearrange for temperature difference:

$$T - T_c = \Delta T = \frac{\dot{q}}{k}(L - z)$$

The temperature difference between the temperature at $z = 1$ mm and temperature at $z = 0$ is:

$$T - T_c = \Delta T = \frac{\dot{q}}{k}L = \frac{4\text{ W}/(0.005 \times 0.005)}{\left(\frac{150\text{ W}}{\text{m}}\text{K}\right)}(0.001) = 1.067C$$

Note that the heat flux $\dot{q}$ is in the units of W/m^2.

COMSOL Simulation

Start COMSOL. Select "3D simulation." For geometry, select "Block." Add the block geometry as shown in Figure 6.84. For physics, select "Heat transfer in solid." The bottom boundary condition is the general inward heat flux (Figure 6.85). The top boundary condition is the "Temperature" of the coolant. The 3D surface temperature profile is shown in Figure 6.86. The diagram shows that the temperature difference between the bottom and top surfaces of the chip is around 1°C. This value agrees with the manually calculated values.

Take a 3D cutline in the middle of the cube. The temperature line graph is shown in Figure 6.87. The figure shows that the difference in temperature between the bottom and the top surface of the chip is approximately $\Delta T = 1.06°C$.

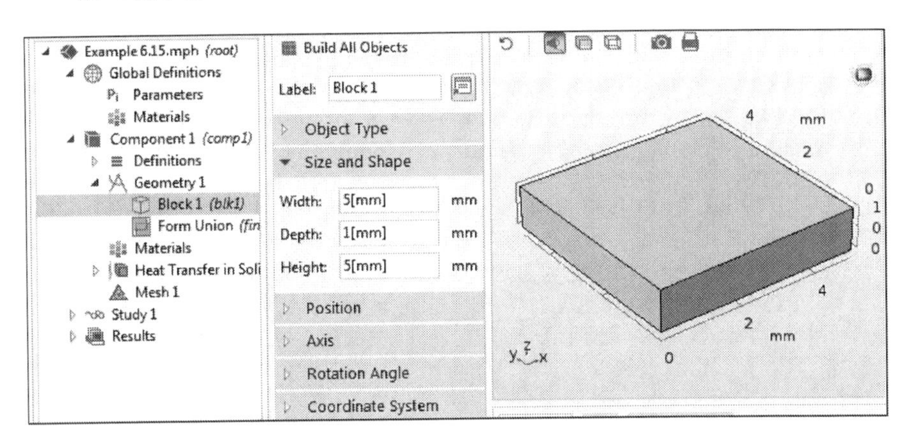

FIGURE 6.84
Specify the dimensions of the block geometry.

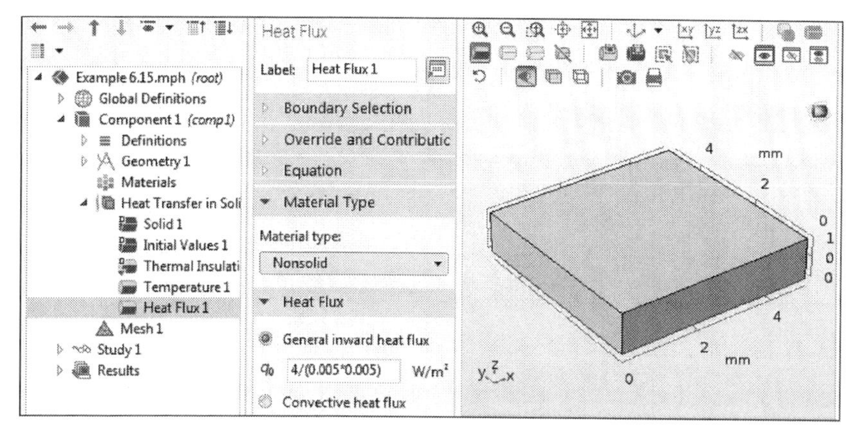

FIGURE 6.85
Heat transfer in solid boundary conditions and heat source.

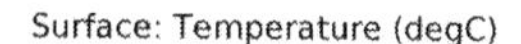

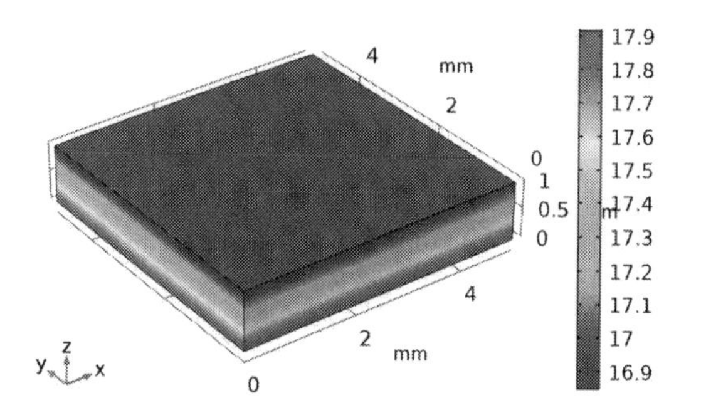

FIGURE 6.86
3D temperature surface plot.

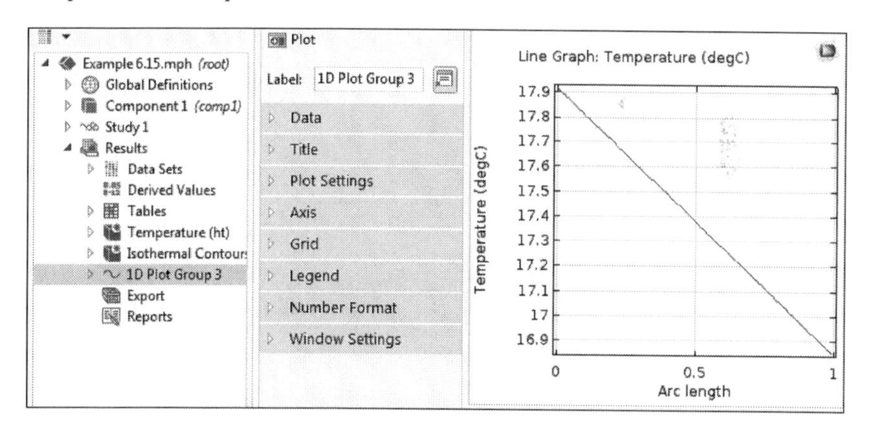

FIGURE 6.87
Temperature distribution curve between the top and bottom chip phases.

Example 6.16: Temperature Distribution in a Cylindrical Rod

A cylindrical steel rod 4 cm in diameter and 50 cm long (Figure 6.88), with thermal conductivity, $k = 13\,\text{W/(m.K)}$, is connected to two steel plates, each at a different temperature, $T_1 = 100°\text{C}$ and $T_2 = 0°\text{C}$. The surface of the rod is cooled by air at $T_a = 20°\text{C}$ passing past the rod. The convective heat transfer coefficient is $h = 11\,\text{W/(m}^2\text{K)}$. How does the temperature vary along the rod [1]?

Solution

Assume the following:

- Steady-state operation
- Constant physical properties

The energy balance equation in cylindrical coordinates is:

$$\rho C_p \frac{\partial T}{\partial t} + \rho C_p \left(v_r \frac{\partial T}{\partial r} + v_\theta \frac{1}{r}\frac{\partial T}{\partial \theta} + v_z \frac{\partial T}{\partial z} \right) = k\left(\frac{1}{r}\frac{\partial}{\partial r}\left(r\frac{\partial T}{\partial r} \right) + \frac{1}{r^2}\frac{\partial^2 T}{\partial \theta^2} + \frac{\partial^2 T}{\partial z^2} \right) + \Phi_H$$

With steady state and no bulk motion, heat is transferred in the x-direction by conduction, and heat is lost to air by convective heat flux:

$$v_r = v_\theta = v_z = 0$$

Steady state — No convective heat $v_r = v_\theta = v_z = 0$ — Temperature gradients in the radial and angular direction is neglected

$$\rho C_p \frac{\partial T}{\partial t} + \rho C_p \left(v_r \frac{\partial T}{\partial r} + v_\theta \frac{1}{r}\frac{\partial T}{\partial \theta} + v_z \frac{\partial T}{\partial z} \right) = k\left(\frac{1}{r}\frac{\partial}{\partial r}\left(r\frac{\partial T}{\partial r} \right) + \frac{1}{r^2}\frac{\partial^2 T}{\partial \theta^2} + \frac{\partial^2 T}{\partial z^2} \right) + \Phi_H$$

Simplify the model to:

$$0 = k\left(\frac{\partial^2 T}{\partial z^2} \right) + \dot{q}$$

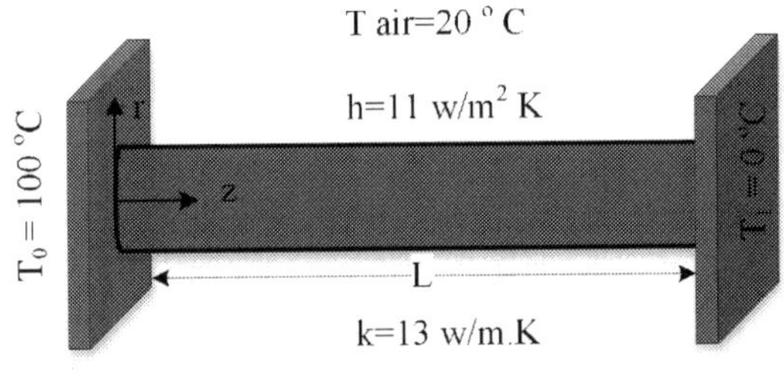

T air=20 °C

h=11 w/m² K

$T_0 = 100$ °C, $T_1 = 0$ °C, r, z, L

k=13 w/m.K

FIGURE 6.88
Schematic diagram of the cylindrical steel rod.

where:

$$\dot{q} = \frac{hA_s(T - T_a)}{V_c}$$

Rearrange:

$$\frac{d^2T}{dz^2} = -\frac{hA_s(T - T_a)}{kV_c} = -\frac{hPL(T - T_a)}{kA_cL} = -\frac{h\pi DL(T - T_a)}{\frac{k\pi D^2 L}{4}} = -\frac{4h(T - T_a)}{kD}$$

where:
- P represents the rod perimeter
- A_c represents the fin cross sectional area
- A_s represents the rod external surface area

$$P = \pi DL, V_c = \pi r^2 L$$

Finally, the second order differential equation is:

$$\frac{d^2T}{dz^2} = -\frac{4h}{kD}(T - T_a)$$

Let $\theta = T - T_a$ and $m^2 = 4hL^2/kD$, $\varphi = z/L$. Substitute and rearrange:

$$\frac{d^2\theta}{d\varphi^2} - m^2\theta = 0$$

The general solution of the above equation [6] is:

$$\theta = C_1\cosh(m\varphi) + C_2\sinh(m\varphi)$$

Boundary conditions:

$$\varphi = 0, \quad \theta = \theta_0, \text{ substitute: } \theta_0 = C_1\cosh(0) + C_2\sinh(0), \rightarrow \theta_0 = C_1 + 0 \rightarrow C_1 = \theta_0$$

$$\varphi = 1, \quad \theta = \theta_1, \text{ substitute: } \theta_1 = C_1\cosh(\mathrm{m}) + C_2\sinh(\mathrm{m})$$

Substitute $C_1 = \theta_0$:

$$\theta_1 = \theta_0\cosh(\mathrm{m}) + C_2\sinh(\mathrm{m})$$

Rearrange to find C_2:

$$\theta_1 - \theta_0\cosh(\mathrm{m}) = C_2 \sinh(\mathrm{m})$$

Accordingly:

$$C_2 = \left[\frac{\theta_1 - \theta_0 \cosh(\mathrm{m})}{\sinh(\mathrm{m})}\right]$$

Substitute C_1 and C_2 in the general equation:

$$\theta = \theta_0 \cosh(m\varphi) + \left[\frac{\theta_1 - \theta_0 \cosh(\mathrm{m})}{\sinh(\mathrm{m})}\right]\sinh(m\varphi)$$

Rewrite the solution in terms of T:

$$T - T_a = (T_0 - T_a)\cosh(m\varphi) + \left[\frac{(T_1 - T_a) - (T_0 - T_a)\cosh(m)}{\sinh(\mathrm{m})}\right]\sinh(m\varphi)$$

where:

$$m = \sqrt{4hL^2 / kD} = \sqrt{\frac{4 \times 11 \times 0.5^2}{13 * 0.04}} = 4.59$$

The plot of the temperature data from the analytical solution compared to those obtained from COMSOL is shown in Figure 6.89. This can be done as follows:

1. Save the excel data in comma delimited (.csv) format.
2. Results > Tables: Add a Table.
3. Click on Import and browse to the file that contains the data.
4. Select the 1D plot group where you want to plot the table data.
5. Add a Table Graph plot and change the value in the Table combo box to the table you just created.

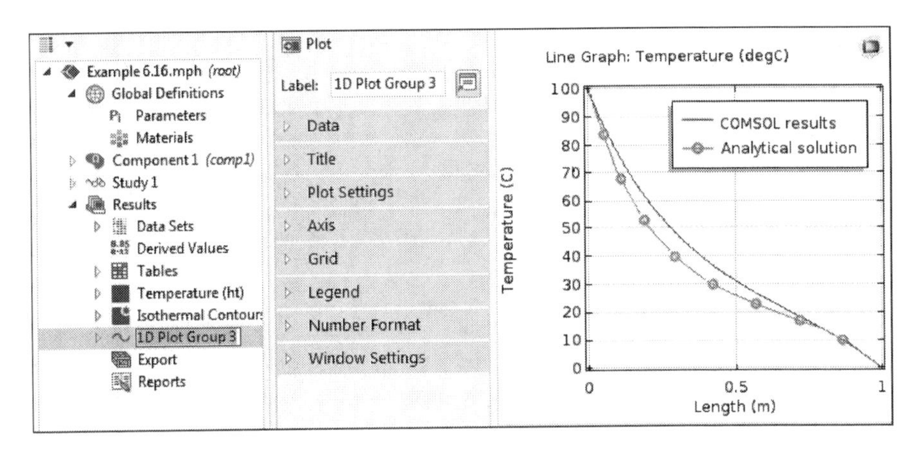

FIGURE 6.89
Comparison between COMSOL results and analytical solution.

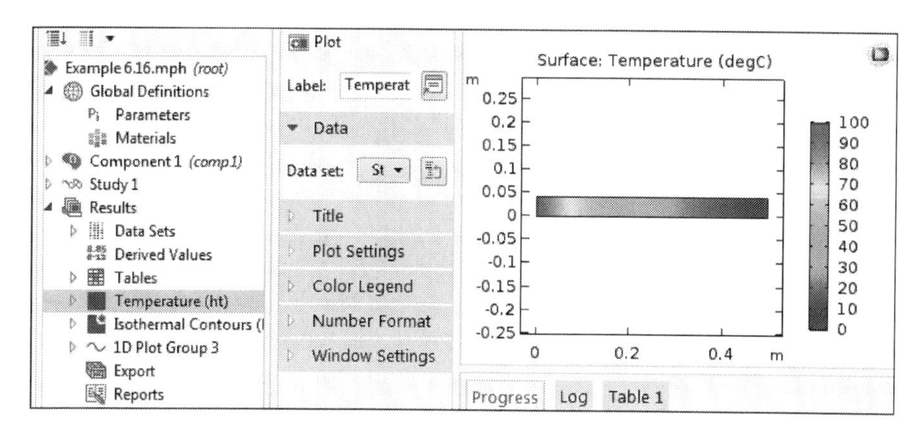

FIGURE 6.90
COMSOL surface temperature plot.

COMSOL Solution

1. Open COMSOL Multiphysics > Model Wizard > select 2D > Heat transfer in Solids > Add > Study > Stationary > Done.
2. Click on "Solid" and enter the thermal conductivity $k = 13\,W/(m.K)$.
3. Right-click on "Heat transfer in solid" to add boundary conditions. Select "Temperatures" two times for left- and right-hand side boundaries, 100°C and 0°C, respectively.
4. Again, right-click on the selected physics and select "Heat flux." Enter $h = 11\,W/(m^2.K)$, and $T_{ext} = 293.15$ K.

Figure 6.90 demonstrates the physics and boundary conditions used, along with the temperature surface plot. The figure shows that temperature decreases along the length of the rod in the x-direction, which is attributed to the low boundary conditions of the right-hand side of the rod.

Example 6.17: Temperature Distribution in a Plane Wall

A large plane wall is subjected to specified temperature on the left surface ($T_1 = 80°C$) and convection on the right surface ($h = 24\,W/(m^2.°C)$, and to an ambient temperature of $T_a = 15°C$. The thermal conductivity of the wall is $k = 2.3\,W/m°C$. The thickness of the wall is 0.4 m. The surface area of the wall is $A = 20\,m^2$. Develop the mathematical model that describes the variation of temperature across the wall (Figure 6.91), and determine the rate of heat transfer for steady one-dimensional heat transfer.

Solution

Consider the following assumptions in the heat transfer in the wall:

- Heat conduction is steady and one dimensional.
- Thermal conductivity is constant.
- There is no heat generation.

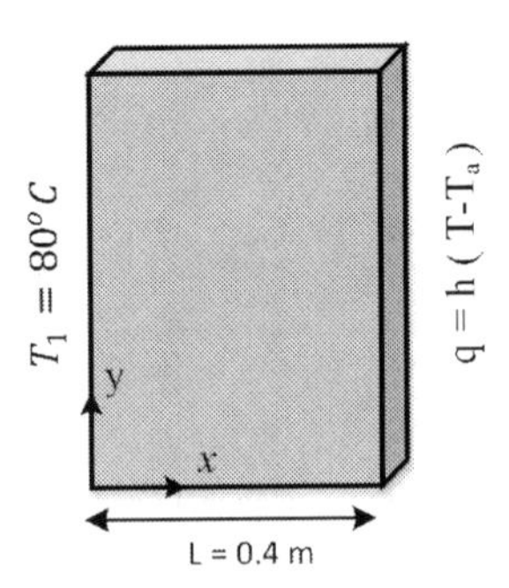

FIGURE 6.91
Heat transfer in a large plane wall.

The direction normal to the surface of the wall is in the x-direction, with $x = 0$ at the left surface. After applying the given assumptions on the heat equations of change in cylindrical coordinates, the mathematical formulation of this problem can be expressed as:

$$\frac{d^2T}{dx^2} = 0$$

The boundary conditions:

$$\text{B.C.1: at } x = 0, \ T(0) = T_1 = 80°\text{C}$$
$$\text{B.C.2: at } x = L \quad -k\,dT(L)/dx = h(T(L) - T_a)$$

Integrating the differential equation twice with respect to x yields:

$$T(x) = C_1x + C_2$$

where C_1 and C_2 are arbitrary constants. Applying the boundary conditions gives:

$$x = 0: \ T(0) = T_1 = C_1 \times 0 + C_2 \rightarrow C_2 = T_1$$

$$x = L: \ -kC_1 = h\left[(C_1L + C_2) - T_a\right] \rightarrow C_1 = -\frac{h(C_2 - T_a)}{k + hL} \rightarrow C_1 = -\frac{h(T_1 - T_a)}{k + hL}$$

Substitute C_1 and C_2 into the general solution to determine the variation of temperature:

$$T(x) = -\frac{h(T_1 - T_a)}{k + hL}x + T_1$$

$$T(x) = -\frac{\left(24\dfrac{\text{W}}{\text{m}^2.°\text{C}}\right)(80-15)°\text{C}}{2.3\dfrac{\text{W}}{(\text{m}.°\text{C})} + \left(24\dfrac{\text{W}}{(\text{m}.°\text{C})}\right)(0.4\ \text{m})}x + 80°\text{C} = 80 - 131.1x$$

The temperature at the right wall is:

$$T = 80 - 131.1 * 0.4 = 27.56°\text{C}$$

The rate of heat conduction through the wall is:

$$q = -kA\frac{dT}{dx} = -kAC_1 = kA\frac{h(T_1 - T_a)}{k + hL}$$

$$q = \left(2.3\frac{\text{W}}{(\text{m}.°\text{C})}20\ \text{m}^2\right)\frac{24\dfrac{\text{W}}{(\text{m}^2.°\text{C})}(80-15)°\text{C}}{2.3\dfrac{\text{W}}{\text{m}.°\text{C}} + \left(24\dfrac{\text{W}}{(\text{m}^2.°\text{C})}\right)(0.4\ \text{m})} = 6030\ \text{W}$$

Note that, under steady conditions, the rate of heat conduction through a plain wall is constant.

COMSOL Solution

The geometrical dimension of the block diagram is shown in Figure 6.92. The 3D surface plot temperature across the large plane wall is shown in Figure 6.93.

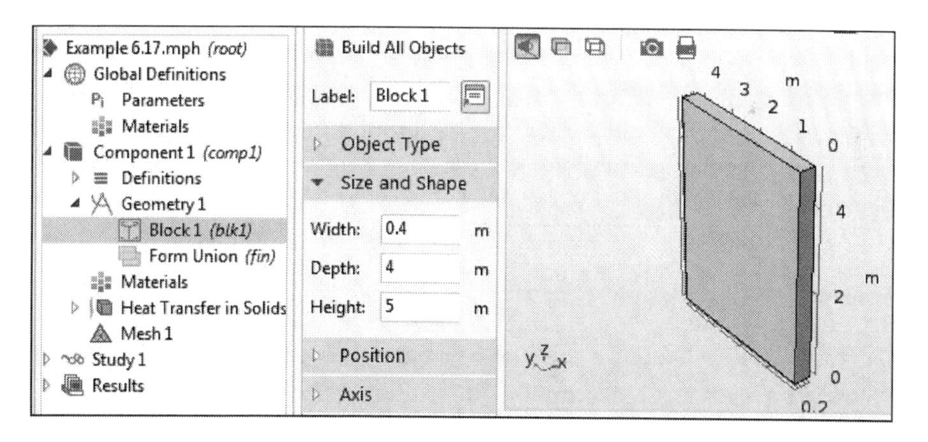

FIGURE 6.92
Simulation of temperature distribution in a large plane wall.

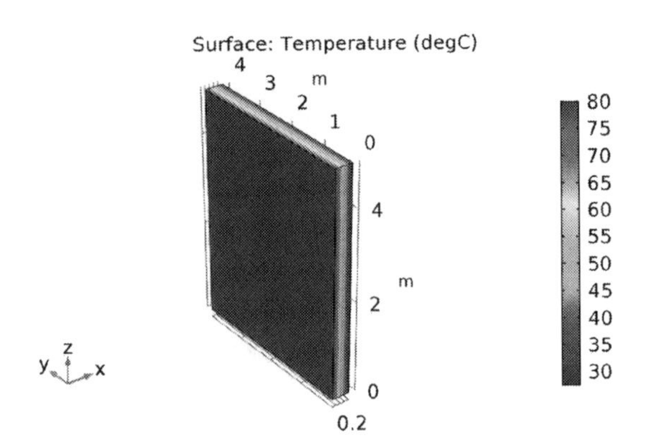

FIGURE 6.93
Surface temperature across large plane wall.

PROBLEMS

6.1 Increasing Automobile Rear Windows Temperature

A very thin transparent heating element is attached to the inner surface of the rear window to defog an automobile. The glass of an automobile rear window is 0.004 m thick and has a thermal conductivity of $k = 1.4\ \mathrm{W/(m \cdot K)}$. The interior air temperature and convection coefficient are 25°C and 10 W/m^{2}K, respectively, and the exterior (ambient) air temperature and convection coefficient are −10°C and 65 W/m^2, respectively. Develop an expression for the temperature distribution $T(z)$ in the window. What is the inner surface temperature of the window?

Answer: –4.6°C

6.2 Heat Transfer in a Wooden Slab

The heat flux through a slab made from wood 0.05 m thick has been determined to be 40 W/m^2. The inner surface temperature of the wooden slab is 40°C. The thermal conductivity of the wood is $k = 0.1$ W/(m.K). Develop an expression for the temperature distribution $T(z)$ in the wooden slab. Determine the outer surface temperature of the wooden slab.

Answer: 20°C

6.3 Temperature Distribution in a Cylindrical Rod With Heat Generation

Develop the steady-state temperature profile for the heat conduction equation inside a cylinder (radius r_0) made of homogeneous material of constant thermal properties undergoing internal energy generation (q) at a constant rate in cylindrical coordinates with azimuthal symmetry and immersed in a medium that keeps the temperature at the surface of the cylinder constant at T_a. Prove that the temperature distribution in the cylinder is described by the following equation:

$$T(r) = T_a + \frac{q}{k}(r_o^2 - r^2)$$

6.4 Heat Transfer in a Shell of Cylinder without Heat Generation

Consider a fluid flowing in a pipe with the inner and outer temperatures T_1 and T_2 at r_1 and r_2, respectively. Show that for r-directed steady heat flow in a cylinder (without heat generation), the temperature of the fluid in the pipe shell is described by the following relation:

$$T = T_1 + (T_2 - T_1)\frac{\ln(r/r_1)}{\ln(r_2/r_1)}$$

6.5 Heat Transfer in a Shell of a Sphere

An important case is a spherical shell, a geometry often encountered in situations where fluids are pumped and heat is transferred. The inner and outer temperatures are T_1 and T_2 at r_1 and r_2, respectively. Show that the temperature distribution in the shell of the sphere is:

$$T = T_1 - (T_1 - T_2)\frac{1-(r_1/r)}{1-(r_1/r_2)}$$

6.6 Temperature Distribution in a Cylindrical Pellet

Uranium dioxide is used as a fuel for a certain nuclear power station. The fuel consists of cylindrical pellets, of 0.005 m radius, made of slightly enriched uranium dioxide. Heat is generated by nuclear fission within these pellets at a uniform rate, 4×10^7 W/m^3, and is conducted out to the reactor coolant. The outer surface of the pellet is maintained at 220°C. Assume steady-state operation. The thermal conductivity of uranium is $k = 29.5\,\text{W/mK}$. The density is $\rho = 10.4\,\text{g/cm}^3$. Develop an expression for the temperature distribution $T(r)$ in the cylindrical pellet. What is the maximum temperature within the pellet [5]?

Answer: 228°C

6.7 Conductive Heat Loss Through a Window Pane

The window glass thickness, L, is 0.0032 m. This is the only window in a room of 25 m^3, and the area of the window is 0.6 m × 0.9 m or 0.54 m^2. The room temperature is 15.56°C. It is winter and the exterior temperature is −6.67°C, and k is $0.71\,\text{W/mK}$. Develop an expression for the temperature distribution in the window and then calculate the heat loss through the window.

Answer: 2.66 kW

6.8 Conduction with Heat Generation

A long resistance electrical wire is generated uniformly at a constant rate because of resistance heating. The wire is of radius $r_1 = 0.003$ m, and thermal conductivity $k_1 = 18$ W/(m.K) generates heat of 4.8×10^6 W/m^3. The electrical wire is covered with a 0.004 m plastic shield whose thermal conductivity is $k_2 = 1.8\,\text{W/(m.K)}$. The outer radius of the plastic cover loses heat by convection to the ambient air at $T_a = 25°\text{C}$, with an average combined heat transfer coefficient of $h = 14\,\text{W/(m}^2\text{.K)}$. Assume one-dimensional heat transfer and develop an expression for the temperature distribution in the wire. Then determine the temperature at the center under steady-state conditions.

Answer: 256°C

6.9 Heater Installed in a Solid Wall

A solid wall has a thickness of 0.08 m and thermal conductivity of $k = 2.5\,\text{W/m.K}$. The wall contains electrical heat wires installed inside. The right-side face of the wall is exposed to a convective environment of $h = 50\,\text{W/(m}^2\text{K)}$ and external temperature of 30°C. The left-side face is exposed to a heat transfer coefficient of $h = 75\,\text{W(m}^2\text{K)}$ and external temperature of $T_{ex} = 50°\text{C}$. Develop an expression for the temperature distribution $T(z)$ in the solid wall. Determine the temperature at the center of the wall using COMSOL if the maximum allowable heat generation rate in the wall is $0.42\,\text{MW/m}^3$

Answer: 450°C

6.10 Computer Silicon Chip

A silicon chip of thickness 0.00675 m and thermal conductivity of 135 W/mK, is encapsulated so that all the power it dissipates is transferred by convection to a fluid, under steady-state conditions. The fluid stream heat transfer coefficient is $h = 1000$ W/m^2.K, and $T_\infty = 25°\text{C}$. The chip is separated from the fluid by a 0.002 m-thick aluminum cover plate (thermal conductivity, $k = 238\,\text{W/m.K}$). The chip surface area is $10^{-4}\,\text{m}^2$. Develop an expression for the temperature distribution in the chip. Find its maximum allowable temperature if the maximum allowable power dissipation in the chip is 5.667 W.

Answer: 85°C

References

1. Incropera, F. P., D. P. DeWitt, 1996. *Introduction to Heat Transfer,* 3rd ed., New York: John Wiley & Sons.
2. Cengel, A. Y., 1998. *Heat Transfer, A Practical Approach,* New York: McGraw-Hill.
3. Welty, J., G. L. Rorrer, D. G. Foster, 2015. *Fundamentals of Momentum, Heat and Mass Transfer,* 6th ed., New York: John Wiley & Sons.
4. Geankoplis, C.J. 2009. *Transport Processes and Separation Processes Principles,* 4th ed., Upper Saddle River, NJ: Prentice Hall.
5. Cengel, Y.A., A. Ghajar, 2014. *Heat and Mass Transfer, Fundamental and Application,* 5th ed., New York: McGraw-Hill.
6. Bird, R. B., W. E. Steward, E. N. Lightfoot, 2002. *Transport Phenomena,* 2nd ed., New York: John Wiley & Sons.

7

Case Studies

Eight modeling and simulation cases are considered in this chapter: The first case is a membrane reactor for hydrogen production, the second is the absorption of carbon dioxide from flue gas in an absorption tower, the third involves packed bed reactors, the fourth is about fluid flow of two immiscible liquids, the fifth shows the production of propylene glycol in an isothermal and adiabatic tubular reactor, the sixth case introduces the coupling between fluid and heat transfer (Multiphysics), the seventh is about the unsteady diffusion of contaminated source from the skin of pipe line, and the eighth and final case describes the use of the Maxwell Stefan method to simulate hydrogen production via steam reforming.

LEARNING OBJECTIVES

- Develop mathematical models for systems with Multiphysics.
- Solve systems using COMSOL Multiphysics.
- Compare model predictions with experimental data when possible.

7.1 Membrane Reactors

Methane gas has been proposed as a source of hydrogen for fuel cells that need hydrogen with high purity. In the process known as steam reforming via membrane reactor [1,2], methane gas and steam are reacted into carbon monoxide and hydrogen at a relatively high temperature (methane steam reforming):

$$CH_4 + H_2O \leftrightarrow 3H_2 + CO \tag{7.1}$$

The carbon monoxide can be converted into hydrogen by using a water gas shift reactor per the following reaction:

$$CO + H_2O \leftrightarrow H_2 + CO_2 \tag{7.2}$$

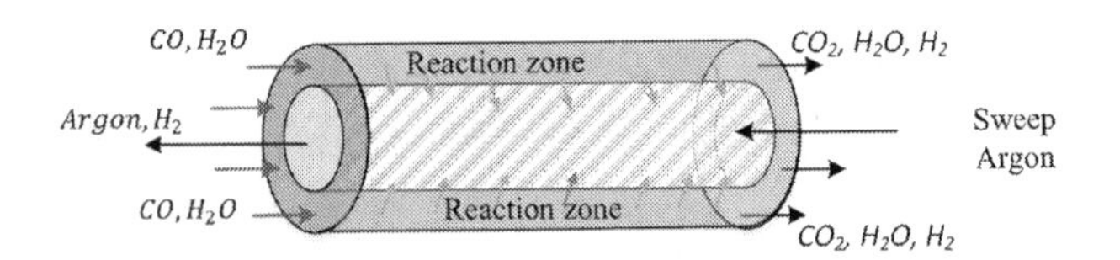

FIGURE 7.1
Schematic of countercurrent membrane reactor.

Figure 7.1 illustrates the case of a water gas shift reaction taking place in an 8-cm-long reactor. The water gas shift reaction using an iron chromium oxide catalyst takes place in the outer (annular) region. The outer reaction zone is separated from the inner reaction zone by a 20-μm-thick palladium membrane that has outstanding hydrogen selectivity. The palladium membrane is typically a metallic tube of a palladium and silver alloy material possessing the unique property of allowing only monatomic hydrogen to pass through its crystal lattice when it is heated above 300°C.

The reaction rate of the water gas shift reaction taking place in the reaction zone is described by the following rate of reaction:

$$r = \frac{k(C_{CO}C_{H_2O} - C_{CO_2}C_{H_2}/K_{eq})}{1 + 2.4 \times 10^5 C_{H_2O} + 7.2 \times 10^5 C_{CO_2}} \tag{7.3}$$

where r represents the reaction rate with the units of mol/(min—cm reactor length) and is given in terms of gas concentration in units of mol/cm^3, where:

$$k = 7.4 \times 10^8 \text{ in units of cm}^6/(\text{mol} \cdot \text{min} \cdot \text{cm reactor length})$$

$$K_{eq} = 11.92 \text{ (dimensionless) at the reaction conditions of } 673\,\text{K}$$

The reaction rates are related by:

$$r = -r_{CO} = -r_{H_2O} = r_{CO_2} = r_{H_2} \tag{7.4}$$

The membrane flux in units of mol/(min · cm reactor length) is given by Equation 7.5 of Uemiya et al. [1]:

$$j = \frac{q}{t}\left(C_{H2,r}^{076} - C_{H2,p}^{076}\right) \tag{7.5}$$

where:

$$q = 1.1 \times 10^2 \frac{\text{mol}^{0.24} \cdot \mu\text{m thickness} \cdot \text{cm}^{2.28}}{\text{min} \cdot \text{cm reactor length}} \tag{7.6}$$

Consider the feed rate of 1.1×10^{-3} mol/min CO and 1.1×10^{-3} mol/min H_2O at 2 atm pressure in a reaction zone. In the separation zone, the feed is 1.8×10^{-2} mol/min Ar at 1 atm pressure, The argon works as a sweeping inert gas. The reactor length is 8 cm, diameter is 8 mm, and catalyst thickness is t = 20 μm. The diffusion coefficients of gas components are assumed to be constant, $D_{AB} = 2\times10^{-5}\,\text{m}^2/\text{s}$. Perform the following studies:

1. Assume that there is no hydrogen flux through the reactor walls ($j = 0$) and determine the equilibrium conversion if there is no membrane separation of hydrogen. Compare the results with actual conversion.
2. Construct a numerical model assuming that $j = 0$ to predict the molar flow rates of CO, H_2O, CO_2, and H_2 as a function of distance if there is no membrane separation of hydrogen. Determine if the equilibrium is reached in this reactor and find the carbon dioxide conversion in this system.
3. Adjust the numerical model to predict the molar flow rates of CO, H_2O, CO_2, and H_2 as a function of distance if hydrogen can permeate through the palladium membrane. Compare your results with the example problem (when there is no membrane). What is the CO conversion in this system (assume $j \neq 0$)?
4. The reaction zone feed is changed to 7.0×10^{-4} mol/min CO and 1.5×10^{-3} mol/min H_2O at 2 atm pressure. Assume there is no membrane separation of hydrogen and thus $j = 0$. Determine the equilibrium conversion.
5. Study the effect of membrane thickness on system performance.

7.1.1 Equilibrium Conversion

The reaction feed stream is equal molar of CO and H_2O, which corresponds to a partial pressure of 1 atm for each gas (total pressure = 2 atm). The corresponding inlet concentrations of carbon monoxide and steam are given as:

$$C_{CO,o} = C_{H_2O,o} = \frac{P_i}{RT} = \frac{1\,\text{atm}}{0.08206\dfrac{\text{L}\cdot\text{atm}}{\text{mol}\cdot\text{K}}\times 673\,\text{K}}\frac{1000\,\text{L}}{1\,\text{m}^3} = 18\,\text{mol/m}$$

Argon concentration in the sweeping section is:

$$C_{Ar,o} = C_{H_2O,o} = \frac{P_{Ar}}{RT} = \frac{1\,\text{atm}}{0.08206\dfrac{\text{L}\cdot\text{atm}}{\text{mol}\cdot\text{K}}\times 673\,\text{K}}\frac{1000\,\text{L}}{1\,\text{m}^3} = 18\,\text{mol/m}^3$$

During the reaction, the product concentration of CO_2 and H_2 will increase linearly with the reactor length, whereas the reactant concentration of CO and H_2O will decrease linearly with the reactor length. If we denote ξ as the extent of the reaction, then to calculate the concentration changes from the feed state to the equilibrium state, we would have to do the following:

$$C_{CO,eq} = C_{H_2O,eq} = C_o - \xi$$

and

$$C_{CO_2,eq} = C_{H_2,eq} = \xi$$

at equilibrium

$$0 = \frac{C_{CO,eq}C_{H_2O,eq} - C_{CO_2,eq}C_{H_2,eq}}{K_{eq}}$$

Rearrange and substitute the extent of reaction, ξ:

$$K_{eq} = 11.92 = \frac{C_{CO_2,eq}C_{H_2,eq}}{C_{CO,eq}C_{H_2O,eq}} = \frac{\xi \times \xi}{(C_o - \xi)^2}$$

Simplify:

$$C_o^2 - 2C_o\xi + \xi^2 = \frac{1}{11.92}\xi^2$$

Rearrange and substitute the value of $C_o = 18$ mol/m^3:

$$0.916\xi^2 - 36\xi + 324 = 0$$

The extent of the reaction is the root of the quadratic equation:

$$\xi = 13.96 \text{ mol/m}^3$$

The equilibrium conversion, X_{eq}:

$$X_{eq} = \frac{\xi}{C_o} = \frac{13.96}{18} = 0.775$$

7.1.2 Numerical Solution of Equilibrium Conversion

The model equation for the reaction zone is obtained using the component balance equation of change after considering the following assumptions:

- Steady state.
- Isothermal operation.
- Transportation of gases takes place by diffusion and convection.
- Diffusion in θ coordinate is negligible.
- Diffusion coefficient of gas is constant.

The component balance equation using the equation of change in cylindrical coordinates is:

$$\frac{\partial C_A}{\partial t}+\left(v_r\frac{\partial C_A}{\partial r}+v_\theta\frac{1}{r}\frac{\partial C_A}{\partial \theta}+v_z\frac{\partial C_A}{\partial z}\right)=D_A\left(\frac{1}{r}\frac{\partial}{\partial r}\left(r\frac{\partial C_A}{\partial r}\right)+\frac{1}{r^2}\frac{\partial^2 C_A}{\partial \theta^2}+\frac{\partial^2 C_A}{\partial z^2}\right)+R_A \tag{7.7}$$

After applying the stated assumptions, the component balance equation is simplified to the following form:

$$0+\left(0+0+v_z\frac{\partial C_A}{\partial z}\right)=D_A\left(\frac{1}{r}\frac{\partial}{\partial r}\left(r\frac{\partial C_A}{\partial r}\right)+0+\frac{\partial^2 C_A}{\partial z^2}\right)+r_A \tag{7.8}$$

Rearrange:

$$v_z\frac{\partial C_A}{\partial z}=D_A\left(\frac{1}{r}\frac{\partial}{\partial r}\left(r\frac{\partial C_A}{\partial r}\right)+\frac{\partial^2 C_A}{\partial z^2}\right)+r_A \tag{7.9}$$

The inlet velocity of gas is calculated from the total inlet molar flow rate:

$$v=2.2\times10^{-3}\frac{\text{mol}}{\text{min}}\frac{22.4\,\text{L}}{\text{mol}}\frac{\text{m}^3}{1000\,\text{L}}\frac{1\,\text{min}}{60\,\text{sec}}\times\frac{1}{\frac{\pi(0.008)^2}{4}}=0.016\,\text{m/s}$$

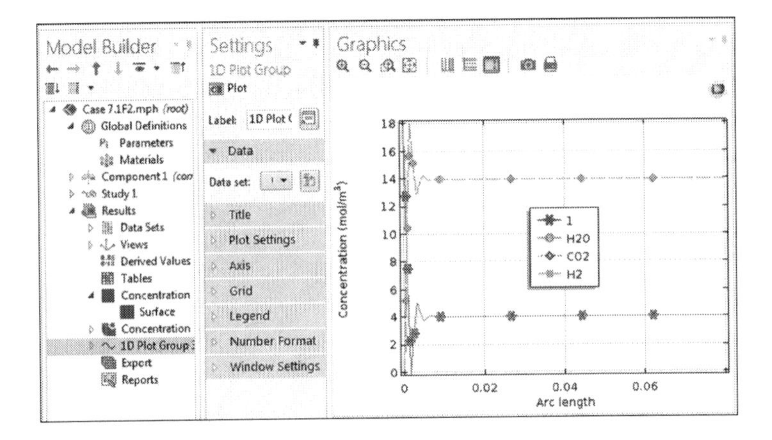

FIGURE 7.2
Component concentration profile as a function of reactor length; hydrogen not allowed to permeate.

The model equations are solved using COMSOL software package as shown below:

The conversion without hydrogen flux

$$x = \frac{C_{CO,o} - C_{CO}}{C_{CO,o}} = \frac{18 - 4.043}{18} = 0.775$$

The equilibrium conversion is achieved rapidly in the reactor. Figure 7.2 shows the component concentration profile for species along the length of the reactor. The figure shows that the reaction reaches equilibrium in the first few mm of the reactor length. The longer reactor length facilitates hydrogen separation into the separation zone.

7.1.3 Numerical Solution in Case of Hydrogen Permeation

Figure 7.3 shows that the numerical model solved by COMSOL is used to predict the molar flow rates of CO, H_2O, CO_2, and H_2 as a function of membrane length: the case with hydrogen permeation through the palladium membrane layer ($j \neq 0$). A comparison with the case when there is no hydrogen dispersion (Figure 7.2) through membrane walls (i.e., $j = 0$) demonstrates that a higher conversion is achieved and even more than the equilibrium conversion. The hydrogen flux inserted in COMSOL is in the unit of mol/m^{2}s; accordingly, the flux is modified as follows:

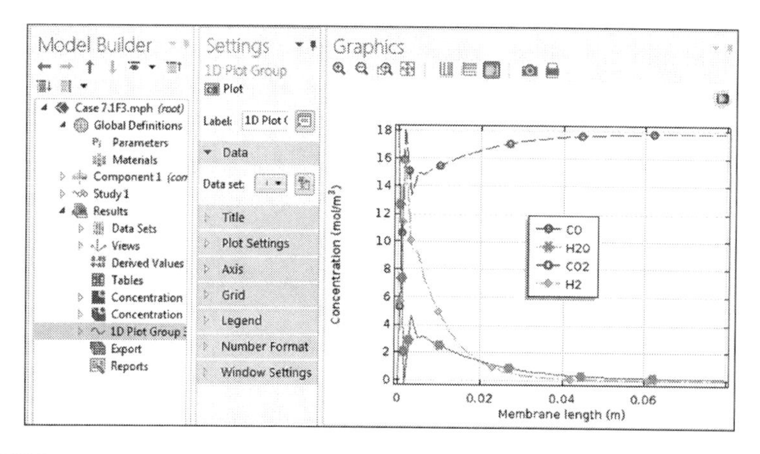

FIGURE 7.3
Gas concentration as a function of reactor length; hydrogen can permeate through the palladium catalyst.

$$\text{Flux} = j = -\left(\frac{q}{t}\right)\left(C_{H_{2,r}}^{0.76} - C_{H_{2,p}}^{0.76}\right)(RT)^{0.76}(7.44 \times 10^{-5})\left(\frac{1}{\pi D}\right)$$

where:

$C_{H_{2,p}}$ is the hydrogen concentration in the permeate zone
$C_{H_{2,r}}$ is the hydrogen concentration inside the reaction zone

The CO conversion (x) in the case of hydrogen permeated through membrane walls is:

$$x = \frac{C_{CO,o} - C_{CO}}{C_{CO,o}} = \frac{18 - 1.19}{18} = 0.934$$

Assume negligible partial pressure of hydrogen in the permeate zone due to the continuous sweeping flow rate of argon. The surface plot for the concentration of CO across the membrane reactor sections is shown in Figure 7.4. The line graph for the concentration of gas component through the membrane reactor is shown in Figure 7.5.

The concentration of reactant decreased along the reactor length, by contrast produced gases, CO_2 while H_2 increased and then decreased due to hydrogen permeation rate. The plot of the hydrogen surface concentration profile in the reaction and in the shell zone of the membrane

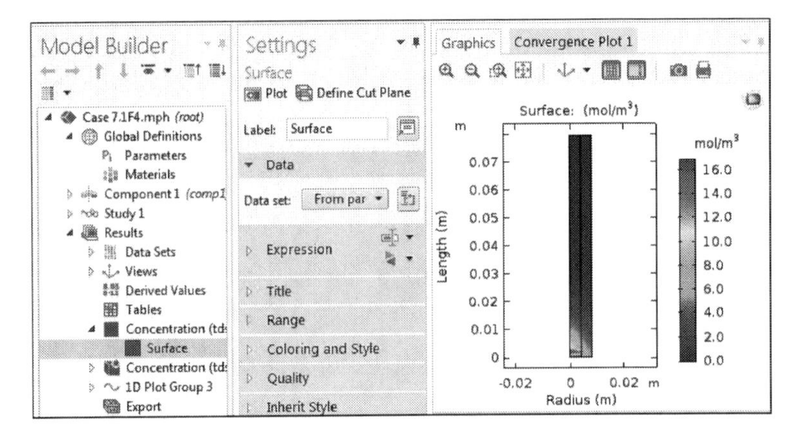

FIGURE 7.4
Surface plot of the concentration of hydrogen through membrane reactor.

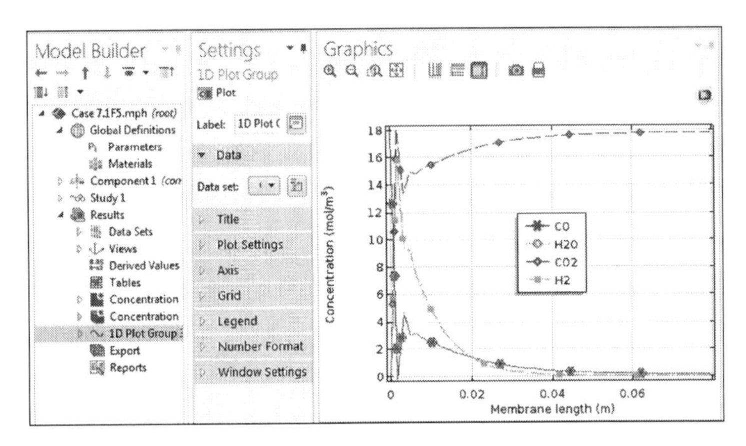

FIGURE 7.5
Concentration of gas in the reaction zone in the case of negligible hydrogen partial pressure in permeate zone.

reactor for the case when the hydrogen partial is equal to the atmospheric pressure in the shell side (the case when there is no sweep gas existing in the shell side) is shown in Figure 7.6. The line graph for component concentration in the reaction zone is depicted in Figure 7.7. Comparison of Figures 7.5 and 7.7 illustrates the case of negligible hydrogen partial pressure in the permeate zone and the case when the hydrogen pressure is 1 atm in the permeate zone; the conversion is higher in the first case relative to the second case.

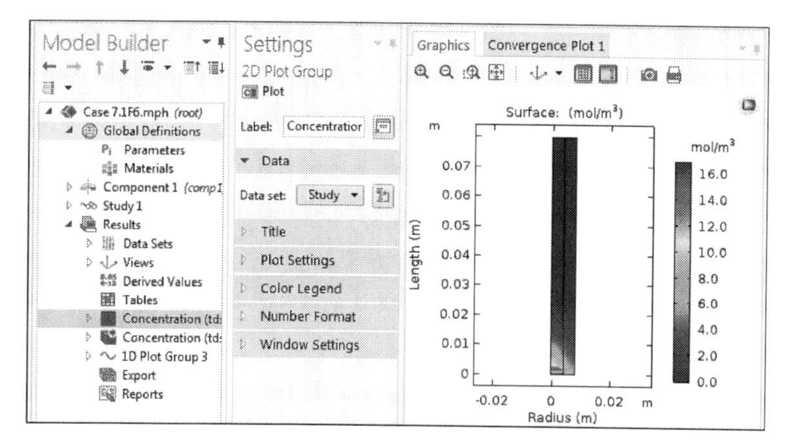

FIGURE 7.6
Hydrogen surface concentration; hydrogen pressure in permeate zone is 1 atm.

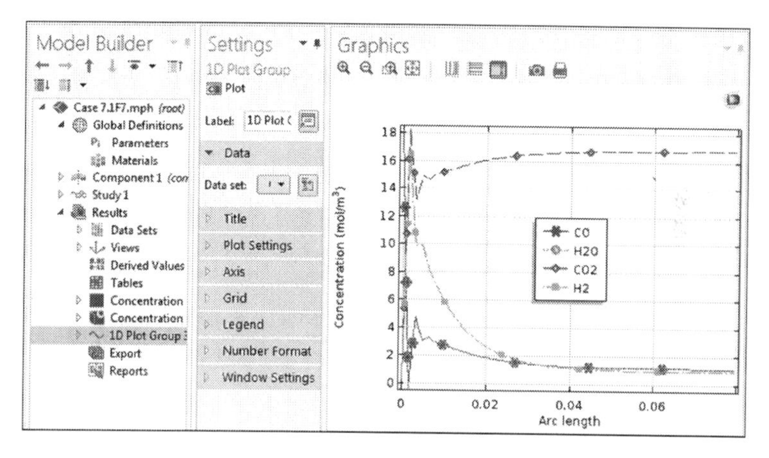

FIGURE 7.7
Component concentration profile; hydrogen pressure in permeate zone is 1 atm.

7.1.4 Variable Feed Concentration

In this case, we will adjust the feed stream to the reaction zone by making the steam flow rate higher than that of carbon dioxide (7.0×10^{-4} mol/min CO and 1.5×10^{-3} mol/min H_2O) at a total inlet pressure of 2 atm and when there is no membrane separation of hydrogen ($j=0$). The inlet concentrations of carbon monoxide and steam are calculated as follows:

$$C_{CO,o}=\frac{y_{CO}P_{tot}}{RT}=\frac{7.4\times10^{-4}}{\left(7.4\times10^{-4}+1.5\times10^{-3}\right)}\frac{2\,\text{atm}}{0.08206\dfrac{\text{L}\cdot\text{atm}}{\text{mol}\cdot\text{K}}673\,\text{K}}\frac{1000\,\text{L}}{1\,\text{m}^3}=11.9\frac{\text{mol}}{\text{m}^3}$$

$$C_{H_2O,o} = \frac{y_{H_2O} P_{tot}}{RT} = \frac{1.5\times10^{-3}}{\left(7.4\times10^{-4}+1.5\times10^{-3}\right)} \frac{2\,\text{atm}}{0.08206\frac{\text{L}\cdot\text{atm}}{\text{mol}\cdot\text{K}}673\,\text{K}} \frac{1000\,\text{L}}{1\,\text{m}^3}$$

$$= 24.1\frac{\text{mol}}{\text{m}^3}$$

Rearrange and substitute the extent of reaction, ξ:

$$K_{eq} = 11.92 = \frac{C_{CO_{2,eq}} C_{H_{2,eq}}}{C_{CO,eq} C_{H_2O,eq}} = \frac{\xi.\xi}{\left(C_{CO,o}-\xi\right)\left(C_{H_2O,o}-\xi\right)} = \frac{\xi^2}{(11.9-\xi)(24.1-\xi)}$$

Rearrange:

$$0.916\xi^2 - 38\xi + 286.79 = 0$$

Solve the quadratic equation, $\xi = 9.92$. Since the limiting reactant in this case is CO, the equilibrium conversion is:

$$X_{eq} = \frac{9.92}{11.92} = 0.83$$

The equilibrium conversion is higher when we have excess steam in the reactor rather than equal molar feed rate. The percentage of excess steam is:

$$\%\text{excess } H_2O = \frac{24.1-11.9}{24.1}\times100\% = 50.6\%$$

The case has feed rate 7.0×10^{-4} mol/min CO and 1.5×10^{-3} mol/min H_2O (thus, in your simulations, assume $j \neq 0$). The surface concentration profile is shown in Figure 7.8.

The concentration profile of all components in the reaction zone in the presence of hydrogen penetration across the membrane is shown in Figure 7.9. The conversion in the case of excess steam and the presence of hydrogen separation shows high conversion.

$$X = \frac{11.9-0.08}{11.9} = 0.99$$

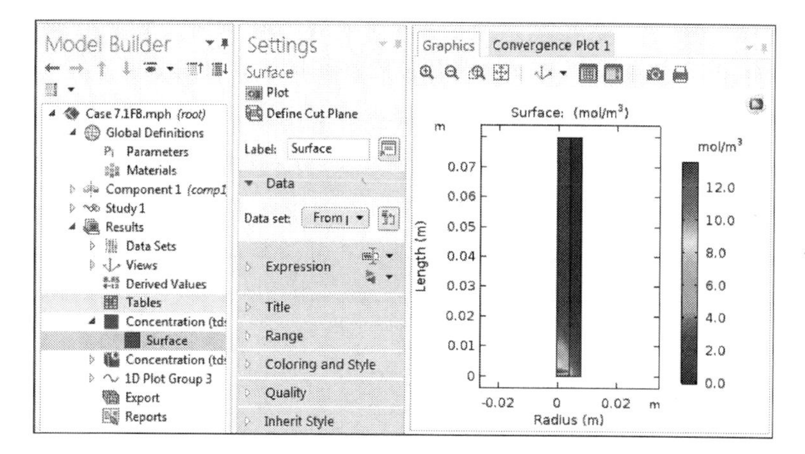

FIGURE 7.8
Hydrogen surface concentration in the case of excess steam and hydrogen separation.

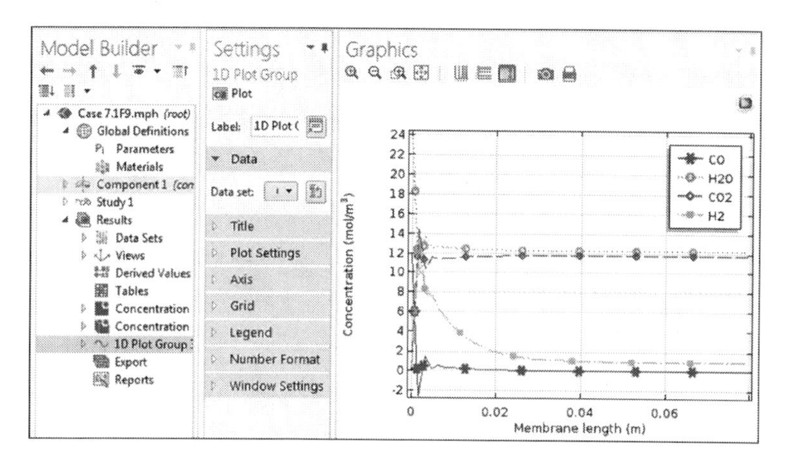

FIGURE 7.9
Component concentration profile in the presence of hydrogen permeate.

7.1.5 Effect of Membrane Thickness

The effect of membrane thickness (50, 100, 200, 500 μm) on the percent conversion of carbon monoxide is depicted in Figure 7.10. The results reveal that the percent CO conversion decreases as the membrane thickness is increased. This is attributed to the fact that, as membrane thickness increases, the membrane resistance increases.

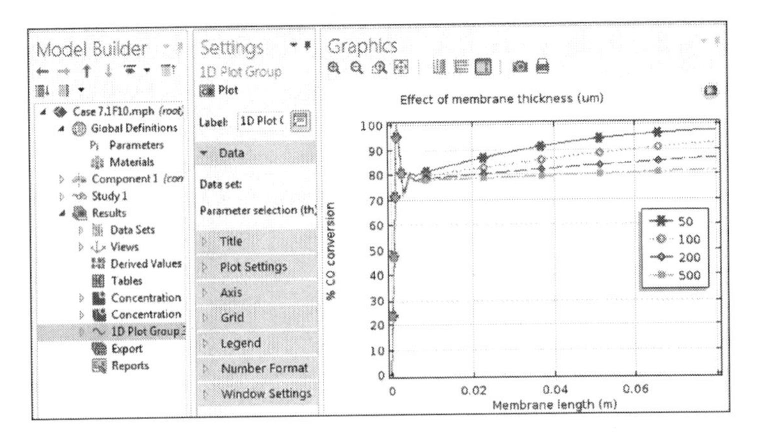

FIGURE 7.10
Effect of membrane thickness (μm) on membrane conversion.

7.2 Absorption of Carbon Dioxide from Flue Gas

A gas in an atmosphere that absorbs and releases radiation within the thermal infrared range is known as a greenhouse gas. CO_2 acts like a blanket in the earth's atmosphere and prevents heat from escaping from the earth's surface, in turn causing a rise in global temperatures. CO_2 is a greenhouse gas because it increases the temperature level of the earth in the same way the heat is generated inside the greenhouse where plants are grown [3–5].

7.2.1 Capture of Carbon Dioxide Using Fresh Water

In this case (CO_2) is absorbed from flue gas by two different absorbents: first, using fresh water and second using 0.1 M aqueous sodium hydroxide in a pilot scale absorption column. The diameter is 0.08 m and the height is 1.83 m. The liquid enters from the top of the column at a rate of $1.05\times10^{-3}\,m^3/min$. The flue gas enters the bottom of the column at a rate of $1.8\times10^{-3}\,m^3/min$. The inlet CO_2 mole fraction in flue gas is 0.185. The diffusivity of CO_2 in flue gas is $1.6\times10^{-5}\,m/s^2$, and the diffusivity of CO_2 in liquid absorbent is $1.6\times10^{-9}\,m/s^2$. The equilibrium constant, k_e is 1400 atm. The mass transfer coefficient is K_ya is $0.214\,mol/m^3s$. Simulate the absorption column using the two absorbents and compare the results. The column is operated at atmospheric pressure. The rate of absorption per unit volume using fresh water is:

$$r = K_ya(y - y^*)$$

where the equilibrium line for the concentration of carbon dioxide in gas and liquid phase is described by the following equilibrium relation:

$$y^* = k_e x$$

where:

y is the mole fraction of CO_2 in gas phase
x is the mole fraction of CO_2 in liquid phase
y^* is the equilibrium mole fraction in the gas phase

7.2.1.1 Model Equations

The mole fraction of the carbon dioxide component is calculated from the CO_2 concentration in the gas phase (C_g, mol/m^3), which is the default unit used for concentration in COMSOL. The relation is extracted from the ideal gas law ($PV = nRT$). Accordingly, the mole fraction of carbon dioxide in gas phase (y) is defined as:

$$y = \frac{C_g}{C_{\text{tot,gas}}} = \frac{C_g}{P/RT} = \frac{C_g \times 8.314 \dfrac{\text{m}^3\text{Pa}}{\text{mol K}} \times (298\,\text{K})}{101325\ \text{Pa}} = 0.025\,C_g$$

The CO_2 liquid molar concentration (C_w) is used to calculate the mole fraction of CO_2 in the liquid phase (x):

$$x = \frac{C_w}{C_{\text{tot,liq}}} = \left(\frac{\text{mol}}{\text{m}^3}\right)\left(\frac{1\,\text{m}^3}{1{,}000{,}000\,\text{g}}\right)\left(\frac{18\ \text{g}}{1\,\text{mol}}\right) = 1.8 \times 10^{-5}\,C_w$$

The initial concentration of carbon dioxide in liquid water, $C_{w,0}$ is set to zero:

$$C_{w,0} = 0$$

The absorption rate (r) is:

$$r = K_y a\left(y - k_e x\right)$$

$K_y a$ is the overall mass transfer coefficient based on the gas phase driving force, 0.214 mol/m^{3}s. The initial concentration of carbon dioxide in feed stream to the absorber, $C_{g,0}$, is:

$$C_{g,0} = \frac{y_{CO_2} P}{RT} = \frac{0.185 \times 101325\ \text{Pa}}{8.314 \dfrac{\text{m}^3\text{Pa}}{\text{mol K}}(298\,\text{K})} = 7.57\,\text{mol/m}^3$$

Inlet liquid and gas velocities (v_l) are calculated from the gas inlet volumetric flow rate (V_{liq}) and inlet cross sectional area (A_c):

$$v_l = \frac{V_{\text{liq}}}{A_c} = -\left(1.05\times10^{-3}\right)\frac{\text{m}^3}{\text{min}} \times \frac{1\,\text{min}}{60\,\text{s}} \times \frac{1}{\frac{\pi(0.08\,\text{m})^2}{4}} = -0.00185\frac{\text{m}}{\text{s}}$$

The negative sign represents the current flow of liquid in the opposite direction of the *z*-axis. The inlet gas velocity, v_g, is:

$$v_g = \frac{V_g}{A} = \left(1.8\times10^{-3}\right)\times\frac{\text{m}^3}{\text{min}} \times \frac{\text{min}}{60\,\text{s}} \frac{1}{\frac{\pi(0.08\,\text{m})^2}{4}} = 0.00634\frac{\text{m}}{\text{s}}$$

The component balance equation in terms of molar concentrations is:

$$\overbrace{\frac{\partial C_A}{\partial t}}^{\text{accum.}} + \overbrace{\left(v_x\frac{\partial C_A}{\partial x} + v_y\frac{\partial C_A}{\partial y} + v_z\frac{\partial C_A}{\partial z}\right)}^{\text{Convection}} - \overbrace{D_{AB}\left(\frac{\partial^2 C_A}{\partial x^2} + \frac{\partial^2 C_A}{\partial y^2} + \frac{\partial^2 C_A}{\partial z^2}\right)}^{\text{Diffusion}} = \overbrace{r_A}^{\text{reaction}} \quad (7.10)$$

A transient term, convective term, diffusive term, and reaction term are the main terms of the component mole balance equation.

7.2.1.2 COMSOL Simulation

The following steps are followed in the simulation process:

1. Start COMSOL Multiphysics 5.3a and then click "Model Wizard."
2. From the Space dimension list, select "Axisymmetric (2D)."
3. From the Select physics list, select Chemical Species transport > Transport of diluted species.
4. Edit the field in the Dependent variables. Type the name of the concentration variable: c_g.
5. Repeat item 3: From the Select physics list, select Chemical Species transport > Transport of diluted species.
6. In the Dependent variables field, type the name of the concentration variable: c_w.
7. Click "Study."
8. Select "Stationary."
9. Right-click "Geometry" and select "rectangle." Specify the following dimensions in meters: width = 0.08, length = 1.83.

10. Under Global definitions, select "Parameters" and enter the parameter as shown in Figure 7.11.

▾ Parameters

Name	Expression	Value
Dg	1.6e-5	1.6E-5
Dl	1.6e-9	1.6E-9
vg	0.0064	0.0064
vl	-0.00185	-0.00185
cg0	7.57	7.57
cl0	0	0
kya	0.214	0.214
ke	1400	1400

FIGURE 7.11
Parameters used in the absorption.

11. Right-click on "Definitions" and select "Variables." Type in the reaction rate and gas mole fraction as shown in Figure 7.12.

▾ Variables

Name	Expression	Unit
rxr	kya*(y-ke*x)	mol/m³
y	cg*0.025	mol/m³
x	cw*1.8e-5	mol/m³

FIGURE 7.12
Variables used in the simulation.

12. Right-click on "Transport of dilute species (tds)" and select "Reaction."
13. Click on "Reaction," and under reaction rate, type "$-rxr$" for the gas phase. Use the minus sign because of the rate of disappearance of reactant gas.
14. Click on "inflow" and below concentration type "cg_0." The inlet concentration of carbon dioxide in the inlet gas is from the bottom.

    Steps 1 to 14 are represented in Figure 7.13.

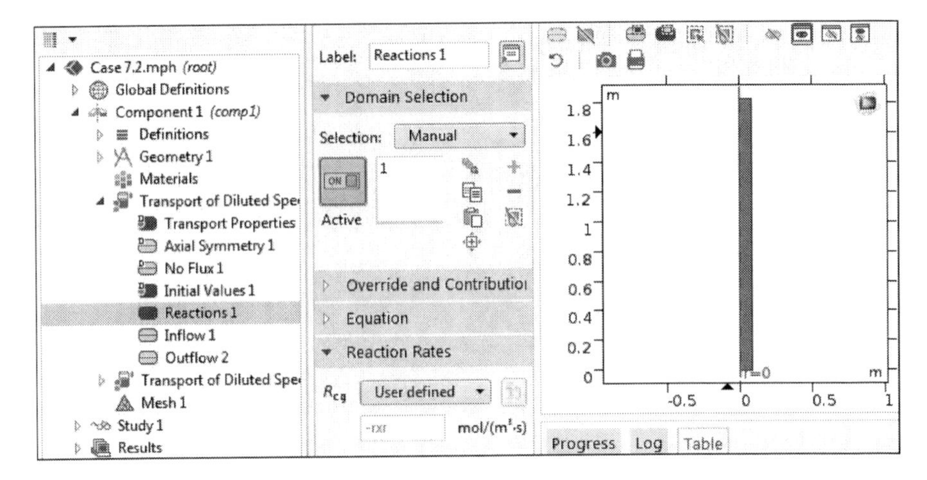

FIGURE 7.13
Reaction rate of CO_2.

15. Right-click on "Transport of dilute species (tds2)" and select "Reaction."
16. Click on "Reaction." Type under reaction rate "*rxr*" for the liquid phase (Figure 7.14).

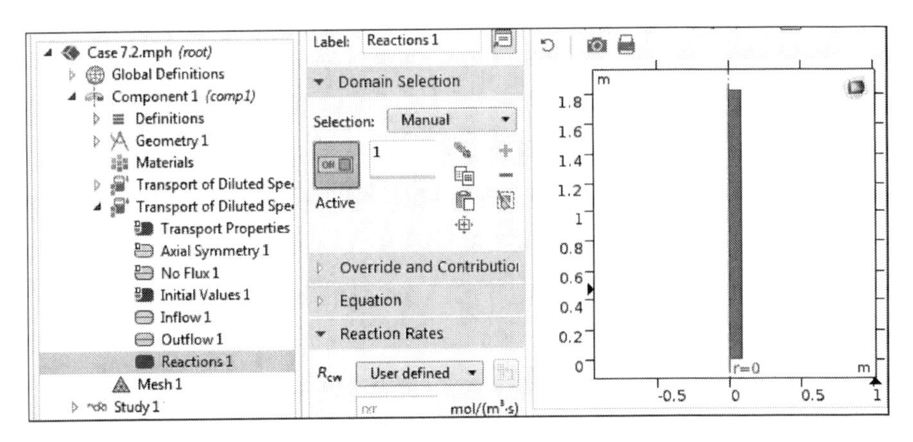

FIGURE 7.14
Absorption reaction rate of water.

17. Click on "Inflow" and below concentration type "0," which is the concentration of carbon dioxide in the inlet liquid stream.

    The default mesh is sufficient and the problem is ready for calculation.

18. Click on "Study" and then click on "Compute." The surface of the mole fraction of carbon dioxide in the gas phase is shown in Figure 7.15. The carbon dioxide concentration profile along the length of the absorber is depicted in Figure 7.16. The carbon dioxide concentration decreased along the length of the absorber.

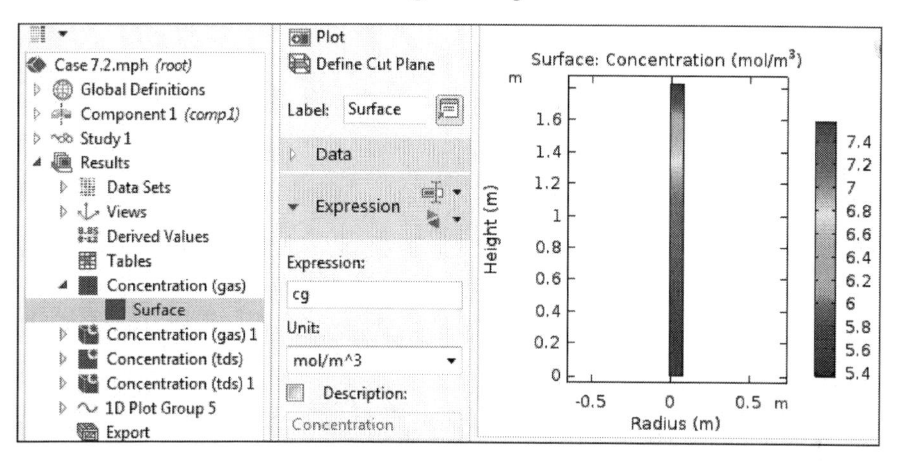

FIGURE 7.15
Surface plot of CO_2 composition across the absorption tower.

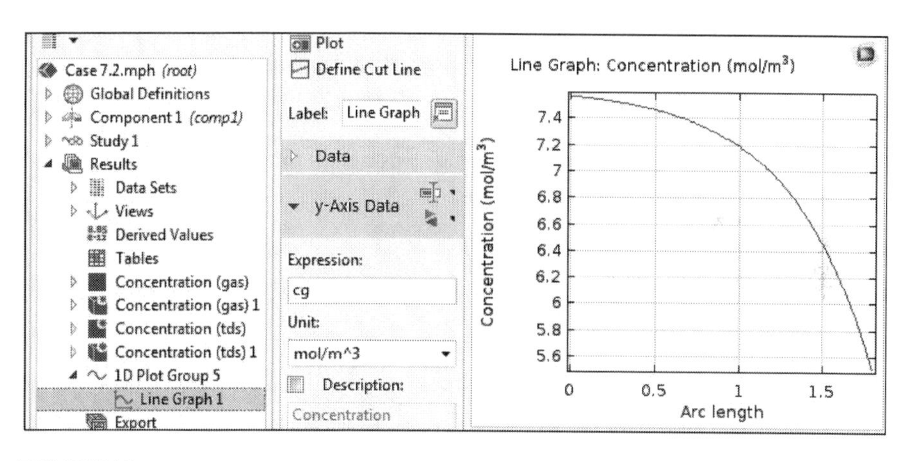

FIGURE 7.16
Carbone dioxide concentration profile across the tower.

7.2.2 Capture of CO_2 Using Aqueous Sodium Hydroxide

When CO_2 is absorbed into an aqueous NaOH solution, Na_2CO_3 is formed accompanied by consumption of NaOH in the pH range of 12 and 10, which is followed by formation of $NaHCO_3$ accompanied with consumption of Na_2CO_3 in the pH range of 10 and 8.

7.2.2.1 *Model Equations*

The reaction taking place between carbon dioxide and aqueous sodium hydroxide is based on the following reaction:

$$CO_2 + 2NaOH \rightarrow Na_2CO_3 + H_2O$$

The reaction rates are:

$$r_{CO_2} = -k_r[CO_2][NaOH]$$

$$r_{NaOH} = -2k_r[CO_2][NaOH]$$

The reaction rate constant k_r is:

$$k_r\left(\frac{m^3}{mol\cdot s}\right) = 10^{11.916-\frac{2382}{T}}/1000$$

The dimensionless Henry's constant H is:

$$H = 2.82\times10^6\exp(-2044/T)/(RT)$$

The mole fraction of carbon dioxide in gas phase is defined as:

$$y = \frac{C_{CO_{2,g}}\times R\times T}{P} = \frac{C_{CO_{2,g}}\times 8.314\frac{m^3\,Pa}{mol\,K}\times(298\,K)}{101325\,Pa} = 0.025\,C_{CO_{2,g}}$$

The mole fraction of CO_2 in the liquid phase x is:

$$x = C_{CO_{2,l}}\left(\frac{mol}{m^3}\right)\times\left(\frac{L}{55.56\,mol}\right)\times\left(\frac{1\,m^3}{1000}\right) = 1.8\times10^{-5}\,C_{CO_{2,l}}$$

The inlet concentration of carbon dioxide in aqueous sodium hydroxide solution ($C_{CO_{2,l,0}}$) is zero ($C_{CO_{2,l,0}} = 0$).

The concentration of carbon dioxide in the gas stream fed to the absorber, ($C_{CO_{2g,0}}$) is calculated as follows:

$$C_{CO_{2g,0}} = \frac{0.185 \times 101325 \text{ Pa}}{8.314 \frac{\text{m}^3 \text{ Pa}}{\text{mol K}} (298 \text{ K})} = 7.57 \frac{\text{mol}}{\text{m}^3}$$

The inlet liquid velocities, v_l, is calculated as follows:

$$v_l = -\left(1.05 \times 10^{-3} \frac{\text{m}^3}{\text{min}}\right) \times \frac{\text{min}}{60 \text{ s}} \frac{1}{\frac{\pi (0.08 \text{ m})^2}{4}} = -0.00185 \frac{\text{m}}{\text{s}}$$

The negative sign of the liquid velocity is due to the flow of liquid opposite to the z-coordinate.

The inlet gas velocity, v_g, is calculated from the volumetric flow rate as follows:

$$v_g = \left(1.8 \times 10^{-3} \frac{\text{m}^3}{\text{min}}\right) \times \frac{1 \text{min}}{60 \text{ s}} \frac{1}{\frac{\pi (0.08 \text{ m})^2}{4}} = 0.00634 \frac{\text{m}}{\text{s}}$$

The following is the component balance equation:

$$\overbrace{\frac{\partial C_A}{\partial t}}^{\text{accum.}} + \overbrace{\left(v_x \frac{\partial C_A}{\partial x} + v_y \frac{\partial C_A}{\partial y} + v_z \frac{\partial C_A}{\partial z}\right)}^{\text{Convection}} - \overbrace{D_{AB}\left(\frac{\partial^2 C_A}{\partial x^2} + \frac{\partial^2 C_A}{\partial y^2} + \frac{\partial^2 C_A}{\partial z^2}\right)}^{\text{Diffusion}} = \overbrace{r_A}^{\text{reaction}} \tag{7.11}$$

The component balance equation consists of the accumulation term, a convective term, a diffusive term, and a reaction.

7.2.2.2 COMSOL Simulation

1. Start COMSOL Multiphysics 5.3a and click "Model Wizard."
2. From the Space dimension list and select "Axisymmetric (2D)."
3. From the Select physics list, select Chemical Species transport > Transport of diluted species.
4. In the edit field of the Dependent variables, type the name of the concentration variable c_g.
5. From the Select physics list, again select Chemical Species transport > Transport of diluted species.

6. In the edit field of the Dependent variables, type the name of the concentration variable, the concentration of the carbon dioxide in liquid phase (c_{gw}), and the concentration of sodium hydroxide in the liquid phase (c_s).
7. Click "Study."
8. Select "Stationary."
9. Right-click on "Geometry" and select two rectangles, one for gas phase and the second for liquid phase. Specify the following dimensions in meters: width = 0.02, length = 1.83, as shown in Figure 7.17.

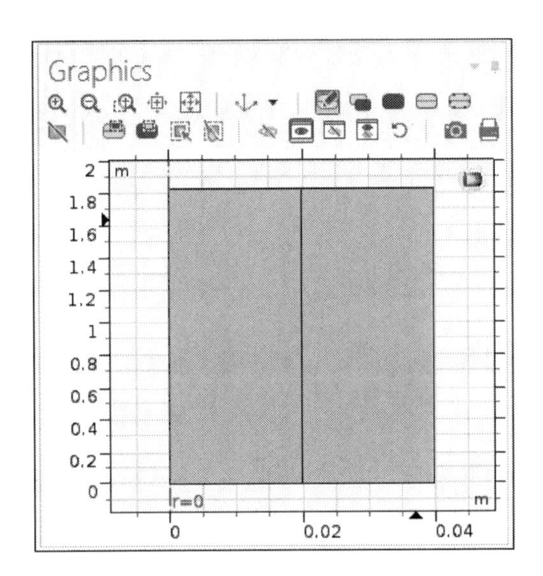

FIGURE 7.17
Gas and liquid section of the absorber column.

10. Select "Parameters" by right-clicking on "Global definitions," then type in the parameters as shown in Figure 7.18.
11. Right-click on "Definitions" and select "Variables," then type in the reaction rate and gas mole fraction as shown Figure 7.19.

Parameters

Name	Expression	Value
Dg	1.6e-5	1.6E-5
Dl	1.6e-9	1.6E-9
vg	0.0064	0.0064
vl	-0.00185	-0.00185
cg0	7.57	7.57
cl0	0	0
kya	0.214	0.214
ke	1400	1400
H	2.82e6*exp(-204...	1.195
T	298	298
m	1/H	0.83681

FIGURE 7.18
Input parameters.

Variables

Name	Expression
rNaOH	2*kr*cgw*cs
rCO2	kr*cgw*cs
kr	(10^(11.916-2382/T))/1000

FIGURE 7.19
Input variables.

12. Right-click on "Transport of dilute species (tds) 2" and select "Reaction."
13. Left-click on "Reaction" and under reaction rate, type "$-r_{NaOH}$" and "$-r_{CO_2}$" in the liquid phase (Figure 7.20).

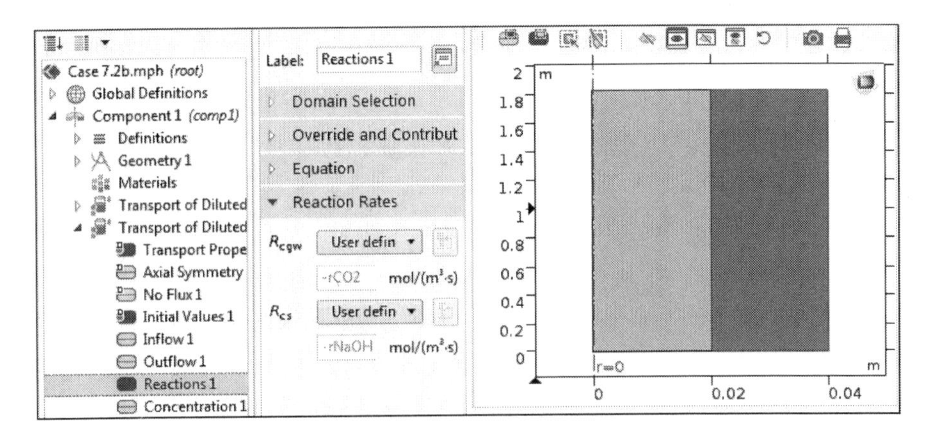

FIGURE 7.20
Reaction rate takes place in the liquid phase.

14. Left-click on "inflow" boundary condition, and below "concentration," type "cg_0," which is the inlet concentration of carbon dioxide in the inlet gas.
15. Left-click on the "concentration" boundary condition and specify the boundary concentration at the gas-liquid interface as shown in Figure 7.21.

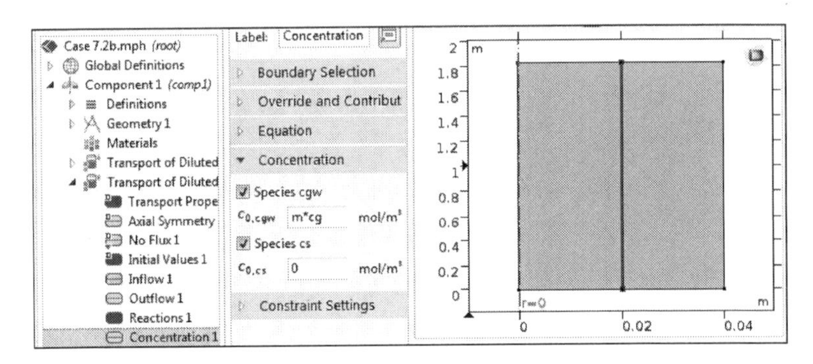

FIGURE 7.21
Concentration as boundary conditions.

16. Choose the "Inflow" boundary condition, and below concentration, type "0," which is the concentration of carbon dioxide in the inlet liquid stream.
17. The default mesh is sufficient and the problem is ready for calculation.
18. Click on "Study" and then click on "Compute." The surface of the mole fraction of carbon dioxide in the gas phase is shown in Figure 7.22.

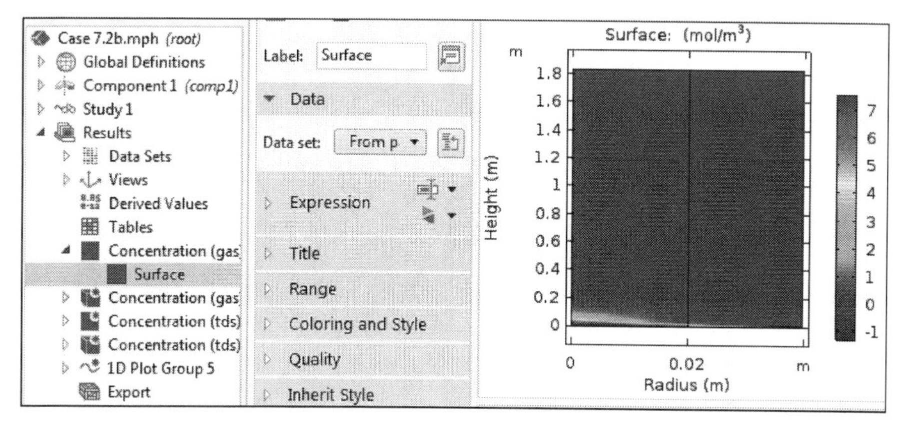

FIGURE 7.22
Surface plot of CO_2 composition across the absorption tower.

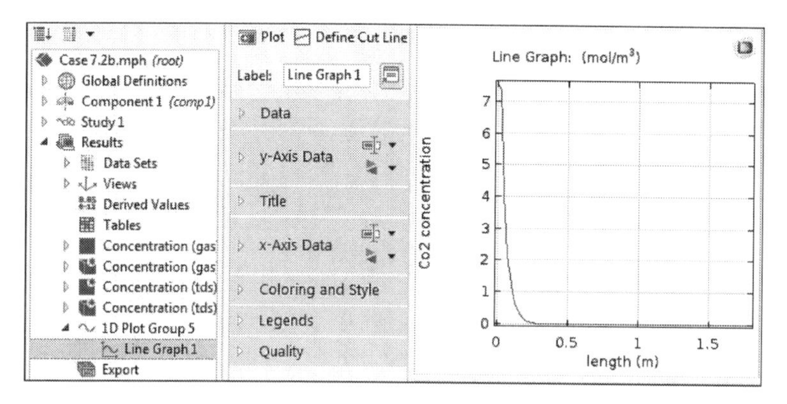

FIGURE 7.23
Concentration of CO_2 in the gas phase zone across the tower.

Figure 7.23 shows a sharp drop in the CO_2 concentration at liquid flow rate of 1 L/min. The diagram reveals that approximately 0.3 m length of the reactor is enough to achieve depletion of the entire carbon dioxide.

7.3 Packed Bed Reactors

Packed bed reactors are used in chemical reactions to catalyze gas reactions. These reactors are tubular and are filled with solid catalyst particles. In chemical processing, a packed bed reactor is a hollow tube or a pipe filled with a packing material. The packing can be randomly filled with small objects like Raschig rings. Packed beds contain adsorbents or catalyst particles such as zeolite pellets or granular activated carbon. Packed beds improve the contact area between two phases in a chemical process and can be used in a chemical reactor, an absorber, and a distillation process. The advantage of using a packed bed reactor is the higher conversion per weight of catalyst than other catalytic reactors. Three main types of reactor configurations are used depending on their function: adiabatic tubular reactor, isothermal tubular reactor, and heat exchanger fixed bed reactor. A tubular heat exchanger type reactor is often used to increase the capacity of production with enhanced yields. In isothermal processes, a change of system occurs at a constant temperature. This case occurs when a system meets an external thermal reservoir and, using heat exchange, the temperature changes gradually to allow the system to continually adjust to the temperature of the reservoir. On the other hand, an adiabatic process is where a system exchanges no heat with its surroundings ($Q = 0$). In other words, in an isothermal process, the value $\Delta T = 0$, and therefore $\Delta U = 0$ (only for an ideal gas), but $Q \neq 0$. While in an adiabatic process, $\Delta T \neq 0$ but $Q = 0$.

7.3.1 Isothermal Packed Bed Reactor

Isothermal means that the system is operated at constant temperature. In this example, steam and natural gas react to form mostly carbon monoxide and hydrogen with some carbon dioxide also produced. There can also be excess water in the reformate stream in a process known as methane steam reforming. The process produces hydrogen used for fuel cells and as a fuel for vehicles. In the reactor, methane (CH_4) and water (H_2O) are fed as reactants, while carbon dioxide (CO_2), carbon monoxide (CO), and hydrogen (H_2) are produced over a nickel catalyst on an alumina support. Consider a packed bed reactor 0.02 m diameter and 0.4 m long. Feed of 10,000 mol/h CH_4, 10,000 mol/h H_2O and 100 mol/h H_2 goes to a steam-reforming reactor that operates at 1000 K and 1 atm feed pressure. The catalyst density is 2 g/cm^3. The diffusivity $D_i = 2 \times 10^{-5} m/s^2$. Determine the molar flow rates of CH_4, H_2O, CO_2, CO, and H_2 as a function of reactor length (catalyst weight up to 382 g). Determine the overall methane conversion [6].

7.3.1.1 Model Development

Three simultaneous reactions take place in a packed bed reactor. The first reaction for methane steam reforming is given as:

$$CH_4 + H_2O \leftrightarrow 3\,H_2 + CO \qquad \Delta H_r^o = +206\ \text{kJ/mol} \tag{7.12}$$

The reaction rate is given by:

$$r_1 = \frac{\left(\frac{k_1}{P_{H_2}^{2.5}}\right)\left[P_{CH_4}P_{H_2O} - \left(\frac{P_{H_2}^3 P_{CO}}{K_{e1}}\right)\right]}{\left(1 - k_{CH_4}P_{CH_4} + k_{CO}P_{CO} + k_{H_2}P_{H_2} + k_{H_2O}\left(\frac{P_{H_2O}}{P_{H_2}}\right)\right)^2} \tag{7.13}$$

The equilibrium constant, K_{e1}, is:

$$K_{e1} = \exp\left(30.42 - \frac{27106}{T}\right) \tag{7.14}$$

The forward reaction rate constant is given by the Arrhenius relationships, k_1:

$$k_1 = 4.22 \times 10^{15} \exp\left(-\frac{240100}{RT}\right) \tag{7.15}$$

The second reaction is the water gas shift reaction also takes place in the steam reformer as follows:

$$CO + H_2O \leftrightarrow H_2 + CO_2 \qquad \Delta H_r^o = -41\ \text{kJ/mol} \tag{7.16}$$

The reaction rate is given by:

$$r_2 = \frac{\left(\frac{k_2}{P_{H_2}}\right)\left[P_{CO}P_{H_2O} - \left(\frac{P_{H_2}P_{CO_2}}{K_{2e}}\right)\right]}{\left(1 - k_{CH_4}P_{CH_4} + k_{CO}P_{CO} + k_{H_2}P_{H_2} + k_{H_2O}\left(\frac{P_{H_2O}}{P_{H_2}}\right)\right)^2} \tag{7.17}$$

The reaction equilibrium constant, K_{e2}:

$$K_{e2} = \exp\left(-3.798 + \frac{4160}{T}\right) \tag{7.18}$$

The forward reaction rate constant is given by the Arrhenius relationships as, k_2:

$$k_2 = 1.9 \times 10^6 \exp\left(-\frac{67130}{RT}\right) \tag{7.19}$$

Adding together the methane-steam reforming and water gas shift reactions gives the overall reaction:

$$CH_4 + 2H_2O \leftrightarrow 4H_2 + CO_2 \qquad \Delta H_r^o = +165 \text{ kJ/mol} \tag{7.20}$$

The reaction rate is given by:

$$r_3 = \frac{\left(\dfrac{k_3}{P_{H_2}^{3.5}}\right)\left[P_{CH_4} * P_{H_2O}^2 - \left(\dfrac{P_{H_2}^4 P_{CO_2}}{K_{e3}}\right)\right]}{\left(1 + k_{CH_4} P_{CH_4} + k_{CO} P_{CO} + k_{H_2} P_{H_2} + k_{H_2O}\left(\dfrac{P_{H_2O}}{P_{H_2}}\right)\right)^2} \tag{7.21}$$

The reaction equilibrium constant, K_{e3}:

$$K_{e3} = \exp\left(34.218 - \frac{31266}{T}\right) \tag{7.22}$$

The forward reaction rate constant, k_3:

$$k_3 = 1.02 \times 10^{15} \exp\left(-\frac{243900}{RT}\right) \tag{7.23}$$

The equilibrium constants can be expressed in terms of partial pressures and temperature (in atm and in degrees Kelvin [K], respectively). Helium (He) is added as an inert nonreacting gas. The reaction rate, r_i is in the unit of $\text{mol}/(g_{cat} \cdot h)$. The coefficients in the equations are given by the Arrhenius relationships:

$$K_{CH_4} = 6.65 \times 10^{-4} \exp\left(\frac{38280}{RT}\right) \tag{7.24}$$

$$K_{H_2O} = 1.77 \times 10^{5} \exp\left(-\frac{88680}{RT}\right) \tag{7.25}$$

$$K_{H_2} = 6.12 \times 10^{-9} \exp\left(\frac{82900}{RT}\right) \tag{7.26}$$

$$K_{CO} = 8.23 \times 10^{-5} \exp\left(\frac{70650}{RT}\right) \tag{7.27}$$

Note the gas constant $R = 8.314\,\text{J/(mol}\cdot\text{K)}$ in these expressions. The component balance equation is:

$$\overbrace{\frac{\partial C_A}{\partial t}}^{\text{accum.}} + \overbrace{\left(v_x \frac{\partial C_A}{\partial x} + v_y \frac{\partial C_A}{\partial y} + v_z \frac{\partial C_A}{\partial z} \right)}^{\text{Convection}} - \overbrace{D_{AB} \left(\frac{\partial^2 C_A}{\partial x^2} + \frac{\partial^2 C_A}{\partial y^2} + \frac{\partial^2 C_A}{\partial z^2} \right)}^{\text{Diffusion}} = \overbrace{r_A}^{\text{reaction}} \tag{7.28}$$

The partial differential equation is to be solved by the COMSOL software package, as shown in the following section.

7.3.1.2 COMSOL Simulation

The steps to solve the model equation via COMSOL software are as follows:

1. Start COMSOL Multiphysics 5.3a and click "Model Wizard."
2. From the Space dimension list, select "Axisymmetric (2D)."
3. From the Select physics list, select Chemical Species transport > Transport of diluted species.
4. Type the name of the six concentration variables in the Dependent variables edit field: cco, ch2o, cco2, ch2, che, cch4.
5. Click on "Study."
6. Select "Stationary."
7. Right-click "Geometry" and select "Rectangle." Once it is selected, click on it and specify the following dimensions in meters: width = 0.02, length = 0.3, as shown in Figure 7.24.
8. Enter the parameters shown in Figure 7.25 in the parameters window.

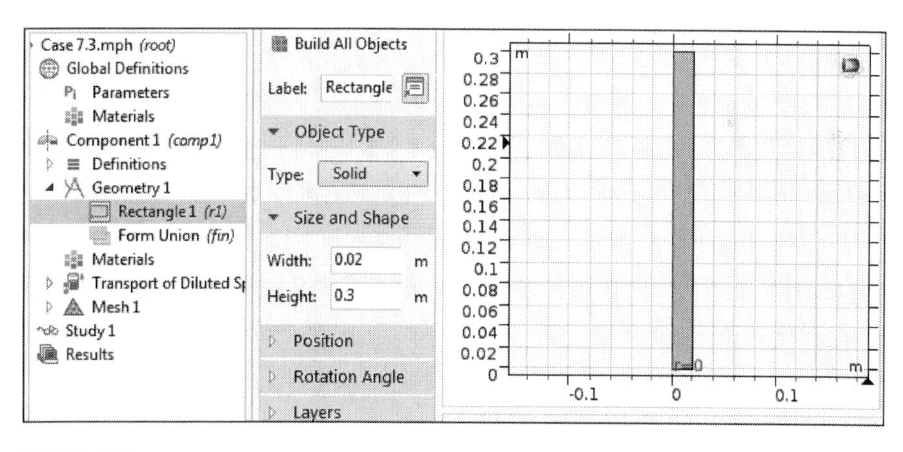

FIGURE 7.24
Reactor dimensions.

Parameters

Name	Expression	Value	Description
T	1000	1000	K
P	1	1	atm
Rg1	8.314	8.314	J/mol/K
Fch40	10000	10000	mol/h
Fh2o0	10000	10000	mol/h
Fh20	100	100	ml/h
Ft0	Fch40+Fh2o0+Fh20	20100	mol/h
ych40	Fch40/Ft0	0.49751	
yh2o0	Fh2o0/Ft0	0.49751	
yh20	Fh20/Ft0	0.0049751	
ch40	ych40*P/(Rg1*T)	5.984E-5	
ch2o0	yh2o0*P/(Rg1*T)	5.984E-5	
ch20	yh20*P/(Rg1*T)	5.984E-7	
vel	vol/(3.14*D^2/4)	3.6958E7	m/s
vol	Ft0*Rg1*T/P/3600	46420	m3/s
vc	(382/2)/1e6/(1-0.3)	2.7286E-4	m3 catalyst (rho=2 g/cc)
L	vc/(3.14*D^2/4)	0.21724	m
D	0.04	0.04	m

FIGURE 7.25
Parameters used in solving the model equations.

9. Right-click on "Definitions." Select "Variables." Once it is selected, click on "Variables" and type in the variables shown in Figure 7.26. These variables include the reaction rates and related parameters.

Variables

Name	Expression	Unit
k1	4.2248e15*exp(-240100/8.314...	
k2	1.955e6*exp(-67130/8.314/T)	
k3	1.0202e15*exp(-243900/8.314...	
kch4	6.65e-4*exp(38280/8.314/T)	
kco	8.23e-5*exp(70650/8.314/T)	
kh2	6.12e-9*exp(82900/8.314/T)	
kh2o	1.77e5*exp(-88680/8.314/T)	
DEN	1+kch4*Pch4+kco*Pco+kh2...	
Ke1	exp(30.420-27106/T)	
Ke2	exp(-3.798+4160/T)	
Ke3	exp(34.218-31266/T)	
r1	k1/Ph2^2.5/DEN^2*(Pch4*P...	
r2	k2/Ph2/DEN^2*(Pco*Ph2o-P...	
r3	k3/Ph2^3.5/DEN^2*(Pch4*P...	
Ph2	ch2*Rg1*T	mol/m³
Pch4	cch4*Rg1*T	mol/m³
Pco	cco*Rg1*T	mol/m³
Ph2o	ch2o*Rg1*T	mol/m³
Pco2	cco2*Rg1*T	mol/m³
Phe	che*Rg1*T	mol/m³
R1	r1*2*1e6/3600	
R2	r2*2e6/3600	
R3	r3*2e6/3600	

FIGURE 7.26
Process variables used in the case study.

10. Right-click on "Transport of Diluted species" under "components" and select "Reactions." Once it is selected, type in the reactions like that shown in Figure 7.27.

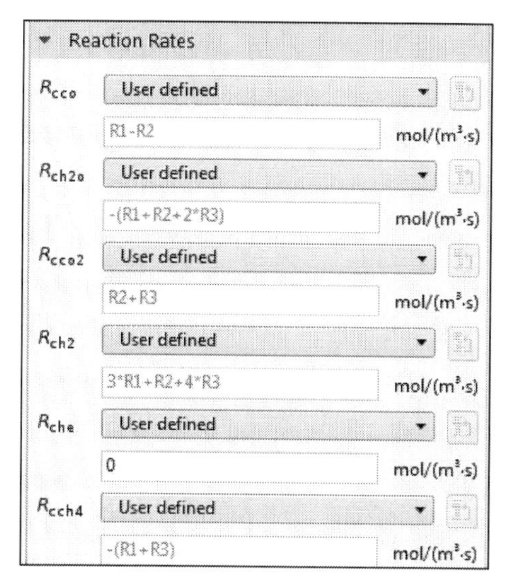

FIGURE 7.27
Components' reaction rates.

11. Right-click on "Transport of Diluted Species (tds)," select the "inflow" boundary condition, then specify the inflow concentration of each component as shown in Figure 7.28.

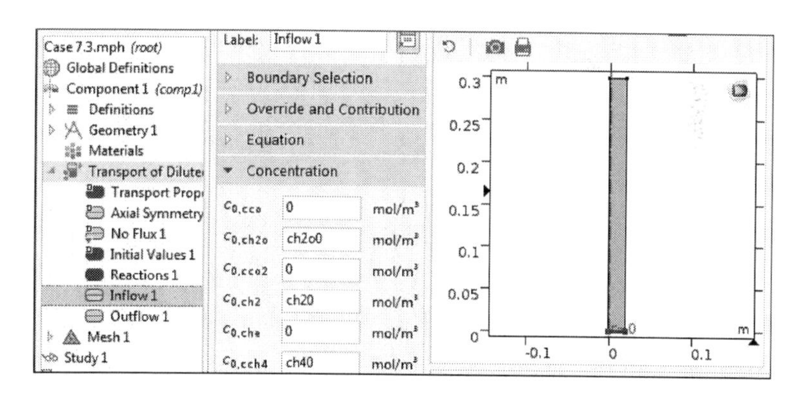

FIGURE 7.28
Reactor feed concentration.

12. For the top boundary, right-click on "Transport of Diluted Species (tds)," then select and click on the "outflow."
13. Set the right boundary for "No flux."
14. By default, the center is the "Axial Symmetry 1."
15. Right-click on "Mesh" and select "Mapped."
16. Right-click on "Mapped" and select "Distribution" twice.
17. Click on "Distribution 1." Select boundary "1, 4" next under "Distributed number of elements" and type 50 under the "Number of elements."
18. Click on "Distribution 2." Select boundary "2, 3" under "Distributed number of elements," then type 50 under the "Number of elements."
19. The system is now supposed to be ready. If it is, click on "Study," then "Compute."
20. The surface concentration diagram of hydrogen production is shown in Figure 7.29.

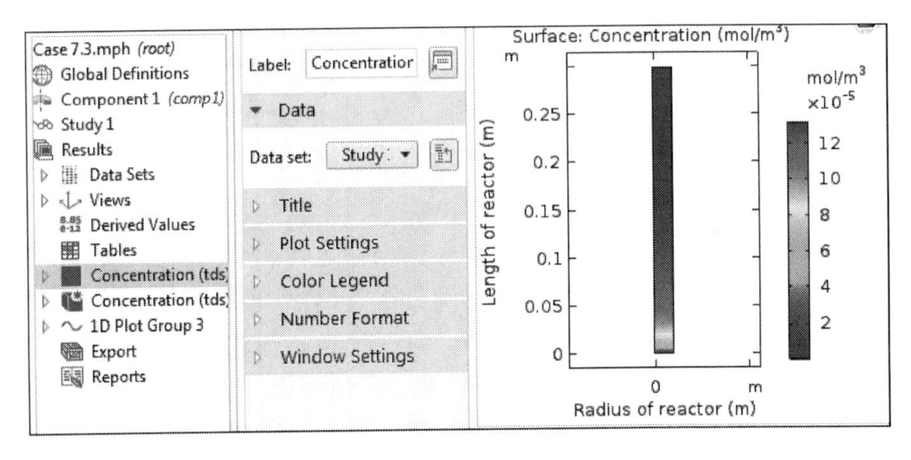

FIGURE 7.29
Surface plot of the concentration of hydrogen in the reactor.

21. Right-click on "Data" under results and select "cut line 2D" and click "Plot." The cutline should look like the line in Figure 7.30. The concentration will be calculated along that line. The cutline is located on the axial symmetry of the rectangular block.

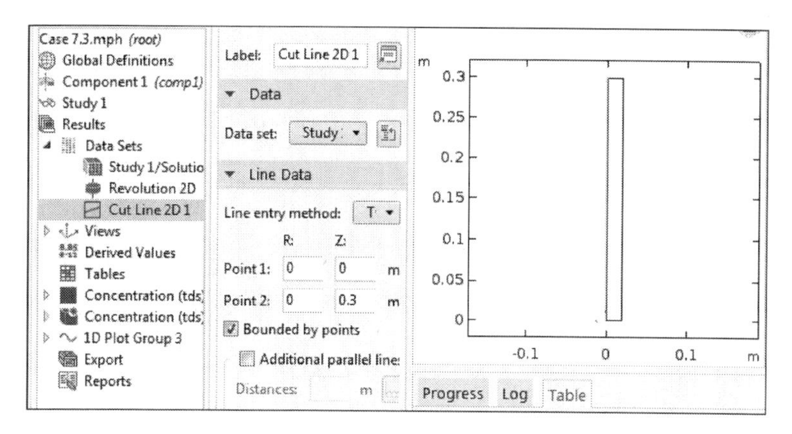

FIGURE 7.30
Generation of line graph using cutline 2D.

22. Right-click on "Results" and select "1D plot group."
23. Right-click on "1D plot group" and select "line graph" six times.
24. Relate each line graph to a specific concentration.
25. To change concentration to moles per hour, multiply each concentration by gas volumetric flow rate and the conversion of hour to seconds, as shown in the Figure 7.31.

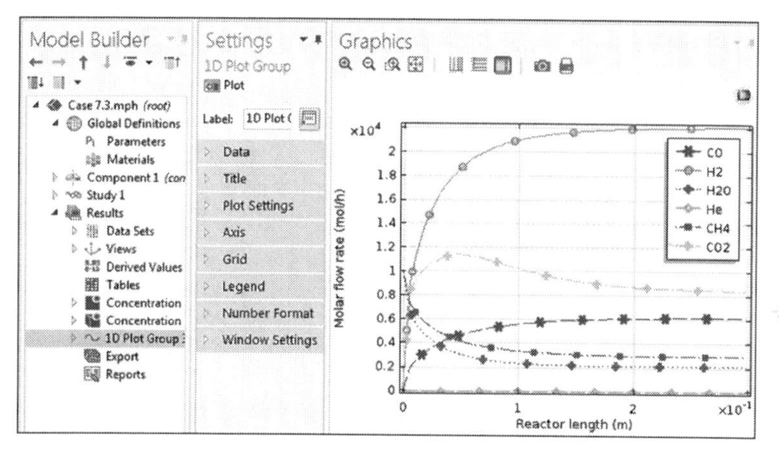

FIGURE 7.31
Line graph molar flow rate.

26. Click on "line graph" and capture the molar flow rate for all components. The results should look like the trend shown in Figure 7.32.

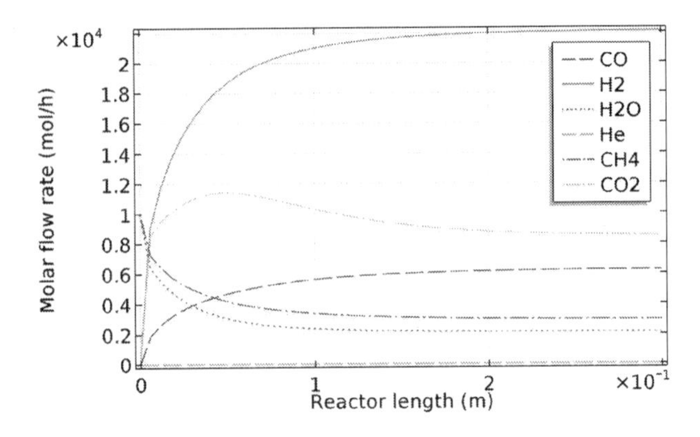

FIGURE 7.32
Species molar flow rates as a function of reactor length.

27. The exiting CH_4 molar flow rate is about 2945 mol/h. This corresponds to a CH_4 conversion of:

$$x = \frac{10,000 - 2945}{10,000} = 0.71$$

7.3.1.2.1 Conclusion

The number of moles of hydrogen produced increases along the length of the reactor. It is notable that, with a negligible pressure drop in the reactor, the gas can be expanded by increasing the volumetric flow rate. The partial pressure of a chemical species can be calculated using the total pressure and the number of moles of that species. The conversion is around 70%.

7.3.2 Adiabatic Packed Bed Reactor

An adiabatic reactor is one that occurs without transfer of heat between a system and its surroundings. In an adiabatic process, energy is transferred to its surroundings only as work. In COMSOL, under the option "heat/energy transfer physics," one should select the option of heat source, then define the heat source as ($\Delta H_{rxn}R$), where R is the reaction and ΔH_{rxn} is the enthalpy of the reaction, which might or might not depend upon temperature. After that, all the rate constants should be defined in terms of variable T (temperature) of the heat transfer physics, which allow the variation of all the rate constants as T varies. The slight decrease in temperature through the reactor is due to the low heat of the reaction (Figure 7.33).

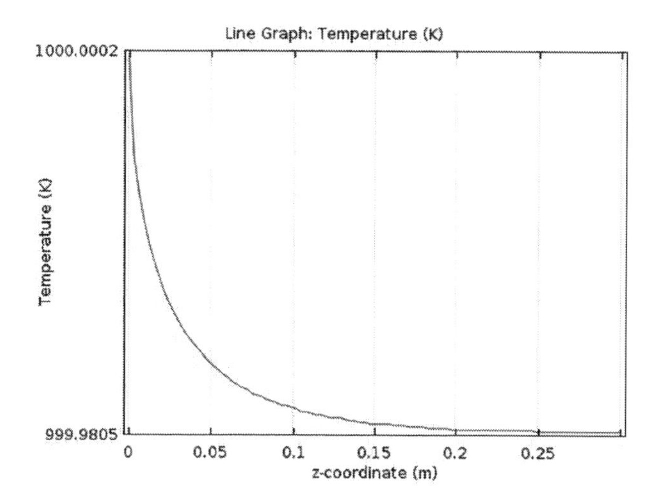

FIGURE 7.33
Temperature as a function of reactor length for adiabatic reactor.

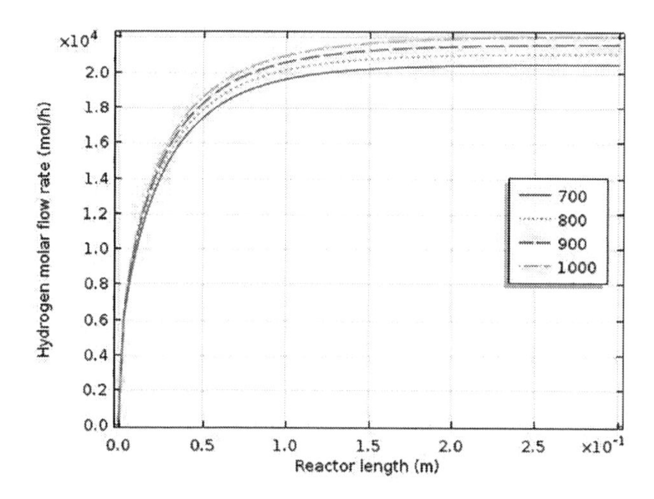

FIGURE 7.34
Hydrogen production molar flow rate versus reactor length at variable reactor feed temperatures.

The effect of feed temperature on hydrogen production rate is shown in Figure 7.34. The figure reveals that, as the reaction temperature increases, production rate of hydrogen also increases, which is credited to the increase in reaction rate. The diagram shows that there is very slight decrease in hydrogen production after the reactor length of 1.5 m. At a fixed reactor length, hydrogen production increases with an increase in reactor temperature.

7.4 Fluid Flow of Two Immiscible Liquids

Immiscible liquids are liquids do not dissolve in each other. It is possible to shake up the liquids and get them to mix temporarily, but they will separate shortly. Separating immiscible liquids is done simply by using a separating funnel. Oil and water are an example. In the following case, water and transformer oil are two immiscible and viscous liquids, flowing at 300 K in the z-direction in a horizontal thin slit of length 0.5 m and width 0.1 m under the influence of a pressure gradient. Both liquids can be considered incompressible. At the inlet, both liquids can be assumed to have the same pressure of 0.02 Pa. Determine the corresponding velocity profile for each fluid.

7.4.1 Model Development

For an incompressible fluid flowing in a pipe (Cartesian coordinates), the momentum balance for the fluids flowing in the z-direction is:

$$\overbrace{\rho\frac{\partial v_z}{\partial t}}^{\text{accum.}}+\overbrace{\rho\left(v_x\frac{\partial v_z}{\partial x}+v_y\frac{\partial v_z}{\partial y}+v_z\frac{\partial v_z}{\partial z}\right)}^{\text{transport by bulk flow}}=\overbrace{\mu\left(\frac{\partial^2 v_z}{\partial x^2}+\frac{\partial^2 v_z}{\partial y^2}+\frac{\partial^2 v_z}{\partial z^2}\right)}^{\text{transport by viscous forces}}-\overbrace{\frac{\partial P}{\partial x}+\rho g_z}^{\text{generation}} \quad (7.29)$$

The model equations are solved using COMSOL, as shown in the following section.

7.4.2 COMSOL Simulation

Start COMSOL. Click on the model wizard. The model wizard will let you specify the dimensions, physics, and study type of the model. Specify the model Space Dimension: 2D. The geometry of the system can be divided into two subregions. Both regions are rectangles of width 0.5 m and height of 0.05 m. One of these regions will contain water, the other will contain oil. First, right-click the geometry tab and then left-click "rectangle" from the menu. Next, enter the dimensions of the rectangle. Duplicate the first rectangle. Define two subdomains or regions. The first subdomain consists of the liquid water phase and the second fluid is the liquid transformer oil phase. The Navier-Stokes equation is the main equation for this problem, which is part of the "laminar flow" physics in COMSOL. The walls should have no-slip boundary conditions, and the outlet can be set to zero pressure. The geometry of the system may be defined as shown Figure 7.35.

Right-click the materials tab and select water and transform oil from the library. Assign water to the top rectangle and transformer oil to the bottom domain. Set the boundary conditions for inlet, outlet, and walls. You also need to input the temperature for the model (300 K) and select the incompressible

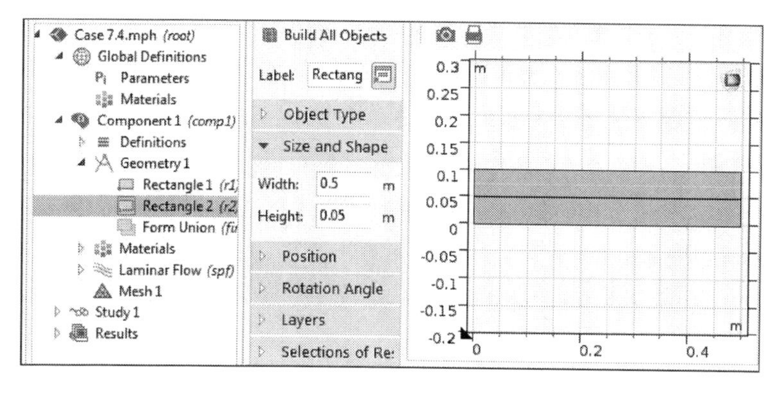

FIGURE 7.35
Selection of rectangular geometry.

form of the Navier-Stokes equation. Left-click on the laminar flow tab to display the governing equation for the model as well as the physical model being used. Physical model is set by default to incompressible flow. Under laminar flow, left-click the Fluid properties tab. In the Temperature field, type 300 [K]. Right-click the Laminar Flow tab and select the inlet boundary condition. Apply this boundary condition to the leftmost surfaces. Change the boundary condition to pressure by left-clicking on the drop-down menu. Enter the inlet pressure to 0.02 Pa (Figure 7.36). Right-click the laminar flow tab and select the outlet boundary condition.

Apply this boundary condition to the rightmost surfaces. By default, this sets the pressure at this surface to 0, as wanted. Apply the no-slip condition to both the upper and lower walls (top- and bottommost surfaces); this has been done by default in COMSOL. To confirm, you may click the subtab Wall 1

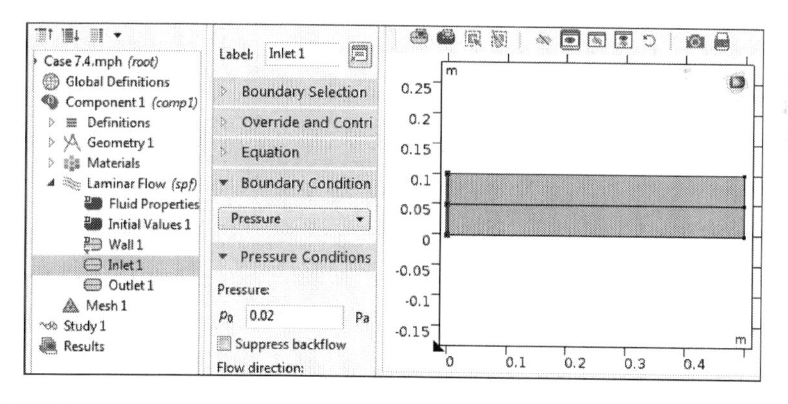

FIGURE 7.36
Inlet boundary condition.

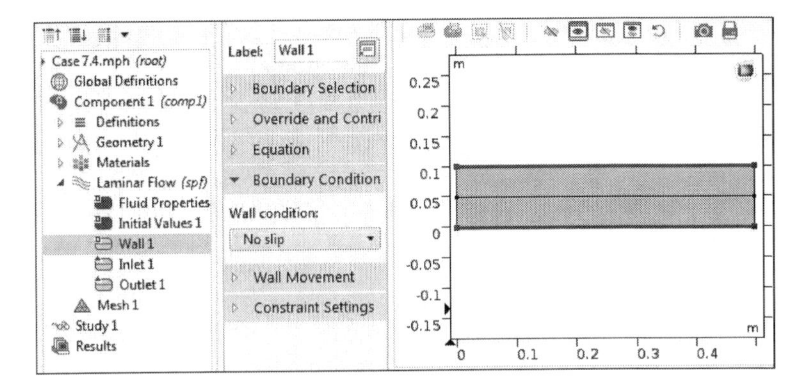

FIGURE 7.37
Boundary conditions, top and bottom wall (no slip).

and observe that the no-slip condition has indeed been applied to the top and bottom surfaces (Figure 7.37).

The model is ready to compute. Right-click on "Study" and left-click the "compute" button. The velocity surface plot should be displayed and should look like the one shown in Figure 7.38.

Right-click on the data sets tab and under the results tab, then left-click on "cut line 2D." Input the two points that define the 2D cut line ($x = 0.40\, y = 0$ and $x = 0.40\, y = 0.1$). Right-click on the 1D plot group tab and then left-click "line graph." This will create a subtab to the 1D plot group tab named line graph 1. Left-click this to bring up the line graph interface. In the data set pull-down menu, select "cut line 2D 1." To change the x axis from arc length to y, simply open the drop-down menu in the line graph interface and

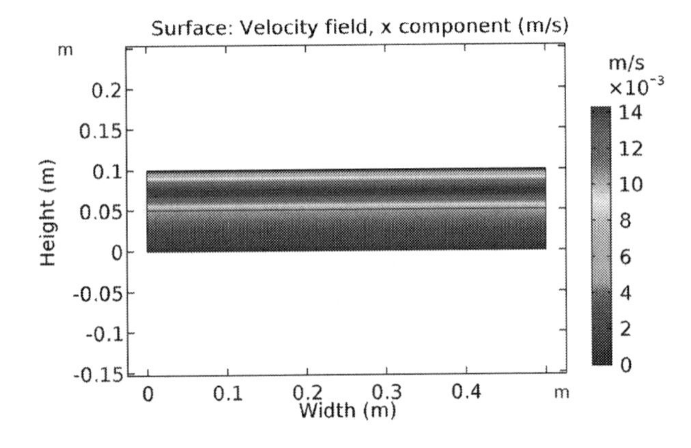

FIGURE 7.38
Surface velocity a cross the tube.

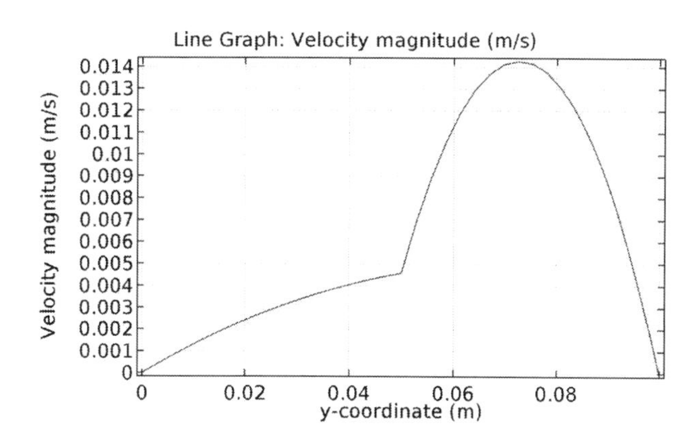

FIGURE 7.39
Velocity profile along the y-coordinate.

change it from "arc length" to expression under x axis, then type y in the field that appears. Pressing the plot button again will update the graph with the changed x axis (Figure 7.39).

7.5 Production of Propylene Glycol in Adiabatic Tubular Reactor

Propylene glycol is a synthetic liquid substance that absorbs water and is used to make polyester compounds, in the production of polymers, and in food processing and as a process fluid in low temperature heat exchange applications. It is produced via catalytic reaction of sulfuric acid of propylene oxide (A) with water (B) to form propylene glycol (C). The hydrolysis of propylene oxide takes place at room temperature when catalyzed. The first-order reaction takes place in an excess of water. The tubular reactor is 1.0 m long, and its diameter is 0.2 m. The feed to the reactor consists of two streams: One has an equal volumetric mixture of propylene oxide and methanol, and the other stream has water containing 0.1 wt% sulfuric acid. The water is fed at a volumetric rate 2.5 times larger than the propylene oxide-methanol feed. The molar flow rate of propylene oxide fed to the tubular reactor is 0.1 mol/s.

The water-propylene oxide-methanol mixture undergoes a slight neglected decrease in volume upon mixing. The heat of mixing causes an immediate temperature to rise upon mixing the two feed streams. The inlet temperature of both streams is set to 312 K to account for heat of mixing. The operation temperature should not exceed 325 K because propylene oxide has a rather low boiling point and too much propylene oxide will be lost by vaporization.

Using a tubular reactor can tackle this problem by slightly increasing the pressure to prevent a temperature rise and a loss of reactant. To make this case clear, the problem is solved at an atmospheric pressure resulting in temperatures higher than the recommended temperature. The reaction is an apparent zero-order reaction due to excess water and a first-order reaction in propylene oxide concentration and with the specific reaction rate [7].

The reaction rate law is:

$$-r_A = kC_A$$

with:

$$k(h^{-1}) = 16.96\times 10^{12} e^{-\frac{75362\,\mathrm{J/mol}}{RT}}$$

The thermal conductivity of the reaction mixture is $k_e = 0.599$ W/m/K, and the diffusivity is $D_e = 1\times 10^{-9}\,\mathrm{m^2/s}$. The overall heat-transfer coefficient is 1300 $\mathrm{W/m^2K}$, and the temperature of the cooling jacket is assumed to be constant and is set to 273 K. The density ($\mathrm{kg/m^3}$) of ethylene oxide, methanol, water, and ethylene glycol are 830, 791.3, 1000, 1040, respectively, and the heat capacity, C_p (J/(mol K)), is 146.54, 81.095, 75.36, 192.59, respectively. The heat of reaction is $\Delta H_{rxn} = -84666$ J/mol.

7.5.1 Model Development

The equation below formulates the general mass balance that takes into consideration radial changes in a tubular reactor:

$$\frac{\partial C_A}{\partial t} + \left(v_r \frac{\partial C_A}{\partial r} + v_\theta \frac{1}{r}\frac{\partial C_A}{\partial \theta} + v_z \frac{\partial C_A}{\partial z} \right) = D_e \left(\frac{1}{r}\frac{\partial}{\partial r}\left(r\frac{\partial C_A}{\partial r} \right) + \frac{1}{r^2}\frac{\partial^2 C_A}{\partial \theta^2} + \frac{\partial^2 C_A}{\partial z^2} \right) + r_A \quad (7.30)$$

where:

C_A is the concentration of species i, ($\mathrm{mole/m^3}$)

D_e is the effective diffusivity, ($\mathrm{m^2/s}$)

v_r and v_z are the superficial velocity in the radial and axial directions, respectively, (m/s)

r_i is the reaction rate of species i

$$r_i\left(\mathrm{mol/(m^3\cdot s)}\right) = v_i(r)$$

v_i is the stoichiometric coefficient.

Accumulation is presented by the term on the left-hand side. The first two terms on the right-hand side represent the difference in flux over a volume element. The total flux consists of two parts: flux by diffusion and convection both in the radial direction and axially. If the flux of one species out of the specific volume element is less than the flux going into the same element, that implies that this species is accumulating or disappearing through reaction. Therefore, the sum of the accumulation and the net flux equals the reaction rate.

Because the reaction is studied under steady-state conditions, the equation is simplified by neglecting the time dependent accumulation term:

$$0+\left(v_r\frac{\partial C_A}{\partial r}+v_\theta\frac{1}{r}\frac{\partial C_A}{\partial \theta}+v_z\frac{\partial C_A}{\partial z}\right)=D_A\left(\frac{1}{r}\frac{\partial}{\partial r}\left(r\frac{\partial C_A}{\partial r}\right)+\frac{1}{r^2}\frac{\partial^2 C_A}{\partial \theta^2}+\frac{\partial^2 C_A}{\partial z^2}\right)+r_A \quad (7.31)$$

The convective flux in the radial direction is much smaller than the diffusive flux and can therefore be neglected; i.e., v_r is approximately 0. Therefore, the radial flux will only consist of the diffusive term:

$$0=D_A\frac{1}{r}\frac{\partial}{\partial r}\left(r\frac{\partial C_A}{\partial r}\right)+D_A\frac{\partial^2 C_A}{\partial z^2}-\left(v_z\frac{\partial C_A}{\partial z}\right)+r_A \quad (7.32)$$

Expanding the derivative of the radial diffusivity yields the final form of the general mass balance being used in this exercise:

$$0=D_A\frac{\partial^2 C_A}{\partial r}+\frac{D_A}{r}\frac{\partial C_A}{\partial r}+D_A\frac{\partial^2 C_A}{\partial z^2}-v_z\frac{\partial C_A}{\partial z}+r_A \quad (7.33)$$

The velocity profile in the z-direction v_z is calculated using the momentum change equation for the z-component of the velocity profile:

$$\rho\frac{\partial v_z}{\partial t}+\rho\left(v_r\frac{\partial v_z}{\partial r}+\frac{v_\theta}{r}\frac{\partial v_z}{\partial \theta}+v_z\frac{\partial v_z}{\partial z}\right)=\mu\left(\frac{1}{r}\frac{\partial}{\partial r}\left(r\frac{\partial v_z}{\partial r}\right)+\frac{1}{r^2}\frac{\partial^2 v_z^2}{\partial \theta^2}+\frac{\partial^2 v_z}{\partial z^2}\right)-\frac{\partial P}{\partial z}+\rho g_z \quad (7.34)$$

The general energy balance that takes into consideration radial variations in a tubular reactor is:

$$\rho C_p\frac{\partial T}{\partial t}+\rho C_p\left(v_r\frac{\partial T}{\partial r}+v_\theta\frac{1}{r}\frac{\partial T}{\partial \theta}+v_z\frac{\partial T}{\partial z}\right)=k\left(\frac{1}{r}\frac{\partial}{\partial r}\left(r\frac{\partial T}{\partial r}\right)+\frac{1}{r^2}\frac{\partial^2 T}{\partial \theta^2}+\frac{\partial^2 T}{\partial z^2}\right)+Q \quad (7.35)$$

The first term on the left-hand side of Equation 7.35, represents the accumulation of energy; the second term represents the convective heat fluxes in the radial and axial directions. The first term on the right-hand side represents the conductive heat fluxes in the radial and axial directions. The last term represents the heat production through the heat of reaction. The difference in the energy flux over the volume element is either due to the accumulation of heat or the production or consumption of it. Finally, we expand the radial conduction term and change the axial convection term slightly. If ρC_p is constant, the energy balance becomes:

$$0 = \frac{k_e}{r}\frac{\partial T}{\partial r} + k_e\frac{\partial^2 T}{\partial r^2} + k_e\frac{\partial^2 T}{\partial z^2} - v_z\rho C_p\frac{\partial T}{\partial z} + Q \tag{7.36}$$

where:
C_p is the heat capacity of the mixture
T is the temperature
k is the thermal conductivity of the reaction mixture
Q is the heat of the reaction

$$Q = \Delta H_{Rx} r_A$$

where:
ΔH_{Rx} is the heat of reaction
$-r_A$ is the reaction rate based on the limiting species A

For example, the reaction rate for an elementary first-order reaction can be expressed as:

$$-r_A = kC_A$$

7.5.1.1 Boundary Conditions

The differential equations require boundary conditions and one initial value. For both the mass and energy balance, we know the temperature and the concentration at the inlet boundary, and these values are also the initial values. The boundary conditions for mass, temperature, and momentum can be expressed, respectively, as follows:

at $r = 0$, $\partial C_i/\partial r = 0$ and $\partial T/\partial r = 0$, $\partial v_z/\partial r = 0$, axial symmetry for all
at $r = R$, $\partial C_i/\partial r = 0$ and $k_e\partial T/\partial r = 0$, no flux, thermal insulation, wall (no slip)
at $z = 0$, then $C_i = C_{io}$ and $T = T_o$, $v = v_0$, inlet concentration, temperature, and velocity

Solving the mass and energy balance using the boundary and initial conditions stated above produces Figure 7.46, demonstrating the temperature profile in the reactor. The temperature surface plot displays a cross-section of the temperature profile. It clearly shows the temperature rise along the length of the reactor, presenting the effect of the exothermic reaction. The radial temperature profiles plot displays the radial profile at three different locations: at the inlet, at the outlet, and halfway through the reactor.

7.5.2 COMSOL Simulation

1. Start COMSOL Multiphysics 5.3a and click on the Model Wizard.
2. Select "Axisymmetric (2D)" from the Space dimension list.
3. From the "Select physics" list, select Chemical Species transport > Transport of diluted species.
4. Go to the dependent variables and in the edit field, type the name of the three concentration variables: C_A, C_B, C_C.
5. Select the "Heat Transfer in Fluids" option.
6. Select the "Laminar Flow."
7. Click on "Study."
8. Select "Stationary."
9. Right-click "Geometry" and select "Rectangle." Once it is selected, click on it and specify the following dimensions in meters: width = 0.1, length = 1, as shown in Figure 7.40.

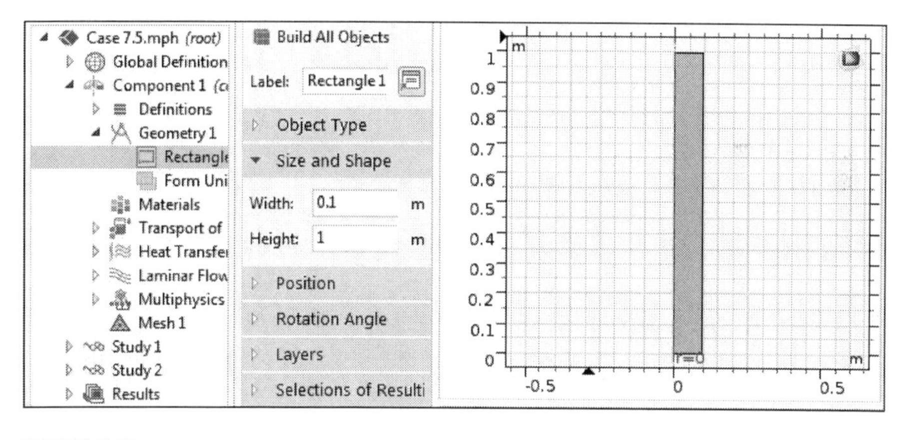

FIGURE 7.40
Reactor dimensions.

10. Enter the parameters shown in Figure 7.41.

Parameters

Name	Expression	Value	Description
E	75362[J/mol]	75362 J/mol	Activation energy
A	16.96e12[1/h]	4.7111E9 1/s	Frequency factor
ke	0.559[W/m/K]	0.559 W/(m·K)	Thermal conductivity
Diff	1e-9[m^2/s]	1E-9 m²/s	Diffusion coefficient
dHrx	-84666[J/mol]	-84666 J/mol	Heat of reaction
T0	312[K]	312 K	Inlet temperature
Mpo	58.095[g/mol]	0.058095 kg/...	Molar weight, propylene...
Mw	18[g/mol]	0.018 kg/mol	Molar weight, water
rhopop	830[kg/m^3]	830 kg/m³	Propylene oxide (po) Den...
rhowp	1000[kg/m^3]	1000 kg/m³	Water density,
rhomp	791.3[kg/m^3]	791.3 kg/m³	density, methanol
Cppo	146.54[J/mol/K]	146.54 J/(mol·K)	Propylene oxide specific...
Cpw	75.36[J/mol/K]	75.36 J/(mol·K)	Water specific heat, w
Cppg	192.59[J/mol/K]	192.59 J/(mol·K)	Propylene glycol specific...
Cpm	81.095[J/mol/K]	81.095 J/(mol·K)	specific heat, methanol
vratio	3.5	3.5	water / (prop+me)
npo0	0.1[mol/s]	0.1 mol/s	Molar flow rate po in
Ra	0.1[m]	0.1 m	radium of reactor
Mm	32.042[g/mol]	0.032042 kg/...	Mw methanol

FIGURE 7.41
Model parameters.

The variables are shown in Figure 7.42.

Variables

Name	Expression	Unit	Description
u0	v0/(pi*Ra^2)	m/s	inlet velocity
xA	(cA0-cA)/cA0		Conversion species A
rA	-A*exp(-E/R_const/T)*cA	mol/(m...	rate of reaction
Q	(-rA)*(-dHrx)	W/m³	Heat due to reaction
cpm	(Cppo*cA+Cpw*cB+Cp...	J/(kg·K)	Mixture specific heat
v0	vw0+vpo0+vm0	m³/s	Total inlet volumetric flo...
cA0	npo0/v0	mol/m³	intial conc. of A
cB0	nw0/v0	mol/m³	initial conc. of B
cMe0	nm0/v0	mol/m³	inlet conc. of methanol
Cp0	(Cppo*cA0+Cpw*cB0+C...	J/(kg·K)	specific heat of feed
rho0	(cA0*Mpo+cMe0*Mm+...	kg/m³	density of feed stream
vpo0	npo0*Mpo/rhopop	m³/s	vol. feed of A
vm0	vpo0	m³/s	vol. feed rate of methanol
vw0	vratio*(vpo0+vm0)	m³/s	inlet water feed rate
nw0	rhowp*vw0/Mw	mol/s	moles of water in feed
nm0	vpo0*rhomp/Mm	mol/s	moles of methanol in feed

FIGURE 7.42
Parameters used in the simulation.

The reaction rate is to be set as shown in Figure 7.43.

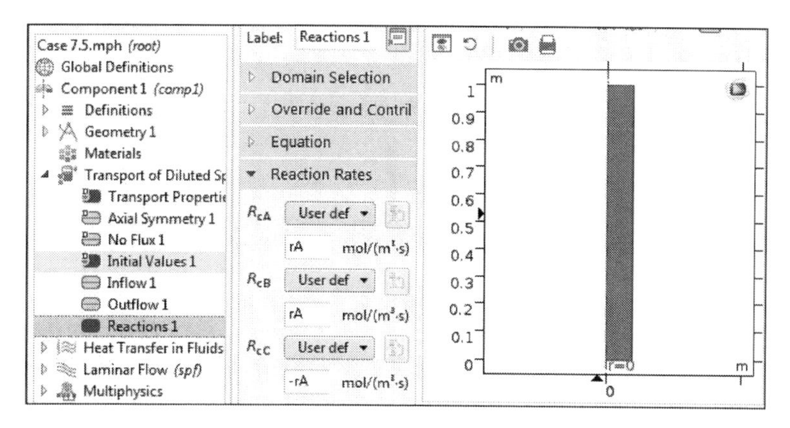

FIGURE 7.43
Reaction rates of each component.

The heat source generated from the exothermic reaction is shown in Figure 7.44. The surface conversion profile of A (propylene oxide) across the reactor is shown in Figure 7.45.

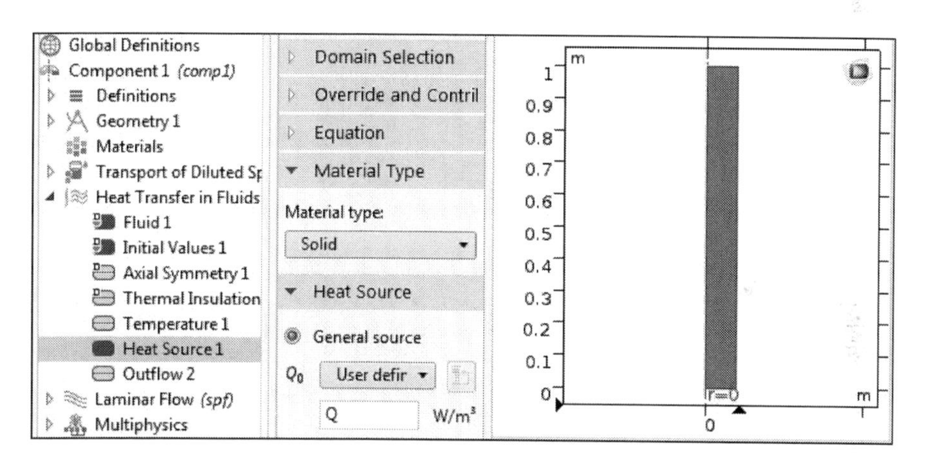

FIGURE 7.44
Setting of heat source.

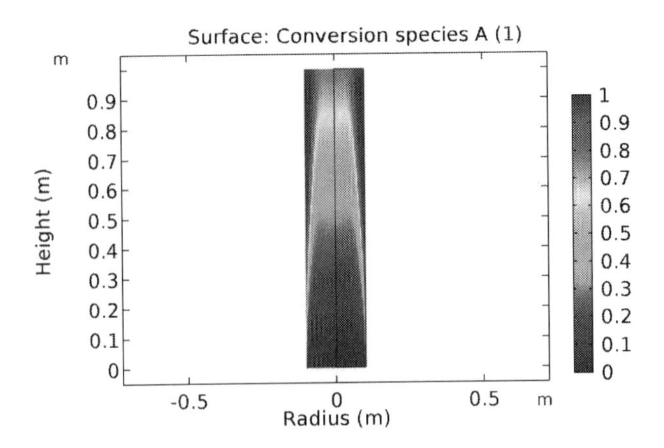

FIGURE 7.45
Surface plot of conversion.

The temperature profile is shown in Figure 7.46.

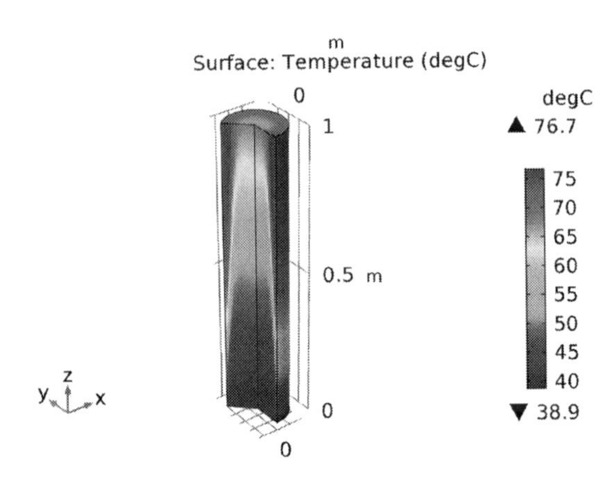

FIGURE 7.46
Temperature 3D surface plot.

The surface velocity profile distribution is shown in Figure 7.47. The length of the arrows is proportional to the magnitude of the velocity in the z-direction. The velocity of the fluid is not fully developed before the length of reactor of approximately 0.5 m. The velocity is maximum at the center of the pipe and zero near the borders, which agrees with the bed boundary conditions.

Conversion of propylene oxide along the length of the reactor is shown in Figure 7.48. The figure illustrates that the conversion increased along the length of reactor.

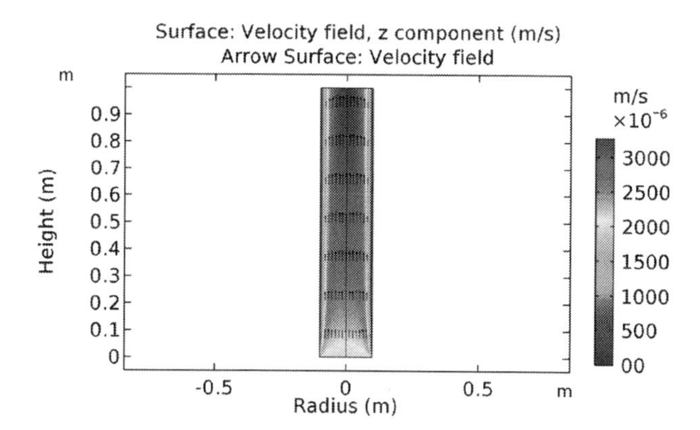

FIGURE 7.47
Velocity distribution inside the reactor.

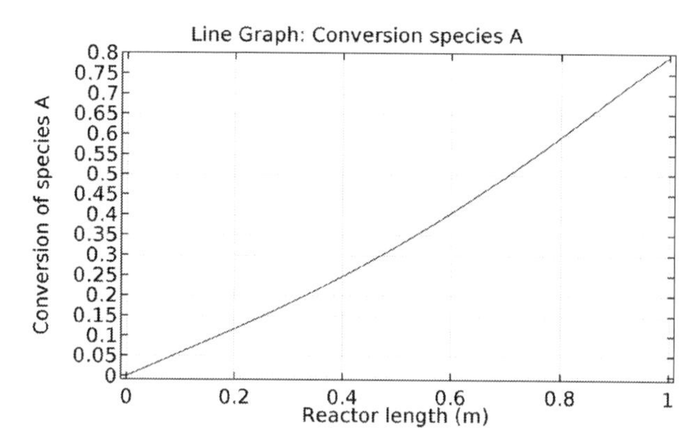

FIGURE 7.48
Propylene oxide conversion along the reactor length.

The final conversion is $x = 0.8$. Conversion of the reactor could be increased if the length of the reactor increased until it reaches equilibrium conversion.

7.6 Coupling of Fluid and Heat Transfer (Multiphysics)

An incompressible fluid is flowing upward between two parallel and vertical plates. The distance between the plates is 0.1 m; the plates are maintained at 80°C. In this example, we couple laminar fluid flow (continuity equation and

Navier-Stocks equation) and heat transfer (energy equation) physics through the heat transfer convective term, as follows:

1. Start COMSOL and then select 2D.
2. From the "Select physics" menu, select "Laminar flow" and "Heat transfer in fluids," as shown in Figure 7.49.

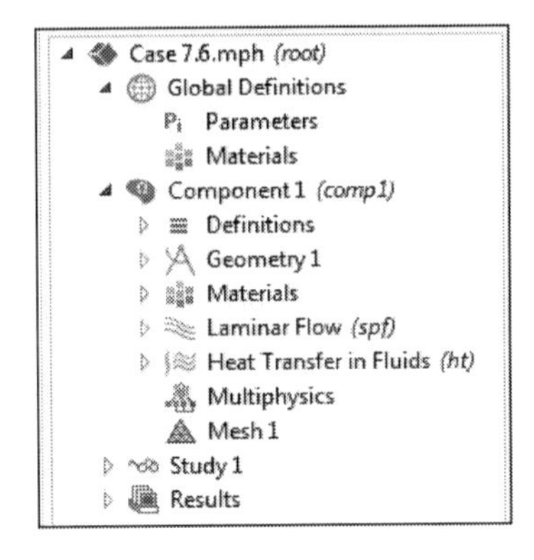

FIGURE 7.49
Selection of physics interfaces.

3. Click on "Study" on the bottom right corner, and select "stationary" and then the "Done" button at the bottom of the study page to go in the Model Builder page.
4. While in the Model Builder page, right-click on "Geometry" and select "Rectangle." Set 01 m for width and 1 m for height, then click on the "Build all Object" button (Figure 7.50).

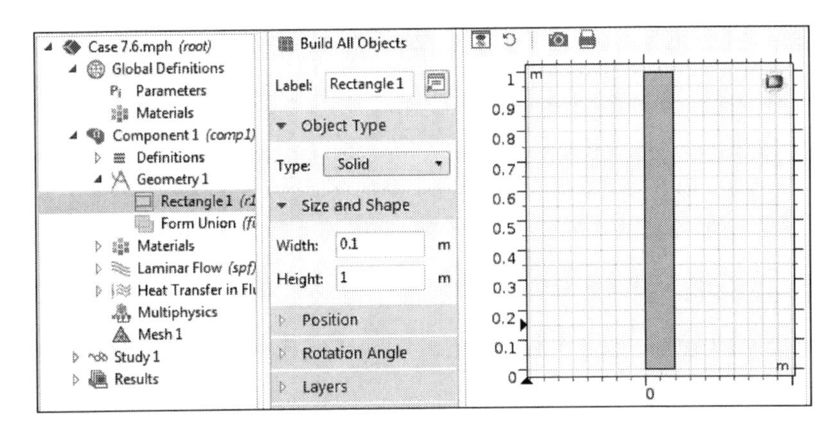

FIGURE 7.50
Dimensions of rectangle geometry.

5. The fluid flowing between the two plates is water. Accordingly, right-click on "Material," left-click on "Add material from library," and select "water" under liquids.
6. First we will complete the Laminar flow (spf) physics, the boundary conditions and the necessary data. Right-click on "Fluid flow" and select "inlet" and "outlet." Click on "Inlet" boundary condition, select boundary 2 the bottom line, and set the velocity to 0.0.01 m/s. For the outlet, click on the top line and leave the pressure value at $p_0 = 0$, which is the gauge pressure. Click on "Fluid property." For temperature, change "User defined" to Temperature (ht), so that the temperature will be calculated from the heat transfer. Your final laminar flow setting screen should look like Figure 7.51.

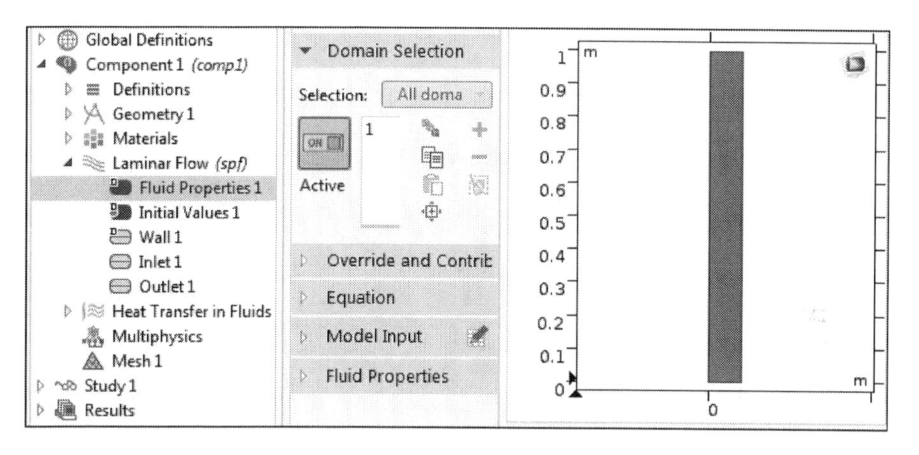

FIGURE 7.51
Laminar flow physics.

7. The second most important step while in the Model Builder is to complete the required information for the Heat transfer in fluids (ht) physics. In the energy balance equation, there is a convective term and there is a velocity term u associated with the convective term. This u is basically coming from laminar flow; this is the place where the two equations are coupled (the laminar fluid flow and heat transfer). In the laminar fluid flow, the velocity must be coupled with heat transfer equation. The coupling must come from the laminar fluid flow physics. Click on Fluid, then under the model input, change the Absolute pressure from User defined to Absolute pressure (spf). Now the coupling is done. The temperature of the walls is 80°C and the water is entering at 25°C. Two temperature boundary conditions are to be selected, and an Outlet boundary is also needed (Figure 7.52).

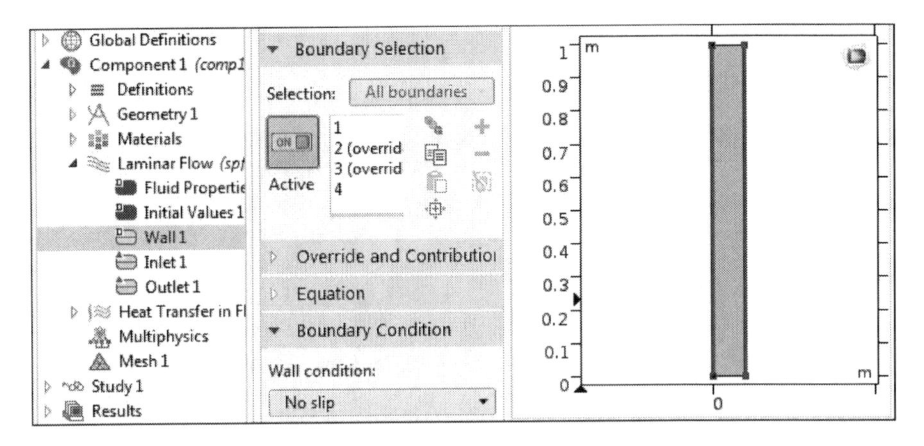

FIGURE 7.52
Properties and boundary conditions of the physics heat transfer in fluids.

8. The model is supposed to be ready for running; the default mesh by COMSOL is sufficient. Click on "Compute."
9. The resulting surface velocity profile is presented in Figure 7.53.

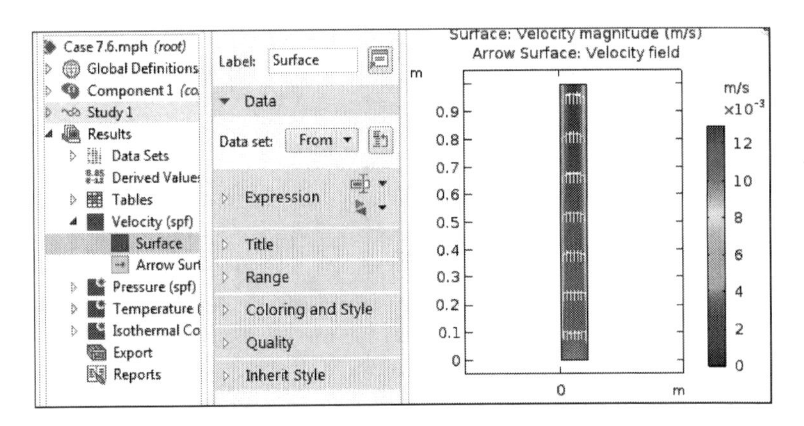

FIGURE 7.53
Velocity surface profile.

10. Under "Results," click on temperature. Under "Coloring and Style," change Color table from Thermal light to Rainbow because most students are used to relating high temperature to red and cold temperature to blue. The results of the surface temperature should look like Figure 7.54.

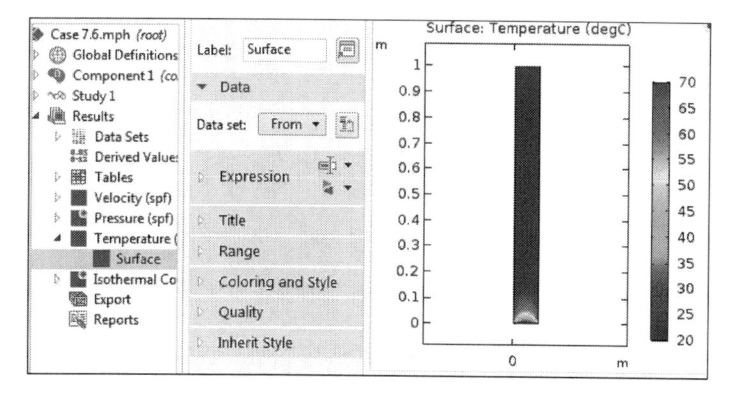

FIGURE 7.54
Surface temperature profile.

7.7 Unsteady Diffusion of Contaminated Source from the Skin of a Pipe Line

A horizontal pipe line has a length of 1 m and a diameter of 0.1 m, and pure water is flowing inside the pipe due to a pressure gradient of 10 Pa. The diffusion coefficient is $1 \times 10^{-3} m^2/s$. The contaminant is 0.05 m long and starts from the entrance of the pipe stuck to the pipe bottom wall.

Solution

This is a coupled problem where laminar flow and transport of diluted species modules are used. Use the following procedure and COMSOL to solve the problem.

Start COMSOL and select the two physics: laminar flow and transport of diluted species (Figure 7.55).

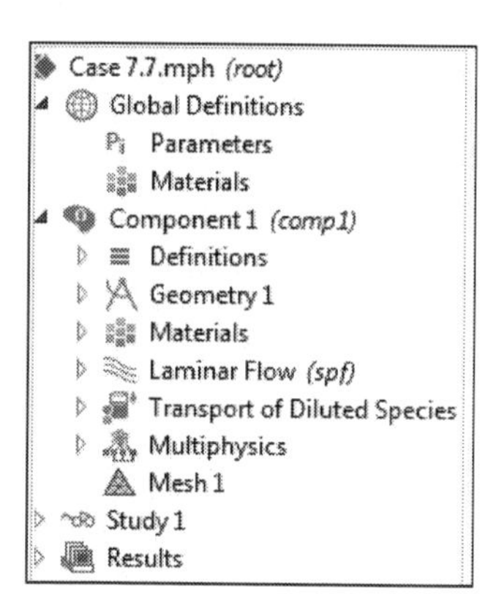

FIGURE 7.55
Physics used in the problem.

From "Study," select "Time dependent" and click the "Done" button. Right-click on "Geometry" and select "Rectangle." Specify 1 m for width and 0.1 m for height (the diameter of the pipe). Right-click again on "Geometry" and select "Point," as shown in Figure 7.56.

In the laminar flow model, go for the incompressible flow model. Right-click on the Transport of diluted species and select the two boundary conditions Inlet and Outlet. For inlet boundary conditions, select pressure instead of velocity and set the inlet pressure to 10 Pa gauge pressure and the outlet pressure to ambient pressure zero pressure gauge. Next go to the Transport of Diluted module. Click on Transport Properties, change the velocity, and use the one from the laminar flow section (coupling). Add Inflow and Outflow boundary conditions. Set the concentration of contaminant in the inlet water to zero concentration. Now add the source of the contaminant. Right-click on Transport of diluted species and select concentration. Click on the line boundary at the inlet of the pipe, as shown in Figure 7.57.

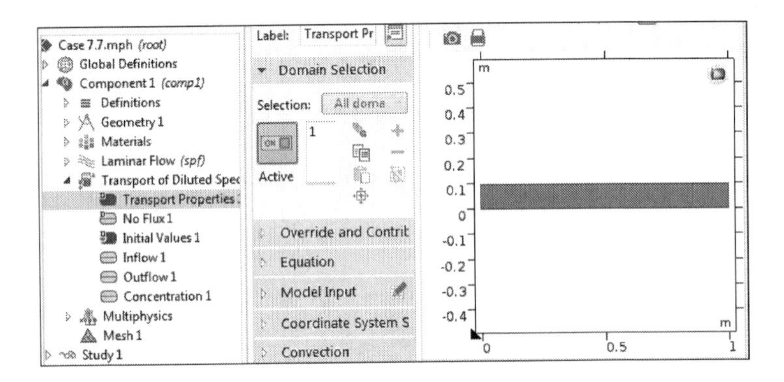

FIGURE 7.56
Setting of contaminant location inside the pipe.

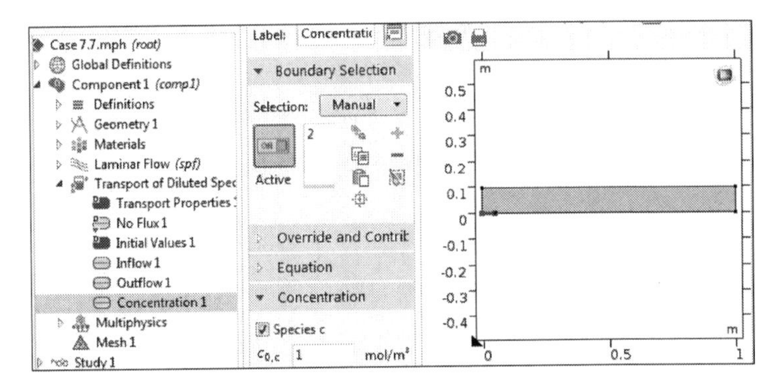

FIGURE 7.57
Pollutant source at the first bottom edge of the pipe.

Accomplish the Multiphysics coupling by right-clicking on Multiphysics and then left-clicking "Flow Coupling," as shown in Figure 7.58. This is important.

Click on "Mesh" and select the default mesh or, for fast computation, select courser mesh. Then click on "Time dependent" and change the time to 100 seconds, as shown in Figure 7.59.

The surface plot of the concentration profile after 100 seconds is shown in Figure 7.60.

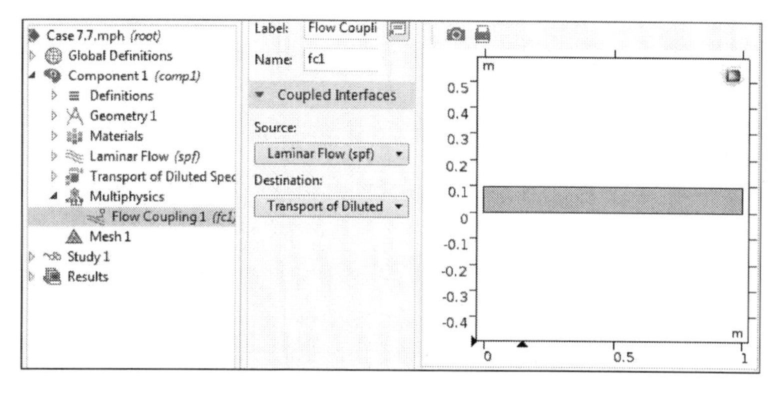

FIGURE 7.58
Coupling of laminar flow and transport of dilute species physics.

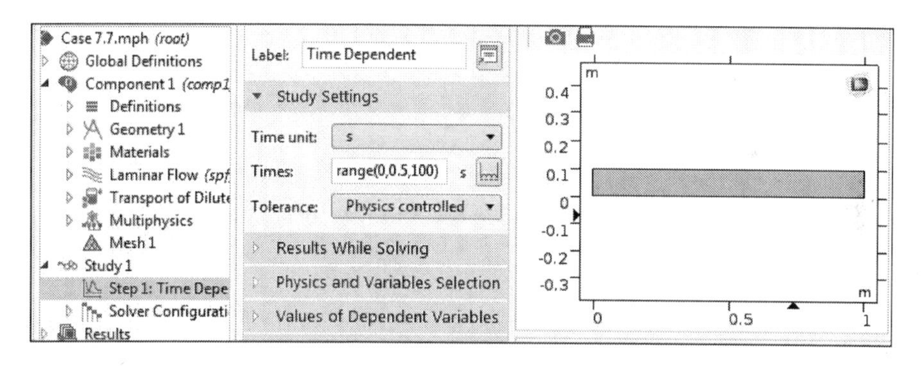

FIGURE 7.59
Setting of time-dependent limit and step size.

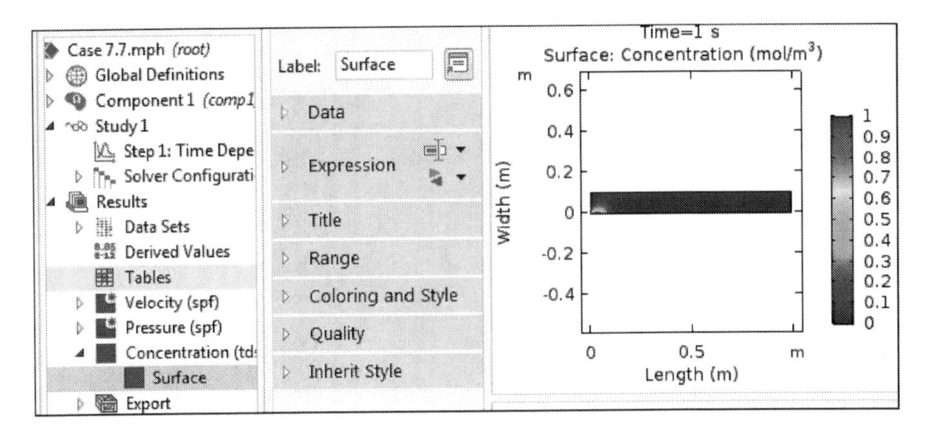

FIGURE 7.60
Surface concentrating profile versus pipe dimensions.

7.8 Maxwell-Stefan Diffusion

The diffusion coefficient cannot be treated as a constant or as composition-independent when there is diffusion of multicomponents in gas mixtures or when more than one chemical species is present in substantial mass fractions in concentrated solutions. Consequently, the diffusion coefficient becomes a tensor, and the equation for diffusion is altered to relate the mass flux of one chemical species to the concentration gradients of all chemical species present. In this case, the Maxwell-Stefan description of diffusion is considered; it is often applied to describe gas mixtures. The dependent variables in this case are the mole or mass fractions (x_i and w_i, respectively), not the concentration [8]. The mole fractions are related to the concentrations:

$$x_i = \frac{c_i}{\sum c_j}$$

The mass fractions are related to concentrations:

$$w_i = \frac{M_i c_i}{\sum M_j c_j}$$

Mole fractions are related to mass fractions:

$$x_i = \frac{w_i}{M_i \sum w_i / M_j}$$

where M_i is the relative molar mass ($kg\,mol^{-1}$) of species i.

The resulting mass balance included in the chemical engineering module yields the following expression:

$$\frac{\partial}{\partial t}\rho w_i + \nabla.\left[-\rho w_i \sum_{j=1}^{N} D_{ij}\left\{\frac{M}{M_j}\left(\nabla w_j + w_j \frac{\nabla M}{M}\right) + \left(x_j - w_j\right)\frac{\nabla p}{p}\right\} + w_i \rho u + D_i^T \frac{\nabla T}{T}\right] = R_i \tag{7.37}$$

where M denotes the total mol mass of the mixture, M_j the molar mass (molecular weight) of species j, x_j is the mole fraction of j, and R_j is the reaction rate of species j.

M and x_j can also be expressed in term of the mass fractions w_j. This implies that the only dependent variable in the application is the mass fraction w, while the temperature field T, pressure field p, thermal diffusivity D^T, and velocity u, are obtained in combination with the energy, momentum, and continuity equations. The application combines the mass balances for each species except one, which is eliminated through: $\sum w_i = 1$.

The average molecular weight of the mixture (M_{mix}) is defined from:

$$M_{mix} = \left(\sum \frac{w_i}{M_i}\right)^{-1}$$

The mixture gas density (ρ) is:

$$\rho = \frac{p.M_{mix}}{R\ .T}$$

The mole fraction of component i (x_i) in the mixture is:

$$x_i = \frac{w_i M_{mix}}{M_i}$$

The partial pressure of component i (p_i) in the gas mixture is:

$$p_i = x_i P$$

7.8.1 Hydrogen Production

The steam reforming of isooctane is used to produce hydrogen in a monolithic reactor of square channels ($0.001 \times 0.001 \times 0.02$ m). The walls of the channel are coated with a thin layer of catalyst. The feed velocity to the reactor at

0.1 m/s consists of 40 wt.% isooctane, 50 wt.% steam, and 10 wt.% nitrogen. The total mass of catalyst inside the reactor is 0.01 kg. The reactor is maintained at 350°C and at a pressure of 100 kPa. The stream reforming of isooctane is described by the following reaction:

$$C_8H_{18} + 8H_2O \rightarrow 8CO + 17H_2$$

The species-independent reaction rate of steam reforming over a N_i/Al_2O_3 catalyst is:

$$r = \frac{k\,p_{C_8H_{18}}\;p_{H_2}}{\left(1 + K_a\,p_{C_8H_{18}} + K_b\,p_{H_2O}\right)^2} \tag{7.38}$$

The reaction rate constant is $k = 2.9\times10^{-12}\,\text{mol/kg}\cdot\text{cat}\cdot\text{s}\cdot\text{Pa}^2$. The equilibrium constant is $k_a = 2.25\times10^{-3}\,\text{Pa}^{-1}$, $K_b = 4.60\times10^{-5}\,\text{Pa}^{-1}$.

Solution

COMSOL Multiphysics 3.5a is used for the solution of this case study. Open a new case in COMSOL, click on the Model Wizard, and select a 2D coordinate system. Under Fluid flow, select Laminar flow and select Transport of concentrated species under Chemical species transport. After selecting Multiphysics, click "next," click on "stationary" (steady-state operation), and then click on "Done" to enter the Model Builder menu. Right-click on "Geometry," select "Rectangle," and add the width at 0.02 m and the height at 0.001 m. While in the Model Builder, click on "Parameters" and add the constants shown in Figure 7.61.

Under Component 1, right-click on "Variables," select variables and enter the variables shown in Figure 7.62. The average molecular weight of the mixture (M_{mix}) is:

$$M_{mix} = \left(\frac{w_{C_8H_{18}}}{M_{C_8H_{18}}} + \frac{w_{H_2O}}{M_{H_2O}} + \frac{w_{CO}}{M_{CO}} + \frac{w_{H_2}}{M_{H_2}} + \frac{w_{N_2}}{M_{N_2}}\right)^{-1} \tag{7.39}$$

▾ Parameters

Name	Expression	Value	Description
T	623 [K]	623 K	Temperature (K)
P	100000	1E5	Pressure (Pa)
R	8.314 [J/mol*K]	8.314 kg·m²·K...	Gas constant
mu	3e-5	3E-5	Gas viscosity (kg/m.s)
Mc8h18	114e-3	0.114	Mw (mol/kg)
Mh2o	18e-3	0.018	Mw (mol/kg)
Mco	28e-3	0.028	Mw (mol/kg)
Mh2	2e-3	0.002	Mw (mol/kg)
Mn2	28e-3	0.028	Mw (mol/kg)
u0	0.1	0.1	Inlet velocity (m/s)
krxn	2.9e-12	2.9E-12	Rate cons. (mol/kgcat.s.P
Ka	2.25e-3	0.00225	Eqm. constant
Kb	4.6e-5	4.6E-5	Eqm. constant
wc8h18i	0.4	0.4	mass fraction
wh2oi	0.5	0.5	mass fraction
wcoi	0	0	mass fraction
wh2i	0	0	mass fraction
wn2i	0.1	0.1	mass fraction
mc	0.01	0.01	kg catalyst
Area	0.02*0.001	2E-5	m^2

FIGURE 7.61
Constants used in the model.

▾ Variables

Name	Expression	Unit	Description
Mmix	(wc8h18/Mc8h18+w...		Average Mw
rho	(p*Mmix/R/T) [kg/m...	kg·mol/...	Gas density
xc8h18	wc8h18*Mmix/Mc8h18		mole fraction
xh2o	wh2o*Mmix/Mh2o		mole fraction
xco	wco*Mmix/Mco		mole fraction
xh2	wh2*Mmix/Mh2		mole fraction
xn2	wn2*Mmix/Mn2		mole fraction
pc8h18	xc8h18*P		Partial pressure
ph2o	xh2o*P		Partial pressure
pco	xco*P		Partial pressure
ph2	xh2*P		Partial pressure
pn2	xn2*P		Partial pressure
r	krxn*(pc8h18*ph2o)/(...		Reaction rate

FIGURE 7.62
Variable used in the solutions.

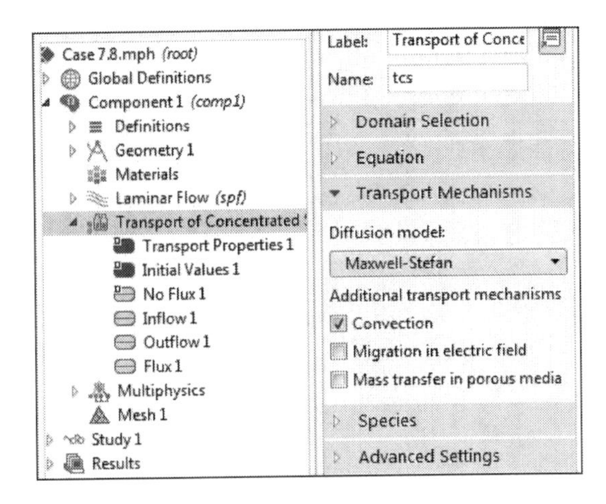

FIGURE 7.63
The selection of Maxwell-Stefan diffusion model.

While in the Model Builder, click on Transport of concentrated species and select Maxwell-Stefan from the pulldown menu under diffusion model (Figure 7.63).

The binary diffusivities (m^2/s) is shown in Table 7.1.

Right-click on the physics Fluid flow and select inlet and outlet as inlet and exit boundary conditions, respectively. The inlet fluid velocity boundary conditions are shown in Figure 7.64.

The steam reforming of isooctane is described by the following reaction:

$$C_8H_{18} + 8H_2O \rightarrow 8CO + 17H_2$$

The rate of reaction given in this case study is species independent:

$$r = \frac{k p_{C_8H_{18}} \; p_{H_2}}{\left(1 + K_a p_{C_8H_{18}} + K_b p_{H_2O}\right)^2} \tag{7.40}$$

TABLE 7.1
Binary Diffusivities of Gas Mixture

$D_{C_8H_{18}-H_2O} = 2.8 \times 10^{-5}$	$D_{H_2O-H_2} = 1.8 \times 10^{-4}$
$D_{C_8H_{18}-CO} = 2.1 \times 10^{-5}$	$D_{H_2O-N_2} = 6.6 \times 10^{-5}$
$D_{C_8H_{18}-H_2} = 7.8 \times 10^{-5}$	$D_{CO-H_2} = 1.6 \times 10^{-4}$
$D_{C_8H_{18}-N_2} = 2. \times 10^{-5}$	$D_{CO-N_2} = 5.1 \times 10^{-5}$
$D_{H_2O-CO} = 6.4 \times 10^{-5}$	$D_{H_2-N_2} = 1.6 \times 10^{-5}$

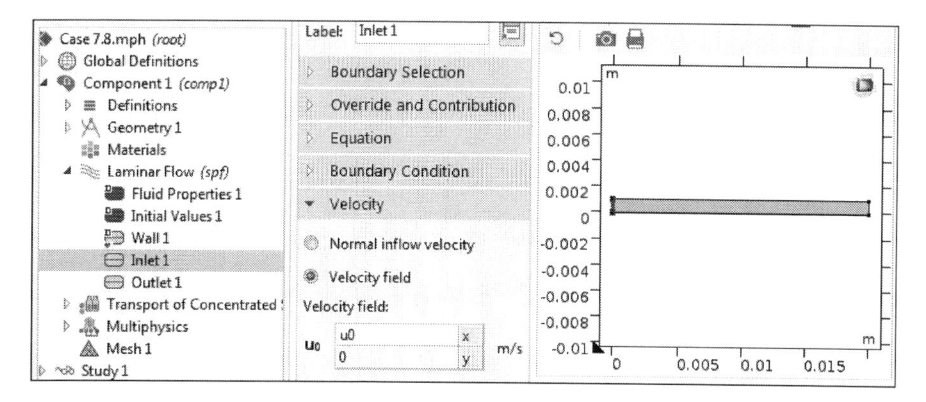

FIGURE 7.64
Inlet boundary condition for laminar flow physics.

Accordingly:

$$r = -\frac{r_{C_8H_{18}}}{1} = -\frac{r_{H_2O}}{8} = \frac{r_{CO}}{8} = \frac{r_{H_2}}{17}$$

The reaction takes place on the walls of the channel in the monolith; the walls are coated with a thin layer of catalyst. For simplicity, in the 2D model, the catalyst layer thickness is considered negligible. The bottom boundary condition represents the catalyst-gas interface and can be defined by the flux of each component due to chemical reaction, as shown in Figure 7.65.

The hydrogen mass fraction is shown in Figure 7.66. The diagram reveals that the mass fraction of hydrogen is generated along the length of the reactor.

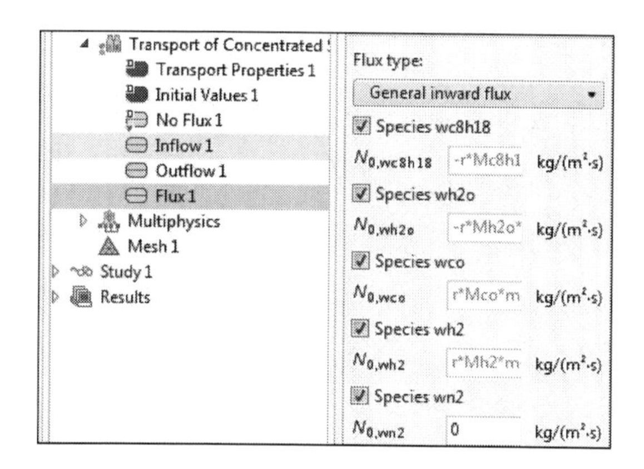

FIGURE 7.65
The boundary conditions where reaction is taking place.

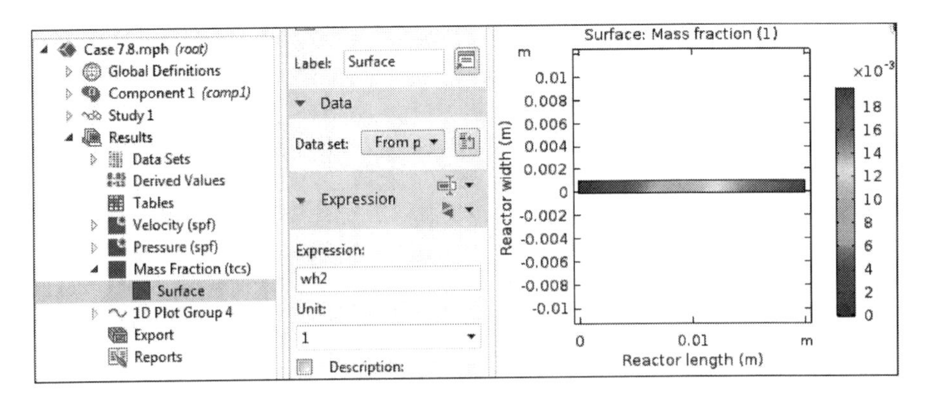

FIGURE 7.66
Hydrogen surface concentration.

References

1. Uemiya, S., N. Sato, H. Ando, E. Kikuchi, 1991. The water gas shift reaction assisted by a palladium membrane reactor. *Industrial & Engineering Chemistry Research* 30: 585.
2. Folger, S., 2006. *Elements of Chemical Reaction Engineering*, 4th ed., Upper Saddle River, NJ: Prentice Hall.
3. Versteeg, G. F., W. P. M. Van Swaaij, 1988. On the kinetics between CO_2 and alkanol amines both in aqueous and non-aqueous solutions. I. Primary and secondary amines. *Chemical Engineering Science* 43: 573–585.
4. Barth, D., C. Tondre, J. Delpuech, 1986. Stopped-flow investigation of the reaction kinetics of carbon dioxide with some primary and secondary alkanol amines in aqueous solutions. *International Journal of Chemical Kinetics* 18: 445–457.
5. Zanfir, M., A. Gavriilidis, C. Wille, V. Hessel, 2005. Carbon dioxide absorption in a falling film microstructure reactor: Experimental and modelling. *Indistrial and Engineering Chemistry Research* 44: 1742–1751.
6. Vielstich, W., A. Lamm, H. T. A. Gasteiger, 2003. *Handbook of Fuel Cells: Fundamentals, Technology, and Applications*, 4th Vol., New York: John Wiley & Sons.
7. Fogler, S., 2005. *Elements of Chemical Reaction Engineering*, 4th ed., Upper Saddle River, NJ: Prentice Hall.
8. Taylor, R., R. Krishna, 1993. *Multicomponent Mass Transfer*, New York: John Wiley & Sons.

8

Computing Solutions of Ordinary Differential Equations

Numerical methods for ordinary differential equations (ODEs) are methods used to find numerical approximations to the solutions of ODEs. Their use is also known as numerical integration. Many differential equations are generated because of engineering problems analysis, most of which cannot be easily solved. These equations can be solved numerically using a computer. There is a close relationship between solving initial value problems and a finite integral using an approximation method; this inspires the development of the techniques outlined in this chapter. Several differential equations cannot be solved using symbolic computation. However, a numeric approximation to the solution is often sufficient for practical purposes. The problems that contain first-order ODEs are solved by employing methods of Euler, modified Euler, midpoint, Heun, and Runge-Kutta second, third, and fourth order.

LEARNING OBJECTIVES

- Solve first-order ODEs using the methods of Euler, modified Euler, midpoint, Heun, and Runge-Kutta.
- Solve coupled first-order ODEs.
- Use MATLAB® to solve various ODEs via the numerical solution methods.

8.1 Introduction

There are many types of differential equations, as well as a wide variety of solution techniques, even for equations of the same type. This chapter introduces some terminology that aids in the classification of equations and in the selection of their solution techniques. While differential equations have three basic types, ordinary differential equations (ODEs), partial differential equations (PDEs), and differential-algebraic equations (DAEs), they can be further described by attributes such as order, linearity, and degree. An ODE is an equation that depends on one or more derivatives of functions of a single variable. A PDE is an equation that depends on one or more partial derivatives of functions of several variables. In many cases, PDEs are solved by reducing to multiple ODEs [1, 2]. One example is the following heat transfer equation:

$$\frac{\partial T}{\partial t} = k\frac{\partial^2 T}{\partial x^2} \tag{8.1}$$

where k is a constant. Equation 8.1 is an example of a PDE, as its solution, $T(x, t)$ is a function of two independent variables (x, t), and the equation includes partial derivatives with respect to both variables. The order of a differential equation can be classified as the order of the highest derivative of any unknown function in the equation. The following is an example of a first-order differential equation:

$$\frac{dy}{dt} = ay - b \tag{8.2}$$

where a and b are constants. The equation is a first-order differential equation, as only the first derivative dy/dt appears in the equation. The following equation, on the other hand, is a second-order differential equation:

$$\frac{d^2y}{dx^2} + \frac{3dy}{dx} + 2y = 0 \tag{8.3}$$

In a differential equation, when the variables and their derivatives are only multiplied by constants, the equation is linear. The variables and their derivatives must always appear as a simple first power. An example of nonlinear differential equation is equation (8.4).

$$\frac{d^3y}{dx^3} + \left(\frac{dy}{dx}\right)^2 = kx \tag{8.4}$$

"Linear equation" means that the variable in an equation appears only with a power of one. So x is linear, but x^2 is nonlinear. Any function like $\cos(x)$ is nonlinear. Nonlinear equations are usually very difficult to solve, so in many cases a linear equation is used to approximate a nonlinear equation to solve the problem. This approximation is called a linearization. The simplest form of ODE is:

$$y' = \frac{dy}{dx} = f(x) \tag{8.5}$$

where the right-hand side is a function of x only. The problem can be easily solved by a direct method:

$$y(x) = \int_a^x f(x)dx + C \tag{8.6}$$

Constant C is evaluated from some initial condition, $y(a)$. For example, the solution of the linear first-order ordinary deferential equation:

$$\frac{dy}{dt} = ay, \ y(0) = y_0 \tag{8.7}$$

where a is a constant, has an analytical solution:

$$y = y_0 e^{at} \tag{8.8}$$

8.2 Numerical Solution of Single Ordinary Equation

The following is a first-order ODE:

$$\frac{dy}{dx} = f(x,y) \tag{8.9}$$

The initial condition can be taken as:

$$y(x_0) = y_o \tag{8.10}$$

Then we could use a Taylor series about $x = x_0$ and obtain the complete solution:

$$y(x) = y(x_0) + y'(x_0)(x - x_0) + \frac{1}{2!}y''(x_0)(x - x_0)^2 + \ldots + \frac{1}{n!}y^n(x_0)(x - x_0)^n \tag{8.11}$$

The most general first-order differential equation can be written as:

$$\frac{dy}{dt} = f(y,t) \tag{8.12}$$

We shall now consider systems of simultaneous linear differential equations which contain a single independent variable and two or more dependent variables. In general, the number of equations will be equal to the number of dependent variables; that is, if there are n dependent variables, there will be n equations.

8.2.1 Euler Method

The Euler method is a practical numerical method. The Euler method is a first-order numerical procedure for solving ODEs with a given initial value. Divide the region of interest $[a,b]$ into discrete values of $x = nh$, $n = 0,1....N$, spaced at intervals $h = (b-a)/N$. Use the forward difference approximation for the differential coefficient:

$$f(x_n, y_n) \approx \frac{(y_{n+1} - y_n)}{h} \tag{8.13}$$

Rearranging the equation will give the following equation:

$$y_{n+1} \approx y_n + hf(x_n, y_n) \tag{8.14}$$

The MATLAB function that describes the Euler method is shown in Figure 8.1.

The numerical solution of the ODE $dy/dt = 0.06y$ is as follows. First, write the equation (F), then enter the values of all the parameters required by the function, as shown in Figure 8.2.

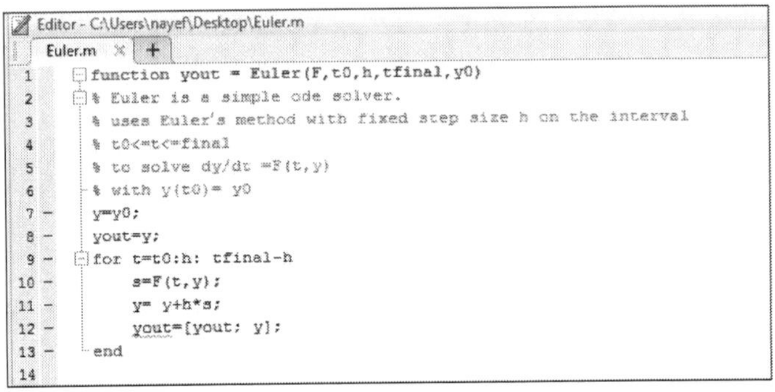

```
Editor - C:\Users\nayef\Desktop\Euler.m
Euler.m
1    function yout = Euler(F,t0,h,tfinal,y0)
2    % Euler is a simple ode solver.
3    % uses Euler's method with fixed step size h on the interval
4    % t0<=t<=final
5    % to solve dy/dt =F(t,y)
6    % with y(t0)= y0
7 -  y=y0;
8 -  yout=y;
9 -  for t=t0:h: tfinal-h
10 -      s=F(t,y);
11 -      y= y+h*s;
12 -      yout=[yout; y];
13 -  end
14
```

FIGURE 8.1
MATLAB function of the Euler method.

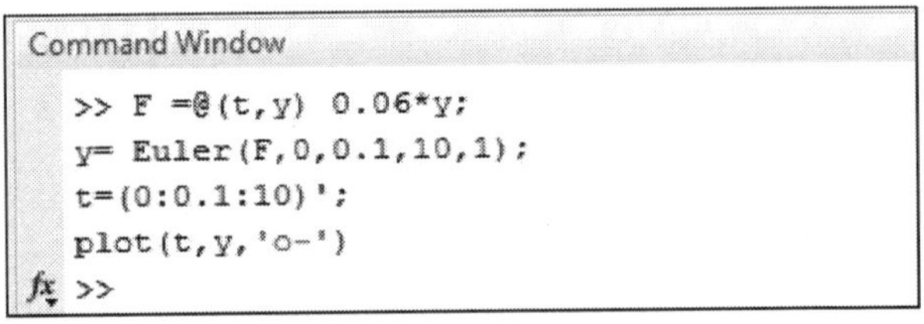

```
Command Window
>> F =@(t,y) 0.06*y;
y= Euler(F,0,0.1,10,1);
t=(0:0.1:10)';
plot(t,y,'o-')
fx >>
```

FIGURE 8.2
Method of solution via the Euler method.

Example 8.1: Euler Method

Solve the following first-order ODE using the Euler method:

$$\frac{dy}{dt} = 2y,\ y(0) = 10, \quad 0 \le t \le 3$$

Solution

To solve the ODE using the Euler method, just type the steps in the MATLAB command windows as shown in Figure 8.3. The plot of time versus y is shown in Figure 8.4.

```
Command Window
>> y=Euler(@(t,y) 2*y, 0, 1, 3,10)

y =

    10
    30
    90
   270

fx >>
```

FIGURE 8.3
Numerical solution using the Euler method.

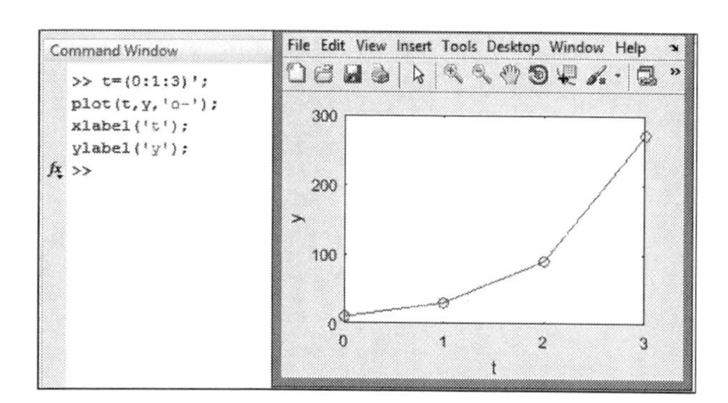

FIGURE 8.4
Numerical solution using the Euler method of the ODE: $dy/dt = 2y,\ y(0) = 1.$

Example 8.2: Euler Method

Solve the following first-order differential equation using the Euler method for $x = [0 2]$ with initial condition $y(0) = 0$. The first-order differential equation is:

$$\frac{dy}{dx} = y + x + 1$$

Compare the solution obtained from employing the Euler method and the exact analytical solution ($y = 2e^x - x - 2$) with $h = 1, 0.2$:

$$f(x,y) = y + x + 1$$

$$y_{n+1} = y_n + hf(x,y)$$

Solution

The numerical solutions via the Euler method are summarized in Tables 8.1 and 8.2 for step sizes $h = 1$ and $h = 0.2$, respectively.

TABLE 8.1

Solution for Step Size $h = 1$

n	x_n	y_n	**True, y_n**
0	0	0	0
1	1	1	2.437
2	2	4	10.778

TABLE 8.2

Solution for Step Size $h = 0.2$

n	x_n	y_n	**True, y_n**
0	0	0.00	0.00
1	0.2	0.20	0.24
2	0.4	0.48	0.58
3	0.6	0.86	1.04
4	0.8	1.35	1.65
5	1.0	1.98	2.44
6	1.2	2.77	3.44
7	1.4	3.77	4.71
8	1.6	5.00	6.31
9	1.8	6.52	8.30
10	2.0	8.38	10.78

=F2+0.2*(F2+D2+1)

D	E	F
x	y,analytical	Y, Euler
0	0.00	0.00
0.2	0.24	0.20
0.4	0.58	0.48
0.6	1.04	0.86
0.8	1.65	1.35
1	2.44	1.98
1.2	3.44	2.77
1.4	4.71	3.77
1.6	6.31	5.00
1.8	8.30	6.52
2	10.78	8.38

FIGURE 8.5
Excel screenshot of solution of the ODE: $dy/dx = x + y + 1$ using the Euler method.

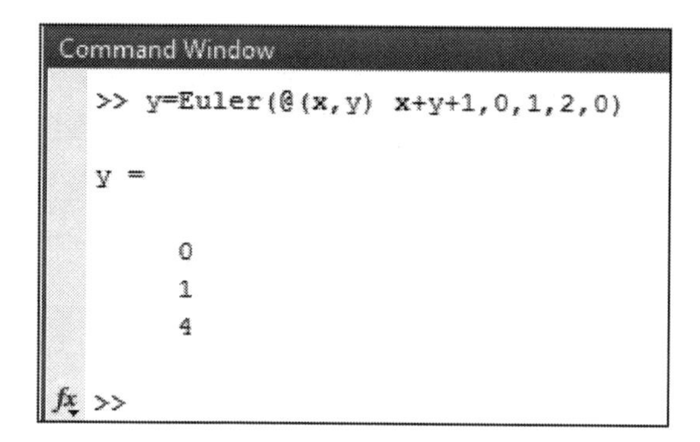

FIGURE 8.6
Numerical solution using the Euler method and MATLAB function.

The Euler solution rapidly diverges from the true solution. The smaller the step size, the less discrepancy between the analytical and numerical solution (Figure 8.5).

The MATLAB solution of the first-order ODE, $dy/dx = x + y + 1$, obtained for $h = 1$ and time from 0 to 2, is shown in Figure 8.6.

Example 8.3: Euler Method

Use the Euler method and $h = 0.02$ to solve the following ODE and plot the numerical and exact analytical solution:

$$\frac{dy}{dx} = x + y, \qquad y(0) = 1$$

Solution

The Euler method is:

$$y_{(i+1)} = y_i + hf\left(x_i, y_i\right),\ y_0 = y\left(x_0\right)$$

Use the Euler method and $h = 0.02$:

$$y_0(0) = 1,\ x_0 = 0 \text{ and } x_1 = 0.02$$

Note that $f(x_i, y_i) = x_i + y_i$; then:

$$y_{(i+1)} = y_i + hf\left(x_i, y_i\right) => y_1 = y_0 + hf(0,1) = 1 + 0.02(0+1) = 1.02$$

The rest of the values can be calculated using Microsoft Excel. The exact solution can easily be obtained employing the online Wolfram Alpha solver.

Solve $y' = x + y,\ y(0) = 1$

The analytical solution of the differential equation is:

$$y(x) = -x + 2e^x - 1$$

The numerical and analytical (Exact) solution using the Euler method is summarized in Table 8.3.

The relative error is calculated as follows:

$$\Delta y_{\text{Euler}} = y_5 - y_0 = 1.13 - 1.00 = 0.13$$

$$\Delta y_{\text{Exact}} = y_5 - y_0 = 1.11 - 1.00 = 0.11$$

$$\text{Relative Error} = \frac{\left|\Delta y_{\text{Euler}} - \Delta y_{\text{Exact}}\right|}{\frac{1}{2}\left(\Delta y_{\text{Euler}} + \Delta y_{\text{Exact}}\right)} = \frac{0.02}{0.12} = 0.17$$

Comparison of the Euler method and the exact analytical solution is shown in Figure 8.7.

TABLE 8.3

Exact Solution and Numerical Solution Using the Euler Method ($h = 0.02$)

i	x_i	y_{Euler}	y_{exact}
0	0.00	1.00	1.00
1	0.02	1.04	1.02
2	0.04	1.06	1.04
3	0.06	1.08	1.06
4	0.08	1.11	1.09
5	0.10	1.13	1.11

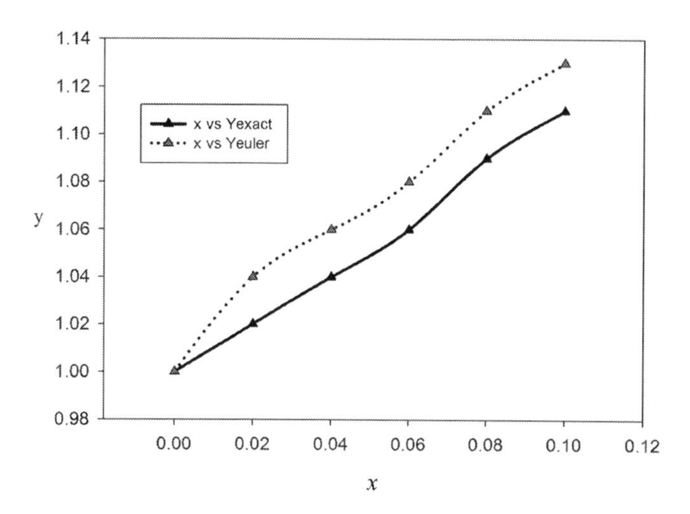

FIGURE 8.7
Comparison between the Euler method and exact solution.

8.2.2 Modified Euler Method

The Euler forward structure required a very small step size for a reasonable result. The Euler method might be very simple to implement, but it cannot give precise solutions. An upgrade over Euler is to take the arithmetic average of the slopes at x_i and x_{i+1}. The scheme so obtained is called modified or improved Euler method. It works first by relating a value to y_{i+1} and then refining it by making use of average slope. Recall the Euler governing equation:

$$y_{n+1} \approx y_n + hf(x_n, y_n) \tag{8.15}$$

A better estimation of the slope from (x_n, y_n) to (x_{n+1}, y_{n+1}) would be:

$$y_{n+1} \approx y_n + \frac{h}{2}\left(f(x_n, y_n) + f(x_{n+1}, y_{n+1})\right) \tag{8.16}$$

The value of y_{n+1} is not known; however, we can estimate it by using the Euler method to give a two-stage predictor-corrector algorithm that is called the Modified Euler algorithm:

Step 1. Predictor

$$y^*_{n+1} = y_n + hf(x_n, y_n) \tag{8.17}$$

Step 2. Corrector

$$y^*_{n+1} = y_n + \frac{h}{2}\left(f(x_n, y_n) + f(x_n, y^*_n)\right) \tag{8.18}$$

```
Euler.m  MEuler.m  +
1    function yout = MEuler(F,t0,h,tfinal,y0)
2    % The Modified Euler method for ODE is a simple ode solver.
3    % uses MEuler method with fixed step size h on the interval
4    % t0<=t<=final
5    % to solve dy/dt =F(t,y)
6    % with y(t0)= y0
7 -  y=y0;
8 -  yout=y;
9 -  for t=t0:h: tfinal-h
10 -     s1=F(t,y);
11 -     s2=F(t+h,y+h*s1);
12 -     y= y+(h/2)*(s1+s2);
13 -     yout=[yout;y];
14 -  end
```

FIGURE 8.8
MATLAB function of the modified Euler method.

Both can be combined in one equation:

$$y_{i+1} = y_i + \frac{h}{2}\left[f\left(x_i, y_i\right) + f\left(x_i + h, y_i + hf\left(x_i,\ y_i\right)\right)\right] \tag{8.19}$$

Figure 8.8 is the MATLAB code for the function created for the modified Euler method.

Example 8.4: Modified Euler Method

Find $y(1)$ for the first-order differential equation: $y' = 3x + y$, when $y(0) = 5$, $h = 0.5$ using modified Euler method.

Solution

Modified Euler method states that:

$$y_{n+1} = y_n + \frac{h}{2}(f(x_n, y_n) + f(x_{n+1}, y^*_{n+1})), \text{ where}$$

$$y^*_{n+1} = y_n + hf(x_n, y_n) \text{ and } x_{n+1} = x_n + h$$

We have: $h = 0.5,\ x_o = 0,\ y_o = 5$

The right-hand side of the equation is:

$$f(x, y) = 3x + y$$

Step 1:

$$x_1 = x_o + h = 0 + 0.5 = 0.5$$

$$y^*_1 = y_0 + hf\left(x_0, y_0\right) = 5 + hf(0,5) = 5 + 0.5(5) = 7.5$$

$$y_1 = y_o + \frac{h}{2}\left(f\left(x_o, y_o\right) + f\left(x_1, y^*_1\right)\right) = 5 + \frac{h}{2}(f(0,5) + f(0.5,7.5))$$

$$y_1 = 5 + \frac{0.5}{2}(5 + 9) = 8.5$$

```
Command Window
>> y=MEuler(@(x,y) 3*x+y,0,0.5,1,5)

y =

    5.0000
    8.5000
   15.1250

fx >>
```

FIGURE 8.9
Solution of Example 8.4 using the modified Euler method.

Step 2:

$$x_2 = x_1 + h = 0.5 + 0.5 = 1$$

$$y_2^* = y_1 + h.f(x_1, y_1) = 8.5 + h.f(0.5, 8.5) = 8.5 + 0.5(10.0) = 13.5$$

$$y_2 = y_1 + \frac{h}{2}\left(f(x_1, y_1) + f(x_2, y_2^*)\right) = 8.5 + \frac{h}{2}(f(0.5, 8.5) + f(1, 13.5))$$

$$y_2 = 8.5 + \frac{0.5}{2}(10.0 + 16.5) = 15.125$$

The MATLAB solution using the modified Euler method is shown in Figure 8.9.

Example 8.5: Modified Euler Method

Solve the following ODE using the modified Euler method for the interval $x = [0\,2]$, the initial condition $y(0) = 0$, and a step size, $h = 1$:

$$y' = \frac{dy}{dx} = y + x + 1$$

Compare the results obtained from the modified Euler scheme solution with the exact analytical solution: $y = 2e^x - x - 2$.

Solution

The solution is shown in Table 8.4. The solution of the modified Euler method is much more accurate than that for the simple Euler method (Example 8.2).

The numerical solution using the modified Euler method of ODE, $dy/dx = y + x + 1$, $y(0) = 0$, is shown in Figure 8.10.

TABLE 8.4

Solution Using the Modified Euler Scheme

n	x_n	$y_{n,\text{MEuler}}$	True, y_n
0	0	0	0
1	1	2	2.437
2	2	8.5	10.78

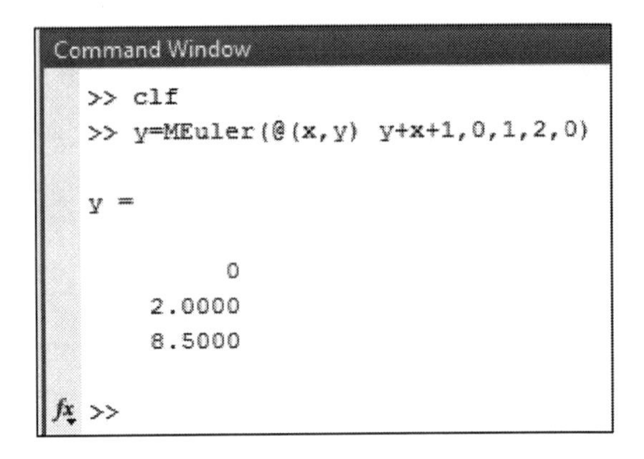

FIGURE 8.10
Numerical solution using the modified Euler method.

Example 8.6: Modified Euler Method

Solve the following ODE using the modified Euler method with $h = 0.2$ and calculate the value of $y(1.2)$:

$$\frac{dy}{dx} = x + y^{0.5}, \; at\, x = 1,\; y = 2$$

Solution

Use the modified Euler method. The right hand-side of the given ODE is:

$$f(x_i, y_i) = x_i + y_i^{0.5},$$

Substitute $x_i = 1$ and $y_i = 2$:

$$f(x_i, y_i) = 1 + 2^{0.5} = 2.4142$$

Use the modified Euler method:

$$y_{i+1} = y_i + \frac{h}{2}\left[f(x_i, y_i) + f\left(x_i + h, y_i + hf(x_i, y_i)\right)\right]$$

```
Command Window
>> clf
>> y=MEuler(@(x,y) x+y^0.5,1,0.2,1.2,2)

y =

    2.0000
    2.5190
```

FIGURE 8.11
Solution of the ODE $dy/dx = x + y^{0.5}$, via the modified Euler methods.

Substitute x and y values:

$$y_{i+1} = 2 + \frac{h}{2}\left[f(1,2) + f(1+0.2, 2+0.2f(1,\ 2))\right]$$

Simplify:

$$y_{i+1} = 2 + \frac{h}{2}\left[2.414 + f(1.2, 2+0.2(2.414))\right]$$

The value of y_{i+1} is:

$$y_{i+1} = 2 + \left(\frac{0.2}{0.5}\right)\ [2.414 + 2.776] = 2.519$$

The MATLAB solution can be obtained while in the MATLAB command window, as shown in Figure 8.11.

8.2.3 Midpoint Method

The midpoint method is a one-step method for numerically solving ODEs [3]. The midpoint method is an improvement over the Euler method. It is an explicit technique for approximating the solution of the initial value problem $y' = f(x,y);\ y(x_0) = y_0$ *at* x for a given step size h. The midpoint method is:

$$s_1 = f(t_n, y_n) \tag{8.20}$$

$$s_2 = f\left(t_n + \frac{h}{2}, y_n + \frac{h}{2}s_1\right) \tag{8.21}$$

$$y_{n+1} = y_n + hs_2 \tag{8.22}$$

```
Editor - C:\Users\nayef\Desktop\Midpnt.m
Euler.m | MEuler.m | Midpnt.m
1      function yout = Midpnt(F,t0,h,tfinal,y0)
2      % Midpoint is a simple ode solver.
3      % uses Midpoint's method with fixed step
4      % size h on the interval
5      % t0<=t<=final
6      % to solve dy/dt =F(t,y)
7      % with y(t0)= y0
8  -   clf
9  -   y=y0;
10 -   i=0;
11 -   yout=y;
12 -   for t=t0:h: tfinal-h
13 -       i=i+1;
14 -       s1=F(t,y);
15 -       s2=F(t+h/2,y+h*s1/2);
16 -       y= y+h*s2;
17 -       ymd(i)=y;
18 -       time(i)= t;
19 -       plot(time,ymd,'o-');
20 -       hold on
21 -       xlabel ('time');
22 -       ylabel ('y');
23 -       yout = [yout; y];
24 -   end
```

FIGURE 8.12
Midpoint function in MATLAB.

The midpoint method coded as a MATLAB function is shown in Figure 8.12.

The midpoint method uses the derivative at the starting point to approximate the solution at the midpoint. The midpoint method can be summarized as follows:

- The Euler method is used to estimate the solution at the midpoint.
- The value of the rate function $f(x, y)$ at the midpoint is calculated.
- This value is used to estimate y_{i+1}.

Example 8.7: Solve a Nonlinear ODE Using the Midpoint Method

Solve the following nonlinear first-order ODE using the midpoint method:

$$\frac{dy}{dt} = \sqrt{1 - y^2}$$

Initial conditions:

$$y(0) = 0$$

For the range, $0 < t < \frac{\pi}{2}$, and a step size, $h = \pi/32$.

```
Command Window
>> y=Midpnt(@(x,y) sqrt(1-y^2),0,pi/32,pi/2,0);
>> t=(0:pi/32:pi/2)';
plot(t,y,'o-')
xlabel('time')
ylabel('y')
>>
```

FIGURE 8.13
Solution in MATLAB command window via the midpoint method.

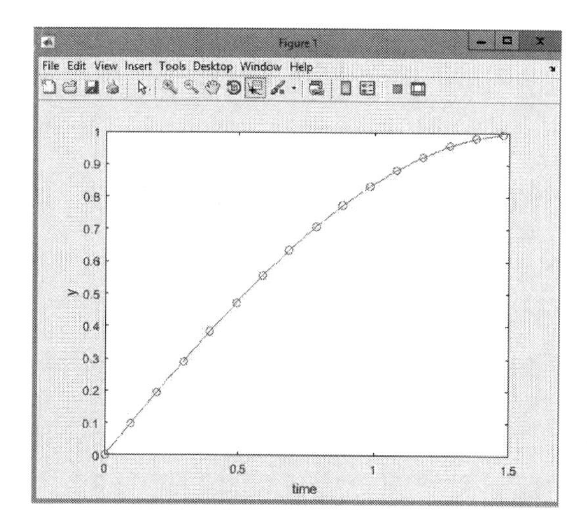

FIGURE 8.14
Plot to y versus x using the midpoint method in MATLAB.

The MATLAB solution using the midpoint method can be achieved using the MATLAB code while in the MATLAB working area (Figure 8.13). The result is plotted in Figure 8.14 for time versus y.

Example 8.8: Solve ODE Using the Midpoint Method

Use the midpoint method to solve the following ODE:

$$\frac{dy}{dx} = 1 + x^2 + y$$

With the initial condition $y(0) = 1$ and a step size, $h = 0.1$, determine $y(0.1)$ and $y(0.2)$. The analytical solution is:

$$y = -x^2 - 2x + 4e^x - 3$$

Solution

Let the right-hand side of the ODE be as follows:

$$f(x,y)=1+x^2+y,\quad y_0=y(0)=1,\quad h=0.1$$

$i=0$

$$y_{0+\frac{1}{2}}=y_0+\frac{h}{2}f\left(x_0,y_0\right)=1+\frac{0.1}{2}\left(1+0^2+1\right)=1.1$$

$$y_1=y(0.1)=y_0+hf\left(x_{0+\frac{1}{2}},y_{0+\frac{1}{2}}\right)=1+0.1\left(1+0.05^2+1.1\right)=1.210$$

The analytical solution, $y_1=1.211$:

$i=1$

$$y_{1+\frac{1}{2}}=y_1+\frac{h}{2}f\left(x_1,y_1\right)=1.211+0.05\left(1+0.1^2+1.21\right)=1.32$$

$$y_2=y(0.2)=y_1+hf\left(x_{1+\frac{1}{2}},y_{1+\frac{1}{2}}\right)=1.211+0.1(2.34)=1.446$$

The analytical solution of $y_2=1.446$. The MATLAB solution using function of the midpoint method is shown in Figure 8.15.

```
Command Window
>> y=Midpnt(@(x,y) 1+x^2+y,0,0.1,0.2,1)

y =

    1.0000
    1.2103
    1.4446
```

FIGURE 8.15
Numerical solution of: $dy/dx=1+x^2y$ in MATLAB.

Example 8.9: Numerical Integration of a Nonlinear ODE with the Midpoint Method

Solve the following ODE:

$$\frac{dy}{dx} = yx^2 - 1.2y$$

using the midpoint method with $h = 0.5$ and initial condition, $y(0) = 1$, over the interval $[0, 2]$. Solve for $y(2)$.

The analytical solution of the differential equation is:

$$y(x) = e^{0.333x^3 - 1.2x}, \text{ and } y(2) = 1.305$$

Solution

The midpoint equations used in the solution are:

$$y_{i+\frac{1}{2}} = y_i + \frac{h}{2} f(x_i, y_i)$$

$$y_{(i+1)} = y_i + hf\left(x_{i+\frac{1}{2}}, y_{i+\frac{1}{2}}\right)$$

The solution using the midpoint method to find $y(2)$ with $h = 0.5$ is as follows. The right-hand side: $f(x, y) = yx^2 - 1.2y$

$i = 0$

$$y_{0+\frac{1}{2}} = 1 + \frac{0.5}{2}\left(1 \times 0^2 - 1.2 \times 1\right) = 0.7$$

$$y_1 = y(0.5) = 1 + 0.5\left(0.7 \times 0.25^2 - 1.2 \times 0.7\right) = 0.6019$$

$i = 1$

$$y_{\left(1+\frac{1}{2}\right)} = 0.6019 + \frac{0.5}{2}\left(0.6019 \times 0.5^2 - 1.2 \times 0.6019\right) = 0.4589$$

$$y_2 = y\left(1\right) = 0.601875 + 0.5\left(0.4589 \times 0.75^2 - 1.2 \times 0.4589\right) = 0.4556$$

```
Command Window
>> y=Midpnt(@(x,y) y*x^2-1.2*y,0,0.5,2,1)

y =

    1.0000
    0.6019
    0.4556
    0.5340
    1.1619
```

FIGURE 8.16
MATLAB midpoint numerical solution of: $dy/dx = yx^2 - 1.2y$, $y(0) = 1.0$.

$i = 2$

$$y_{2+\frac{1}{2}} = 0.4556 + \frac{0.5}{2}\left(0.4556 \times 1^2 - 1.2 \times 0.4556\right) = 0.4328$$

$$y_3 = y(1.5) = 0.4556 + 0.5\left(0.4328 \times 1.25^2 - 1.2 \times 0.4328\right) = 0.5340$$

$i = 3$

$$y_{3+\frac{1}{2}} = 0.5340 + \frac{0.5}{2}\left(0.5340 \times 1.5^2 - 1.2 \times 0.5350\right) = 0.6742$$

$$y_4 = y(2) = 0.5340 + 0.5\left(0.6742 \times 1.75^2 - 1.2 \times 0.6742\right) = 1.1619$$

The MATLAB solution using the function of the midpoint method is shown in Figure 8.16.

8.2.4 Heun Predictor-Corrector Method

The Heun method is a predictor-corrector method. The Heun method uses average derivative for the entire interval.

Predictor step: Use the Euler method to find a first estimate for y_{i+1}:

$$y_{i+1}^{*} = y_i + hf\left(x_i, y_i\right)$$

Corrector step: Take the average of slopes at x_i and x_{i+1}:

$$y_{i+1} = y_i + \frac{h}{2}\left\{f\left(x_i, y_i\right) + f\left(x_{i+1}, y_{i+1}^{*}\right)\right\}$$

Figure 8.17 is the developed MATLAB function of the Heun method.

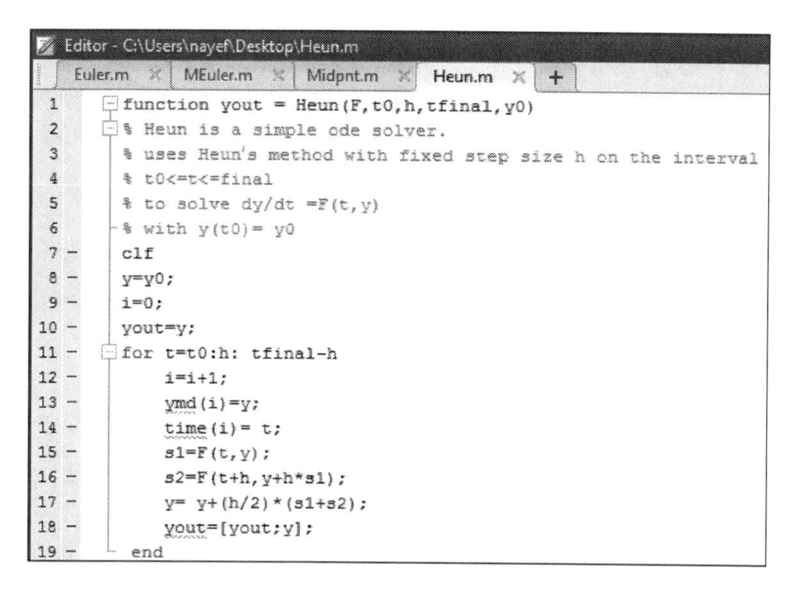

```
Editor - C:\Users\nayef\Desktop\Heun.m
Euler.m  MEuler.m  Midpnt.m  Heun.m  +
1      function yout = Heun(F,t0,h,tfinal,y0)
2      % Heun is a simple ode solver.
3      % uses Heun's method with fixed step size h on the interval
4      % t0<=t<=final
5      % to solve dy/dt =F(t,y)
6      % with y(t0)= y0
7  -   clf
8  -   y=y0;
9  -   i=0;
10 -   yout=y;
11 -   for t=t0:h: tfinal-h
12 -       i=i+1;
13 -       ymd(i)=y;
14 -       time(i)= t;
15 -       s1=F(t,y);
16 -       s2=F(t+h,y+h*s1);
17 -       y= y+(h/2)*(s1+s2);
18 -       yout=[yout;y];
19 -    end
```

FIGURE 8.17
MATLAB function of the Heun method.

Example 8.10: Heun Method

Use the Heun method to solve the following ODE:

$$y'(x) = 1 + x^2 + y,\ y(0) = 1, \text{ use a step size, } h = 0.1$$

Determine $y(0.1)$ and $y(0.2)$.

Solution

This example illustrates the use of the Heun method in solving first-order ODEs. The right-hand side of the ODE is:

$$f(x,y) = 1 + y + x^2$$

$$y_0 = y(0) = 1,\ h = 0.1$$

Step 1:

$$y(0) = 1.0$$

$$\text{Predictor: } y_1^* = y_0 + hf(x_0, y_0) = 1 + hf(0,1)$$

$$y_1^* = 1 + 0.1(1 + 1 + 0^2) = 1.2$$

$$\text{Corrector: } y_1 = y_0 + \frac{h}{2}\left(f(x_0, y_0) + f(x_1, y_1^*)\right)$$

$$y_1 = y(0.1) = 1 + \frac{0.1}{2}\left(\left(1 + 1 + 0^2\right) + \left(1 + 1.2 + 0.1^2\right)\right) = 1.2105$$

```
Command Window
>> y=Heun(@(x,y) y+x^2+1,0,0.1,0.2,1)

y =

    1.0000
    1.2105
    1.4452
```

FIGURE 8.18
MATLAB solution of: $dy/dx = y + x^2 + 1$ via the Heun method.

Step 2:

$$y(0.1) = 1.2105$$

$$\text{Predictor: } y_2^* = y_1 + h(x_1, y_1) = 1.2105 + 0.1(1 + 1.2105 + 0.1^2) = 1.4326$$

$$\text{Corrector: } y_2^1 = y_1 + \frac{h}{2}\left(f(x_1, y_1) + f(x_2, y_2^*)\right)$$

$$y_2 = y(0.2) = 1.2105 + \frac{0.1}{2}\left(1 + 1.2105 + 0.1^2\right) + \left(1 + 1.4326 + 0.2^2\right)$$

$$y_2 = y(0.2) = 1.2105 + \frac{0.1}{2}(2.2205) + (2.4726) = 1.4452$$

The MATLAB solution using the Heun method is shown in Figure 8.18.

8.2.5 Runge-Kutta Method

The objective of this section is to understand the Runge Kutta methods of different orders to solve first-order ODEs. Higher-order Runge-Kutta methods are available as second-order, third-order, and fourth-order Runge-Kutta. Higher-order methods are more accurate but require more calculations. The most popular numerical methods are the second-, third-, and fourth-order Runge-Kutta methods.

8.2.5.1 Second-Order Runge-Kutta (RK2)

The second-order Runge-Kutta method, known as RK2, is equivalent to the Heun method with a single corrector.

$$k_1 = f\left(x_i, y_i\right) \tag{8.23}$$

$$k_2 = f\left(x_i + h, y_i + k_1 h\right) \tag{8.24}$$

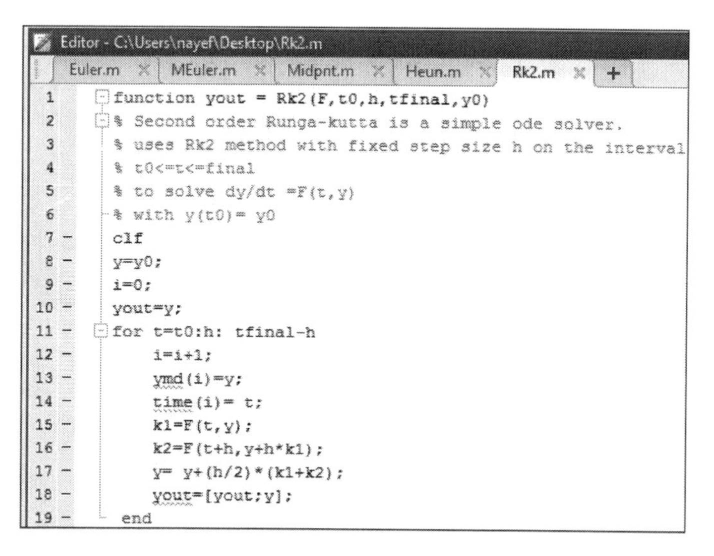

FIGURE 8.19
MATLAB function of the second-order Runge-Kutta method.

$$y_{i+1} = y_i + \frac{h}{2}\left(k_1 + k_2\right) \tag{8.25}$$

The developed MATLAB function of the second-order Runge-Kutta method is shown in Figure 8.19.

8.2.5.2 *Third-Order Runge-Kutta (RK3)*

The third-order Runge-Kutta method uses weighted-average value for three slopes, k_1, k_2, k_3. The third-order Runge-Kutta method is used to approximate solutions of the initial value problem $y'(x) = f(x, y)$; $y(x_0) = y_0$, which evaluates the integral, $f(x, y)$, three times per step. For step $i+1$ and $x_i = x_0 + ih$:

$$k_1 = f\left(x_i, y_i\right) \tag{8.26}$$

$$k_2 = f\left(x_i + \frac{h}{2}, y_i + \frac{1}{2}k_1 h\right) \tag{8.27}$$

$$k_3 = f\left(x_i + h, y_i - k_1 h + 2k_2 h\right) \tag{8.28}$$

The third-order Runge-Kutta method for ODE $y' = f(x, y)$ can be written as:

$$y_{i+1} = y_i + \frac{h}{6}\left(k_1 + 4k_2 + k_3\right) \tag{8.29}$$

The MATLAB function of the third-order Runge-Kutta method is shown in Figure 8.20.

```
Editor - C:\Users\nayef\Desktop\Rk3.m
Euler.m | MEuler.m | Midpnt.m | Heun.m | Rk2.m | Rk3.m | +
1      function yout = Rk3(F,t0,h,tfinal,y0)
2      % The third order Runge-Kutta method for ODE is a simple ode solver.
3      % uses Rk3 method with fixed step size h on the interval
4      % t0<=t<=final
5      % to solve dy/dt =F(t,y)
6      % with y(t0)= y0
7 -    y=y0;
8 -    i=0;
9 -    yout=y;
10 -   for t=t0:h: tfinal-h
11 -       i=i+1;
12 -       ymd(i)=y;
13 -       time(i)= t;
14 -       k1=F(t,y);
15 -       k2=F(t+h/2,y+h*k1/2);
16 -       k3=F(t+h,y-h*k1+2*k2*h);
17 -       y= y+(h/6)*(k1+4*k2+k3);
18 -       yout=[yout;y];
19 -   end
```

FIGURE 8.20
MATLAB function of the third-order Runge-Kutta method.

8.2.5.3 Fourth-Order Runge-Kutta

The fourth-order Runge-Kutta method is popular and uses several predictive steps instead of just one. It involves calculating four auxiliary quantities k_1, k_2, k_3 and k_4. The fourth-order Runge-Kutta algorithm is:

$$k_1 = f(x_i, y_i) \tag{8.30}$$

$$k_2 = f\left(x_i + \frac{h}{2}, y_i + \frac{h}{2}k_1\right) \tag{8.31}$$

$$k_3 = f\left(x_i + \frac{h}{2}, y_i + \frac{h}{2}k_2\right) \tag{8.32}$$

$$k_4 = f\left(x_i + h, y_i + k_3 h\right) \tag{8.33}$$

The updated value for y can be found from the following equation:

$$y_{i+1} = y_i + \frac{h}{6}\left(k_1 + 2k_2 + 2k_3 + k_4\right) \tag{8.34}$$

Note that, while all the algorithms presented are for first-order equations with one dependent variable, they can be readily extended to systems of higher-order differential equations. The MATLAB function, ODE4, implements the

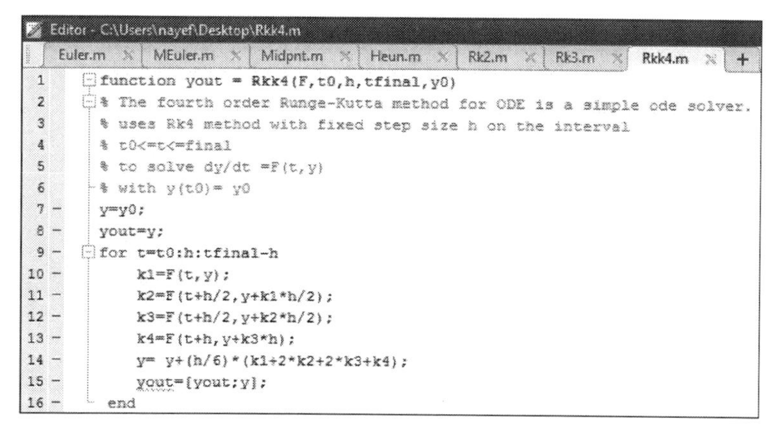

FIGURE 8.21
MATLAB function of the fourth-order Runge-Kutta method.

classic Runge-Kutta method, which is the most widely used numerical method for ODEs. The MATLAB function for the classical fourth-order Runge-Kutta method is shown in Figure 8.21.

Example 8.11: Using Various ODE Solver Methods

Use the Euler method; modified Euler method; midpoint method; Heun method; and second-, third-, and fourth-order Runge-Kutta methods to solve the following ODE with $h = 0.2$, (solve for y at $x = 1.2$, i.e., $y(1.2)$):

$$\frac{dy}{dx} = x + y^{0.5}, \text{ at } x = 1,\ y = 2$$

The exact analytical solution of $y(1.2) = 2.52009$.

Solution

Solve the following ODE using the Euler method:

$$\frac{dy}{dx} = x + y^{0.5},$$

at $x = 1$, $y = 2$.

Euler Method

$$y_{(i+1)} = y_i + hf(x_i, y_i)$$

Solve for $f(x, y)$:

$$f(x,y) = x + y^{0.5} = 1 + 2^{0.5} = 2.4142$$

Solve using the Euler method:

$$y_{(i+1)} = y(1.2) = y_i + hf(x_i, y_i) = 2 + 0.2 \times (2.4142) = 2.4828$$

```
Command Window
>> y=Euler(@(x,y)x+y^0.5,1,0.2,1.2,2.0)

y =

    2.0000
    2.4828
```

FIGURE 8.22
Numerical solution using the Euler method.

Figure 8.22 is the Euler method MATLAB solution obtained via the MATLAB code while in MATLAB command window.

Solve the following ODE using the modified Euler methods:

$$\frac{dy}{dx} = x + y^{0.5},$$

at $x = 1, y = 2$.

The modified Euler method:

$$y_{n+1} = y_n + 0.5h\left[f(x_n, y_n) + f\left(x_n + h, y_n + hf(x_n, y_n)\right)\right]$$

Solve for function $f(x_n, y_n)$:

$$f(x_n, y_n) = 1 + 2^{0.5} = 2.4142$$

$$y_n + hf(x_n, y_n) = 2 + (0.2)(2.4142) = 2.4828$$

$$f\left(x_n + h, y_n + hf(x_n, y_n)\right) = 2.7757$$

$$y_{n+1} = y(0.2) = 2 + (0.5)(0.2)[2.4142 + 2.7757] = 2.5190$$

The MATLAB solution obtained via MATLAB code while in the command windows of the first-order differential equation, $dy/dx = x + y^{0.5}$, $y(0) = 2.0$ is depicted in Figure 8.23.

```
Command Window
>> y=MEuler(@(x,y)x+y^0.5,1,0.2,1.2,2.0)

y =

    2.0000
    2.5190
```

FIGURE 8.23
Numerical solution using the modified Euler method.

Midpoint Method

Solve the following ODE using the midpoint method:

$$\frac{dy}{dx} = x + y^{0.5},$$

at $x = 1$, $y = 2$.

The midpoint method:

$$y_{n+1} = y_n + h\left[f\left(x_n + \frac{h}{2}, y_n + \frac{h}{2} f(x_n, y_n)\right)\right]$$

The value of $f(x_n, y_n)$:

$$f(x_n, y_n) = 1 + 2^{0.5} = 2.4142$$

The value of the function after h:

$$f\left(x_n + \frac{h}{2}, y_n + \frac{h}{2} f(x_n, y_n)\right) = 2.5971$$

Calculate the value of y_{n+1}:

$$y_{n+1} = 2 + (0.2)(2.5971) = 2.5194$$

The MATLAB solution is simply obtained following the steps shown in Figure 8.24.

Heun Method

The solution of the following ODE uses the Heun method:

$$\frac{dy}{dx} = x + y^{0.5},$$

at $x = 1$, $y = 2$.

```
Command Window
>> y=Midpnt(@(x,y)x+y^0.5,1,0.2,1.2,2.0)

y =

    2.0000
    2.5194
```

FIGURE 8.24
Numerical solution using the midpoint method.

The Heun method:

$$y_{n+1} = y_n + 0.25k_1 + 0.75k_2$$

Calculate the value of k_1:

$$k_1 = hf(x_n, y_n) = hf(1,2) = h\left(1 + 2^{0.5}\right) = (0.2)(2.4142) = 0.4828$$

Calculate the value of k_2:

$$k_2 = hf\left(x_n + \frac{2}{3}h, y_n + \frac{2}{3}k_1\right) = (0.2)(2.6471) = 0.5294$$

$$k_2 = 0.2f\left(1 + \frac{2}{3}0.2, 2 + \frac{2}{3}0.4828\right) = 0.2f(1.13, 2.32) = (0.2)(2.6471) = 0.53$$

Substitute in the main equation:

$$y_{n+1} = 2 + \left[(0.25)(0.4828) + (0.75)(0.53)\right] = 2.5190$$

The MATLAB solution is obtained using the MATLAB code while in the MATLAB working area (Figure 8.25).

Second-Order Runge-Kutta

The second-order Runge-Kutta is known as RK2. It is equivalent to the Heun method with a single corrector:

$$\frac{dy}{dx} = x + y^{0.5},$$

at $x = 1$, $y = 2$.

From the previous examples:

$$k_1 = f(x_i, y_i) = 1 + 2^{0.5} = 2.4142$$

```
Command Window
>> y=Heun(@(x,y)x+y^0.5,1,0.2,1.2,2.0)

y =

    2.0000
    2.5190
```

FIGURE 8.25
Numerical solution using the Heun method.

Solve for k_2:

$$k_2 = f\left(x_i + h, y_i + k_1 h\right) = f(1 + 0.2, 2 + 0.2 * 2.4142)$$

Then:

$$k_2 = 2 + 0.2\left(1.2 + 0.4828^{0.5}\right) = 2.7757$$

The new value of y:

$$y_{i+1} = y_i + \frac{h}{2}(k_1 + k_2) = 2 + \frac{0.2}{2}(2.4142 + 2.7757) = 2.5190$$

The MATLAB solution via the second-order Runge-Kutta method is shown in Figure 8.26.

Third-Order Runge-Kutta

Solve the following ODE:

$$\frac{dy}{dx} = x + y^{0.5},$$

at $x = 1$, $y = 2$.

The third-order Runge-Kutta method estimates for the changes are

$$i = 0, x = 1, y = 2$$

$$k_1 = f\left(x_i, y_i\right) = f(1, 2) = 1 + 2^{0.5} = 2.41$$

$$k_2 = f\left(x_i + \frac{h}{2}, y_i + \frac{1}{2}k_1 h\right) = f\left(1 + \frac{0.2}{2}, 2 + \frac{1}{2}(2.414 * 0.2)\right) = f(1.1, 2.241)$$

$$k_2 = f(1.1, 2.241) = 1.1 + 2.241^{0.5} = 2.6$$

$$k_3 = f\left(x_i + h, y_i - k_1 h + 2k_2 h\right) = f(1 + 0.2, 2 - 2.41 * 0.2 + 2 * 2.6 * 0.2)$$

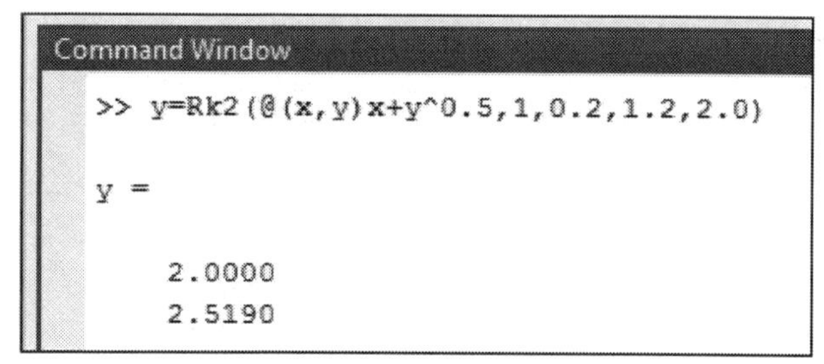

FIGURE 8.26
Numerical solution using the second-order Runge-Kutta method.

```
Command Window
>> y=Rk3(@(x,y)x+y^0.5,1,0.2,1.2,2.0)

y =

    2.0000
    2.5201
```

FIGURE 8.27
Numerical solution using the third-order Runge-Kutta.

$$k_3 = f(1+0.2,2-2.41*0.2+2*2.6*0.2) = f(1.2,2.558) = 1.2+2.558^{0.5} = 2.8$$

$$y_{i+1} = y_i + \frac{h}{6}(k_1 + 4k_2 + k_3)$$

$$y_{i+1} = 2 + \frac{0.2}{6}(2.41 + 4(2.6) + 2.8) = 2.52$$

The MATLAB numerical solution using the third-order Runge-Kutta method is shown in Figure 8.27.

Fourth-Order Runge-Kutta

Solve the following ODE:

$$\frac{dy}{dx} = x + y^{0.5},$$

at $x = 1, y = 2$.

The fourth-order Runge-Kutta method estimates for the changes are:

$$k_1 = f(x_i, y_i) = f(1,2) = 1 + 2^{0.5} = 2.414$$

$$k_2 = f\left(x_i + \frac{h}{2}, y_i + \frac{1}{2}k_1 h\right)$$

$$k_2 = f\left(1 + \frac{0.2}{2}, 2 + \frac{1}{2}(2.414*0.2)\right) = f(1.1, 2.241) = 2.5971$$

$$k_3 = f\left(x_i + \frac{h}{2}, y_i + \frac{1}{2}k_2 h\right)$$

$$k_3 = f\left(1 + \frac{0.2}{2}, 2 + \frac{1}{2} \times 2.5971 \times 0.2\right) = f(1.2, 2.6) = 1.1 + 2.597^{0.5} = 2.603$$

```
Command Window
>> y=Rkk4(@(x,y)x+y^0.5,1,0.2,1.2,2.0)

y =

    2.0000
    2.5201
```

FIGURE 8.28
Numerical solution using the fourth-order Runge-Kutta method.

$$k_4 = f\left(x_i + h, y_i + k_3 h\right)$$

$$k_4 = f(1+0.2, 2+2.603\times 0.2) = f(1.2, 2.52) = 1.2 + 2.52^{0.5} = 2.78764$$

and the updated value for y is found from:

$$y_{i+1} = y_i + \frac{h}{6}\left(k_1 + 2k_2 + 2k_3 + k_4\right)$$

$$y_{i+1} = 2 + \frac{0.2}{6}(2.41 + 2(2.5971) + 2(2.603) + 2.78764) = 2.52009$$

The MATLAB numerical solution using the fourth-order Runge-Kutta method is shown in Figure 8.28. The analytical solution is $y(1.2) = 2.52009$.

Example 8.12: Second-Order Runge-Kutta Method

Use the second-order Runge-Kutta method (RK2) to solve the following ODE:

$$\frac{dy}{dx} = 2x + y^2 x$$

where $y(0) = 2$
Solve for $h = 0.1$ and compute $y(0.1)$.
The exact analytical value of $y(0.1) = 2.0303$.

Solution

The second-order Runge-Kutta method (RK2) is equivalent to the Heun method with a single corrector:

$$k_1 = f\left(x_i, y_i\right) = 0 + 0 = 0$$

Solve for k_2:

$$k_2 = f\left(x_i + h, y_i + k_1 h\right) = f(0 + 0.1, 2 + 0.1 * 0)$$

```
Command Window
>> y=Rk2(@(x,y)x+y^0.5,1,0.2,1.2,2.0)

y =

    2.0000
    2.5190
```

FIGURE 8.29
Solving ODE with second-order Runge-Kutta.

Then:

$$k_2 = 2 * 0.1 + 2.0^2 * 0.1 = 0.6$$

The new value of y:

$$y_{i+1} = y_i + \frac{h}{2}(k_1 + k_2) = 2 + \frac{0.1}{2}(0 + 0.6) = 2.03$$

The MATLAB numerical solution via the second-order Runge-Kutta method of $dy/dx = x + x^2$ is shown in Figure 8.29.

Example 8.13: Nonlinear ODE via the Midpoint Method

Solve the following initial value problem using the midpoint method:

$$\frac{dy}{dx} = 3e^{-x} - 0.4y, \qquad y(0) = 5$$

Calculate $y(3)$, using $h = 1.5$, the exact analytical solution of $y(3) = 2.763$.

Solution

The midpoint method:

$$y_{i+1} = y_i + k_2 h$$

$$k_1 = f(x_i, y_i)$$

$$k_2 = f\left(x_i + \frac{1}{2}h, y_i + \frac{1}{2}k_1 h\right)$$

The function is:

$$f(x_i, y_i) = 3e^{-x} - 0.4y$$

Take a step size, $h = 1.5$, at $i = 0$, $x_0 = 0$, $y_0 = 5$:

$$k_1 = f(x_i, y_i) = 3e^{-0} - 0.4(5) = 1$$

$$k_2 = f\left(x_i + \frac{1}{2}h, y_i + \frac{1}{2}k_1 h\right) = f\left(0 + \frac{1}{2} * 1.5, 5 + \frac{1}{2}(1)(1.5)\right) = f(0.75, 5.75)$$

$$k_2 = f(0.75, 5.75) = 3e^{-0.75} - 0.4(5.75) = -0.88$$

Then substitute:

$$y_{i+1} = y_i + k_2 h = 5 + (-0.88) * 1.5 = 3.68$$

For $i = 1$, $x_1 = 1.5$, $y_1 = 3.676$

$$k_1 = f(x_i, y_i) = (1.5, 3.676) = 3e^{-1.5} - 0.4(3.676) = -0.8009$$

$$k_2 = f\left(x_i + \frac{1}{2}h, y_i + \frac{1}{2}k_1 h\right) = f(1.5) + \frac{1}{2} * 1.5, 3.676 + \frac{1}{2} * (-0.8009) * 1.5$$

$$k_2 = f(2.25, 3.075) = 3^{-2.25} - 0.4(3.075) = -0.9138$$

$$y_2 = y_1 + k_2 h = 3.676 + (-0.9138)(1.5) = 2.304 = y(x_2) = y(3)$$

$$x_2 = x_1 + h = 1.5 + 1.5 = 3.0$$

$$\% \text{ Error} = \frac{|2.763 - 2.304|}{2.763} \times 100\% = 16.6\%$$

The solution of $dy/dx = 3e^{-x} - 0.4y, y(0) = 5$ using the midpoint method is shown in Figure 8.30.

```
Command Window
>> y=Midpnt(@(x,y)3*exp(-x)-0.4*y,0,1.5,3.0,5)

y =

    5.0000
    3.6756
    2.3049
```

FIGURE 8.30
Numerical solution using the midpoint method of: $dy/dx = 3e^{-x} - 0.4y$.

Example 8.14: Nonlinear ODE via the Second-Order Runge-Kutta Method

Use the second-order Runge-Kutta method to solve the following ODE:

$$\frac{dy}{dx} = 3e^{-x} - 0.4y, \qquad y(0) = 5$$

Calculate $y(3)$, using $h = 1.5$, the exact analytical solution of $y(3) = 2.763$.

Solution

Use the second-order Runge- Kutta method:

$$y_{i+1} = y_i + \frac{h}{2}(k_1 + k_2)$$

$$k_1 = f(x_i, y_i)$$

$$k_2 = f(x_i + h, y_i + k_1 h)$$

From the initial conditions:
at $i = 0$, $x_0 = 0$, $y_0 = 5$

$$f(x_i, y_i) = 3e^{-x} - 0.4y$$

$$k_1 = f(x_0, y_0) = f(0,5) = 3e^{-0} - 0.4(5) = 1$$

$$k_2 = f(x_i + h, y_i + k_1 h) = f(0 + 1.5, 5 + (1)(1.5))$$

$$= f(1.5, 6.5) = e^{-1.5} - 0.4(6.5) = -1.9306$$

$$y_1 = y_0 + \frac{h}{2}(k_1 + k_2) = 5 + \frac{1.5}{2}(1 + (-1.9306)) = 4.302 = y(1.5)$$

at $i = 1$, $x_1 = 1.5, y_1 = 4.302$

$$k_1 = f(x_1, y_1) = f(1.5, 4.302) = 3e^{-1.5} - 0.4(4.302) = -1.0514$$

$$k_2 = f(x_1 + h, y_1 + k_1 h) = f(1.5 + 1.5, 4.302 + (-1.0514)(1.5))$$

$$= f(3, 2.726) = e^{-3} - 0.4(2.726) = -0.9406$$

$$y_2 = y_1 + \frac{h}{2}(k_1 + k_2) = 4.302 + \frac{1.5}{2}((-1.0514) + (-0.9406)) = 2.808 = y(3.0)$$

```
>> y=Rk2(@(x,y)3*exp(-x)-0.4*y,0,1.5,3.0,5)

y =

    5.0000
    4.3020
    2.8080
```

FIGURE 8.31
Numerical solution using the second-order Runge-Kutta method.

$$\%\text{ Error} = \frac{|2.763 - 2.808|}{2.763} \times 100\% = 1.63\%$$

The MATLAB solution using the second-order Runge-Kutta method of $dy/dx = 3e^{-x} - 0.4y$ is shown in Figure 8.31.

Example 8.15: Solve Using the Fourth-Order Runge-Kutta Method

Use the fourth-order Runge-Kutta method to solve the following initial value ODE:

$$\frac{dy}{dx} = 3e^{-x} - 0.4y, \qquad y(0) = 5$$

Calculate $y(3)$, using $h = 1.5$, the exact analytical solution of $y(3) = 2.76301$.

Solution

The fourth-order Runge-Kutta method:

$$y_{i+1} = y_i + \frac{1}{6}h\left(k_1 + 2k_2 + 2k_3 + k_4\right)h$$

$$k_1 = f\left(x_i, y_i\right)$$

$$k_2 = f\left(x_i + \frac{1}{2}h, y_i + \frac{1}{2}k_1h\right)$$

$$k_3 = f\left(x_i + \frac{1}{2}h, y_i + \frac{1}{2}k_2h\right)$$

$$k_4 = f\left(x_i + h, y_i + k_3h\right)$$

$$i = 0, x_o = 0, y_o = 5, h = 1.5$$

$$k_1 = f(x_0, y_0) = f(0,5) = 3e^{-0} - 0.4 * 5 = 1$$

$$k_2 = f\left(0 + \frac{1}{2}(1.5), 5 + \frac{1}{2}(1)(1.5)\right) = f(0.75, 5.75) = 3e^{-0.75} - 0.4(5.75) = -0.8829$$

$$k_3 = f\left(0 + \frac{1}{2}1.5, y_0 + \frac{1}{2}(-08829)(1.5)\right) = f(0.75, 4.338)$$

$$k_3 = 3e^{-0.75} - 0.4(4.338) = -0.318$$

$$k_4 = f(0 + 1.5, 5 + (-0.318 * 1.5)) = (1.5, 4.523)$$

$$k_4 = 3e^{-1.5} - 0.4(4.523) = -1.140$$

$$y_1 = y_0 + \frac{1}{6}h(k_1 + 2k_2 + 2k_3 + k_4)$$

$$y_1 = 5 + \frac{1}{6}1.5(1 + 2 * (-0.8829) + 2(-0.318) - 1.140) = 4.365$$

$$x_1 = x_0 + h = 0 + 1.5 = 1.5$$

Accordingly:

$$y(x_1) = y(1.5) = 4.365$$

$$x_1 = 1.5, y_1 = 4.365$$

$$i = 1, x_1 = 1.5, y_1 = 4.365, h = 1.5$$

$$k_1 = f(x_1, y_1) = f(1.5, 4.365) = 3e^{-1.5} - 0.4 * 4.35 = -1.076$$

$$k_2 = f\left(x_1 + \frac{1}{2}h, y_1 + \frac{1}{2}k_1 h\right)$$

$$k_2 = f\left(1.5 + \frac{1}{2}(1.5), 4.365 + \frac{1}{2}(-1.076)(1.5)\right)$$

$$k_2 = f(2.25, 3.557) = 3^{-2.25} - 0.4(3.557) = -1.107$$

$$k_3 = f\left(x_1 + \frac{1}{2}h, y_1 + \frac{1}{2}k_2 h\right) = f(2.25, 3.535) = 3e^{-2.25} - 0.4(3.535) = -1.098$$

$$k_4 = f(x_1 + h, y_1 + k_3 h) = f(1.5 + 1.5, 4.365 + (-1.098)(1.5))$$

```
Command Window
>> y=Rkk4(@(x,y)3*exp(-x)-0.4*y,0,1.5,3.0,5)

y =

    5.0000
    4.3646
    2.7588
```

FIGURE 8.32
Numerical solution using the fourth-order Runge-Kutta method.

$$k_4 = f(x_1 + h, y_1 + k_3h) = f(3, 2.718) = 3e^{-3} - 0.4(2.718) = -0.9378$$

$$y_2 = y_1 + \frac{1}{6}h(k_1 + 2k_2 + 2k_3 + k_4)$$

$$y_2 = 4.365 + \frac{1}{6}1.5(-1.076 + 2(-1.107) + 2(-1.098) + (-0.9378)) = 2.759$$

$$x_2 = x_1 + h = 1.5 + 1.5 = 3.0$$

$$y_2 = y(x_2) = y(3) = 2.759$$

$$\%\ \text{Error} = \frac{|2.763 - 2.759|}{2.763} \times 100\% = 0.145\%$$

The MATLAB solution using the fourth-order Runge-Kutta method of $dy/dx = 3e^{-x} - 0.4y$ is shown in Figure 8.32.

Example 8.16: Numerical Solution Using the Heun Method

Solve the following initial value ODE for $y(0.5)$:

$$\frac{dy}{dx} = 8e^{-x} - 3\sin\, y, \qquad y(0) = 0.5$$

Use the Heun method and $h = 0.5$ for the analytical solution of $y(0.5) = 2.3323$.

Solution

Heun method:

$$y_{i+1} = y_i + \frac{h}{2}(k_1 + k_2)$$

```
Command Window
>> y=Heun(@(x,y)8*exp(-x)-3*sin(y),0,0.5,0.5,0.5)

y =

    0.5000
    3.8009
```

FIGURE 8.33
Number solution using the Heun method.

where:

$$k_1 = f(x_i, y_i)$$

$$k_2 = f(x_i + h, y_i + hk_1)$$

$$i = 0, x_0 = 0, y_0 = 0.5$$

$$k_1 = f(x_0, y_0) = f(0,0.5) = 8e^{-0} - 3\sin(0.5) = 6.5417$$

$$k_2 = f(x_0 + h, y_0 + hk_1) = f(0 + 0.5, 0.5 + 0.5 * 4.5417) = f(0.5, 3.78)$$

$$k_2 = 8e^{-0.5} - 3\sin(3.78) = 6.6422$$

Substitute the values of k_1 and k_2:

$$y_1 = y_0 + \frac{h}{2}(k_1 + k_2) = 0.5 + \frac{0.5}{2}(6.5417 + 6.6422) = 3.796$$

Accordingly, $y(0.5) = 3.796$:

$$\% \text{ Error} = \frac{|2.33233 - 3.796|}{2.33233} \times 100\% = 61.34\%$$

The MATLAB solution using the Heun method of the differential equation $dy/dx = 8e^{-x} - \sin(y)$ is shown in Figure 8.33.

8.3 Simultaneous Systems of First-Order Differential Equations

A simultaneous differential equation is one of the mathematical equations for an indefinite function of one or more than one variable that relate the values of the function. Differential equations play an important function in engineering, physics, economics, and other disciplines. This analysis concentrates on linear equations with constant coefficients. A single differential equation of second and higher order can also be converted into a system of first-order differential equation. A system of simultaneous first-order ODEs has the general form [2]:

$$x_1' = f_1 = (x_1, x_2, x_3 \ldots\ldots x_n)$$

$$x_2' = f_2 = (x_1, x_2, x_3 \ldots\ldots x_n)$$

$$\vdots$$

$$x_n' = f_n = (x_1, x_2, x_3 \ldots\ldots x_n)$$

where each x_i is a function of t. If each f_i is a linear function of $x_1, x_2, \ldots, x_n$, then the system of equations is said to be linear; otherwise, it is nonlinear.

Example 8.17: Simultaneous Differential Equations

Using the Euler method, solve the following differential equation by simplify the equation into two initial value ODEs:

$$\frac{d^2y}{dx^2} + 1.5\frac{dy}{dx} + 2.5y - 5.5e^{-x} = 0$$

where $y(0) = 7, dy/dx(0) = 13$

Calculate the value of $y(0.5)$ using step size $h = 0.25$.

The analytical solution can be calculated using the Wolfram Alpha website http://www.wolframalpha.com/:

$$solve\ \ y'' + 1.5\, y' + 2.5{*}y - 5.5{*}e\,\hat{}\,(-x),\ \ y(0) = 7,\ \ y'(0) = 13,\ \ y(0.5) = ?$$

The analytical solution of the differential equation is:

$$y(x) = 2.75e^{-x} + 13.6051e^{-0.75x}\sin(1.39194x) + 4.25e^{-0.75x}\cos(1.39194x)$$

The differential equation solution at point is:

$$y(0.5) = 9.90459$$

Solution

Now we will simplify the equation into two ODEs. Let:

$$\frac{dy}{dx} = z \qquad y(0) = 7$$

$$\frac{d^2y}{dx^2} = \frac{dz}{dx}$$

$$\frac{dz}{dx} = 5.5e^{-x} - 1.5z - 2.5y \qquad z(0) = 13$$

Set the function of each equation:

$$f_1(x,y,z) = z$$

$$f_2(x,y,z) = 5.5e^{-x} - 1.5z - 2.5y$$

Use the Euler method:

$$y_{i+1} = y_i + f_1\left(x_i, y_i, z_i\right)h$$

$$z_{i+1} = z_i + f_2\left(x_i, y_i, z_i\right)h$$

at $i = 0, x_0 = 0, y_0 = 7, z_0 = 13$

Substitute the values in the Euler equation:

$$y_1 = y_0 + f_1\left(x_0, y_0, z_0\right)h = 7 + f_1(0,7,13)(0.25)$$

$$y_1 = 7 + (13)(0.25) = 10.25$$

$$z_1 = z_0 + f_2\left(x_0, y_0, z_0\right)h = 13 + f_1(0,7,13)(0.25)$$

$$z_1 = 13 + [5.5e^{-0} - 1.5(13) - 2.5(7)] = 5.125$$

Second step size:

$$i = 1, x_1 = 0.25, y_1 = 10.25, z_1 = 5.125$$

$$y_2 = y_1 + f_1\left(x_1, y_1, z_1\right)h = 10.25 + f_1(0.25, 10.25, 5.125)(0.25)$$

$$y_1 = 10.25 + (5.125)(0.25) = 11.53$$

Accordingly, $y(0.5) = 11.53$.

Try using a lower step size ($h = 0.125$, $h = 0.0625$).

The numerical solutions via the Euler method for step sizes $h = 0.125$ and $h = 0.0625$ are summarized in Tables 8.5 and 8.6, respectively.

TABLE 8.5

Euler Method and Step Size $h = 0.125$

i	x_n	y_n	z_n	y_{exact}
0	0.000	7.000	13.000	7.00
1	0.125	8.625	9.062	8.383
2	0.250	9.758	5.275	9.300
3	0.375	10.417	1.772	9.791
4	0.500	10.639	−1.343	9.905

TABLE 8.6

Euler Method and Step Size $h = 0.0625$

i	x_n	y_n	z_n	y_{exact}
0	0.0000	7.000	13.000	7.000
1	0.0625	7.8125	11.031	7.751
2	0.1250	8.5020	9.099	8.383
3	0.1875	9.071	7.221	8.897
4	0.2500	9.5220	5.412	9.300
5	0.3125	9.860	3.684	9.596
6	0.3750	10.091	2.050	9.791
7	0.4375	10.219	0.5170	9.892
8	0.5000	10.251	−0.906	9.905

8.4 Summary

The Euler, midpoint, and Heun methods are similar in the following sense:

1. They use different methods for different estimates of the slope.

$$y_{(i+1)} = y_i + h \times \text{slope}$$

2. Both the midpoint and Heun methods are comparable in accuracy to the second-order Taylor series method.
3. The fourth-order Runge-Kutta (RK4) method is more accurate.

PROBLEMS

8.1 Euler Method

Using the Euler method, solve the for following initial value first-order differential equation:

$$\frac{dy}{dx} = x + 2y, \quad y(0) = 0$$

Using step size, $h = 0.25$, find a value for the solution at $x = 1$.

Answer: 0.515625

8.2 Euler Method

Using the Euler method, find the value of $y(0.5)$ in the initial value problem:

$$dy/dt = -2x - y, \; y(0) = -1$$

Use a step length of 0.1. Also find the error in the approximation.

Answer: −0.7715

8.3 Euler Method

Solve the differential equation $y' = x/y$, $y(0) = 1$ by the Euler method to get $y(1)$. Use the step lengths $h = 0.1$ and 0.2, and compare the results with the analytical solution.

Answer: 1.3855, 1.3550

8.4 Euler Method

Use the Euler method to find the numerical solution of the following ODE:

$$\frac{dy}{dt} = t - 2y, \quad y(0) = 1$$

The exact solution is

$$y = \frac{1}{4}\left[2t - 1 + 5e^{-2t}\right]$$

Solve for step size $h = 0.2$ and find the value of $y(0.6)$.

Answer: 0.32

8.5 Midpoint Method

Solve the following initial value problem using the midpoint method:

$$\frac{dy}{dt} = 3e^{-x} - 0.4y, \qquad y(0) = 5$$

Calculate $y(3)$ using $h = 1.5$. The analytical solution

$$y(x) = 10e^{-0.4x} - 0.4y$$

The exact solution: $y(3) = 2.76301$

Answer: 2.304

8.6 Second-Order Runge-Kutta Method

Use the second-order Runge-Kutta method to solve the following ODE:

$$\frac{dy}{dt} = 5e^{-x} - 0.4y, \qquad y(0) = 5$$

Calculate $y(3)$ using $h = 1.5$.
The analytical solution: $y(x) = 13.33e^{-0.4x} - 8.33e^{-x}$
The analytical solution of $y(3) = 3.601$

Answer: 4.493

8.7 Fourth-order Runge-Kutta method

Solve the following initial value ODE using the fourth-order Runge-Kutta method:

$$\frac{dy}{dt} = 5e^{-x} - 0.4y, \qquad y(0) = 5$$

Calculate $y(3)$ using $h = 1.5$.
The analytical solution: $y(x) = 13.33e^{-0.4x} - 8.33e^{-x}$
The analytical solution of $y(3) = 3.601$.

Answer: 3.6

8.8 Heun's Method

Solve the following differential equation using Heun's method
$dy/dx = 8e^{-x} - 3\sin(y), \quad y(0) = 0.5$ and $h = 0.5$ determine $y(0.5)$.

Answer: 3.8

8.9 Heun's Method

Solve the following initial value ODE:
$dy/dx = 8e^{-x} - 3\sin(y), \qquad y(0) = 0.5$ and $h = 0.5$
Solve the following initial value ODE:

$$\frac{dy}{dx} = e^{-x} - 0.4y, \qquad y(0) = 5$$

Using the Heun method and $h = 1.5$, determine the value of $y(1.5)$.

Answer: 3.763

8.10 Initial Value ODE

Solve the following initial value ODE:

$$\frac{dy}{dx} = 3e^{-x} - 0.4y \qquad y(0) = 5$$

Using Heun method and $h = 3.0$, determine the value of $y(3)$. Compare your results with a step size of $h = 1.5$.

Answer: 2.763

References

1. James, G., 2001. *Modern Engineering Mathematics*, 5th ed., Upper Saddle River, NJ: Prentice-Hall.
2. Strauss, W. A., 2008. *Partial Differential Equations: An Introduction*, 2nd ed., New York: John Wiley & Sons.
3. Chapra, S., R. Canale, 2017. *Numerical Methods for Engineers*, 7th ed., New York: McGraw Hill.

9

Higher-Order Differential Equations

Differential equations are designated by their order and are determined by the term with the highest derivatives. An equation containing only first derivatives is a first-order differential equation; an equation containing the second derivative is a second-order differential equation. Higher-order ordinary differential equations (ODEs) are expressions that involve derivatives other than the first order. Classic mathematical models of physical systems are most often second-order or even higher, which is another reason to concentrate on second-order systems. By contrast, a partial differential equation (PDE) is a differential equation that contains unknown multivariable functions and their partial derivatives. In this chapter, we will deal with high-order differential equations. The high-order differential equations will be solved numerically using the COMSOL Multiphysics 5.3a software package.

LEARNING OBJECTIVES

- Solve high-order differential equations by COMSOL software.
- Solve boundary-value and initial value problems using COMSOL software.
- Solve high-order ODE and coupled value problems using COMSOL software.

9.1 Introduction

Differential equations are called ODEs or PDEs according to whether they contain partial derivatives. The order of a differential equation is the highest order derivative occurring. A solution of a differential equation of order n consists of a function defined and n times differentiable on a domain D; the domain has the property that the functional equation obtained by substituting the function and its n derivatives into the differential equation holds for every point in the D domain. We learned in Chapter 8 how to solve first-order ODEs using Euler, midpoint, Heun, and Runge-Kutta methods. The first-order differential equations are of the form [1]:

$$\frac{dy}{dx} = f(x,y), \;\; y(0) = y_0 \tag{9.1}$$

The following equation is higher than first order; it is an nth order differential equation of the form shown in Equation 9.2:

$$a_n \frac{d^n y}{dx^n} + a_{n-1} \frac{d^{n-1} y}{dx^{n-1}} + \ldots + a_1 \frac{dy}{dx} + a_0 y = f(x) \tag{9.2}$$

What do we do to solve simultaneous (coupled) differential equations, or differential equations that are higher than first order?

With $(n-1)$, initial conditions can be solved by assuming:

$$y = z_1 \tag{9.3}$$

$$\frac{dy}{dx} = \frac{dz_1}{dx} = z_2 \tag{9.4}$$

$$\frac{d^2 y}{dx^2} = \frac{dz_2}{dx} = z_3 \tag{9.5}$$

$$\frac{d^{n-1} y}{dx^{n-1}} = \frac{dz_{n-1}}{dx} = z_n \tag{9.6}$$

$$\frac{d^n y}{dx^n} = \frac{dz_n}{dx} = \frac{1}{a_n}\left(-a_{n-1}\frac{d^{n-1} y}{dx^{n-1}} \ldots . -a_1 \frac{dy}{dx} - a_0 y + f(x)\right) \tag{9.7}$$

$$= \frac{1}{a_n}\left(-a_{n-1} z_n \ldots - a_1 z_2 - a_0 z_1 + f(x)\right) \tag{9.8}$$

Equations 9.3 to 9.8 represent n first-order differential equations as follows:

$$\frac{dz_1}{dx} = z_2 = f_1\left(z_1, z_2, \ldots, x\right) \tag{9.9}$$

$$\frac{dz_2}{dx} = z_3 = f_2\left(z_1, z_2, \ldots, x\right) \tag{9.10}$$

$$\frac{dz_n}{dx} = \frac{1}{a_n}\left(-a_{n-1} z_n \ldots . -a_1 z_2 - a_0 z_1 + f(x)\right) \tag{9.11}$$

These first-order ODEs are simultaneous in nature, but they can be solved by the methods used for solving first-order ODEs that we have already learned (Euler, modified Euler, midpoint, Heun, and Runge-Kutta). Each of the n first-order ODEs is accompanied by one initial condition. The first-order differential equation can be linear or nonlinear. A linear first-order differential equation is one that can be written in the form of Equation 9.12:

$$\frac{dy}{dx} + P(x)y = Q(x) \tag{9.12}$$

where P and Q are continuous functions of x. Accordingly the following equation is linear:

$$\frac{dy}{dx} + 2xy = x^2 \tag{9.13}$$

Equation 9.14 is not linear because of the square root.

$$\frac{dy}{dx} + x\sqrt{y} = 2x \tag{9.14}$$

Example 9.1: Change Second-Order ODE to Multiple First-Order ODE

Change the following differential equation to a set of first-order differential equations:

$$3\frac{d^2y}{dx} + 2\frac{dy}{dx} + 5y = e^{-x},\ y(0) = 5,\ y'(0) = \frac{dy}{dx}(0) = 7$$

Solution

The ODE should be rewritten as follows. Assume the following:

$$\frac{dy}{dx} = z$$

Then:

$$\frac{d^2y}{dx^2} = \frac{dz}{dx}$$

Substituting this in the given second-order ODE gives:

$$3\frac{dz}{dx} + 2z + 5y = e^{-x}$$

Rearrange:

$$\frac{dz}{dx} = \frac{1}{3}(e^{-x} - 2z - 5y)$$

The set of two simultaneous first-order ODEs complete with the initial conditions is:

$$\frac{dy}{dx} = z, \quad y(0) = 5$$

$$\frac{dz}{dx} = \frac{1}{3}(e^{-x} - 2z - 5y), \quad z(0) = 7$$

Now one can apply any of the numerical methods used for solving first-order ODEs.

Example 9.2: Solve a Second-Order ODE Using the Euler Method

Given the following second-order nonlinear differential equation:

$$\frac{d^2y}{dt^2} + 2\frac{dy}{dt} + y = e^{-t}, \quad y(0) = 1, \quad \frac{dy}{dt}(0) = 2$$

Find by the Euler method: $y(0.5)$.
Use a step size of $h = 0.25$.

Solution

First, the second-order differential equation is written as two simultaneous first-order differential equations as follows. Assume:

$$\frac{dy}{dt} = z$$

Then:

$$\frac{dz}{dt} + 2z + y = e^{-t}$$

Rearrange:

$$\frac{dz}{dt} = e^{-t} - 2z - y$$

Accordingly, the two simultaneous first-order differential equations are:

$$\frac{dy}{dt} = z = f_1(t, y, z), \quad y(0) = 1$$

$$\frac{dz}{dt} = e^{-t} - 2z - y = f_2(t, y, z), \quad z(0) = 2$$

Using the Euler method, we get:

$$y_{i+1} = y_i + hf_1(t_i, y_i, z_i)$$

$$z_{i+1} = z_i + hf_2(t_i, y_i, z_i)$$

To find the value of $y(0.5)$, and since we are using a step size of 0.25 and starting at $t = 0$, we need to take three steps to find the value of $y(0.5)$:

$i = 0$

For $i = 0$, $t_0 = 0$, $y_0 = 1$, $z_0 = 2$
From the Euler equation:

$$y_1 = y_0 + hf_1(t_0, z_0, y_0)$$

$$y_1 = 1 + 0.25f_1(0, 1, 2)$$

$$y_1 = 1 + 0.25(2) = 1.5$$

The variable y_1 is the approximate value of y at $t = t_1 = t_0 + h = 0 + 0.25 = 0.25$. Accordingly:

$$y_1 = y(0.25) \approx 1.5$$

From the following equation:

$$\frac{dz}{dt} = e^{-t} - 2z - y$$

$$z_1 = z_0 + hf_2(t_o, y_0, z_0)$$

Substitute values:

$$z_1 = 2 + 0.25f_2(0, 1, 2)$$

$$z_1 = 2 + 0.25(e^{-0} - 2(2) - 1) = 1$$

The variable z_1 is the approximate value of z (same as dy/dt) at $t = 0.25$. Accordingly

$$z_1 = z(0.25) \approx 1$$

$i = 1$

For $i = 1$, $t_1 = 0.25$, $y_1 = 1.5$, $z_1 = 1$

From the previous equation:

$$y_2 = y_1 + hf_1(t_1, y_1, z_1)$$

Substitute values:

$$y_2 = 1.5 + 0.25f_1(0.25, 1.5, 1)$$

Solve:

$$y_2 = 1.5 + 0.25(1) = 1.75$$

The variable y_2 is the approximate value of y at t_2:

$$t = t_2 = t_1 + h = 0.25 + 0.25 = 0.50$$

Accordingly:

$$y_2 = y(0.5) \approx 1.75$$

The analytical solution of the second-order ODE is:

$$y = e^{-t} + 3e^{-t}t + \frac{e^{-t}t^2}{2}$$

Accordingly:

$$y(0.5) = 1.592$$

9.2 Initial and Boundary Value Problems

Initial value differential equations are ODEs of the following form:

$$\frac{dy}{dt} = f(t, y) \tag{9.15}$$

where their initial conditions imposed at the same locations, most likely $t = 0$ in time, are of the form $y(0) = y_0$.

Every initial value of the elements of y is specified at the same location in time. For example, the supplementary conditions are at one point of the independent variable:

$$\frac{d^2y}{dt^2} + 2\frac{dy}{dt} + y = e^{-2t} \tag{9.16}$$

With initial conditions:

$$y(0) = 1, \frac{dy}{dt}(0) = 2.5$$

A boundary value problem (BVP) is a differential equation with a set of additional constraints called the boundary conditions. A solution to a BVP is a solution to the differential equation that also satisfies the boundary conditions. In the initial value problems, the initial conditions are being imposed at the same point in the independent variable (in this case, t). By contrast, in BVPs, boundary conditions are imposed at different values of the independent variable. As an example of a BVP, consider the second-order ODE:

$$\frac{d^2y}{dx^2} + ay^2 = 0 \tag{9.17}$$

with boundary conditions given by:

$$y(0) = 0 \text{ and } y(1) = 1$$

This problem cannot be solved using the methods we learned for the initial value problems because the two conditions imposed on the problem are not at coincident locations of the independent variable x.

9.3 Shooting Method

The shooting method in numerical analysis is a method for solving a BVP by reducing it to the solution of an initial value problem. The main idea of the shooting method is to choose the remaining information in one x value so that we can start the integration and observe how the boundary condition in the other x value is satisfied. The following steps of the shooting method are used in solving ODEs:

1. Guess a value for the auxiliary conditions at one point in time.
2. Solve the initial value problem using Euler, Runge-Kutta, and so on.
3. Check if the boundary conditions are satisfied; if they are not, modify the guess and resolve the problem.
4. Use interpolation in updating the guess. It is an iterative procedure and can be efficient in solving the BVP.

The following examples illustrate the use of the Euler and Runge-Kutta methods to solve BVP s using shooting methods.

Example 9.3: BVPs

The boundary conditions are not at one point of the independent variable, and this type of problem is more difficult to solve than initial value problem. Consider the following second-order differential equation:

$$\frac{d^2y}{dx^2}+2\frac{dy}{dx}+y=e^{-2x}$$

With boundary conditions:

$$y(0)=1$$

$$y(2)=1.5$$

Solve using COMSOL software.

Solution

1. In COMSOL, go to the "select physics menu." Select "Mathematics/PDE interfaces Coefficient Form PDE." Click the "Add" button. While in the right-side Review physics interface menu, change the field name u to T and click the "Study" button. Select the "stationary analysis option" and click the "Done" button.

2. Right-click on "geometry" and left-click "interval." Enter the start and end dimensions of the interval to the desired geometry: 0 and 2. This single subdomain represents the one-dimensional geometry for this problem (Figure 9.1).

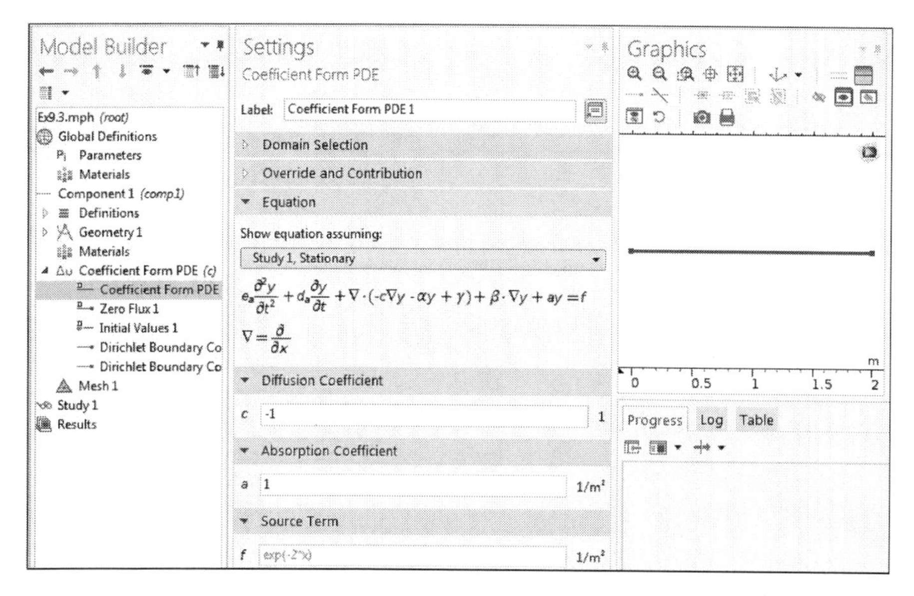

FIGURE 9.1
Coefficient of the general COMSOL PDE form.

The first boundary condition is shown in Figure 9.2 ($y(0) = 1$).

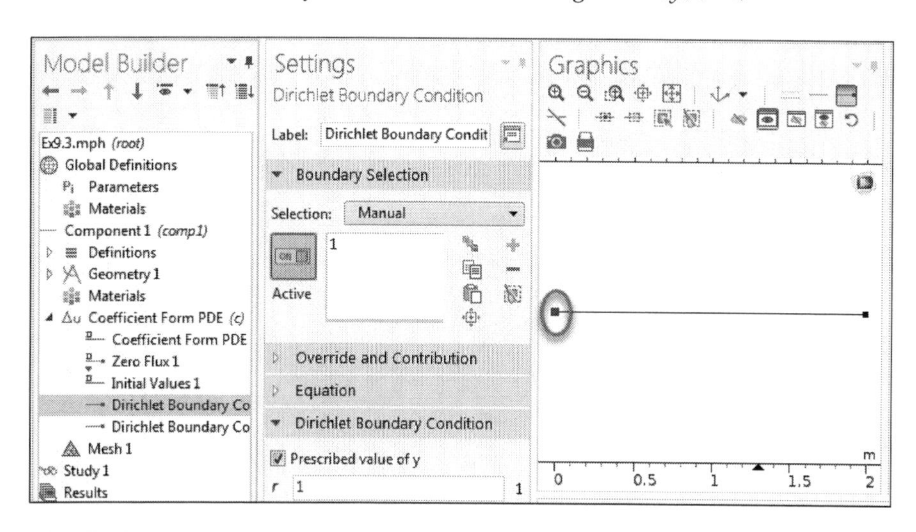

FIGURE 9.2
Initial boundary condition.

The second boundary conditions are shown in Figure 9.3 ($y(2) = 1.5$).

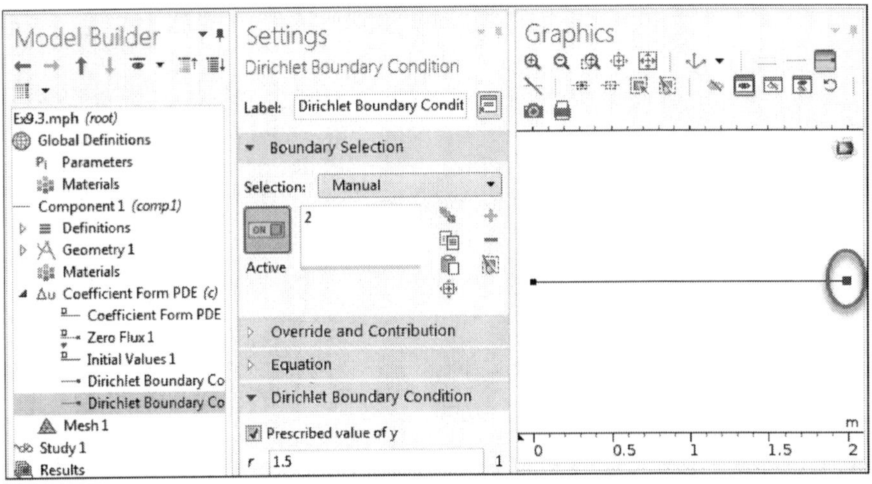

FIGURE 9.3
Second boundary condition.

The solution of the problem is shown in Figure 9.4.

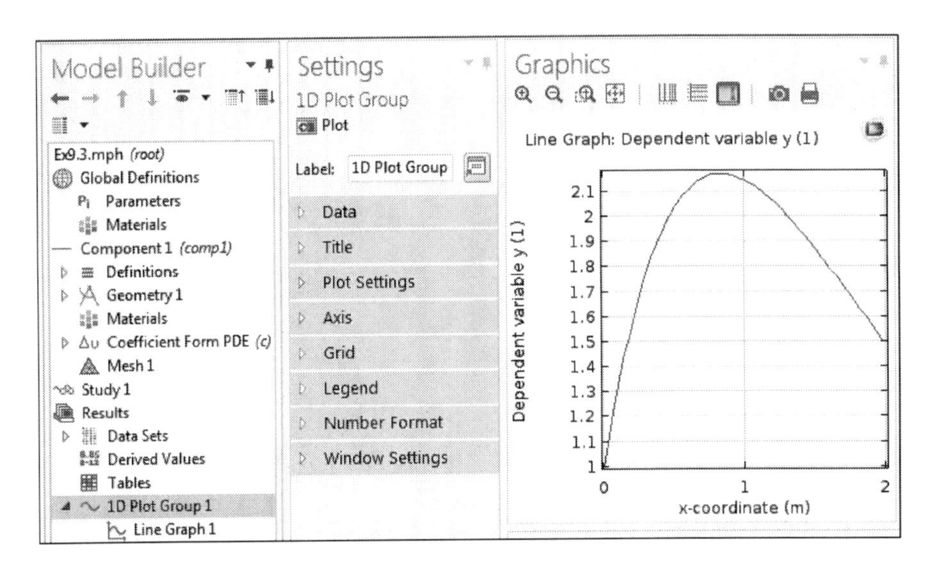

FIGURE 9.4
Solution of the second-order BVP.

Example 9.4: BVP

Solve the following BVP using the Euler and second-order Runge-Kutta (RK2) methods:

$$\frac{d^2y}{dx^2} - 4y + 4x = 0$$

where:

$$y(0) = 0, \quad y(1) = 1$$

Solution

1. Convert to a system of first ODEs:

$$\frac{dy}{dx} = z, \quad y(0) = 0.0$$

$$\frac{dz}{dx} = 4(y - x), \quad z(0) = ?$$

2. Guess the initial conditions that are not available.
3. Solve the initial-value problem using the Euler method with $h = 0.2$:

$$y_{i+1} = y_i + hf_1(x_i, y_i, z_i)$$

where:

$$f_1(x_i, y_i, z_i) = z_i$$

$$z_{i+1} = y_i + hf_2(x_i, y_i, z_i)$$

where:

$$f_2(x_i, y_i, z_i) = 4(y_i - x_i)$$

$i = 0$

Guess 1: $z(0) = 0.5$

$$y_1 = y_0 + hf_1(x_0, y_0, z_0) = hf_1(0.0, 0, 0.5) = 0 + 0.2(0.5) = 0.1$$

$$z_1 = z_0 + hf_2(x_0, y_0, z_0) = 0.5 + 0.2f_2(0.0, 0.0, 0.5) = 0.5 + 0.2(4(0.0 - 0.0)) = 0.5$$

TABLE 9.1

Numerical Solution Using the Euler Method, $z(0) = 0.5$

i	x	y	z
1	0	0.00	0.50
2	0.2	0.10	0.50
3	0.4	0.30	0.42
4	0.6	0.48	0.34
5	0.8	0.67	0.25
6	1.0	0.85	0.14

TABLE 9.2

The Correct Guess is $z(0) = 4.33$

i	x	y	z
1	0	0.00	4.33
2	0.2	0.87	4.33
3	0.4	1.07	4.86
4	0.6	1.37	5.40
5	0.8	1.68	6.01
6	1	2.00	6.72

The solution using the Euler method is summarized in Tables 9.1 and 9.2.

Solution Using the Euler Method

Solution Using the Second-Order Runge-Kutta Method

1. Use the second-order Runge-Kutta (RK2) method for two simultaneous ODEs:

$$k_{11} = f_1(x_i, y_i, z_i)$$

$$k_{12} = f_2(x_i, y_i, z_i)$$

Then k_2:

$$k_{21} = f_1(x_i + h, y_i + k_{11}h, z_i + k_{12}h)$$

$$k_{22} = f_2(x_i + h, y_i + k_{21}h, z_i + k_{21}h)$$

$$y_{i+1} = y_i + \frac{h}{2}(k_{11} + k_{12})$$

$$z_{i+1} = z_i + \frac{h}{2}(k_{21} + k_{22})$$

Substitute values as follows:

$$k_{11} = f_1(x_i, y_i, z_i) = f_1(0,0,0.5) = 0.5$$

$$k_{12} = f_2(x_i, y_i, z_i) = f_2(0,0,0.5) = 4(0-0) = 0$$

$$k_{21} = f_1(x_i + h, y_i + k_{11}h, z_i + k_{12}h) = f_1(0+0.2, 0+0.5*0.2, 0.5+0.5*0.2)$$

$$k_{21} = f_1(x_i + h, y_i + k_{11}h, z_i + k_{12}h) = f_1(0.2, 0.1, 0.6) = 0.6$$

$$k_{22} = f_2(x_i + h, y_i + k_{21}h, z_i + k_{21}h) = f_2(0+0.2, 0+0.6*0.2, 0.5+0.6*0.2)$$

$$k_{22} = f_2(x_i + h, y_i + k_{21}h, z_i + k_{21}h) = f_2(0.2, 0.12, 0.62) = 4(0.12-0.2) = -0.32$$

Then y:

$$y_{i+1} = y_i + \frac{h}{2}(k_{11} + k_{21})$$

$$z_{i+1} = z_i + \frac{h}{2}(k_{21} + k_{22})$$

at $i = 1$, $x_1 = 0.2$, $y_1 = 0$, $z_1 = 0.5$ (first guess).
For simultaneous differential equations:

$$k_{11} = f_1(x_0, y_0, z_0) = f_1(0.2, 0, 0.5) = 0.5$$

$$k_{12} = f_2(x_1, y_1, z_1) = f_2(0.2, 0, 0.5) = 4(0.5-0.2) = 1.2$$

$$k_{21} = f(x_1 + h, y_1 + k_{11}h) = f(0.2+0.2, 0+0.5*0.2) = 0.1$$

$$k_{22} = f(x_1 + h, y_1 + k_{11}h) = f(0.2+0.2, 0+0.5*0.2) = 4(1-0.4) = 2.4$$

Substitute values in RK2:

$$y_2 = y_1 + \frac{h}{2}(k_{11} + k_{21}) = 0 + \frac{0.2}{2}(0.5+1.2) = 0.17$$

$$z_2 = z_1 + \frac{h}{2}(k_{21} + k_{22}) = 0.5 + \frac{02}{2}(0.1 + 2.4) = 0.75$$

COMSOL Multiphysics was designed specifically to solve problems written in the form of PDEs involving one, two, or three spatial coordinates and time. For steady-state problems and one-dimensional geometry, the general problem description reduces to a simple BVP. COMSOL Multiphysics can also be used to study the BVPs that we have been solving with the shooting method and the finite difference method, although the program was designed to handle much more complex situations. The following example illustrated the use of COMSOL in solving BVP problems.

Example 9.5: Temperature Distribution in a Reactor Tube

Consider a tubular reactor of length $L = 0.1\,\text{m}$ and radius $R = 0.005$ m. The volumetric heat source generated due to the exothermic reaction is $Q = 1.410^8\ \text{W/m}^3$. The thermal conductivity of fuel is $k_f = 1.9\,\text{W/(m}\cdot{}^\circ\text{C)}$, and the outer surface temperature of the reactor is $T_s = 650°\text{C}$.

Solution

The following equation and boundary conditions represent the heat transfer in the radial direction of the tubular reactor:

$$\frac{1}{r}\frac{d}{dr}\left(k_f r \frac{dT}{dr}\right) + Q = 0$$

Multiply each term by r and rearrange as follows:

$$\frac{d}{dr}\left(k_f r \frac{dT}{dr}\right) = -Qr$$

With the boundary conditions (B.C.):

B.C. 1: at $r = 0$, $(dT/dr)\big|_{r=0} = 0$ (axial symmetry at $r = 0$)
B.C. 2: at $r = R$, $T(R) = T_s$ (fixed temperature at surface)

A typical set of parameters used to get numerical results for this problem was given as follows:

Volumetric heat source due to the reactions: $Q = 1.410^8\ \text{W/m}^3$
Reactor outer radius: $R = 0.005$ m
Outer surface temperature of tubular reactor: $T_s = 650°\text{C}$
Constant thermal conductivity of reactor fuel: $k_f = 1.9\,\text{W/(m}\cdot{}^\circ\text{C)}$

The equation represents a BVP and is solved with COMSOL Multiphysics 5.3a. To solve this problem within COMSOL Multiphysics, we must accomplish the following steps:

1. Define the geometry of interest.
2. Set the equation constants.
3. Specify the boundary conditions.
4. Set solution parameters and solve the problem.
5. Perform postprocessing and analysis of the results.

The following steps are identified in solving the BVP using COMSOL:

1. Start COMSOL Multiphysics.
2. Click the Model Wizard under "new" and select the 1D geometry.
3. While in the select physics menu, select "Mathematics/PDE interfaces/Coefficient Form PDE." Click the "Add" button. While in the right-side Review physics interface menu, change the field name from u to T and click the "Study" button. Select the stationary analysis option and click the "Done" button.
4. Right-click on "geometry" and left-click "interval." Enter the start and end dimensions of the interval to the desired geometry: 0 and 0.005. This single subdomain represents the one-dimensional geometry (Figure 9.5). This case presents only one dependent variable, $T(x)$, within the following general PDE:

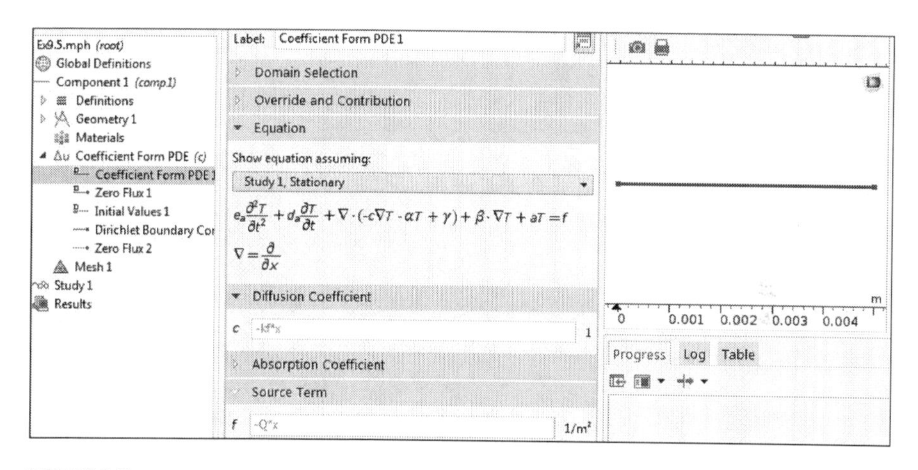

FIGURE 9.5
Diffusion and absorption coefficient and source term.

$$e_a \frac{\partial^2 T}{\partial t^2} + d_a \frac{\partial T}{\partial t} + \nabla.(-c\nabla T - \alpha T + \gamma) + \beta.\nabla T + aT = f \tag{9.18}$$

where e_a and d_a will be set to zero for stationary (time-independent) problems.

Now we must match the ODE of interest to the general PDE form that is available in COMSOL Multiphysics. In the current case, the following correspondence is needed:

independent variable: $r \rightarrow x$
dependent variable: $T \rightarrow u$
coefficients: $\alpha = \beta = \gamma = a = 0$

$$f = -Q * x$$

$$c = -k_f * x$$

Thus, we simply enter these values into the "Value/Expression" field under the "Coefficients" tab on the Coefficient form PDE menu. The options available under the other tabs are either not needed or do not need to be changed for this problem.

Set the Boundary Conditions

For the left boundary condition, right-click on "Coefficient from PDE" and then left-click "zero/flux." For the right boundary condition, right-click on "Coefficient form PDE" and left-click "Dirichlet boundary condition." For the prescribed value of $u = T = 650$, give the desired boundary condition at the right endpoint ($r = 0.005$).

Select Mesh Grid

Once the geometry, equation coefficients, and BCs have been specified, we are now ready to discretize the independent variables, that is, to create a finite element mesh for the geometry of interest. COMSOL Multiphysics also makes this step quite easy for most problems. Simply select the "Initialize Mesh" icon to create the initial grid, then hit the "Refine Mesh" icon as many times as necessary to double the number of mesh uniformly throughout the geometry until we get the desired grid for the problem. For one-dimensional problems, this is relatively straightforward, and a single Refine Mesh for the current problem is quite sufficient.

Compute

Simply hitting the "Compute" button will solve the current problem. Many general PDE problems are difficult to solve, but most one-dimensional BVPs are not. In the simplest form, you can plot the solution $u(x)$ versus x [or $T(r)$ *vs.* r in our case]. This can be visualized in the COMSOL Multiphysics main plot screen, or it can be placed into a new

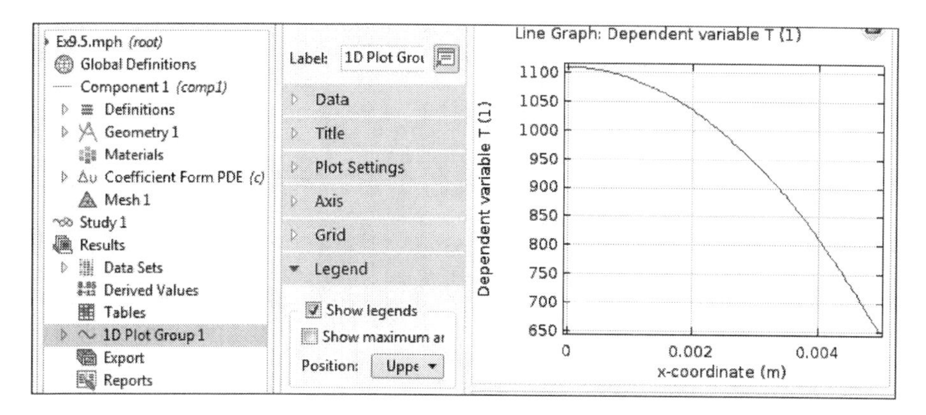

FIGURE 9.6
Solution of BVP using COMSOL PDE form.

figure in COMSOL. For quick access to the primary results, however, COMSOL Multiphysics analysis is easier. The resultant line graph of temperature versus radius of the tube is shown in Figure 9.6. The diagram reveals that the temperature is at a maximum at the center of the tube, where a huge amount of heat is released because of the highly exothermic reaction.

9.4 Simultaneous Ordinary Differential Equation

The following are two simultaneous first-order differential equations:

$$\frac{dy_1}{dt} = f_1(t, y_1, y_2) \tag{9.19}$$

$$\frac{dy_2}{dt} = f_2(t, y_1, y_2) \tag{9.20}$$

k_1:

$$k_{11} = f_1(t, y_1, y_2) \tag{9.21}$$

$$k_{12} = f_2(t, y_1, y_2) \tag{9.22}$$

k_2:

$$k_{21} = f_1\left(\left(t + \frac{h}{2}\right), \left(y_1 + \frac{h}{2}K_{11}\right), \left(y_2 + \frac{h}{2}k_{12}\right)\right) \tag{9.23}$$

$$k_{22} = f_2\left(\left(t+\frac{h}{2}\right),\left(y_1+\frac{h}{2}K_{11}\right),\left(y_2+\frac{h}{2}k_{12}\right)\right) \quad (9.24)$$

k_3:

$$k_{31} = f_1\left(\left(t+\frac{h}{2}\right),\left(y_1+\frac{h}{2}K_{21}\right),\left(y_2+\frac{h}{2}k_{22}\right)\right) \quad (9.25)$$

$$k_{32} = f_2\left(\left(t+\frac{h}{2}\right),\left(y_1+\frac{h}{2}K_{21}\right),\left(y_2+\frac{h}{2}k_{22}\right)\right) \quad (9.26)$$

k_4:

$$k_{41} = f_1\left(\left(t+\frac{h}{2}\right),\left(y_1+\frac{h}{2}K_{31}\right),\left(y_2+hk_{32}\right)\right) \quad (9.27)$$

$$k_{42} = f_2\left(\left(t+\frac{h}{2}\right),\left(y_1+\frac{h}{2}K_{31}\right),\left(y_2+hk_{32}\right)\right) \quad (9.28)$$

Then y_1, y_2:

$$y_{1,j} = y_{1,(j-1)} + h\left(\frac{k_{11}}{6}+\frac{k_{21}}{3}+\frac{k_{31}}{3}+\frac{k_{41}}{4}\right) \quad (9.29)$$

$$y_{2,j} = y_{2,(j-1)} + h\left(\frac{k_{1,2}}{6}+\frac{k_{2,2}}{3}+\frac{k_{3,2}}{3}+\frac{k_{4,2}}{4}\right) \quad (9.30)$$

9.5 Solving High-Order Differential Equations Using COMSOL

In this section, high-order differential equations will be solved using the COMSOL built-in module Mathematics/Coefficient Form PDE. Examples one, two, and three dimensions will be illustrated.

Example 9.6: Solving One-Dimensional ODE Using COMSOL

Solve the following high-order differential equation using COMSOL for the interval from 0 to 2π:

$$\frac{d^2y}{dx^2}+\frac{dy}{dx}+y = 2\cos(x)+x+1$$

The boundary conditions are $u(0) = 0, (2\pi) = 2\pi$.

Solution

1. Start COMSOL and select space dimension, 1D.
2. Then for physics, choose "Coefficient Form PDE." Select the "Mathematics/PDE interfaces/Coefficient Form PDE" and then click the "Add" tab, as shown in Figure 9.7.

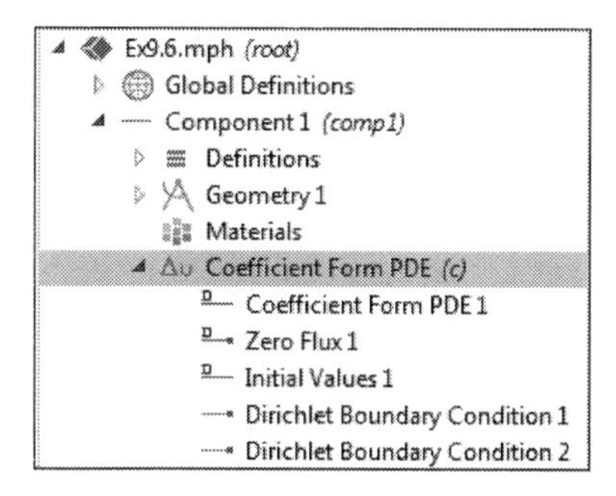

FIGURE 9.7
Selecting of Coefficient Form PDE.

3. Click the "Study" button and choose "Stationary study." Then click the "Done" button. The problem does not contain a time-dependent term.
4. Right-click on "geometry" and select interval $(0, 2\pi)$, as described in Figure 9.8.

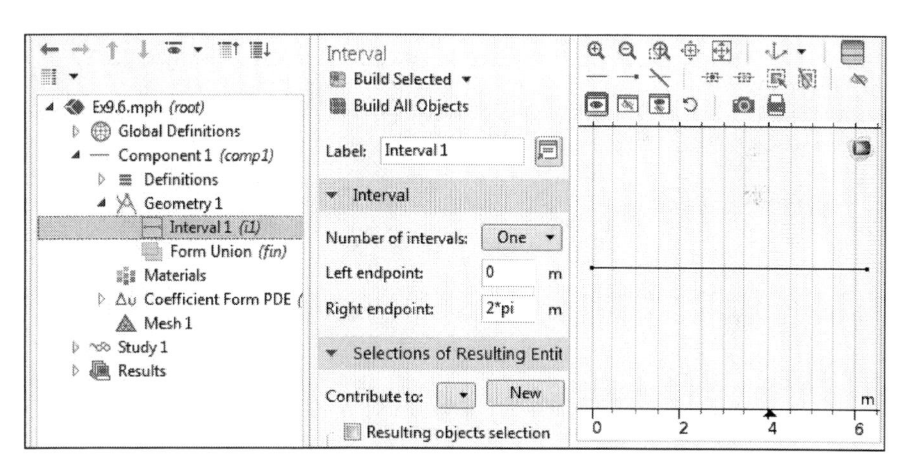

FIGURE 9.8
Selection of interval geometry.

5. Left-click on "Coefficient Form PDE" and change the coefficient values of the equation to get the equation to be solved. Set the following values: $c = -1$, $a = 1$, $f = 2 * \cos(x) + x + 1$, $e_a = 0$, $d_a = 0$, $\alpha = -1$, $\beta = 0$, $\gamma = 0$ (Figure 9.9).

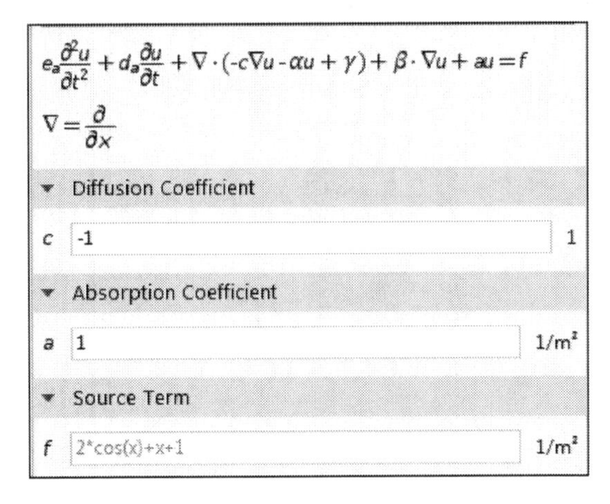

FIGURE 9.9
Coefficients of PDE; the rest are: $e_a = 0$, $d_a = 0$, $\alpha = -1$, $\beta = 0$, $\gamma = 0$.

6. The PDE is now defined. It is time to define the boundary conditions. Right-click "Coefficient Form PDE" tab on the right and select "Dirichlet boundary condition." For the first boundary condition $u(0) = 0$; accordingly, r should be left as zero, and point 1 on the line should be selected on the interval (Figure 9.10).

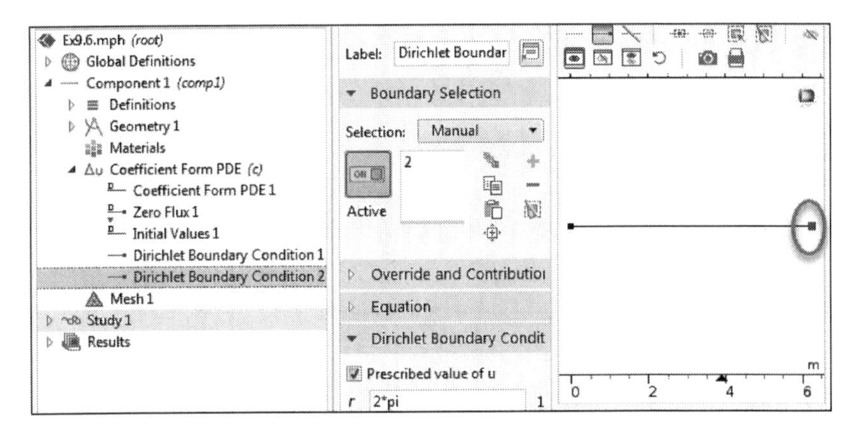

FIGURE 9.10
Setting of boundary conditions, $u(0) = 0$, $u(2\pi) = 2 * pi$.

7. Repeat the same procedure to set the second boundary condition, $u(2pi) = 2pi$. So r should be set as $2pi$, and point 2 on the line should be selected.
8. The required information to solve the differential equation is now completed, and the result can be computed. Go to the "Study" tab and left-click the "Compute" button (Figure 9.11).

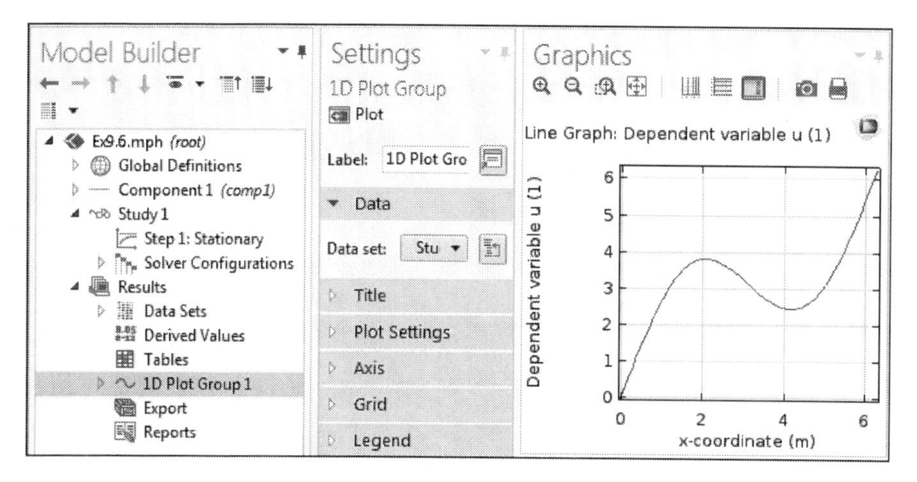

FIGURE 9.11
COMSOL generated results.

Example 9.7: Solving a Two-Dimensional Heat Transfer Problem

A hole of diameter $D = 0.25$m is drilled through the center of a solid copper block of square cross section with $w = 1$ m on aside (Figure 9.12). The hole is drilled along the length, $L = 2$ m, of the block, which has a thermal conductivity of $k = 150$ W/m.K. Hot oil flowing through the hole is characterized by $T_1 = 300$°C and $h_1 = 450$W/m.K. The outer surfaces are exposed to ambient air, with $T_2 = 25$°C and $h_2 = 4$ W/m^{2}K. Determine the corresponding heat rate and surface temperature.

Solution

The solution of this example will be done using two methods:

1. Use the built-in COMSOL Mathematics module (Coefficient Form PDE) as follows.

Start COMSOL. Select "2D" and "Coefficient Form PDE Module," and enter the parameters as shown in Figure 9.13.

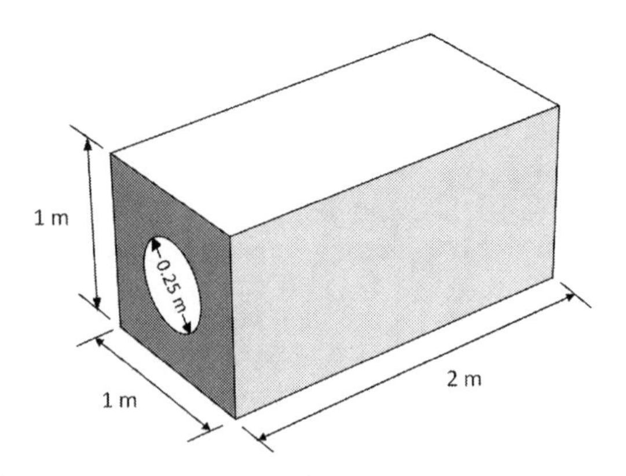

FIGURE 9.12
Schematic diagram of two-dimensional heat transfer problem.

Parameters

Name	Expression
k	150
h1	50
h2	4
T1	573.15
T2	298.15

FIGURE 9.13
Parameter used in solving the problem.

The model builder inner boundary condition is shown in Figure 9.14.
The outer surface boundary conditions are shown in Figure 9.15.
After running the program, you should see the results shown in Figure 9.16.

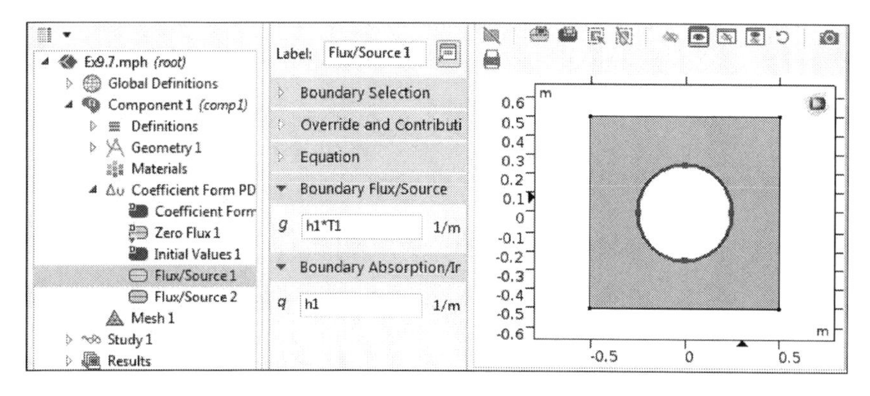

FIGURE 9.14
Inner boundary conditions.

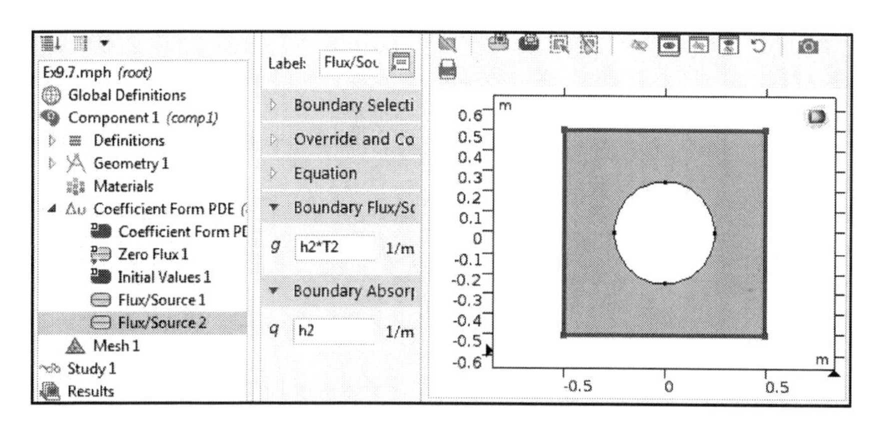

FIGURE 9.15
Model builder and setting of Flux/Source 2.

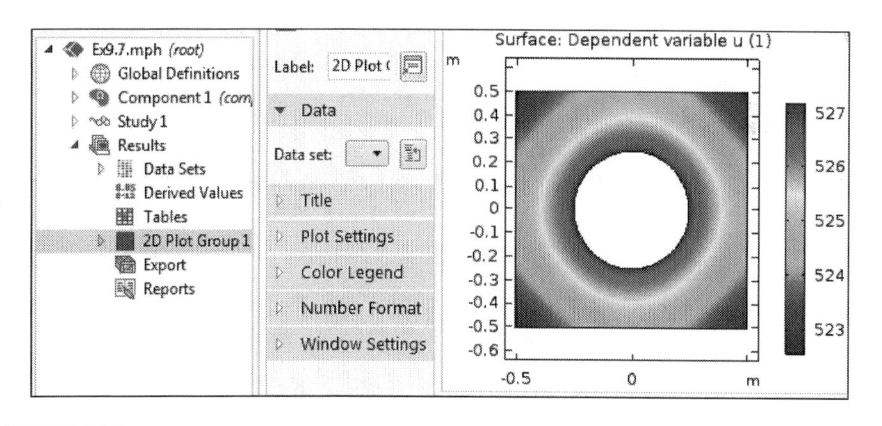

FIGURE 9.16
Two-dimensional surface temperature profile.

2. This simulation uses the built-in COMSOL Heat transfer module. The COMSOL simulation of the block with the hole using the Heat transfer module and the inner hole heat flux are shown in Figure 9.17. The heat flux of the outer surface is shown in Figure 9.18. The predicted 2D surface profile is shown in Figure 9.19.

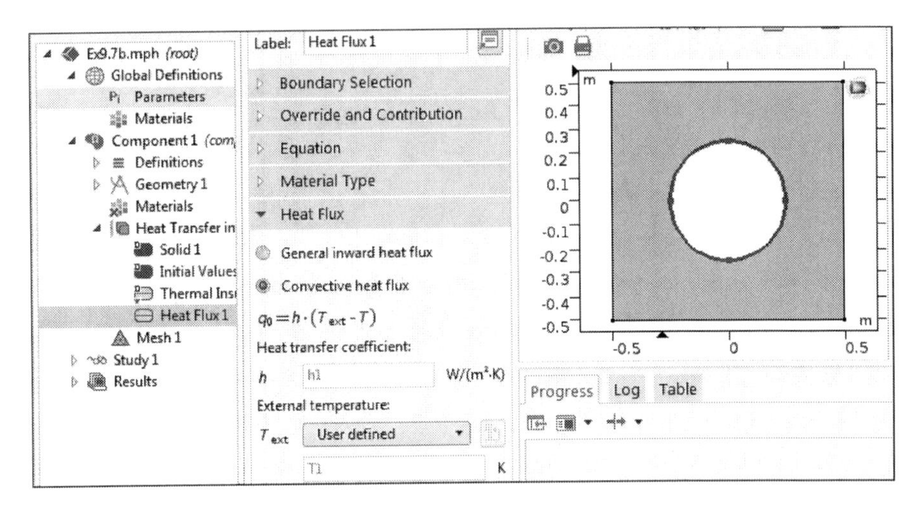

FIGURE 9.17
Heat flux of the inner surface of the cylindrical hole.

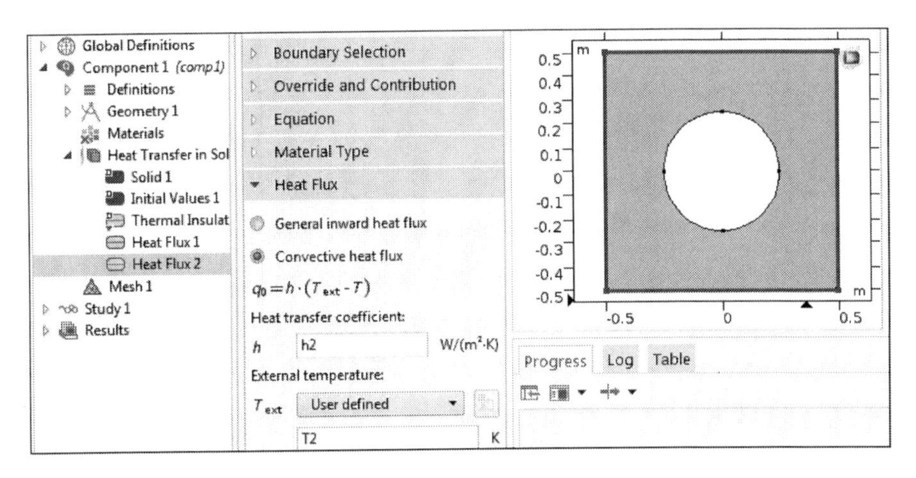

FIGURE 9.18
Heat flux of the block outer surface area.

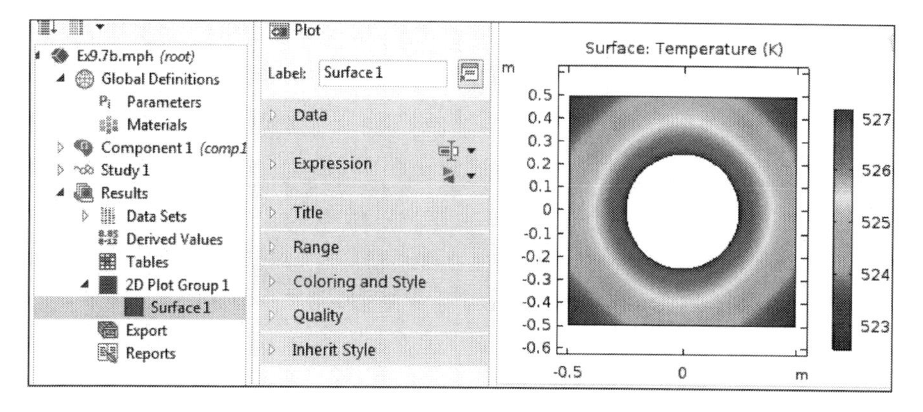

FIGURE 9.19
Surface temperature simulated via heat transfer module.

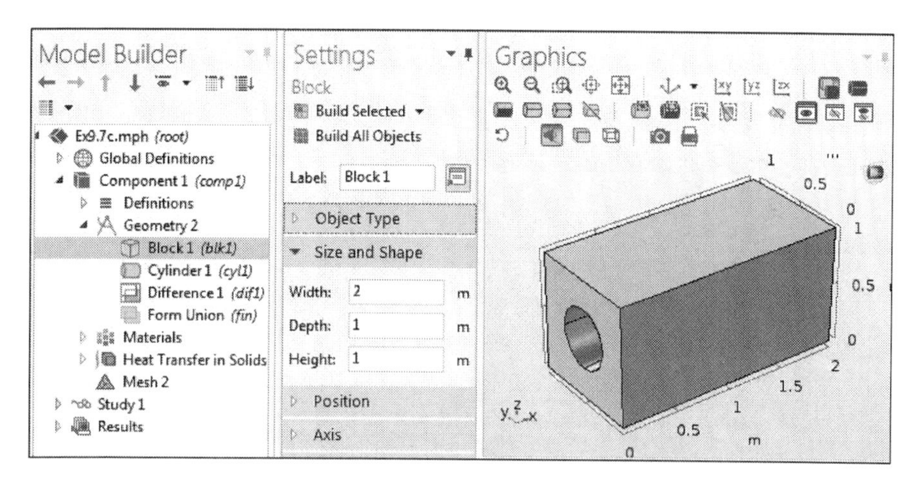

FIGURE 9.20
Three-dimensional surface temperature profile.

The three-dimensional surface temperature profile, after altering the geometry from 2D to 3D, is shown in Figure 9.20. The transparent surface temperature profile is shown in Figure 9.21.

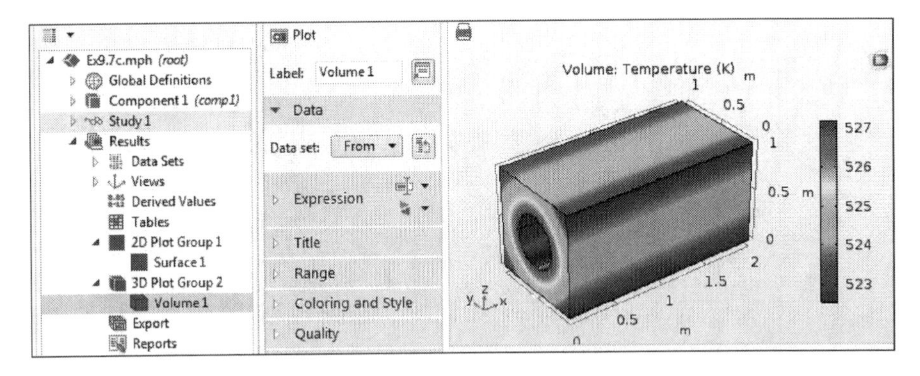

FIGURE 9.21
Transparent surface temperature profile.

Example 9.8: Solving a Three-Dimensional Transient Heat Transfer Problem

Simulate the heat transfer across the rod and the heat source shown in the Figure 9.22. The length of the cross section of the rod is 0.1 m and the length is 1 m. There is a heating source that is cylindrical in shape, with radius 0.05 m and height 0.1 m, placed at the top of the rectangular rod (copper alloy). The heat source is 100 kW/m^3.

Solution

1. Start COMSOL, and select "3D," and "Time dependent."
2. Under Select physics, choose "Mathematics/PDE interfaces." Then select "coefficient Form PDE" and "time dependent" from "select study."

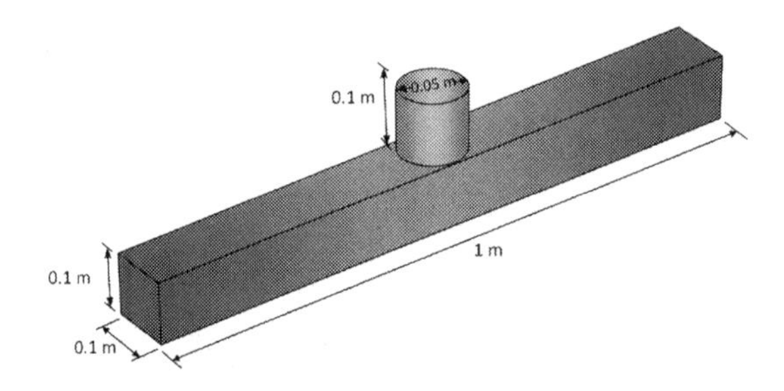

FIGURE 9.22
Schematic of problem setup.

3. Right-click on "Geometry" and select "Block (0.1,1,1)."
4. Right-click on "Geometry" and select "Cylinder."
5. Right-click on "Geometry/Booleans" and "Geometry/union." The geometry is now done (Figure 9.23).

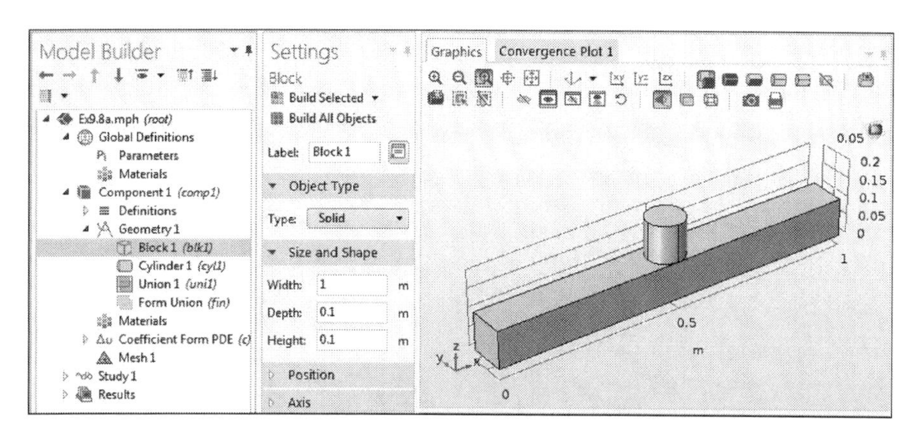

FIGURE 9.23
Dimensions of the block and cylinder geometry.

6. Left-click on "Coefficient Form PDE" and adjust the PDE to match the heat transfer equation of Cartesian coordinates.

   The equation of energy in Cartesian coordinates for Newtonian fluids of constant density has the heat source term Q. The source could be electrical energy due to current flow, chemical energy, and so on. The microscopic energy balance and constant thermal conductivity, in Cartesian coordinates, is:

$$\rho C_p \frac{\partial T}{\partial t} + \rho C_p \left(v_x \frac{\partial T}{\partial x} + v_y \frac{\partial T}{\partial y} + v_z \frac{\partial T}{\partial x} \right) - k \left(\frac{\partial^2 T}{\partial x^2} + \frac{\partial^2 T}{\partial y^2} + \frac{\partial^2 T}{\partial z^2} \right) = Q$$

   The convective term is deleted since the heat is transferred in a solid cube. The heat source is the cylinder above the cube. Accordingly, the energy balance equation is reduced to:

$$\rho C_p \frac{\partial T}{\partial t} + 0 - k \left(\frac{\partial^2 T}{\partial x^2} + \frac{\partial^2 T}{\partial y^2} + \frac{\partial^2 T}{\partial z^2} \right) = Q$$

Enter the values of thermal conductivity k, and $\rho * C_p$. The rest of coefficients are zero (Figure 9.24).

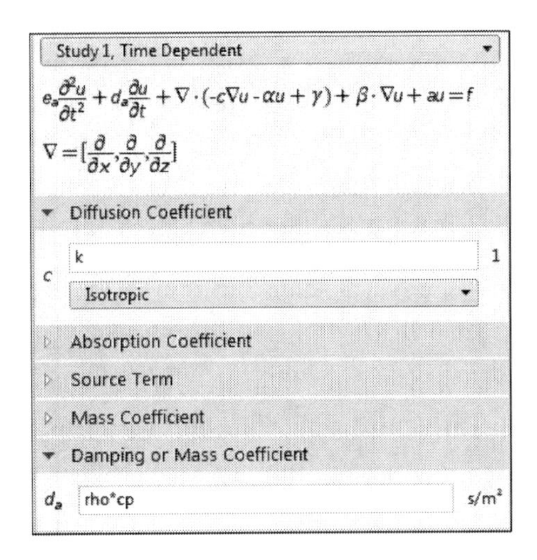

FIGURE 9.24
Setting the coefficient of PDE.

7. Right-click "Coefficient Form PDE" and add "Source" (Figure 9.25).

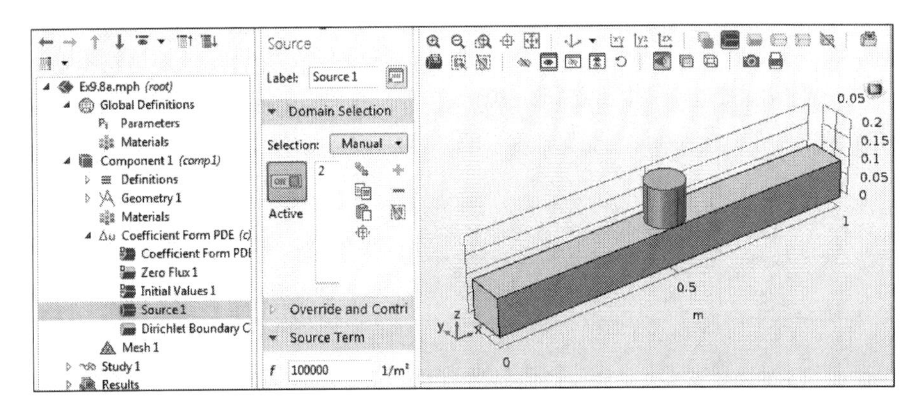

FIGURE 9.25
Value of heat source.

8. The surface temperature profile is shown in Figure 9.26.

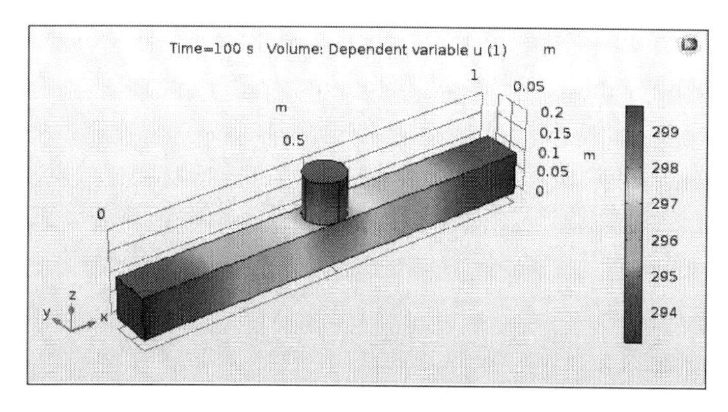

FIGURE 9.26
Surface temperature profile.

9. Using the COMSOL built-in Mathematics/Coefficients Form PDE module, the parameters used in solving the heat transfer problem are shown in Figure 9.27.

▾ Parameters

Name	Expression	Value
k	388	388
rho	8.933e3	8933
cp	0.385	0.385

FIGURE 9.27
Parameters used in the heat transfer problem.

To start the simulation, use the COMSOL built-in physics. Select the COMSOL built-in Heat transfer in solid module, and add the temperature boundary condition at both ends of the rod (293.15 K). Add the value of the heat source by right-clicking the Heat transfer in solid module, add its value, and proceed as shown in Figure 9.28.

The resultant 3D temperature surface plotted after 100 seconds is shown in Figure 9.29.

The results of the COMSOL Mathematics/Coefficient Form PDE and the Heat transfer modules are identical.

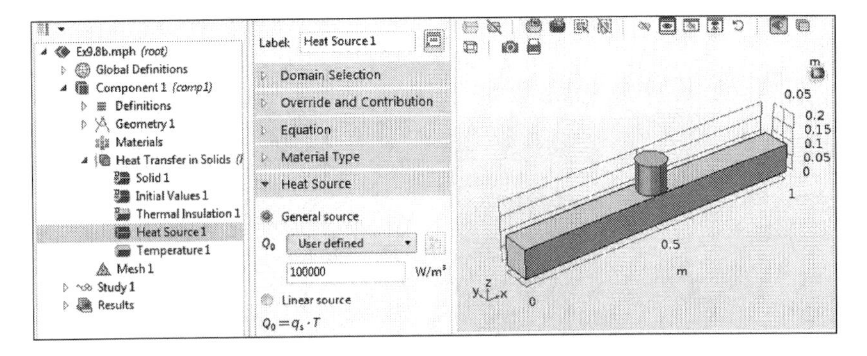

FIGURE 9.28
Model builder using the COMSOL heat transfer in solid module.

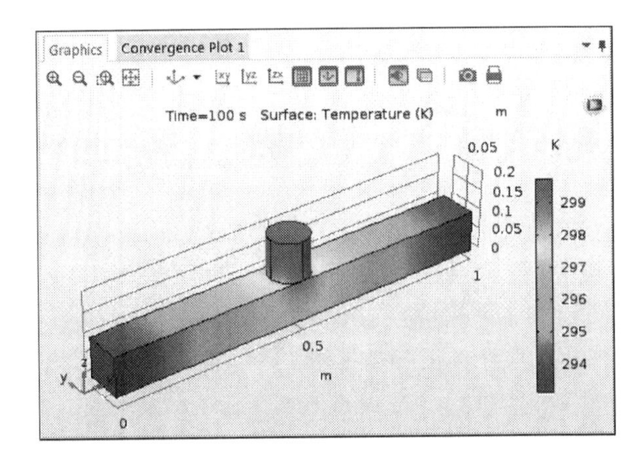

FIGURE 9.29
Two-dimensional temperature surface profile using the heat transfer module.

PROBLEMS

9.1 Second-Order BVP

Consider the following boundary value equation. Solve using the given initial and boundary values conditions:

$$\frac{d^2y}{dx^2} + y = 0$$

subject to $y'(0) = 1$ and $y(\pi) = 0$.

The exact solution is $y = \sin(x)$.

To solve this numerically, we first need to reduce the second-order equation to a system of first-order equations:

$$\frac{dy}{dx} = z$$

$$\frac{dz}{dx} = -y$$

with $z(0) = 1$ and $y(\pi) = 0$.

9.2 BVP

Solve the following BVP. Consider the following differential equation:

$$\frac{d^2y}{dx^2} - 3\frac{dy}{dx} + 2y = 0$$

with boundary conditions:

$$y(0) = 0,\ y(1) = 10$$

9.3 Initial Value Differential Equation

Solve the following initial value differential equation using COMSOL:

$$\frac{dy}{dx} = \sin(x * y)$$

with initial condition $y(0) = \pi$, on the interval [0, 1].

9.4 High-Order Differential Equation

Solve the following high-order boundary value differential equation using COMSOL for the interval from 0 to 2π:

$$\frac{d^2y}{dx^2} + \frac{dy}{dx} + y = 2\cos(x) + x + 1$$

The boundary conditions $y(0) = 0$,

$$\text{at } x = 2\pi,\ \frac{dy}{dx} = 2\pi.$$

9.5 Nonlinear Initial Value Problem

Solve the initial value problem using COMSOL:

$$\frac{dy}{dx} = -2x + \sin(4x) - 4xy, \qquad y(0) = 1$$

9.6 Temperature Distribution in a Rock

Determine the subsurface temperature of a rock. As the surface is heated or cooled, the heat diffuses through the soil and rock. Mathematically, diffusion may be represented by the following equation:

$$\frac{\partial T}{\partial t} = K \frac{\partial^2 T}{\partial x^2}$$

where:
 $T(t, z)$ is the temperature at time t and depth z
 K is the thermal conductivity of the rock

A typical value for K is 2×10^{-6} m²/s. The surface temperature of the rock is 15°C, and the depth of the rock is 100 m. The change in temperature at the bottom of the rock can be neglected assuming that no heat reaches the bottom from the top (i.e., $\partial T / \partial x = 0$). How quickly does heat move through the rock?

9.7 Catalytic Reaction with Diffusion

The catalytic reaction where the diffusion of A into a catalyst follows a first-order irreversible chemical reaction is shown by $A \rightarrow B$. The process is described by the following ODE:

$$\frac{d^2 C_A}{dz^2} - \frac{k}{D_{AB}} C_A = 0$$

with the following boundary conditions

$$at\, z = 0 \qquad C_A = C_{A0} = 0.2 \text{ mol/l}$$

$$at\, z = L = 0.001 \text{ m} \quad \frac{dC_A}{dz} = 0$$

The reaction rate constant is $k = 1 \times 10^{-3}$ 1/s, and the diffusivity is $D_{AB} = 1.2 \times 10^{-9}\,\text{m}^2/\text{s}$. Calculate the concentration of A at the $L = 0.001$ m. Solve using COMSOL.

9.8 Heat Transfer in a Block

The heat transfer in a block of length $l = 1$ m is described to be the following unsteady-state PDE [1]:

$$\frac{\partial T}{\partial t} - \frac{k}{\rho C_p}\frac{\partial^2 T}{\partial x^2} = 0$$

The block is initially at 100°C ($T(0) = 100$), and the boundary condition is:

$$x = 0, \quad T_1 = 0$$

The second boundary is the insulated boundary at:

$$x = L, \quad \frac{dT}{dx} = 0$$

Calculate the temperature profile and the temperature at the insulated boundary of the block. Assume the value of $k/\rho C_p = 2 \times 10^{-5}$. Find the temperature at 0.1 m of the block after 10 minutes.

9.9 Diffusion Problem

Solving the following diffusion problem using COMSOL. Initially the concentration of A is 0 (i.e., $C_A(x,0) = 0$).

$$\frac{\partial C_A}{\partial t} - \frac{\partial}{\partial x}\left(e^{0.5C_A}\frac{\partial C_A}{\partial x}\right) = 0$$

with the following boundary conditions [2]:

$$C_A(0,t) = 0 \quad and \quad C_A(1,t) = 0$$

9.10 Conversion versus Reactor Volume

Determine and plot the conversion as a function of reactor volume for the feed stream of pure A of 1.0 mol/l. The resultant ODE is:

$$\frac{dx}{dV} = 0.08\frac{(1-x)}{(1+2x)}$$

References

1. Cultip, M. B., M. Shacham, 2008. *Problem Solving in Chemical and Biochemical Engineering with Polymath, Excel and* MATLAB, 2nd ed., Boston, MA: Prentice Hall.
2. Finlayson, B. A., 2012. *Introduction to Chemical Engineering Computing*, 2nd ed., Hoboken, NJ: Wiley.

Index

Note: Page numbers in italic and bold refer to figures and tables respectively.